Environmental Science

Series editors: R. Allan · U. Förstner · W. Salomons

Springer

Berlin
Heidelberg
New York
Barcelona
Hong Kong
London
Milan
Paris
Tokyo

Antonio Gianguzza · Ezio Pelizzetti
Silvio Sammartano (Eds.)

Chemistry of Marine Water and Sediments

With 249 Figures and 106 Tables

Springer

Editors

Prof. Antonio Gianguzza
Dipartimento di Chimica Inorganica
Università di Palermo
Via delle Scienze
I-90128 Palermo, Italy

Prof. Ezio Pelizzetti
Dipartimento di Chimica Analitica
Università di Torino
Via Pietro Giuria 5
I-10125 Torino, Italy

Prof. Silvio Sammartano
Dipartimento di Chimica Inorganica, Chimica Analitica e Chimica Fisica
Università di Messina
Salita Sperone 31
I-98166 Messina, Italy

ISSN 1431-6250
ISBN 3-540-42055-X Springer-Verlag Berlin Heidelberg New York

Die Deutsche Bibliothek - CIP-Einheitsaufnahme

Chemistry of marine water and sediments : with 106 tables / Antonio Gianguzza ... (ed.). - Berlin ; Heidelberg ; New York ; Barcelona ; Hong Kong ; London ; Milan ; Paris ; Tokyo : Springer, 2002
(Environmental science)
ISBN 3-540-42055-X

Springer-Verlag Berlin Heidelberg New York
a member of the BertelsmannSpringer Science+Business Media GmbH

Printed in Germany

Cover Design: Struve & Partner, Heidelberg

Dataconversion: Büro Stasch (*www.stasch.com*) · Uwe Zimmermann, Bayreuth

SPIN: 10934449 30/3111 – 5 4 3 2 1 – Printed on acid-free paper

Preface

This book is a collection of all the lectures by the professors attending the 3rd "International School on Marine Chemistry" held in Ustica (Palermo, Italy, September 2000), under the auspices of the United Nations and the Italian Chemical Society. The School was organized by the University of Palermo in co-operation with the Natural Marine Reserve of Ustica Island.

The Organising Committee of the School wishes to thank the University of Messina, the University of Roma "La Sapienza," the Italian University Consortium of Environmental Chemistry, and the Marine Reserve of Ustica Island for their financial support to the School.

This book has been printed with the financial support of the Environmental Research Centre CIRITA of the University of Palermo.

The editors thank all the professors whose outstanding scientific contributions have made it possible to publish this book.

Professor Antonio Gianguzza
Professor Ezio Pelizzetti
Professor Silvio Sammartano

Contents

Contributors

Eric P. Achterberg
(Dr.)
Department of Environmental Sciences
Plymouth Environmental Research Centre
University of Plymouth
Plymouth PL4 8AA, UK

Renato Barbieri
(Professor of Inorganic Chemistry)
Dip. di Chimica Inorganica, Università di Palermo
Viale delle Scienze, I-90128 Palermo, Italy
Phone: +39 091 590578
E-mail: bioinorg@unipa.it

Andrew R. Bowie
(Dr.)
Department of Environmental Sciences
Plymouth Environmental Research Centre
University of Plymouth
Plymouth PL4 8AA, UK

Paola Calza
(Researcher of Analytical Chemistry)
Dip. di Chimica Analitica, Università di Torino
Via Pietro Giuria 5, I-10125 Torino
Phone: +39 011 6707630, Fax: +39 011 6707615
E-mail: pelizzet@chem.unito.it

Vincenzo Cannizzaro
(Mr.)
Department of Environmental Sciences
Plymouth Environmental Research Centre
University of Plymouth
Plymouth PL4 8AA, UK
E-mail: v.cannizzaro@plymouth.ac.uk

Silvio Capri
(Dr.)
Istituto di Ricerca sulle Acque
Consiglio Nazionale delle Ricerche
Via Reno 1, I-00198 Rome, Italy
Phone: +39 06 8841451
E-mail: capri@irsa.rm.cnr.it

Concetta De Stefano
(Professor of Analytical Chemistry)
Dip. di Chimica Inorganica, Chimica Analitica e Chimica Fisica, Università di Messina
Salita Sperone 31, I-98166 Messina, Italy
Phone: +39 090 391354, Fax +39 090 392827
E-mail: destefano@chem.unime.it

Roberta Di Stefano
(Ph.D. Chemistry)
Dip. di Chimica Inorganica, Università di Palermo
Viale delle Scienze, I-90128 Palermo, Italy
E-mail: bioinorg@unipa.it

Tiziana Fiore
(Ph.D. Chemistry)
Dip. di Chimica Inorganica, Università di Palermo
Viale delle Scienze, I-90128 Palermo, Italy
E-mail: bioinorg@unipa.it

Claudia Foti
(Professor of Analytical Chemistry)
Dip. di Chimica Inorganica, Chimica Analitica e Chimica Fisica, Università di Messina
Salita Sperone 31, I-98166 Messina, Italy
Phone: +39 090 391354, Fax: +39 090 392827
E-mail: foti@chem.unime.it

Roberto Frache
(Professor of Analytical Chemistry)
Dip. di Chimica e Chimica Industriale
Università di Genova
Via Dodecaneso 31, I-16146 Genova, Italy
Phone: +39 010 3536186
E-mail: frache@chimica.unige.it

Paulo Gardolinski
(Mr.)
Department of Environmental Sciences
Plymouth Environmental Research Centre
University of Plymouth
Plymouth PL4 8AA, UK

Antonio Gianguzza
(Professor of Analytical Chemistry)
Centro Interdipartimentale di Ricerche sulla Interazione Tecnologie-Ambiente (CIRITA)
Università di Palermo
Viale delle Scienze, I-90128 Palermo, Italy
Phone: +39 091 489409, Fax: +39 091 427584
E-mail: giang@unipa.it

Ingmar Grenthe
(Professor of Inorganic Chemistry)
Department of Chemistry, Inorganic Chemistry
Royal Institute of Technology
S-10044 Stockholm, Sweden
E-mail: ingmarg@kth.se

Massimo Guardigli
(Dr.)
Dip. di Scienze Farmaceutiche
Università di Bologna
Via Belmeloro 6, I-40126 Bologna, Italy
Phone: +39 051 343398, Fax: +39 051 343398
E-mail: mguardig@alma.unibo.it

John I. Hedges
(Dr.)
School of Oceanography
University of Washington
Box 357940, Seattle, WA 98195-7940, USA
Phone: +1 (0)206 5430744, Fax: +1 (0)206 5436073
E-mail: jihedges@u.washington.edu

Carmela Ianni
(Ph.D., Researcher of Analytical Chemistry)
Dip. di Chimica e Chimica Industriale
Università di Genova
Via Dodecaneso 31, I-16146 Genova, Italy
Phone: +39 010 3536180, Fax: +39 010 3536190
E-mail: ianni@chimica.unige.it

Cindy Lee
(Dr.)
Marine Science Research Center
State University of New York
Stony Brook, NY 11794-5000, USA
Phone: +1 (0)631 6328741, Fax: +1 (0)631 6328820
E-mail: cindylee@notes.cc.sunysb.edu

Marco Mascini
(Professor of Analytical Chemistry)
Dip. di Chimica
Università di Firenze, Polo Scientifico
Via della Lastruccia 3
I-50019 Sesto Fiorentino (Firenze), Italy
Phone: +39 055 4573283, Fax: +39 055 4573384
E-mail: mascini@unifi.it

Frank J. Millero
(Professor of Marine and Physical Chemistry)
Rosenstiel School of Marine and Atmospheric Science, University of Miami
4600 Rickenbacker Causeway
Miami, Florida 33149, USA
Phone: +1 (0)305 3614707, Fax: +1 (0)305 3614144
E-mail: fmillero@rsmas.miami.edu

John W. Morse
(Scherck Professor of Oceanography)
Texas A&M University, College Station
Texas 77843-3146, USA
Phone: +1 (0)409 8459630, Fax: +1 (0)409 8459631
E-mail: morse@ocean.tamu.edu

Patrizia Pasini
(Ph.D. Chemistry)
Dip. di Scienze Farmaceutiche, Università di Bologna
Via Belmeloro 6, I-40126 Bologna, Italy
Phone: +39 051 343398, Fax: +39 051 343398
E-mail: pasini@alma.unibo.it

Luisa Patrolecco
(Dr.)
Istituto di Ricerca sulle Acque
Consiglio Nazionale delle Ricerche
Via Reno 1, I-00198 Rome, Italy
Phone +39 06 8841451
E-mail: patrolecco@irsa.rm.cnr.it

Ezio Pelizzetti
(Professor of Analytical Chemistry)
Dip. di Chimica Analitica
Università di Torino
Via Pietro Giuria 5, I-10125 Torino, Italy
Phone: +39 011 6707630, Fax: +39 011 6707615
E-mail: pelizzet@chem.unito.it

Claudia Pellerito
(Ph.D. Chemistry)
Dip. di Chimica Inorganica, Università di Palermo
Viale delle Scienze, I-90128 Palermo, Italy
E-mail: bioinorg@unipa.it

Lorenzo Pellerito
(Professor of Inorganic Chemistry)
Dip. di Chimica Inorganica,
Centro Interdiparti-mentale di Ricerche sulla Interazione Tecnologie-Ambiente (CIRITA)
Università di Palermo
Viale delle Scienze, I-90128 Palermo, Italy
Phone : +39 091 590367, Fax: +39 091 427584
E-mail: bioinorg@unipa.it

Valery S. Petrosyan
(Professor of Organic Chemistry)
Department of Organic Chemistry
M.V. Lomonosov University
Moscow, 119 899, Russia
Phone: +7 (0)95 9395643
E-mail: petros@org.chem.msu.ru

Maurizio Pettine
(Dr.)
Istituto di Ricerca sulle Acque
Consiglio Nazionale delle Ricerche
Via Reno 1, I-00198 Rome, Italy
Phone: +39 06 8841451
E-mail: pettine@irsa.rm.cnr.it

Daniela Piazzese
(Ph.D. Analytical Chemistry)
Dip. di Chimica Inorganica, Università di Palermo
Viale delle Scienze, I-90128 Palermo, Italy
Phone: +39 091 489409, Fax: +39 091 427584
E-mail: danielap@unipa.it

Denis Pierrot
(Mr.)
Rosenstiel School of Marine and Atmospheric Science
University of Miami
4600 Rickenbacker Causeway
Miami, Florida 33149, USA
Phone: +1 (0)305 3614680, Fax: +1 (0)305 3614144
E-mail: dpierrot@rsmas.miami.edu

Martin R. Preston
(Dr., B.Sc., Ph.D., MRSC, Chem.)
Oceanography Laboratories
University of Liverpool
Liverpool L69 3BX, UK
Phone: +44 (0)151 7944093, Fax: +44 (0)151 7944099
E-mail: preston@liv.ac.uk

Joseph M. Prospero
(Professor)
Cooperative Institute of Marine and Atmospheric Sciences (CIMAS), Rosenstiel School of Marine and Atmospheric Sciences
University of Miami
4600 Rickenbacker Causeway
Miami, Florida 33149-1098, USA
Phone: +1 (0)305 3614159, Fax: +1 (0)305 3614457
E-mail: jprospero@rsmas.miami.edu

Paola Rivaro
(Ph.D. in Marine Science)
Dip. di Chimica e Chimica Industriale
Università di Genova
Via Dodecaneso 31, I-16146 Genova, Italy
Phone: +39 010 3536178, Fax: +39 010 3536190
E-mail: rivaro@chimica.unige.it

Aldo Roda
(Professor of Analytical Chemistry)
Dip. di Scienze Farmaceutiche
Università di Bologna
Via Belmeloro 6, I-40126 Bologna, Italy
Phone: +39 051 343398, Fax: +39 051 343398
E-mail: roda@alma.unibo.it

Nicoletta Ruggieri
(Dr.)
Dip. di Chimica e Chimica Industriale
Università di Genova
Via Dodecaneso 31, I-16146 Genova, Italy
Phone: +39 010 3536173, Fax: +39 010 3536190
E-mail: nicolettaruggieri@iol.it

Silvio Sammartano
(Professor of Analytical Chemistry)
Dip. di Chimica Inorganica, Chimica Analitica e Chimica Fisica, Università di Messina
Salita Sperone 31, I-98166 Messina, Italy
Phone: +39 090 393659, Fax +39 090 392827
E-mail: sammartano@chem.unime.it

Richard Sandford
(Mr.)
Department of Environmental Sciences
Plymouth Environmental Research Centre
University of Plymouth
Plymouth PL4 8AA, UK

Manuel Sastre de Vicente
(Professor of Physical Chemistry)
Dep. de Química Fisica e Enxeñería Química
Universidad de La Coruña
C/ Alejandro de la Sota 1, E-15071 La Coruña, Spain
Phone: +34 (0)981 167000, Fax: +34 (0)981 167065
E-mail: eman@udc.es

Michelangelo Scopelliti
(Mr.)
Dip. di Chimica Inorganica, Università di Palermo
Viale delle Scienze, I-90128 Palermo, Italy
E-mail: bioinorg@unipa.it

Fabio Triolo
(Ph.D. Chemistry)
Mount Sinai School of Medicine
New York, NY 10029, USA *or*
Dip. di Chimica Inorganica, Università di Palermo
Viale delle Scienze, I-90128 Palermo, Italy
E-mail: bioinorg@unipa.it

Teresa Vilariño
(Assistant Professor)
Dep. de Química Física e Enxeñería Química, Universidad de La Coruña
C/ Alejandro de la Sota 1
E-15071 La Coruña, Spain
Phone: +34 (0)981 167000
Fax: +34 (0)981 167065

Stuart Wakeham
(Dr.)
Skidaway Institute of Oceanography
10 Ocean Science Circle, Savannah, GA 31411, USA
Phone: +1 (0)912 5982310, Fax: +1 (0)912 5982310
E-mail: stuart@skio.peachnet.edu

Paul Worsfold
(Professor of Analytical Chemistry)
Department of Environmental Sciences
Plymouth Environmental Research Centre
University of Plymouth
Plymouth PL4 8AA, UK
E-mail: pworsfold@plymouth.ac.uk

Part I

Biogeochemical Processes at the Air-Water and Water-Sediment Interface

Chapter 1

Sea Water as an Electrolyte

F. J. Millero

1.1 Introduction

The composition of the major components of sea water has been measured by a number of researchers over the years. The relative molar concentration of the major cations (Na^+, Mg^{2+}, Ca^{2+}, K^+, Sr^{2+}) and anions (Cl^-, SO_4^{2-}, HCO_3^-, Br^-, CO_3^{2-}, $B(OH)_4^-$, F^-) in the major oceans has been shown to be constant. These major components of sea water contribute to the physical chemical properties of the oceans. Since the major components of sea water are constant throughout the oceans (The Marcet Principle), it is possible to treat ocean waters as an electrolyte solution (sea salt) with a dash of the non-electrolyte boric acid. This simplifies the physical chemistry of sea water solutions and other natural waters. Some minor components (SiO_2, NO_3^- and PO_4^{3-}) that are added to the oceans from the bacterial oxidation of plant material can also have a minor effect on the properties of deep waters. In this chapter, I will review how one treats sea water as a multi-component solution, and how the major components of sea water contribute to its physical and chemical properties. First we will examine the composition of sea water and the development of salinity.

1.1.1 Composition of Average Sea Water

The composition of the major components of sea water has been measured by a number of researchers over the years (Culkin 1965; Culkin and Cox 1966; Morris and Riley 1966; Riley and Tongudai 1967; Warner 1971; Carpenter and Manella 1973; Wilson 1975). The relative molar composition of the major cations (Na^+, Mg^{2+}, Ca^{2+}, K^+, Sr^{2+}) in the major oceans is shown in Fig. 1.1. Within the experimental error of the measurements, the relative composition of cations and anions is constant in the surface waters of the oceans. The relative composition of the cations in surface and deep waters is shown in Fig. 1.2. All the cations except Ca^{2+} are independent of the depth. This is shown more clearly in Fig. 1.3, where the concentration of Ca^{2+} normalized to a constant salinity is shown as a function of depth. This increase in Ca^{2+} is the result of the dissolution of $CaCO_3$ in deep ocean waters due to the effect of pressure on the solubility.

All of these measurements were made relative to the chlorinity (*Cl*) (the mass of halides in a given mass of sea water). It is determined by titrating sea water with $AgNO_3$

$$\text{Sea water} + AgNO_3 \rightarrow AgCl(s) + AgBr(s) \quad (1.1)$$

Fig. 1.1. The ratio of the concentration of major cations to the chlorinity in different oceans

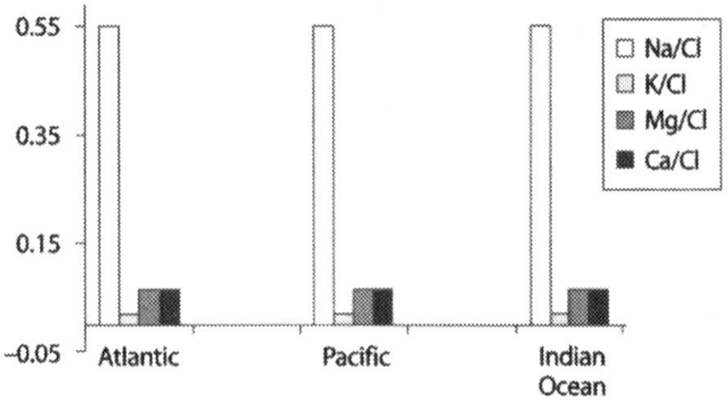

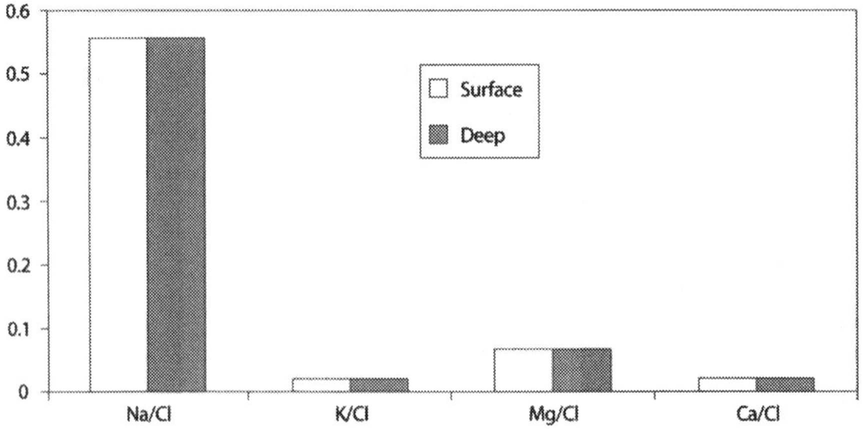

Fig. 1.2. The ratio of the concentration of major cations to the chlorinity in surface and deep waters

which precipitates all the major halides except F^-. Average sea water has a chlorinity of 19.374‰ (parts per thousand). To be consistent over the years, the silver used in the titrations on standard sea water (provided for calibrations) has come from the same bar used originally by Knudsen (1901), who set up the protocol. The relative composition (in grams per chlorinity, $g_i/Cl(‰)$) of average sea water at 25 °C and pH = 8.1 is given in Table 1.1 (Millero 1996). These values of $g_i/Cl(‰)$ can be used to determine the stoichiometry of the components of sea water at a given Cl as well as average sea water having a $Cl(‰) = 19.374$. The values of the total grams ($g_T = \Sigma g_i$), total moles ($n_T = 1/2\Sigma n_i + n_B$), total equivalents ($e_T = 1/2\Sigma n_i Z_i$) and ionic strength ($I = 1/2\Sigma n_i Z_i^2$) can be determined from the values of $g_i/Cl(‰)$ (Table 1.1). This leads to the total molality given by:

$$m_T = 28.903\ Cl(‰)\ /\ [1\,000 - 1.8154\ Cl(‰)] \quad (1.2)$$

and total molal ionic strength ($I = 1/2\Sigma m_i Z_i^2$) given by:

$$I_T = 35.99\ Cl(‰)\ /\ [1\,000 - 1.8154\ Cl(‰)] \quad (1.3)$$

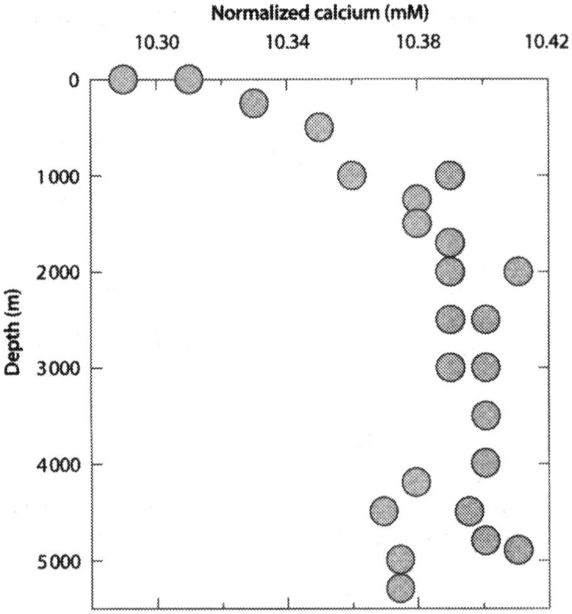

Fig. 1.3. The normalized concentration (to $S = 35$) of Ca^{2+} in the Pacific Ocean as a function of depth (Millero 1996)

Table 1.1. Composition of one kilogram of natural sea water[a] and with a $Cl = 19.374$‰ and pH = 8.1 (Millero 1996)

Species	Molality ($Cl = 19.374$)					
	g_i/Cl	M_i	g_i	m_i	e_i	$n_iZ_i^2/Cl$
Na^+	0.556614	22.9898	10.7838	0.46907	0.46907	0.46907
Mg^{2+}	0.066260	24.3050	1.2837	0.05282	0.10563	0.21127
Ca^{2+}	0.021270	40.0780	0.4121	0.01028	0.02056	0.04113
K^+	0.020600	39.0983	0.3991	0.01021	0.01021	0.01021
Sr^{2+}	0.000410	87.6200	0.0079	0.00009	0.00017	0.00035
Cl^-	0.998910	35.4527	19.3529	0.54588	0.54588	0.54588
SO_4^{2-}	0.140000	96.0636	2.7124	0.02824	0.05648	0.11295
HCO_3^-	0.005524	61.0171	0.1070	0.00175	0.00175	0.00175
Br^-	0.003470	79.9040	0.0672	0.00084	0.00084	0.00084
CO_3^{2-}	0.000830	60.0092	0.0161	0.00027	0.00054	0.00107
$B(OH)_4^-$	0.000407	78.8404	0.0079	0.00010	0.00010	0.00010
F^-	0.000067	18.9984	0.0013	0.00007	0.00007	0.00007
OH^-	0.000007	17.0034	0.0001	0.00000	0.00000	0.00000
$1/2\Sigma =$	0.028895		35.1515	0.55981	0.60565	0.69735
$B(OH)_3$	0.000996	61.8322	0.0193	0.00031	0.00031	
$\Sigma =$	1.815402		35.171	0.56012	0.60596	0.69735

[a] For average sea water $S = 35$, $Cl = 19.374$, $pH_{SWS} = 8.1$, $TA = 2.400$ mmol kg^{-1}, and $t = 25$ °C.

The mean molecular weight (M_T) of sea salt is given by:

$$M_T = \Sigma n_i M_i = 62.793 \quad (1.4)$$

These equations can be converted into functions of the salinity (S) using the approximate relationship:

$$S\,(‰) = 1.80655\,Cl(‰) \quad (1.5)$$

The composition of the major components of sea water is summarized in Fig. 1.4. The major sea salts include NaCl, Na_2SO_4, $MgCl_2$ and $MgSO_4$. The concept of salinity is discussed in more detail in the next section.

1.1.2
The Concept of Salinity

Salinity (S) was originally conceived as a measurement of the mass of dissolved salts in a given mass of sea water (the weight fraction in parts per thousand, ppt, ‰). The experimental determination of the salt content of sea water by drying and weighing presents some difficulties. At the temperatures necessary to drive off the last traces of H_2O, the bicarbonates and carbonates are decomposed to oxides (MO_2, where M = Na or K), and some halides are lost when heating to dryness (HCl and HBr). One can prevent the loss of HCl by adding NaF before evaporation (Morris and Riley 1964). This led earlier researchers to use indirect methods to measure the salinity. A complete chemical analysis of sea water is the only reliable way to determine the true salinity of sea water (S_T). This method, however, requires too much time, and cannot be used for routine work. The early work related the true salinity to chlorinity:

$$S_T = a\,Cl(‰) \quad (1.6)$$

where a = 1.8056 (Dittmar 1884) and 1.8148 (Lyman and Fleming 1940), which can be compared to the values of 1.8154 (Table 1.1). Earlier researchers suggested that Cl(‰) could be used as a measure of salinity. Measurements of the chlorinity and evaporation salinity gave:

$$S\,(‰) = 0.030 + 1.805\,Cl(‰) \quad (1.7)$$

For approximately 65 years, this formula was used in oceanography to determine salinity to an accuracy of 0.01‰ in S. "Normal" sea water of known Cl(‰) (prepared for years in Copenhagen and now in Wormley, England) was used to calibrate the titration methods use to determine Cl. Measurements of the physical properties of sea water such as density as a function of Cl(‰) could be used to calculate physical properties from Cl measurements made at sea. The intercept was due to the use of Baltic Sea waters that have an input of river salts ($Ca(HCO_3)_2$) and little chloride. Since the salts of different rivers can vary, the intercept can vary for each estuarine system. The salinity of sea water can be determined by measuring a number of physical properties (listed below along with the estimated errors in salinity).

Fig. 1.4. The major cations and anions in sea water (Millero 1996)

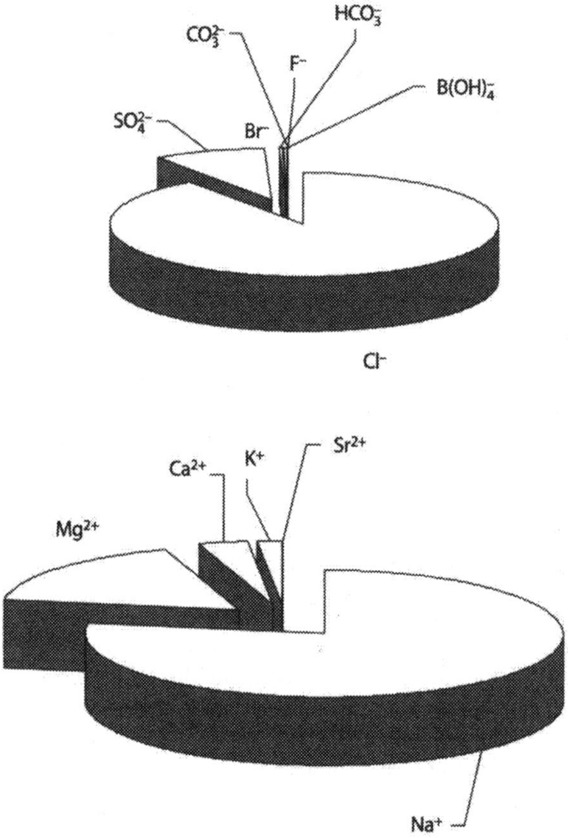

1. Refractive Index ±0.05
2. Sound Speed ±0.03
3. Evaporation ±0.01
4. Composition ±0.01
5. Density ±0.004
6. Chlorinity ±0.0002
7. Conductivity ±0.0004

Due to the high precision, the salinity is presently determined from conductivity measurements. These measurements are made relative to a sample of known conductivity ($R = C_{Sample} / C_{Std}$). Since the $S = 35.000$ at a $Cl = 19.374$ (Eq. 1.7), the earlier conductivity measurements as a function of Cl were converted using $S = 1.80655\ Cl$. Since the composition data for $Cl = 19.374$ gives $S = 35.171$, 0.17 kg of carbonates and boric acid are lost during the evaporation.

The *Practical Salinity Scale* of 1978 was developed from measurements made of the conductivity of sea water of known Cl (19.374) and S (35.000) relative to the conductivity of a given mass of KCl. This new scale breaks the Cl-S relationship in favour of a salinity-conductivity ratio relationship. All waters with the same conductivity ratio have the same salinity (even though the composition may differ). Since salinity is normally used to determine a physical property like density, this was thought to be the best method for determining the effect of changes in ionic composition. This is not always the case since non-electrolytes like SiO_2 are not detected by conductivity. The final equation is:

$$S = a_0 + a_1 R_T^{1/2} + a_2 R_T + a_3 R_T^{3/2} + a_4 R_T^2 + a_5 R_T^{5/2} + \Delta S \quad (1.8)$$

where

$$\Delta S = [(t-15)/(1+k(t-15))] b_0 + b_1 R_T^{1/2} + b_2 R_T + b_3 R_T^{3/2} + b_4 R_T^2 + b_5 R_T^{5/2} \quad (1.9)$$

and $R_T = C(S, t, 0) / C(35, t, 0)$ at atmospheric pressure ($p = 0$). The coefficients are given in Table 1.2. The scale is valid from $S = 2$ to 42 and $t = 0$ to 40 °C. Hill et al. (1986) have formulated equations that can be used in more dilute solutions. The practical salinity has no units.

1.1.3 Causes of Major Components Not Being Conservative

Although the major components of sea water are relatively constant, a number of factors can cause the waters to be non-conservative. They include processes that occur in (1) estuaries, anoxic basins, sediments, hydrothermal vents, and evaporated basins, and (2) by precipitation, dissolution, evaporation, freezing, and oxidation. Some examples will be briefly discussed.

The finding of hydrothermal vents has led to the discovery that a number of elements can be added (Ca, Cu, Zn, Mn, Si) and taken out (Mg, SO_4) of ocean waters. The loss of Mg for hydrothermal vent waters is shown in vent fluids of high SiO_2 in Fig. 1.5. The waters coming out of the vent at high temperatures are devoid of Mg. This is related to the formation of Mg silicates when the sea water reacts with molten basalt. As shown in Fig. 1.6, this deficiency of Mg changes the amount of Mg in deep Pacific waters.

Table 1.2. Coefficients needed to calculate the practical salinity of sea water from conductivity measurements

Parameter	Value	Parameter	Value
a_0	0.0080	b_0	0.0005
a_1	−0.1692	b_1	−0.0056
a_2	25.3851	b_2	−0.0066
a_3	14.0941	b_3	−0.0375
a_4	−7.0261	b_4	0.0636
a_5	2.7081	b_5	−0.0144
$\Sigma a_i =$	35.000	$\Sigma b_i =$	0.0000
$k =$	0.0162		

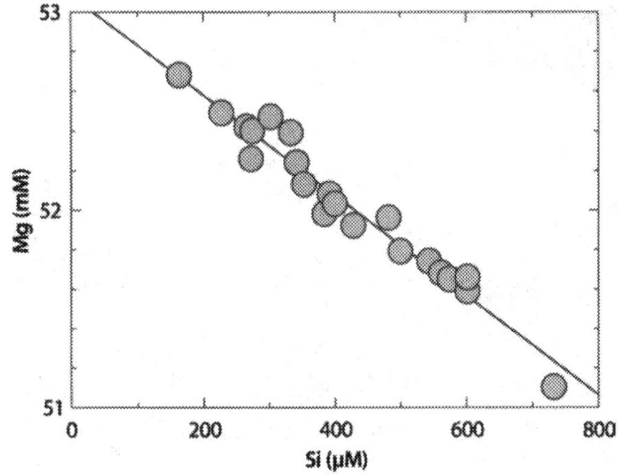

Fig. 1.5. The concentration of Mg^{2+} vs. SiO_2 for hydrothermal vent waters (Millero 1996)

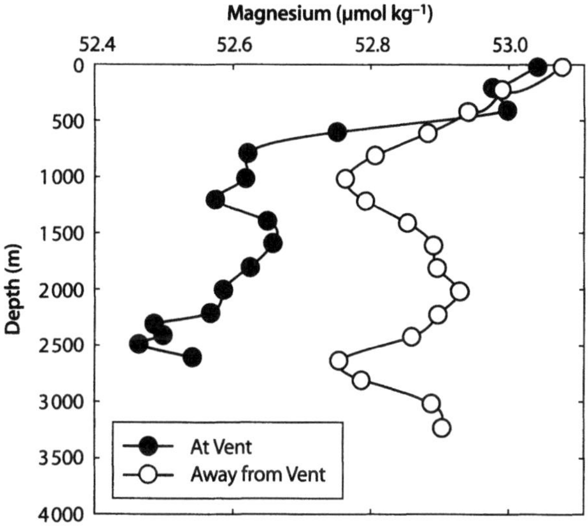

Fig. 1.6. The concentration of Mg^{2+} in deep waters of the Pacific

The concentrations of salts in rivers are controlled by the nature of the rocks being weathered and the soil types yielding ground waters, which differ in chemical composition. The total solids in most rivers are less than 200 ppm or $S = 0.2‰$, and are mostly composed of Mg^{2+}, Ca^{2+} and HCO_3^-. The composition of average world water is compared with normal sea water in Fig. 1.7. The major river cation is Ca^{2+}, and HCO_3^- is the major anion. Most of the NaCl in river waters is recycled from sea salt aerosols. The SiO_2 is predominantly in the unionized form $Si(OH)_4$ at the pH of most rivers (7.3 to 8.0). The major components of world river water (Ca^{2+} and HCO_3^-) come from the weathering of $CaCO_3$. The mixing of river waters with a different composition of sea salts can result in an estuary composition different from sea water. This mixing

Fig. 1.7. A comparison of the composition of average river water with sea water (Millero 1996)

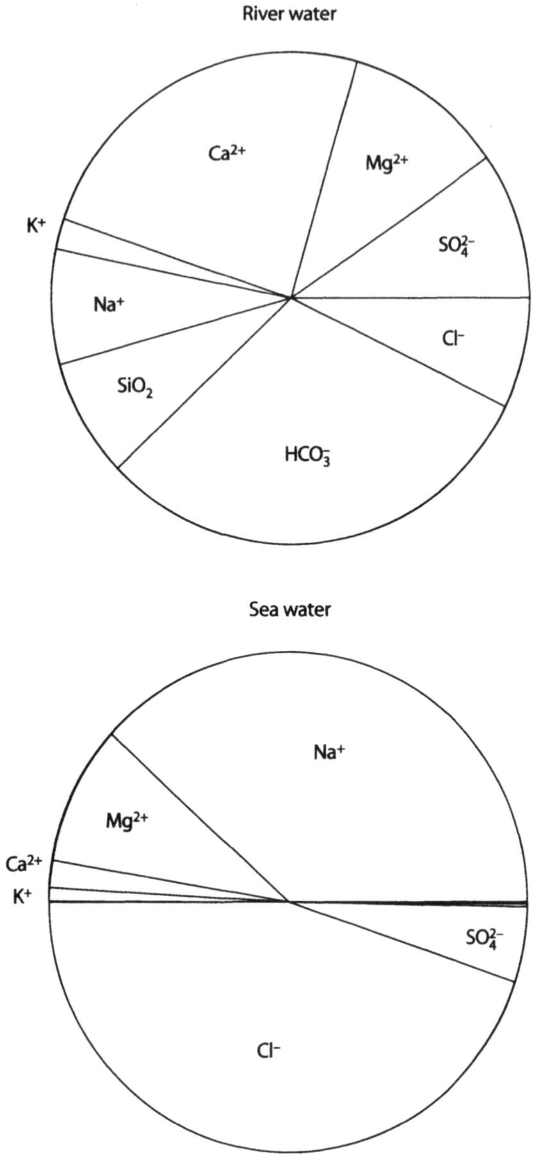

will result in a linear equation with intercepts equal to the values for sea salts as shown for the Baltic estuary in Figs. 1.8 and 1.9 (Millero 1978). The total grams of salts (g_{Est}) in an estuary are given by:

$$g_{Est} = g_R + [(35.171 - g_R) / 19.374] Cl(‰) \tag{1.10}$$

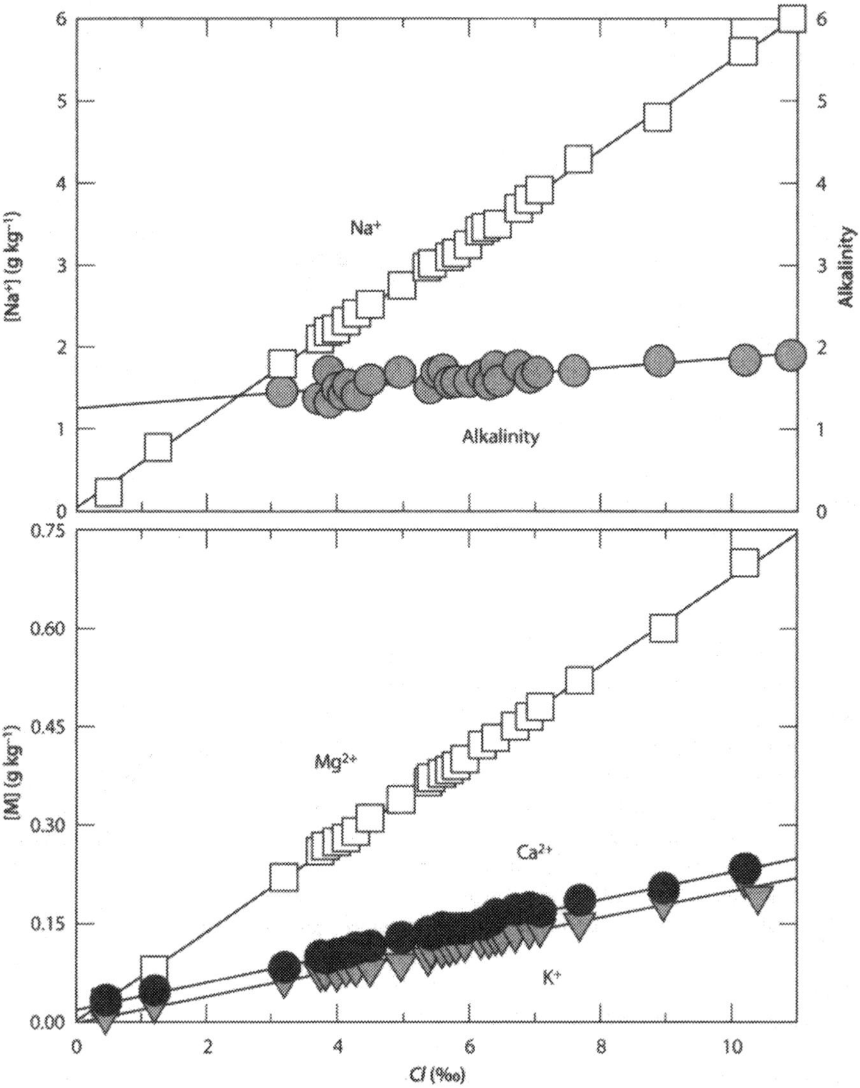

Fig. 1.8. The concentration of cations in the baltic estuary as a function of chlorinity (Millero 1996)

where g_R is the grams of river salts. The values of g_i for conservative elements can be obtained from the measured values in the estuary (g_E):

$$g_i = \frac{(g_E - g_{SW})19.374}{19.374 - Cl(‰)} \tag{1.11}$$

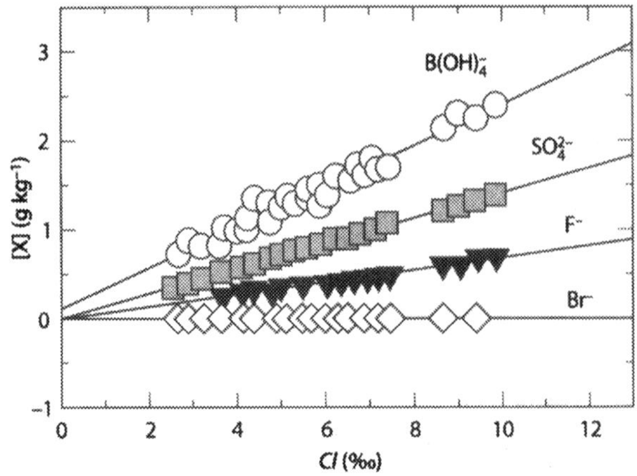

Fig. 1.9. The concentration of anions in the baltic estuary as a function of chlorinity (Millero 1996)

where $g_{SW} = k_i \, Cl(‰)$ and k_i is g_i/Cl for sea water (Table 1.1). For the Baltic, the present total salinity is given by

$$S = 0.044 + 1.8039 \, Cl(‰) \tag{1.12}$$

which is different from the Knudsen (1901) relationship (Eq. 1.7). This is related to the increase of river salts due to a decrease in the dilution of rainwater or the increased weathering of $CaCO_3$ due to acid rain. Since the composition of brines (Fig. 1.10) can be different from sea water, its mixture with sea water will change the composition of the resulting mixture. The evaporation of sea water in isolated basins can also change the composition due to the precipitation of a number of salts. This is shown in Figs. 1.10 and 1.11 for the evaporation of a Mexican lagoon (Fernandez et al. 1982). Ca^{2+}, K^+, HCO_3^- and SO_4^{2-} are lost from the solution during the initial evaporation at values of Cl near 40 ($S = 74$). This is due to the initial precipitation of $CaCO_3$ and later precipitation of $CaSO_4$. The loss of K^+ may be related to its coprecipitation with $CaCO_3$ or $CaSO_4$.

1.1.4
Physical Properties of Natural Waters

The physical properties of natural waters can vary over a wide range of temperatures (0 to 400 °C), salinities (0 to 350) and pressures (0 to 1 000 bar). The most widely studied natural water is sea water (Millero 1982, 1983, 2000b). Many of the physical properties of sea water are available as a function of t, S and P (Millero 2000b; 2001). It has been shown in a number of studies that the physical properties of many dilute rivers, lakes and estuaries are the same as sea water diluted to the same salinity. This is due to the fact that the changes in the physical properties of dilute solutions are not a strong function of the added electrolyte. This is demonstrated for density in Fig. 1.12. The relative density ($\rho - \rho^0$) of NaCl, Na_2SO_4, $MgCl_2$, and $MgSO_4$ is the same in dilute solutions. As the concentrations are increased, the relative densities of Na_2SO_4, $MgCl_2$, and $MgSO_4$ show positive deviations, while the values of NaCl are slightly lower than sea

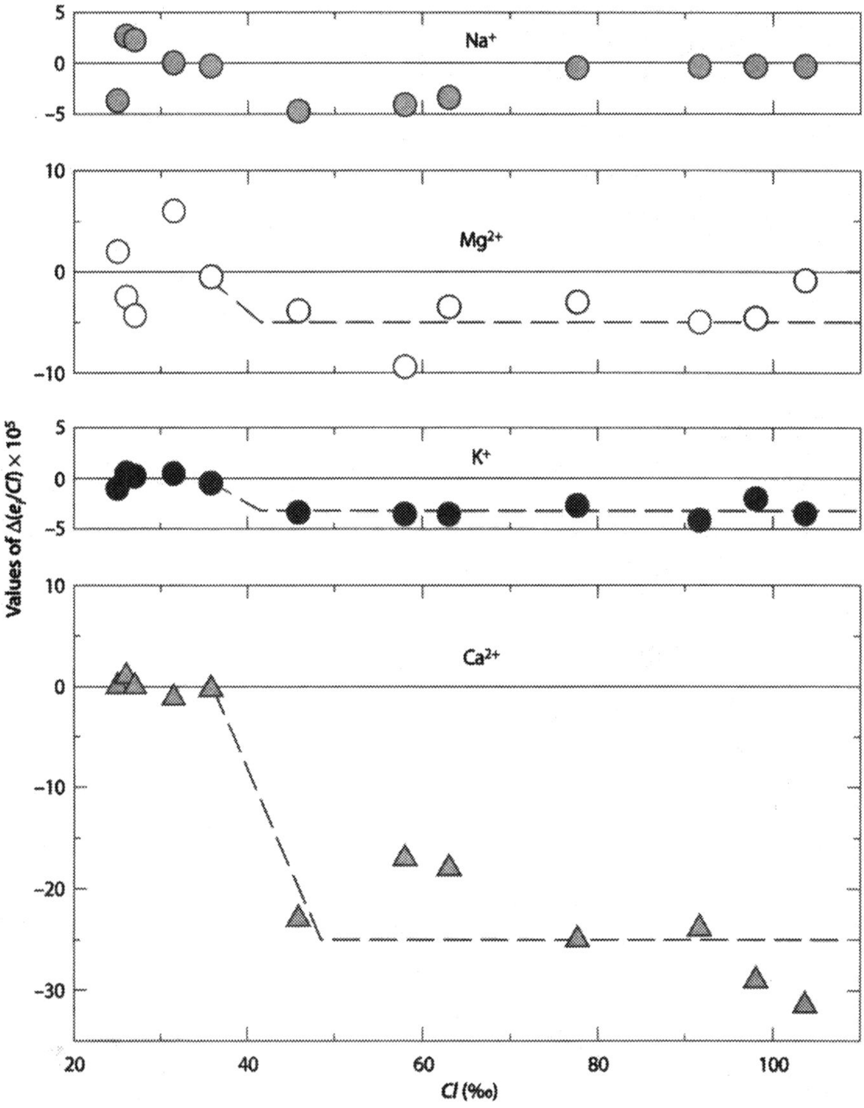

Fig. 1.10. The changes in the cations during the evaporation of Mexican lagoon waters

water. Since most natural waters have NaCl as their major salt, the properties of this salt can be useful as a model for many of these waters.

The reliability of this dilution principle is demonstrated (for the measured relative density) for an estuarine system and sea water diluted to the same salinity (Fig. 1.13). The values of the density of sea water diluted to the same salinity are equal to within 3 ppm. More recently the density of Lake Tanganyika waters has been measured

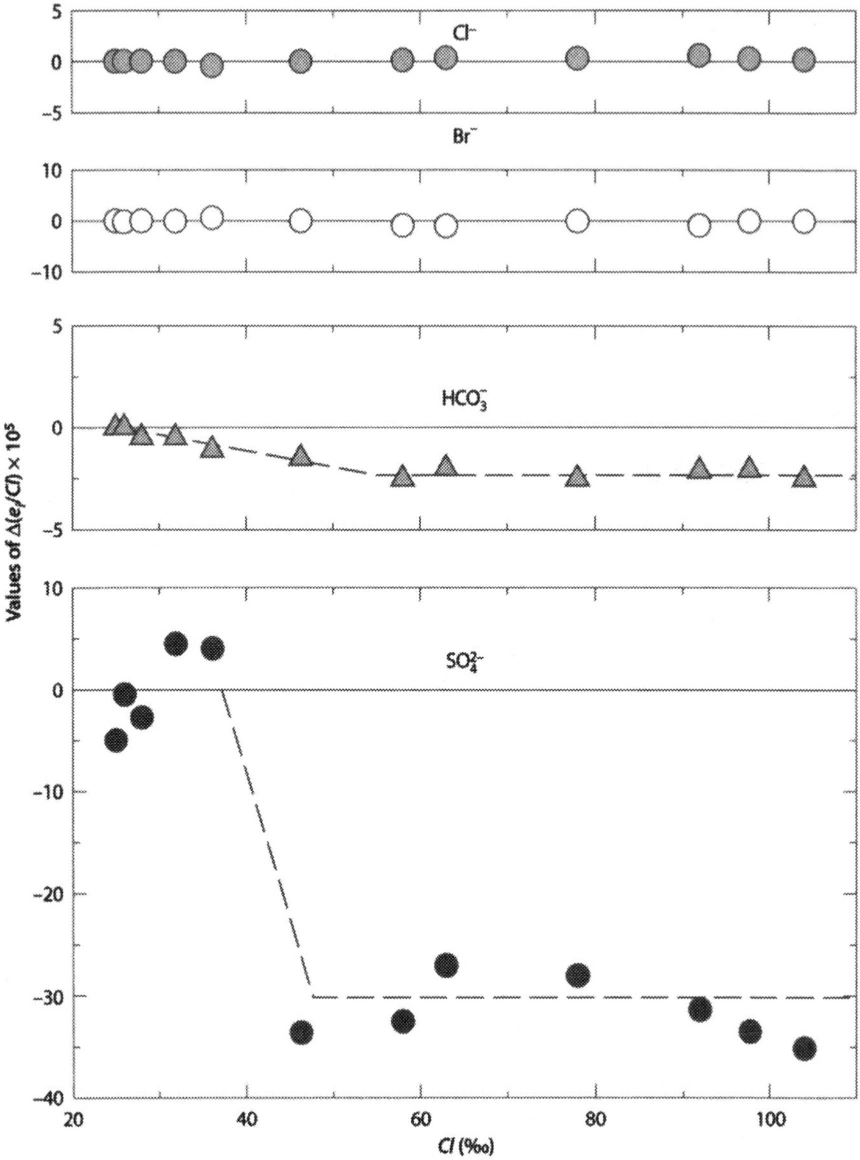

Fig. 1.11. The changes in the anions during the evaporation of Mexican lagoon waters

(Millero 2000a). The results as a function of temperature for the differences in the measured densities and compressibilities and the calculated values shown in Table 1.3 agree to within 2 ppm and 0.006 ppm, respectively, in density and compressibility. The measurements as a function of depth are compared to the calculated values in Fig. 1.14. The increases in the density in the deep waters are due to the addition of nutrients

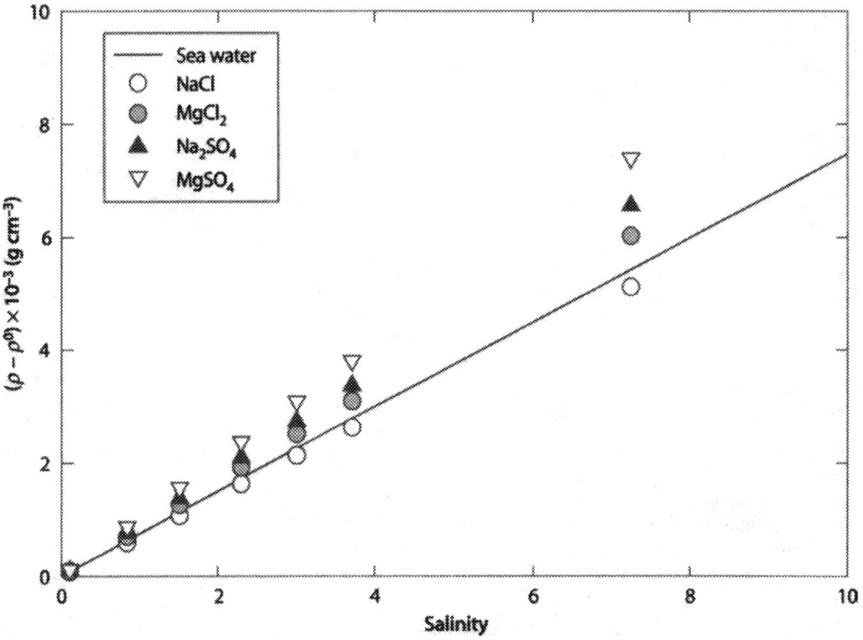

Fig. 1.12. A comparison of the relative density of the major sea salts to sea water as a function of salinity

due to the mineralization of organic material. The increase in the salinity due to the addition of these elements can be adequately accounted for by increasing the salinity of the sea water.

The composition of deep waters in the oceans can change as the result of the mineralization of plant material and the dissolution of $CaCO_3$ (Connors and Weyl 1968; Brewer and Bradshaw 1975; Millero et al. 1976a,b; Millero 1978; Millero and Kremling 1976; Poisson et al. 1980, 1981). The increases in NO_3, PO_4, SiO_2, and alkalinity can change the physical properties of sea water (e.g. density and conductivity). Although these changes are small over small spatial scales, the changes can be important for large-scale and ocean-to-ocean studies. The limitations of the Practical Salinity Scale for estuarine system (Parsons 1982; Sharp and Culberson 1982; Gieskes 1982; Millero 1984) have also been discussed. These studies point out the need to adjust the conductivity salinity for changes in the composition. Recently, Millero (2000b) has examined how these changes can be examined for the density of sea water and the results are briefly discussed below.

Brewer and Bradshaw (1975) were the first to estimate how changes in the composition of sea water would change the calculated density of ocean waters. They estimated that the changes in salinity could be 0.015 and the changes in density of 12×10^{-6} g cm^{-3}. Using partial molar conductance (Poisson et al. 1979) and volume changes they found:

$$\Delta\rho = 53.7\,\Delta TA - 9.6\,\Delta TCO_2 + 42\,\Delta SiO_2 \tag{1.13}$$

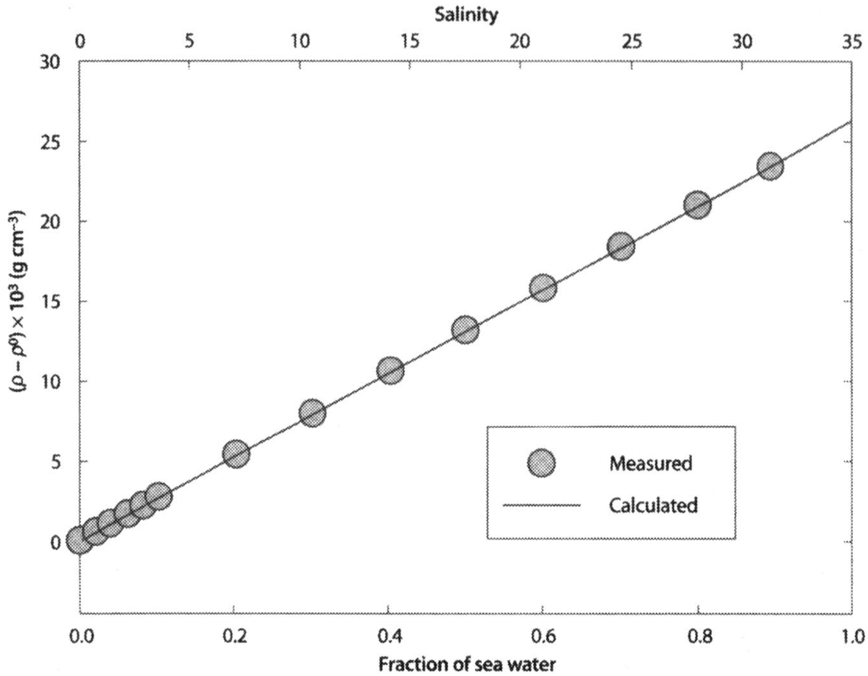

Fig. 1.13. A comparison of the measured densities for estuarine waters with those calculated at the same salinity

Table 1.3. Comparisons of the measured and calculated density of sea water at 25 °C as a function of ionic strength and salinity

Ionic strength	Salinity	$\Delta\rho$(Meas) $-\rho$(Calc) (ppm)
0.11	5	−4
0.21	10	−7
0.31	15	−7
0.41	20	−5
0.51	25	1
0.61	30	11
0.72	35	24
0.83	40	34

where ΔTA, ΔTCO_2, and ΔSiO_2 are respectively the changes in total alkalinity, total carbon dioxide, and silica (mmol kg^{-1}). It should be pointed out that the changes in TA and TCO_2 are normalized to the values for the surface sea water used to determine the equation of state of sea water ($TA = 2.332$ and $TCO_2 = 2.226$ mmol kg^{-1} when $S = 35$). This equation was modified by Millero et al. (1976b), using more reliable partial molar volume for SiO_2 and considering the effect of added NO_3 as HNO_3. They obtained:

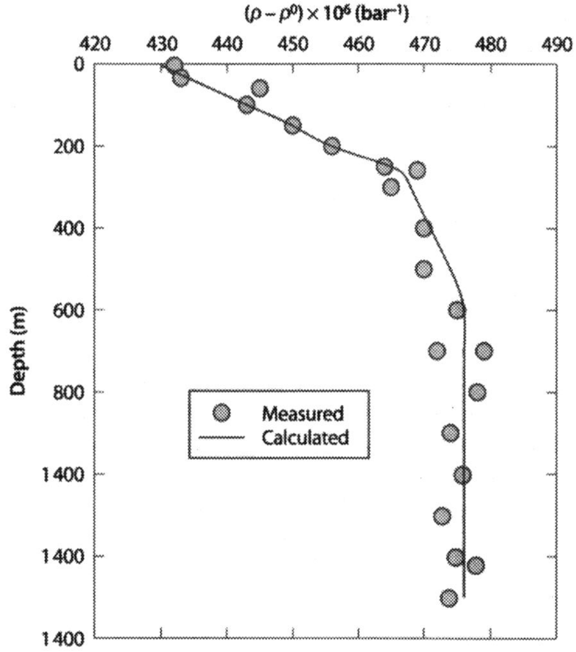

Fig. 1.14. A comparison of the measured densities for Lake Tanganyika waters with those calculated at the same salinity

$$\Delta\rho = 53.7\,\Delta TA - 9.6\,\Delta TCO_2 + 45\,\Delta SiO_2 + 24\,\Delta NO_3 \tag{1.14}$$

This equation was examined by Millero et al. (1976b), using directly measured density and conductivity of ocean waters. They found the values of $\Delta\rho$ were 5 ±1.5 ppm in the North Atlantic and 16 ±3.6 in the North Pacific. These values differed with the calculated densities by ±2.7 ppm in the North Atlantic and ±4.0 in the North Pacific. Millero et al. (1978) made further density measurements on North Pacific waters and found that the measured values were in good agreement with their earlier measurements (Millero et al. 1976b). They found that the measured results could be accounted for by assuming that the increase in density was due to the effect of the added solids on the density ($\Delta\rho = 757\,\Delta S$). They found:

$$\Delta\rho = 37.9\,\Delta TA + 72.8\,\Delta SiO_2 + 47.7\,\Delta NO_3 \tag{1.15}$$

This equation yields results that agree with the measured values to ±4.3 ppm. These results indicate that the density changes in sea water due to changes in the composition can be accounted for by changes in the true salinity due to the mass of added dissolved solids:

$$\Delta S = \Sigma M_i \Delta n_i \tag{1.16}$$

where M_i is the molecular weight and n_i is the change in moles of solute i in 1 kg of sea water. This relationship holds only if the added solute has partial specific properties similar to those of sea salt and the concentrations of added solutes are low (Poisson et al. 1980; Millero 1984).

More recently Millero (2000b) has shown that empirical relationships can also be used to fit the experimental measurements:

$$\Delta\rho = 10.2 + 43.9\,\Delta TCO_2 \qquad (\sigma = 4.2\text{ ppm}) \tag{1.17}$$

$$\Delta\rho = 6.0 + 112\,\Delta TA \qquad (\sigma = 4.4\text{ ppm}) \tag{1.18}$$

$$\Delta\rho = 1.9 + 100.5\,\Delta\mathrm{SiO_2} \qquad (\sigma = 4.1\text{ ppm}) \tag{1.19}$$

$$\Delta\rho = 1.1 + 396\,\Delta\mathrm{NO_3} \qquad (\sigma = 4.1\text{ ppm}) \tag{1.20}$$

The intercept is close to zero except for *TA* and TCO_2, which is due to the difference in the surface values in the Atlantic and Pacific oceans. The individual slopes are larger than the theoretical values, because they include the changes due to all the constituents in the solution. Since nutrient data are more available than carbonate data, the equations using SiO_2 and NO_3 may be more useful. The changes in the density of estuarine waters may be different because of changes in the input of various chemicals from a given river (Poisson et al. 1980, 1981; Millero 1984) and the precipitation of minerals such as $CaCO_3$.

1.2 Modelling the Physical Properties of Natural Waters

The ionic interactions in a mixed electrolyte solution like sea water can affect the physical properties (density, heat capacity, etc.) of natural waters. Since the composition of natural waters can be quite different, it is useful to have models that can be used to describe how the ionic components affect the physical properties. This requires knowledge of ionic interactions in the solutions of interest. Over the years, a great deal of progress has been made in interpreting and modelling the physico-chemical properties of mixed electrolyte solutions (Millero 2001). This has led to the development of models that can be used to estimate the properties of natural waters of known composition. These models consider the changes that occur due to ion-water interactions in dilute solutions and the resultant ion-ion interactions as one moves to more concentrated solutions. The ion-water interactions can be examined using the following models:

1. Continuum Model
2. Ion-Dipole Model
3. Ion Quadrupole Model
4. Ion-Water Structure Model
5. Hydration Model

The continuum model examines the interactions between an ion in a continuous dielectric medium. This can be represented by the transfer of an ion from a vacuum to a solution that has no structure (Fig. 1.15). The free energy, enthalpy and entropy of transferring an ion for a vacuum to the solution can be determined from the equations:

$$\Delta G_h\,(\text{kcal mol}^{-1}) = -(\text{N}e^2Z^2 / 2r)(1 - 1/\text{D}) = -163.9Z^2/r \qquad (1.21)$$

$$\Delta H_h\,(\text{kcal mol}^{-1}) = (\text{N}e^2Z^2 / 2r)[1 - 1/\text{D} - T/\text{D}(\partial \ln \text{D} / \partial T)_P] = -166.8Z^2/r \qquad (1.22)$$

$$\Delta S_h\,(\text{cal mol}^{-1}\,\text{K}^{-1}) = (\text{N}e^2Z^2 / 2r)(\partial \ln \text{D} / \partial T)_P = -9.65Z^2/r \qquad (1.23)$$

where D is the dielectric constant, N is Avogadro's number, r is the radius (Å $= 1 \times 10^{-8}$ cm), Z is the charge, e is the electrostatic charge, T is the absolute temperature and P is the pressure. The free energy of hydration for a number of cations as a function of Z^2/r is shown in Fig. 1.16. The agreement is reasonable, but is better if the radius is increased by 0.85 Å. This can be attributed to the average inter-sphere radius of the firmly bound waters of hydration. The ion-dipole and quadrupole models attempt to account for the interactions of water molecules with individual ions. A more simplistic model can be developed by examining the differences in the properties of the water molecules in the electrostricted region (Fig. 1.17) and the waters in the bulk solution. One can look at the electrostriction as the region where the volume is decreased due to the interactions of the water molecules with a given ion. If one uses the continuum model, the volume of electrostriction (cm^3 mol^{-1}) is given by:

$$V(\text{elect}) = (\text{N}e^2Z^2 / 2\text{D}r)(\partial \ln \text{D} / \partial P)_T = -4.2Z^2/r \qquad (1.24)$$

One can model the partial molal volume of an ion in water as being composed of two components:

$$V(\text{ion}) = V(\text{int}) + V(\text{elect}) = \text{a}r^3 + \text{b}Z^2/r \qquad (1.25)$$

where the $V(\text{int})$ is related to the size of the ion in space $= (4/3)\text{N}\pi r^3 = 2.52\,r^3$, with r in Å and $V(\text{elect}) = -4.2Z^2/r$. The fit of the measured values of $V(\text{ion})$ as a function of Z^2/r in Fig. 1.18 gives values of a = 4.48 and b = –8. These values are larger than those

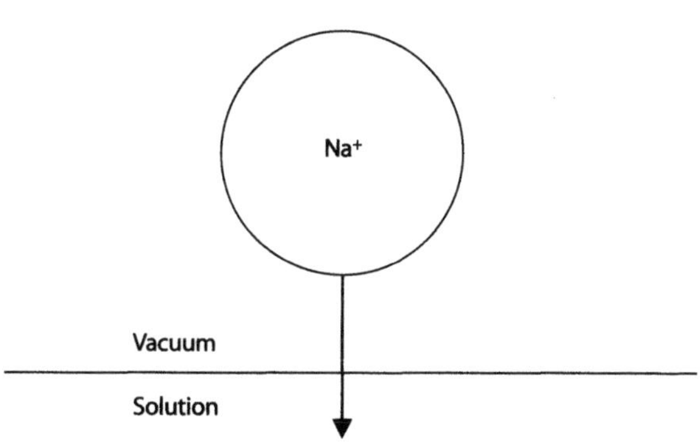

Fig. 1.15. The hydration of an ion

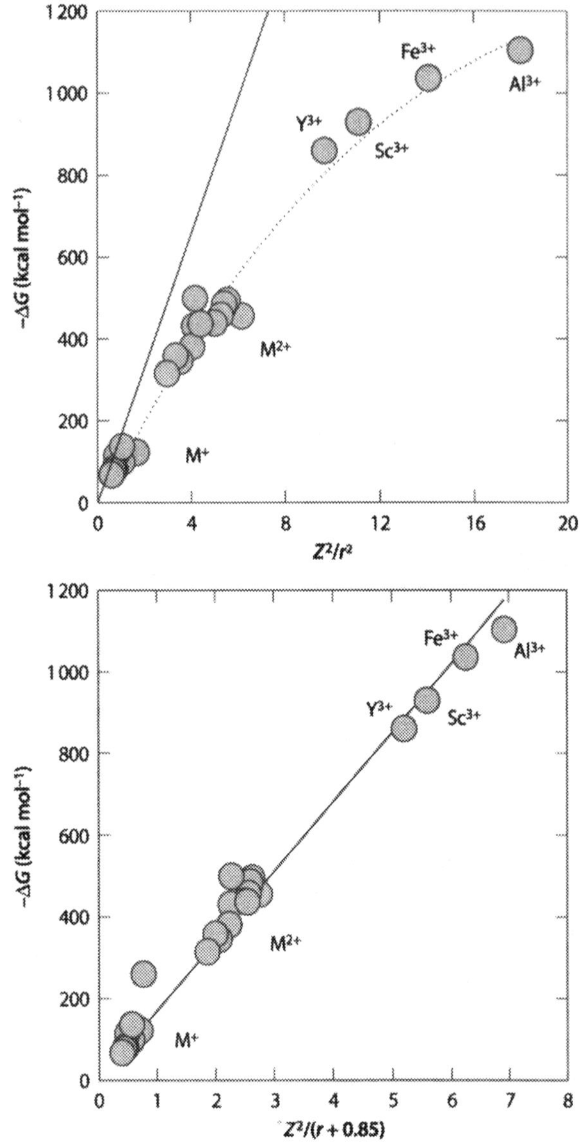

Fig. 1.16. The free energy of hydration for ions in water

expected. This can be attributed to the void volume and dielectric saturation. The addition of a sphere to water increases the volume due to the inefficient packing of the water molecules around the ion. Since the water molecules firmly attached to an ion are not mobile, the dielectric constant is lower near the ion and is much smaller than the value in the bulk solution (increasing a). As with the solution properties, the addition of 0.085 to the radius improves the fit and gives a slope nearer to the continuum value. If one assumes that the water molecules in the electrostricted region are not

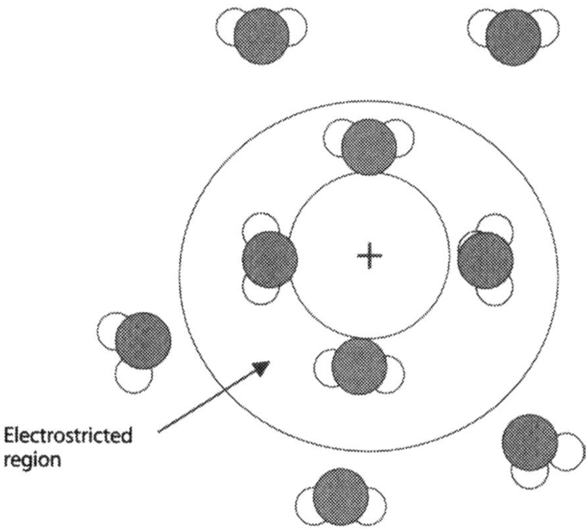

Fig. 1.17. The volume of electrostriction for ions in water

compressible ($\partial V(\text{int}) / \partial P = -K(\text{int}) = 0$), one can determine the number of water molecules hydrated to a given ion by:

$$V(\text{elect}) = h(V_E - V_B) \tag{1.26}$$

where h is the hydration number, V_E is the volume of water in the electrostricted region and V_B is the volume of bulk water (18.015 $cm^3\,mol^{-1}$).

The differentiation of Eq. 1.26 yields the compressibility of electrostriction:

$$K(\text{elect}) = K(\text{ion}) = -\partial V(\text{elect}) / \partial P = h(\partial V_B / \partial P) = -hV_B\beta_B \tag{1.27}$$

where $\beta_B = -(1/V_B)(\partial V_B / \partial P)$ (45.25×10^{-6} bar^{-1}). By rearranging Eq. 1.27, one has:

$$h = -K(\text{ion}) / V_B\beta_B \tag{1.28}$$

Combining equations we have:

$$V(\text{elect}) = -[V_E - V_B) / V_B\beta_B]K(\text{ion}) = -kK(\text{ion}) \tag{1.29}$$

A plot of V(elect) as a function of K(ion) is shown in Fig. 1.19 and yields a value of $k = 5\,000$ bar, which is in reasonable agreement with the continuum model (4 800 bar). This value of k yields a value of $V_E - V_B = -3.9$ $cm^3\,mol^{-1}$. Using $V_B = 18$ $cm^3\,mol^{-1}$, we obtain $V_E = 14$ $cm^3\,mol^{-1}$ for the volume of waters in the hydrated region. This is larger than the crystal volume of 6.6 $cm^3\,mol^{-1}$ or the volume corrected for packing of 11.8 $cm^3\,mol^{-1}$. This means that the water molecules in the electrostricted region are not tightly packed. Hydration numbers calculated from Eq. 1.26 yield values of 3 to 4 for monovalent ions, 6 to 9 for divalent ions and 15 for trivalent ions (Millero 1996).

Fig. 1.18. A sketch of the electrostriction region around an ion

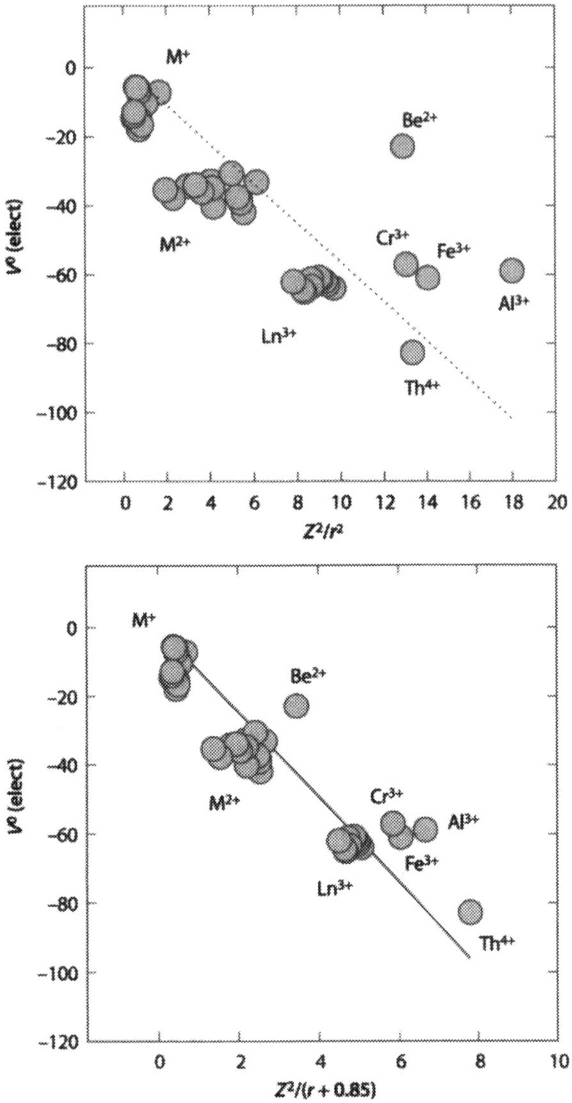

The ion-ion interactions can be examined using

1. Debye-Hückel Theory
2. Ion Pairing Theory
3. Friedman Cluster Expansion Theory

The Debye-Hückel Theory as with the Born Model considers the solution to be a dielectric medium. The free energy or activity coefficient (γ) changes are related to the changes in the ionic strength:

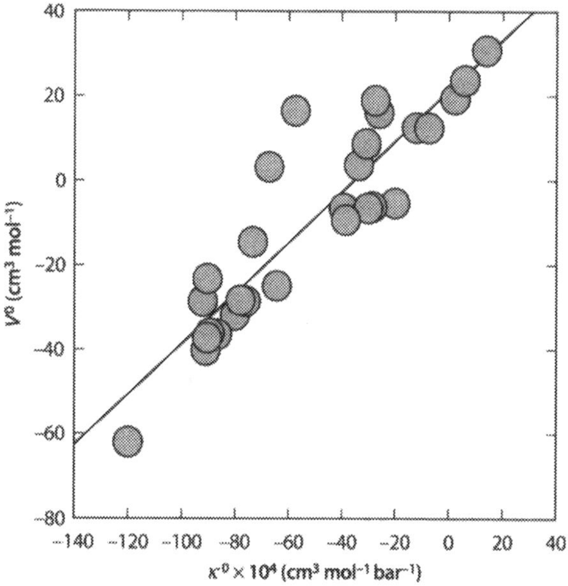

Fig. 1.19. Correlation of the molal volume and compressibility of ions

$$\text{Log}\ \gamma = -AZ^2 I^{1/2} / (1 + BI^{1/2}) \tag{1.30}$$

where A = 0.51 and B = 0.33 at 25 °C. The ion pairing model assumes that deviations from the Debye-Hückel theory are due to the formation of interactions between ions of an opposite sign. The ion pairing model assumes that only the interactions between cations and anions are important in the solution and that these interactions can be strong enough to form a new ion-paired species. The Friedman cluster expansion model attempts to consider interactions of an opposite sign and those of the same sign (Fig. 1.20). The interactions in a mixed electrolyte solution like sea water are accounted for by examining the properties of single (NaCl) and binary electrolyte solutions with a common ion ($NaCl + MgCl_2$). This latter model has proved to be useful in estimating the properties of mixed electrolyte solutions like sea water. The use of these methods is described in more detail in the next section.

1.3 Estimating the Properties of Mixed Electrolytes

This is done by using the apparent molal properties (Φ) of the solution (Fig. 1.21). The apparent molal property is related to the change that occurs when a salt is added to water. The apparent molal property is defined by:

$$\Phi = \Delta\rho / n = (P - P^0) / n \tag{1.31}$$

where n is the number of moles or equivalents of added salt, P is the property of the solution, and P^0 is the property of water. The apparent molal property for a mixed elec-

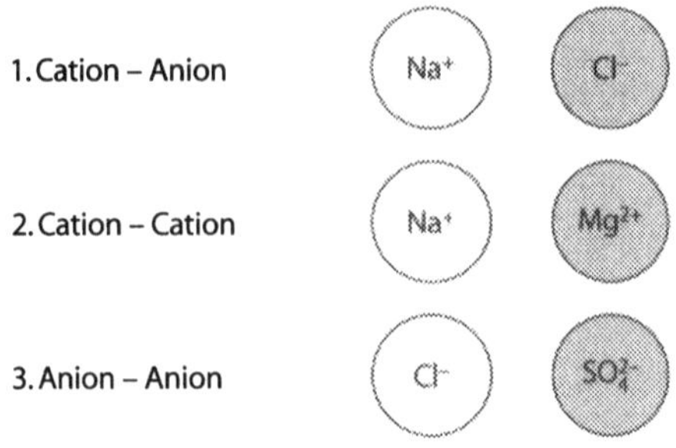

Fig. 1.20. Types of ion-ion interactions

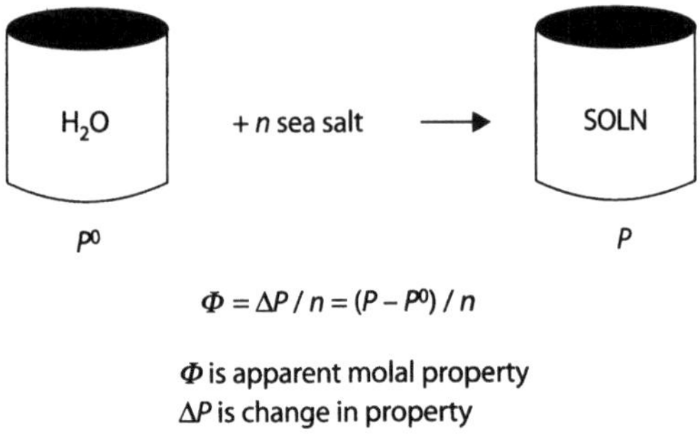

Fig. 1.21. Apparent molal properties of solutes in aqueous solutions.

trolyte solution is nearly equal to the weighted sum of the component electrolytes. This additivity, called Young's rule (Young and Smith 1954), is given by:

$$\Phi = \Sigma E_i \phi_i \tag{1.32}$$

where $E_i = n_i / n_T$ ($n_T = \Sigma n_i$) is the equivalent fraction of electrolyte i in the mixture and ϕ_i is the molal property of i at the ionic strength of the mixture. In terms of the ionic components of a mixed electrolyte solution, the equation becomes:

$$\Phi = \Sigma_M \Sigma_X E_M E_X \phi(MX) \tag{1.33}$$

where E_M and E_X are the equivalent fractions of cations (M) and anions (X) and ϕ(MX) is the apparent property of electrolyte MX at the ionic strength of the mixture. For the major components of sea water, this sum can be made three ways:

$$\Phi(SW) = E_{Na}E_{Cl}\phi(NaCl) + E_{Mg}E_{SO4}\phi(MgSO_4)$$

$$\Phi(SW) = E_{Na}E_{SO_4}\phi(Na_2SO_4) + E_{Mg}E_{Cl}\phi(MgCl_2)$$

$$\Phi(SW) = E_{Na}E_{Cl}\phi(NaCl) + E_{Na}E_{SO_4}\phi(MgSO_4) + E_{Mg}E_{Cl}\phi(MgCl_2) + E_{Mg}E_{SO_4}\phi(MgSO_4)$$

Experimentally, it is found that the third summation works best because it considers the weighted sum of all the possible cation-anion interactions in the solution. These plus-minus interactions represent the major ionic interactions that occur in the mixture. Once the Φ for the mixture is estimated, a given physical property can be determined from:

$$P = P^0 + \Phi n_T \tag{1.34}$$

For sea water, Eq. 1.33 can be broken down into terms for the individual major cations in the solution:

$$\Phi(SW) = E_{Na}\phi(Na\Sigma X_i) + E_{Mg}\phi(Mg\Sigma X_i) + E_{Ca}\phi(Ca\Sigma X_i) + E_K\phi(K\Sigma X_i) + E_{Sr}\phi(Sr\Sigma X_i) \tag{1.35}$$

The individual terms are given by:

$$\phi(Na\Sigma X_i) = E_{Cl}\,\phi(NaCl) + E_{SO_4}\phi(Na_2SO_4) + E_{HCO_3}\,\phi(NaHCO_3) + E_{Br}\phi(NaBr) + E_{CO_3}\phi(Na_2CO_3) + E_{B(OH)_4}\phi(Na(BOH)_4) + E_F\phi(NaF) \tag{1.36}$$

Similar equations can be written for $E_{Mg}\phi(Mg\Sigma X_i)$, $E_{Ca}\phi(Ca\Sigma X_i)$, etc.

Since the apparent molal properties of individual electrolytes have been fitted to equations of the form (Millero 1974, 1975, 2001):

$$\phi(NaCl) = \phi^0(NaCl) + aI^{1/2} + bI + \ldots \tag{1.37}$$

(a, b etc. are empirical constants), the values for sea water or other natural waters have the same form

$$\Phi(SW) = \Phi^0(SW) + AI^{1/2} + BI + \ldots \tag{1.38}$$

where the individual terms are given by:

$$\Phi^0(SW) = \Sigma_M\Sigma_X E_M E_X \phi^0(MX) \tag{1.39}$$

$$A = \Sigma_M\Sigma_X E_M E_X a(MX) \tag{1.40}$$

$$B = \Sigma_M\Sigma_X E_M E_X b(MX) \tag{1.41}$$

The superscript zero is to denote the values at infinite dilution or extrapolations to pure water. These extrapolations yield parameters that are only due to the ion-water interactions of the solution. The terms A and B are related to the ion-ion interactions of the solution. Combining Eqs. 1.34 and 1.38 gives:

$$P = P^0 + A'I + B'I^{3.2} \tag{1.42}$$

where $A' = AkI$ and $B' = BkI$ (since $n_T = kI$ for a mixed electrolyte solution of fixed composition). This equation has been shown to reliably represent a number of physical properties of sea water (Millero 1973a,b, 1975, 1982; Millero and Lepple 1973; Millero and Poisson 1981):

$$P = P^0 + \Sigma\text{ion-water} + \Sigma\text{ion-ion interactions} \tag{1.43}$$

Thus, any physical property of sea water at a given ionic strength is equal to the property of pure water plus a term related to the weighted ion-water and ion-ion interactions. The second term is completely additive for the components of an electrolyte mixture.

These equations have been used to estimate the properties of sea water (Millero 1973a,b, 1974, 1975, 1978; Millero and Lepple 1973; Millero et al. 1977), seas (Millero 1978; Millero and Chetirkin 1980; Millero et al. 1982), lakes (Effler et al. 1986; Millero 2000a), rivers (Millero 1975; Millero et al. 1976c, 1978); estuaries (Millero 1975; Millero and Kremling 1976; Millero et al. 1976a,c) and brines (Millero et al. 1982). This includes the estimates of sound speeds (Millero et al. 1977), heat capacities (Millero et al. 1973a), enthalpies (Millero 1974), freezing points (Millero 1974), densities (Millero 1975), expansibilities (Millero 1973b), and compressibilites (Millero 1973b). Examples of the estimates of the densities and compressibilities of sea water using this simple additivity method are shown in Table 1.3. The calculated values are in good agreement with the measured values. At higher ionic strengths, the estimates are not as good. This is shown in more detail for the density comparisons of Red Sea Brine waters in Table 1.4. These larger errors at higher ionic strengths are related to excess mixing parameters due to the interactions of ions of the same sign (Na^+-Mg^{2+}, Cl^--SO_4^{2-}). These excess-mixing properties can be studied by mixing two electrolyte solutions at a constant ionic strength. For the mixing of the major sea salts, there are six possible mixtures that can be represented by the so-called cross square diagram Fig. 1.22)

The mixing of the salts around the sides of this diagram has either a common cation or anion during mixing. A number of studies by Young et al. (1957) have shown the excess mixing properties ΔP_{E_X} follow some simple rules. They are:

1. The values of ΔP_{E_X} in dilute solutions are not very large and can be assumed to be zero. This leads to the additivity of ϕ or Young's first rule.
2. The values of ΔP_{E_X} for mixtures with a common anion or cation are not strongly affected by the common ion. For example, the ΔP_{E_X} for mixing NaCl and KCl is nearly the same as mixing NaBr and KBr.

   $$\Delta P_{E_X}(\text{NaCl-KCl}) = \Delta P_{E^X}(\text{NaBr-KBr}) \tag{1.44}$$

   Since this mixing process is largely related to cation-cation and anion-anion interactions, this means that as a first approximation, plus-plus and minus-minus interactions are independent of the other ions in the solutions.
3. The third rule is called the cross square rule and is given by:

   $$\Sigma\square = \Sigma X \tag{1.45}$$

   which states the sum of the excess mixing properties around the sides of the diagram given above is equal to the sum of the excess properties of the cross mixtures. For the major sea salts, this gives:

Table 1.4. Comparisons of the measured and calculated density of Red Sea Brines at 25 °C as a function of ionic strength

Ionic Strength	$\Delta\rho$(Meas) $-\rho$(Calc) (ppm)	
	Without ΔV_{EX}	With ΔV_{EX}
0.47	−18	−22
1.22	81	8
1.75	134	13
2.61	199	11
3.56	211	−29
4.23	235	−25
5.33	297	20
6.11	444	188

Fig. 1.22. Cross square diagram for the mixing of the major sea salts

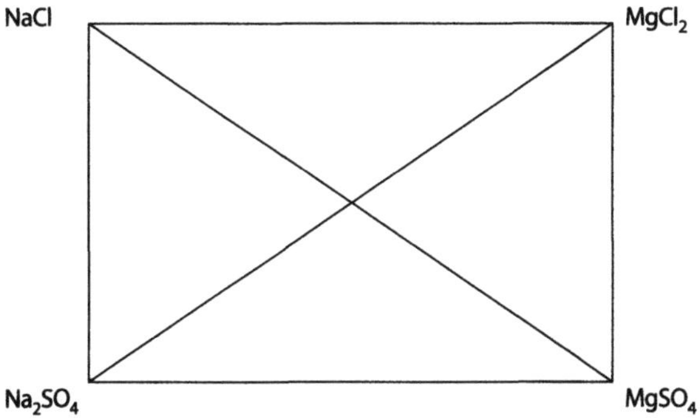

$$\Delta P_{E_X}(NaCl + Na_2SO_4) + \Delta P_{E_X}(MgCl_2 + MgSO_4) + \Delta P_{E_X}(NaCl + MgCl_2)$$
$$+ \Delta P_{E_X}(MgSO_4 + Na_2SO_4) = \Delta P_{E_X}(NaCl + MgSO_4) + \Delta P_{E_X}(Na_2SO_4 + MgCl_2) \quad (1.46)$$

This simplifies the estimation of the ΔP_{E_X} for a complicated mixture like sea water. The addition to Young's simple rule gives:

$$\phi = \Sigma_M \Sigma_X E_M E_X \phi(MX) + \Delta P_{E_X} / n_T \quad (1.47)$$

where $\Delta P_{E_X} / n_T$ is given by:

$$\Delta P_{E_X} / e_T = (n_T / 4)[\Sigma E_X E_N E_M (Z_M + Z_X)(Z_N + Z_X) \Delta P(M,N)^X$$
$$+ \Sigma E_X E_Y E_M (Z_M + Z_X)(Z_M + Z_Y) \Delta P(X,Y)^M] \quad (1.48)$$

where Z_i is the absolute charge on ion i, $\Delta P(M,N)^X$ is the excess mixing properties for MX + NX, and $\Delta P(X,Y)^M$ is the excess mixing properties for MX + MY. This equation attempts to account for cation-cation and anion-anion interactions in the mixture by neglecting triplicate interaction (Young's third rule). Since the values of ΔP_{E_X} are symmetrical around the ionic strength fraction (see Fig. 1.23), the values of ΔP needed to estimate ΔP_M for the mixture can be the value near $y = 0.5$.

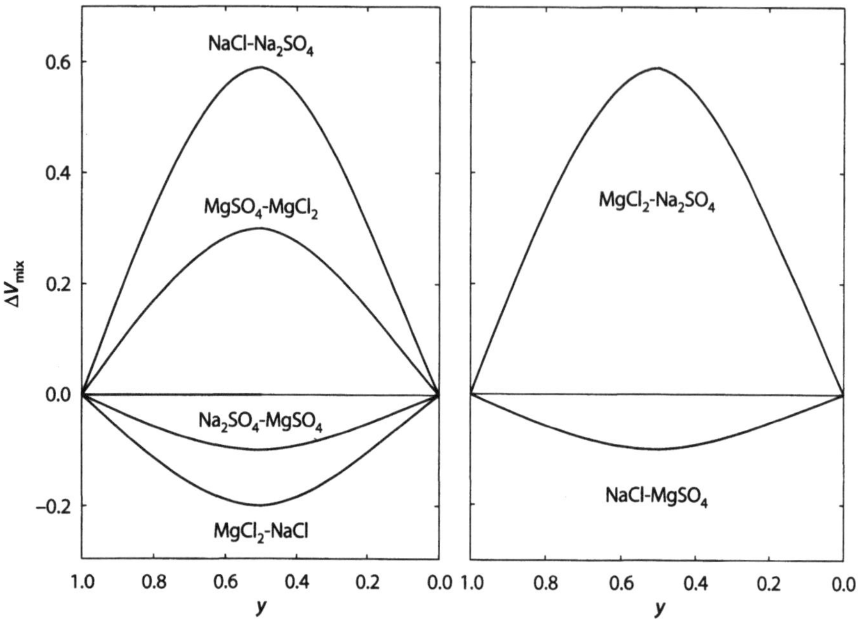

Fig. 1.23. The volume change of mixing the major sea salts

The importance of using these mixing terms to estimate the physical properties of a mixture is demonstrated in Table 1.4. The estimated densities are in better agreement with the measured values when the excess mixing terms are considered (ΔV_{EX} is in this case, the volume of mixing the major sea salts).

For most physical properties in dilute solutions, the estimates can be made without the ΔP_{EX} term. When adding the excess mixing terms to the equations, one is dividing the Σ ion-ion into three terms:

$$\Sigma \text{ ion-ion} = \text{D.H.} + \Sigma \text{ Binary} + \Sigma \text{ Ternary} \tag{1.49}$$

where D.H. is a Debye-Hückel contribution, the Σ Binary term is related to the interactions of ions of opposite (Na-Cl, Mg-Cl) and like sign (Na-Na, Cl-Cl, Na-Mg) and the Σ Ternary term is related to triplet interactions (Na-Mg-Cl, Cl-SO_4-Na). The so-called Pitzer equations (Pitzer 1991) attribute the binary (Na-Na, Cl-Cl) and ternary interactions (Na-Na-Cl) for single electrolytes (NaCl) to three terms β^0, β^1, and C^ϕ. The binary (Na-Mg) interactions for mixtures (NaCl + $MgCl_2$) are related to Θ, and the ternary interactions (Na-Mg-Cl) are related to Ψ. The Pitzer's equations thus incorporate Young's rule and are embodied in the formulation of the Pitzer equations. This general approach, although somewhat complicated, can account for all the possible interactions in a stepwise manner. Computer codes have been written that can be used to estimate the physical chemical properties of natural waters over a wide range of temperature (0 to 50 °C) and ionic strength (0 to 6 m).

1.4 Estimating Transport Properties

All of the properties discussed above were concerned with thermodynamic properties. It has been shown some time ago that the additivity principles discussed above can also be used to estimate the transport properties of mixed electrolyte solutions. For example, the viscosity of sea water has been estimated using these methods and the results are in good agreement with the experimental measurements (Millero 1974). More recently I have examined additivity methods to estimate the conductivity of natural waters (Millero 2000a). This is now necessary since the salinity of lakes and seas are frequently estimated from conductivity measurements made with the CTD (conductivity, temperature and depth) systems used in the oceans (Mcmanus et al. 1992). Since it has been shown above and elsewhere that the properties of lakes and seas are the same as sea water at the same salinity, it would be useful to be able to use these conductivity measurements to determine the total salinity of the lakes, etc. This salinity could be used to determine the physical chemical properties of the lake from equations for sea water. It is thus appropriate to briefly discuss the effect of composition on the conductivity of lakes and other natural waters.

The conductivity, unlike the density, responds only to the ionic components of the solution. The conductivity salinity is thus a strong function of the composition of the waters. This has been demonstrated (Millero 1984, 2000a) by using the equivalent conductances of the major components of sea water and river waters of different composition. The conductance of sea water in dilute solutions is larger than the values for "World" and St. Lawrence River waters. This is because the conductances of Na^+ and Cl^- are larger than Mg^{2+}, Ca^{2+} and HCO_3^- : the major components of river water.

Since the effect of temperature and pressure on the physical chemical properties are not strong functions of the composition, the relationship between the conductivity and the density salinity can be approximated by measurements or calculations at 25 °C. The easiest way to determine this relationship is to make direct conductivity and density measurements on samples from the lake. Recently this has been done (Jellison et al. 1999) on samples from Mono Lake. This requires measurements of both conductivity and density as a function of temperature and composition of the lake waters of variable salinity. This is quite time consuming and essentially defines an equation of state for the lake independent of sea water.

A more useful approach is to make density and conductivity measurements at 25 °C on the waters collected from the lake and sea water of known salinity. This will allow one to use a semi-empirical correlation to relate the sea water salinity determined by conductivity to the lake conductivity:

$$S(\text{Lake})_{\text{Cond}} = S(\text{Sea water})_{\text{Cond}} + \Delta_{\text{Cond}} \qquad (1.50)$$

where Δ_{Cond} is an empirical function of the differences between the calculated salinity using the Practical Salinity Scale for the lake and sea water. If density measurements are also made, one can develop a relationship between the conductance salinity and total salinity derived from density measurements using the equation of state of sea water (Millero and Poisson 1981)

$$S(\text{Lake})_T = S(\text{Lake})_{\text{Cond}} + \Delta(\text{Dens} - \text{Cond}) \quad (1.51)$$

where $\Delta(\text{Dens} - \text{Cond})$ is a constant for function $S(\text{Lake})_{\text{Cond}}$.

The methods outlined above can be used without any detailed knowledge of the composition of the lake. If it is not possible to make conductivity and density measurements on lake waters, the next best thing to do is to make measurements on artificially made lake waters. As we have shown, measurements of artificial waters can provide reliable densities and conductivities of river, estuarine, brine and sea water of known composition (Millero 1973a,b; Millero and Lepple 1973; Millero et al. 1976a, 1982; Millero and Chetirkin 1980).

If only composition data are available, one can determine the salinity from partial molal volume data as described above. The conductivity of the lake can be estimated from the conductivity of the components of the lake (Sorensen and Glass 1987; Wüest et al. 1996) and dilute solution measurements on pure electrolytes (Robinson and Stokes 1959). Wüest et al. (1996) have recently given equations for the equivalent conductivity of the major components of lakes as a function of temperature and composition. They have used these equations to estimate the conductivity of the waters from Lake Malawi. The validity of these methods needs to be examined by making direct measurements on waters of known composition as described above. To examine the approximate relationships between the conductivity of lakes and sea water, I have determined the equivalent conductance of sea water and the lakes considered in this paper using the infinite dilution ionic conductance taken from Robinson and Stokes (1959) and Wüest et al. (1996). The equivalent conductance of the mixture (Λ_0, μS cm^{-1} (eq/l)$^{-1}$) at infinite dilution is determined from:

$$\Lambda_0 = \Sigma_i E_i \lambda_0(i) \quad (1.52)$$

where E_i is the equivalent fraction and $\lambda_0(i)$ is the equivalent conductance of ionic component i. The values of Λ_0 at 25 °C determined from this equation are given in Table 1.5. The lake values are lower largely due to the decrease in the concentration of Na^+ and Cl^-. The ratio of $\Lambda_0(\text{Lake})/\Lambda_0(\text{SW})$ at different temperatures are shown in Fig. 1.24. The ratios are second degree functions of temperature and vary from 0.80 to

Table 1.5. The cross-square rule for the volume and compressibi-lity of the major sea-salts at 25 °C

	ΔV_m	$\Delta 10^4 K_m$
Common ion mixtures		
$NaCl$-Na_2SO_4	0.56	1.65
Na_2SO_4-$MgSO_4$	−0.11	−0.67
$MgSO_4$-$MgCl_2$	0.27	0.65
$MgCl_2$-$NaCl$	−0.22	−1.03
Total	0.50	0.60
Uncommon ion mixtures		
$NaCl$-$MgSO_4$	−0.06	−1.34
$MgCl_2$-Na_2SO_4	0.59	1.98
Total	0.53	0.64

0.84. The conductance ratios for the lakes (0.82) can be compared to 0.88 for rivers (Millero 1984).

The specific conductances of the solutions are related to the equivalent conductances by:

$$\kappa_0\ (\mu\mathrm{S\ cm^{-1}}) = \Lambda_0 e_\mathrm{T} \tag{1.53}$$

where $e_\mathrm{T} = \frac{1}{2}\Sigma Z_i M_i$ If one assumes that the value of κ_0 is proportional to the salinity, one would expect that the values for the lake should be proportional to the ratio of κ_0. The values of κ_0 give ratios for κ_0(Malawi) / κ_0(Baikal) = 2.06 and κ_0(Tanganyika) / κ_0(Malawi) = 2.77, which are in reasonable agreement with the salinity ratios, S(Malawi) / S(Baikal) = 2.13 and S(Tanganyika) / S(Malawi) = 2.77. This simple comparison suggests that one can make a direct link between the conductivity and the total salinity that can be used to determine the *PVT* properties (Wüest et al. 1996). The calculated values of the total salinity can be related to κ by:

$$S_\mathrm{T} = k\kappa \tag{1.54}$$

where k is a function of temperature and κ has been corrected for changes in the composition ($\kappa = f\kappa_0$, where f is a reduction coefficient, Wüest et al. 1996). Since the properties of sea water are accurately known as a function of the practical salinity, it is useful to make a direct link between the measured conductivity on this scale to the total sa-

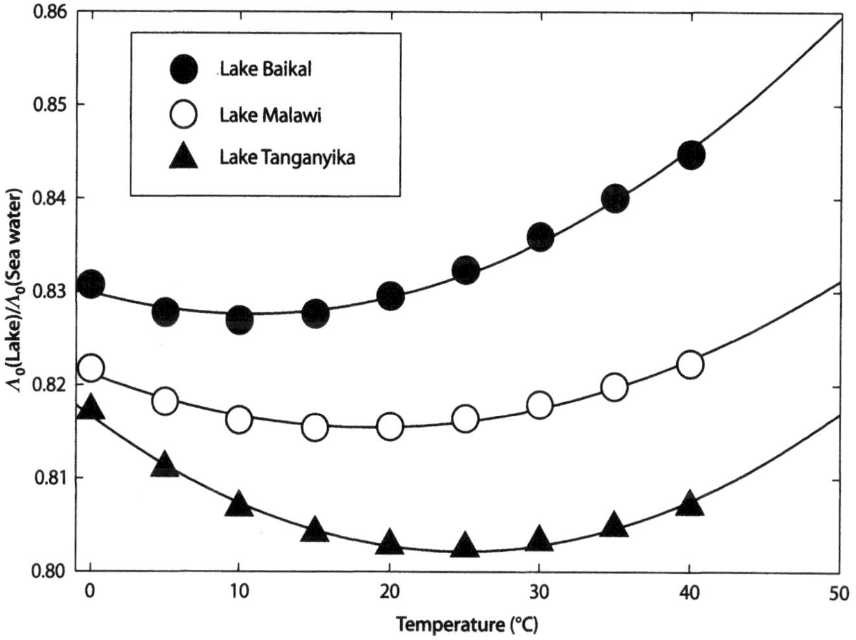

Fig. 1.24. The ratio of the infinite dilution equivalent conductance of Lake Baikal, Malawi and Tanganyika and sea water, Λ_0(lake)/Λ_0(sea water), as a function of temperature (Millero 2000a)

Table 1.6. The equivalence conductance of sea water and lake waters at 25 °C. Calculated using the composition data for the lakes from the literature: Lake Baikal (Callender and Granina 1997; Falkner et al. 1991); Lake Tanganyika (Degens et al. 1971; Edmond 1974); Lake Malawi (Wüest et al. 1996)

Ion	$\lambda_0(i)$	$E_i\lambda_0(i)$			
		Sea water	Baikal	Malawi	Tanganyika
Na^+	49.92	38.66	6.30	16.90	18.55
Mg^{2+}	53.04	9.25	10.89	12.62	23.91
Ca^{2+}	60.96	2.07	39.92	21.34	4.50
K^+	73.56	1.24	1.44	4.55	8.24
Sr^{2+}	59.45	0.02	0.00	0.00	0.00
Cl^-	76.41	68.87	0.77	4.14	6.47
SO_4^{2-}	79.97	7.46	7.48	4.80	1.31
HCO_3^-	44.50	0.13	39.62	39.75	39.64
Br^-	78.14	0.11			
CO_3^{2-}	69.83	0.06		0.30	
F^-	44.50	0.01			
$B(OH)_4^-$	55.40	0.01			
$\Lambda_0 = \Sigma E_i\lambda_0(i)$		127.87	106.43	104.39	102.62
$\kappa_0 = \Lambda_0 e_T$		80.26	0.131	0.270	0.751
$\kappa_0(SW)$[a]			0.221	0.444	1.303
$\kappa_0(SW)/\kappa_0(Lake)$			1.68	1.64	1.73

[a] Seawater diluted to same total salinity of the lake (at $S_T = 35$, $e_T = 0.6277$).

linity determined from the composition. If sea water is diluted to the same total salinity (Table 1.6), the conductivity of sea water at 25 °C is 1.68 ±0.03 larger than the values for the lakes. Direct measurements of the conductance are needed to determine if these methods are reliable.

Acknowledgements

The author would like to acknowledge the support of the Oceanographic Section of the National Science Foundation and the National Oceanic and Atmospheric Administration for supporting this work.

References

Brewer PG, Bradshaw A (1975) The effect of non-ideal composition of sea water on salinity and density. J Mar Res 33:157–175.

Callender E, Granina L (1997) Geochemical mass balances of major elements in Lake Baikal. Limnol Oceanogr 42:148–155

Carpenter JH, Manella ME (1973) Magnesium to chlorinity ratios in seawater. J Geophys Res 78:3621–3626

Connors DN, Weyl PK (1968) The partial equivalent conductance of salt in seawater and the density-conductance relationship. Limnol Oceanogr 13:39–50

Culkin F (1965) The major constituents of sea water. In: Riley P, Skirrow G (eds) Chemical oceanography, vol I. Academic Press, New York, pp 121–161

Culkin F, Cox RA (1966) Sodium, potassium, magnesium, calcium and strontium in seawater. Deep-Sea Res 13:789–804

Degens ET, Von Herzen RP, Wong H-K (1971) Lake Tanganyika: Water chemistry, sediments, geological structure. Naturwissen 58:229–234

Dittmar W (1884) Report on researches into the composition of ocean water collected by H.M.S. *Challenger* 1873–76. Sci. Res. Voyage "Challenger", 1873–76. Chem Phys 1:1–211

Edmond J (1974) Lake chemistry. In: Craig H (ed) Lake Tanganyika geochemical and hydrographic study. University of California, San Diego (SIO Ref. Ser. 75-5, pp 1–64)

Effler SW, Schimel K, Millero FJ (1986) Salinity, ionic strength, and chloride relationships in ion polluted Onondaga Lake N.Y. Wat Air Soil Pollut 27:169–180

Falkner KK, Measures CI, Herbelin SE, Edmond JM, Weiss RF (1991) The major and minor element geochemistry of Lake Baikal. Limnol Oceanogr 36:413–423

Fernandez H, Vazquez F, Millero FJ (1982) The density and composition of hypersaline waters of a Mexican lagoon. Limnol Oceanogr 27:315–321

Gieskes JM (1982) The practical salinity scale 1978. A reply to comments by T. R. Parsons. Limnol Oceanogr 27:387–389

Hill KD, Dalphinee TM, Woods DJ (1986) The extension of the practical salinity scale 1978 to low salinities. IEEE J Oceanic Eng OE-11:109

Jellison R, Macintyre S, Millero FJ (1999) Density and conductivity properties of Na-CO_3-Cl-SO_4 brine from Mono Lake, USA. Int J Salt Lake Res 8:41–53

Knudsen M (1901) Hydrographical tables according to the measurings of Carl Forch, P. Jacobsen, Martin Knudsen and S. P. L. Sorensen, G. E. C. Gad: Copenhagen. Williams Norgate, London

Lyman J, Fleming RH (1940) Composition of seawater. J Mar Res 3:134–146

Mcmanus J, Collier RW, Chen CA, Dymond J (1992) Physical properties of Crater Lake, Oregon: A method for the determination of a conductivity- and temperature-dependent expression for salinity. Limnol Oceanogr 37:41–53

Millero FJ (1973a) Theoretical estimates of the isothermal compressibility of seawater. Deep-Sea Res 20:101–105

Millero FJ (1973b) Seawater – a test of multicomponent electrolyte solution theories. I. The apparent equivalent volume, expansibility and compressibility of artificial seawater. J Solution Chem 2:1–22

Millero FJ (1974) Seawater as a multicomponent electrolyte solution. In: Goldberg E (ed) The sea. Wiley, New York, pp 3–80

Millero FJ (1975) The physical chemistry of estuaries. In: Church TM (ed) Marine chemistry in the coastal environment. ACS, Washington, D.C. (ACS Symp. Ser. 18, pp 25–55)

Millero FJ (1978) The physical chemistry of Baltic Sea waters. Thalassia Jugosl 14:1–46

Millero FJ (1982) The thermodynamics of seawater. Part 1: The PVT properties. Ocean Sci Eng 7:403–460

Millero FJ (1983) The thermodynamics of seawater. Part II: Thermochemical properties. Ocean Sci Eng 8:1–40

Millero FJ (1984) The conductivity-density-salinity-chlorinity relationship for estuarine waters. Limnol Oceanogr 29:1317–1321

Millero FJ (1996) Chemical oceanography. CRC Press, Boca Raton, FL

Millero FJ (2000a) The equation of state of lake waters. Aquatic Geochem 6:1–17

Millero FJ (2000b) Effect of changes in the composition of seawater on the density-salinity relationship. Deep-Sea Res I 47:1583–1590

Millero FJ (2001) The physical chemistry of natural waters. Wiley Scientific, N.Y.

Millero FJ, Chetirkin PV (1980) The density of Caspian Sea waters. Deep-Sea Res 27:265–271

Millero FJ, Kremling K (1976) The densities of Baltic Sea waters. Deep-Sea Res 23:1129–1138

Millero FJ, Lepple FK (1973) The density and expansibility of artificial seawater solutions from 0 to 40 °C and 0 to 21‰ chlorinity. Mar Chem 1:89–104

Millero FJ, Poisson A (1981) International one atmosphere equation of state of seawater. Deep-Sea Res 28:625–629

Millero FJ, Perron G, Desnoyers JE (1973a) Heat capacity of seawater solutions from 5 to 35 °C and 0.5 to 22‰ chlorinity. J Geophys Res 78:4499–4507

Millero FJ, Hansen LD, Hoff EV (1973b) The enthalpy of seawater from 0 to 30 °C and 0 to 40‰ salinity. J Mar Res 31:21–39

Millero FJ, Gonzalez A, Ward GK (1976a) The density of seawater solutions at one atmosphere as a function of temperature and salinity. J Mar Res 34:61–93

Millero FJ, Gonzalez A, Brewer PG, Bradshaw A (1976b) The density of North Atlantic and North Pacific deep waters. Earth Planet Sci Lett 32:468–472
Millero FJ, Lawson D, Gonzalez A (1976c) The density of artificial river and estuarine waters. J Geophys Res 81:1177–1179
Millero FJ, Ward GK, Chetirkin PV (1977) Relative sound velocities of sea salts at 25 °C. J Acoust Soc Am 61:1492–1498
Millero FJ, Forsht D, Means D, Gieskes J, Kenyon K (1978) The density of North Pacific Ocean waters. J Geophys Res 83:2359–2364
Millero FJ, Mucci A, Zullig J, Chetirkin P (1982) The density of Red Sea brines. Mar Chem 11:463–475
Morris AW, Riley JP (1964) The direct gravimetric determination of the salinity of seawater. Deep-Sea Res 11:899–904
Morris AW, Riley JP (1966) The bromide/chlorinity and sulphate/chlorinity ratio in seawater. Deep-Sea Res 13:699–705
Parsons TR (1982) The new physical definition of salinity: Biologists beware. Limnol Oceanogr 27:384–385
Pitzer KS (1991) Ion interaction approach: Theory and data collection. In: Pitzer KS (ed) Activity coefficients in electrolyte solutions. CRS, Boca Raton, FL, pp 75–153
Poisson A, Perie M, Perie J, Chemla M (1979) Individual equivalent ionic conductances of the major ions in seawater. J Solution Chem 8:377–394
Poisson A, Lebel J, Brunet C (1980) Influence of local variations in the ionic ratios on the density of seawater in the St. Lawrence area. Deep-Sea Res 27:763–781
Poisson A, Lebel J, Brunet C (1981) The densities of western Indian Ocean, Red Sea and eastern Mediterranean surface waters. Deep-Sea Res 28:1161–1172
Riley JP, Tongudai M (1967) The major cation/chlorinity ratios in seawater of seawater. Chem Geol 2:263–269
Robinson RA, Stokes RH (1959) Electrolyte solutions, 2nd edn. Butterworths, London, pp 455
Sharp JH, Culberson CH (1982) The physical definition of salinity: A chemical evaluation. Limnol Oceanogr 27:385–387
Sorensen JA, Glass GE (1987) Ion and temperature dependence of electrical conductance for natural waters. Anal Chem 59:1594–1597
Warner TB (1971) Normal fluoride content of seawater. Deep-Sea Res 18:1255–1263
Wilson TRS (1975) Salinity and the major element of sea water. In: Riley JP. Skirrow G (eds) Chemical oceanography, vol 1, 2nd edn. Academic Press, New York, pp 365–413
Wüest A, Piepke G, Halfman JD (1996) Combined effects of dissolve solids and temperature on the density stratification of Lake Malawi. In: Johnson TC, Odada EO (eds) The limnology, climatology and paleoclimatology of the East African lakes. Gordon and Breach Scientific Pubs., pp 183–202
Young TF, Smith MB (1954) Thermodynamic properties of mixtures of electrolytes in aqueous solutions. J Phys Chem 58:716–724
Young TF, Wu Y-C, Krawetz AA (1957) Thermal effects of interaction between ions of like charge. Discuss Faraday Soc 24:37,77

Chapter 2

The Chemical and Physical Properties of Marine Aerosols: An Introduction

J. M. Prospero

2.1 Introduction

The atmosphere is an important pathway for the transport of particulate matter from the continents to the oceans. Although the magnitude of these wind-borne fluxes is not accurately known, there is evidence that some could be large enough to have a significant impact on chemical and biological processes in the oceans. In addition, wind-borne particles (i.e. aerosols) play an important role in climate-related processes which in turn can have a great impact on physical, chemical, and biological processes taking place in the oceans. Aerosols can affect climate directly by scattering and absorbing both solar radiation and that which is re-emitted from the surface of the Earth to the atmosphere. Aerosols play a critical role in water vapour nucleation processes. In this way, they can affect the concentration and size distribution of cloud droplets, which in turn can alter the radiate properties of clouds. The role of aerosols in cloud processes also affects the nature and distribution of rainfall and the subsequent distribution of clouds.

The role of aerosols in climate forcing has only recently received attention. Aerosols were largely ignored in the first Intergovernmental Panel on Climate Change (IPCC) assessment. It was only in the 1995 IPCC report (IPCC 1996) that the importance of aerosols was acknowledged and an effort was made to estimate the forcing due to anthropogenic particles. The report concluded that because of the many unknowns about the chemical, physical, and radiative properties of aerosols and their temporal and spatial variability, only very crude estimates could be made. The IPCC report and other studies have served to stimulate a broad research focus on aerosols and climate-related processes, both in the laboratory and in the field. We have learned much from these programmes, and it is now generally accepted that aerosols play an important role in climate. We are now better able to constrain the estimates of forcing or, at least, to better identify those parameters that must be better defined. Nonetheless, despite these advances, we still do not know enough about many aspects of aerosol properties and their role in climate to make a good assessment – indeed, our knowledge of aerosol processes is much worse than that of most greenhouse gases.

Another major change in the field is that we now have a better appreciation of the types of aerosols that are important. For many years, the aerosol community has focused largely on sulphur (and related) chemistry because of concerns about the impact of sulphur pollution emissions on the atmospheric sulphur cycle. It was not until the mid-90s that other aerosol species began to receive much attention. Today many "natural" species are included in climate assessments: sea salt, mineral dust, and biogenic organics. The inclusion of these species is necessary for two reasons: so that we can assess the relative impacts of anthropogenic and natural species and thereby to be able to partition changes that might occur in both through time; so that we can assess any changes in natural emissions that might occur as a result of human-induced climate change. Knowledge of these natural climate-related species also helps us to understand the factors that played a role in pre-human climate change. Motivated by these concerns, a very substantial part of the community effort is now devoted to a wide range of aerosol species.

The study of the role of aerosols is complicated by the fact that the physical and chemical properties of aerosols are highly variable in both time and space. In contrast, most common "greenhouse" gases (e.g. CO_2, CH_4, halocarbons) have relatively long lifetimes (months to years) in the atmosphere, and as a consequence, their distribution is relatively uniform over the Earth, and changes occur relatively slowly. Aerosols, on the other hand, have a residence time in the troposphere that ranges from a few days to a few weeks. Consequently, the concentration, composition and physical properties of particles in the atmosphere can vary greatly, depending on the distribution of sources, the meteorological processes in the source regions, the large scale circulation systems that subsequently control long-range transport, and finally, the various removal processes that act on the particles. In order to assess the transport of continental materials to the oceans and the general impact on climate, we must have a good understanding of all these processes.

This paper is intended to be a brief review of the chemical and physical properties of aerosol species over the oceans. The emphasis is on those aerosol species that play an important role in marine biogeochemical and climate processes:

- Sea salt aerosols: Sea salt aerosols are produced by the bursting of bubbles at the surface of the ocean. In remote ocean regions, sea salt aerosol is often the primary light scattering aerosol component, and it is an important source of cloud-forming nuclei. Consequently, sea salt has a role in climate. Furthermore, over large areas of the ocean, sea salt aerosol offers a large surface area for chemical and physical interactions with other aerosol species and with gases; because of its large size and high settling velocity, sea salt aerosol can serve as a major sink for other atmospheric components.
- Non-sea-salt (nss) sulphate aerosols: Nss-SO_4^{2-} is that fraction of marine SO_4^{2-} aerosol that is not derived from sea water aerosol droplets. Nss-SO_4^{2-} plays a major role in radiation and cloud processes. As such, it is important to understand the factors affecting its distribution over the oceans. Nss-SO_4^{2-} is the major acidic aerosol species in the atmosphere; over large areas of the world, the pH of aerosols and precipitation is largely controlled by the concentration of nss-SO_4^{2-} and ammonium. Nss-SO_4^{2-} aerosol over the oceans has two major sources: the oxidation of dimethyl-sulphide (DMS) emitted by marine organisms and pollution transported from the continents.
- Nitrate and ammonium: The global budgets of both these species have been hugely impacted by human activities. Ammonium is the major neutralizing species in the atmosphere. Both species can play an important role as nutrient inputs to coastal waters and oceans.

- Mineral dust: Mineral dust is a major aerosol species over many ocean regions. It is an important contributor to deep-sea sediments. The iron associated with dust is believed to be a limiting micronutrient over many remote ocean regions; as such, the dust cycle could be a major factor in the global oceanic carbon cycle. Other elements associated with dust could play an important role in the biogeochemistry of the oceans.

This paper can only serve as a brief overview of the field of aerosol chemistry. For those interested in further study, Hobbs (2000) and Jacob (1999) present excellent and concise introductions to the field of atmospheric chemistry. Graedel and Crutzen (1993) offer a broad and easily readable review of the fundamentals of atmospheric chemistry in climate processes, while Hobbs (1993) provides a good introduction to aerosol-cloud-climate interactions; Charlson and Heintzenberg (1995) present broad coverage of the subject. Broader and more detailed coverage of atmospheric chemistry and aerosols can be found in the excellent texts by Finlayson-Pitts and Pitts (2000) and by Seinfeld and Pandis (1998).

2.1.1
Physical Characteristics of Aerosols

An aerosol is defined as a dispersion of solid or liquid particles in a gaseous phase – in this case, the atmosphere. The size of the particles can range from macromolecules to over 100 μm diameter. In practice most of the aerosol mass that is normally of interest over the oceans is found in the size range under 10–20 μm diameter. Larger particles have high settling velocities in the atmosphere, and consequently they are not carried far from their sources. (For example, a 20 μm diameter particle with unit density has a settling velocity at sea level of 1.2 cm s^{-1}, about 1 km per day.) While aerosols are usually thought of as a solid or liquid particle, in reality, in many regions most particles are complex mixtures of soluble and insoluble species, and depending on the relative humidity, they could consist of both solid and liquid phases.

Aerosols can be produced directly ("primary" aerosols) or as a product of the reactions of gases in the atmosphere ("secondary" aerosols). The mode of production affects the size distribution of the resulting aerosol. Figure 2.1 shows a schematic of the distribution of particle surface area of an idealized ensemble of aerosol particles formed by a variety of processes. (Seinfeld and Pandis 1998, Fig. 2.15, p. 101). Examples of primary aerosols are sea salt, mineral dust, soot emitted from smokestacks, diesel exhaust, and particles shed by plants (e.g. plant waxes, fibres). Physical processes (grinding of rocks, bursting bubbles) normally produce relatively large particles; as a result, the size distribution of primary particles places most of the mass above 1 μm diameter as shown in Fig. 2.1; these are normally referred to as "coarse" particles. For example, the mass median diameter (MMD) of sea salt particles over the ocean is generally in the range of about 5–10 μm diameter or greater (depending on wind conditions); nonetheless, a significant and important fraction of the sea salt mass lies below 1 μm (O'Dowd et al. 1997). The MMD of mineral dust particles over the source regions can be extremely large: many 10's of micrometres, but over the oceans it is typically only several μm (Duce 1995).

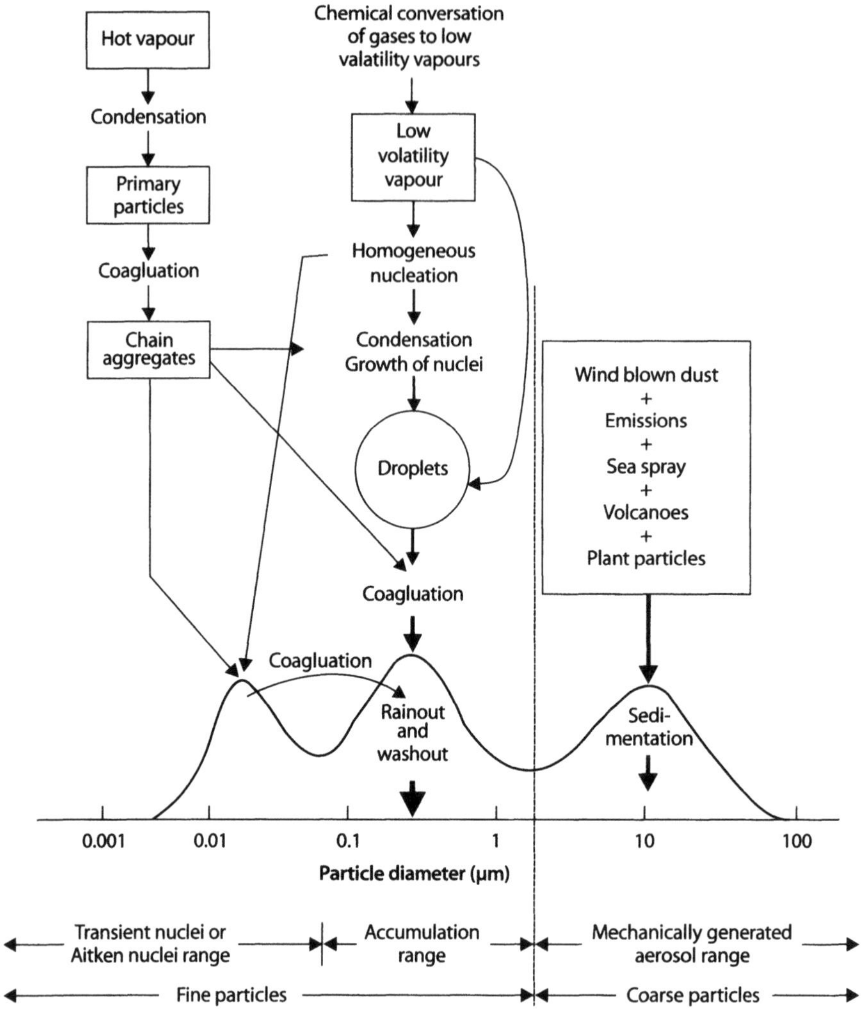

Fig. 2.1. Idealized schematic of surface area as a function of size of an ensemble of aerosol particles formed by a variety of processes. The principal sources and modes of formation are indicated along with mechanisms by which aerosols are transformed and removed from the atmosphere (redrawn from Seinfeld and Pandis 1998, p. 101, Fig. 2.15)

Secondary aerosols are formed from gas phase photochemical reactions in the atmosphere or in heterogeneous reactions of gas phase species in cloud droplets or on the surfaces of existing aerosol particles. Close to the aerosol sources, the newly-formed particles are in a size range that yields a peak in the surface area at around 0.01 μm diameter as shown in Fig. 2.1; these are referred to as Aitken nuclei (after their discoverer) or transient nuclei (because of their short lifetime in the atmosphere – generally a few hours or less). Examples are sulphate particles produced from SO_2 emitted from power plants or the oxidation of DMS emitted from the ocean. Particles in this very fine particle mode are highly mobile. They rapidly coagulate to form larger particles that typically fall in the size range between 0.1 to 1 µm diameter (a size class referred to as the "accumulation" mode) or they can diffuse to the surface of cloud or fog droplets or to larger particles (e.g. sea salt, mineral dust).

Because of the various routes by which particles can be formed, the atmospheric aerosol is extremely complex and dynamic. Some classes of aerosols can be formed by both primary and secondary processes. For example, sulphate particles can be emitted directly from smokestacks; but in advanced technologies, the direct emission of particles contributes a relatively minor source because of emission controls. Furthermore, individual aerosol particles rarely exist as a pure type (e.g. a "pure" sea-salt droplet, a specific sulphate compound). Rather, most particles are comprised of a wide range of compounds with properties and atmospheric lifetimes that differ from those of their individual components. As a result, atmospheric particles often display a wide range of chemical and physical properties.

2.1.2
The Role of Clouds in the Aerosol Cycle

Clouds play a very important role in the aerosol cycle. Cloud droplets are an important pathway for the reaction of gaseous species – for example, the oxidation of gas-phase SO_2 to SO_4^{2-} takes place largely in cloud droplets through the aqueous phase reaction with H_2O_2 (see below). But most clouds do not produce rain. Instead they evaporate, and the cloud droplets are converted to aerosol particles with sizes that typically fall in the range 0.1–1.0 µm diameter (i.e. in the "accumulation" mode; see Fig. 2.1). Nonetheless, the small fraction of clouds that do precipitate play an important role in cleansing the atmosphere. Thus, clouds are important both in the formation of aerosols and in their removal from the atmosphere; the delicate balance between these processes has huge implications from the standpoint of climate. Consequently, cloud processes have the highest priority in climate studies.

The role of clouds in the atmospheric aerosol cycle is paradoxical: clouds are involved in the formation and removal of aerosols from that atmosphere, and yet clouds can not form without the presence of aerosols. At the water vapour supersaturations typically found in the atmosphere (only a few tenths of a percent), water vapour can only condense on hygroscopic aerosol particles (e.g. sulphate and nitrate aerosols, sea salt droplets). Thus, clouds and aerosols are involved in a complex and tightly coupled atmospheric cycle that results in the transformation, transport, and removal of aerosol particles (Wang and Prinn 2000). These processes are depicted schematically in Fig. 2.2: water vapour condensation on the aerosol particle, in-cloud reactions, droplet coalescence, and the formation of the precipitation, which then falls towards the ground. In many cases precipitation will evaporate before it reaches the Earth's surface; the particles so-produced are larger and more chemically complex than those which entered the cloud. Also, as indicated above, the cloud could evaporate before precipitation forms, a process that also yields a transformed aerosol ensemble. Clouds also play an important role in atmospheric nitrogen chemistry in that lighting associated with clouds is a significant source of nitrogen oxides (Wang and Prinn 2000). Finally, clouds can also vent unreacted gases and particles to the middle and upper troposphere (Wang and Prinn 2000), as discussed below.

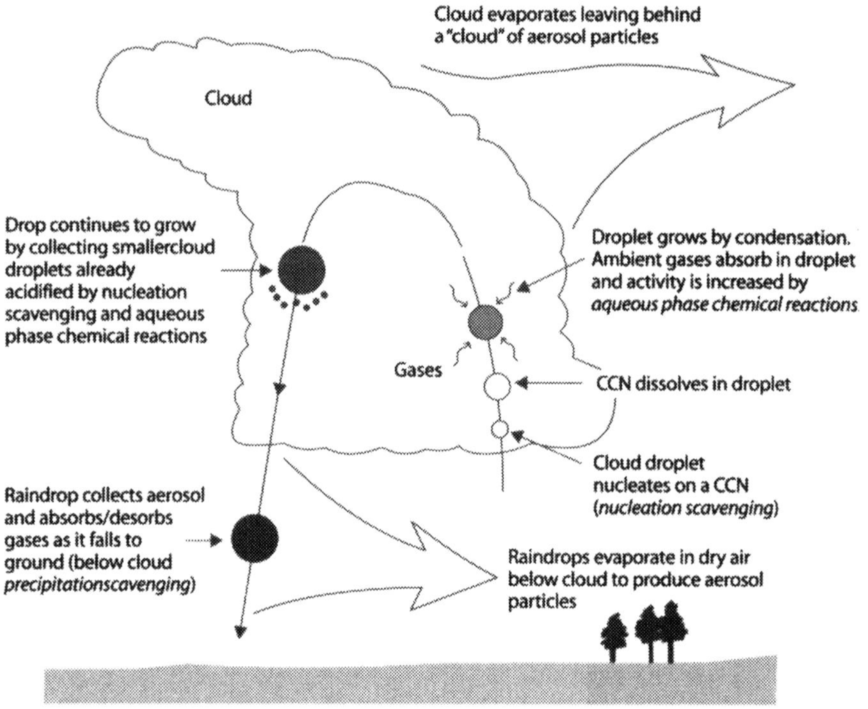

Fig. 2.2. Schematic diagram showing the role of clouds in aerosol processes. Aerosols (cloud condensation nuclei) adsorb water to form cloud droplets, which are subsequently modified by chemical reactions. If the cloud forms precipitation, the chemical species are deposited to the Earth's surface. Alternatively, the cloud can evaporate to produce a "cloud" of transformed aerosol in the middle and upper troposphere; also, the precipitation falling from raining clouds can evaporate to form aerosol in the lower atmosphere. (modified from Hobbs (2000), p. 112, Fig. 7.1)

The role of clouds is also central to the climate issue and the impact of humans on climate. The chemical and physical processes in clouds yield droplets which upon evaporation produce mostly accumulation-mode (0.1–1.0 μm diameter) aerosols – particles that have dimensions comparable to that of visible light. Such particles are efficient scatters of light, and consequently they can have a strong effect on the radiative balance of the atmosphere. It is because of the strong scattering of solar radiation by fine particles that we see dense hazes in polluted atmospheres, hazes that can produce colourful light effects, for example brilliantly coloured sunsets. One of the major challenges in assessing the role of aerosols in climate is understanding how the properties of aerosols (their size distribution, chemical composition, index of refraction, etc.) effect the radiative properties of the atmosphere. This problem is made more difficult because of the extremely complex chemical and physical properties of aerosols that are a consequence of the great variety of sources and processes involved in their formation.

Because of the complexity of the aerosol formation processes and the relatively short residence time of particles in the atmosphere (typically 1–2 weeks in the lower troposphere), their properties and their distribution are highly variable in both time and space. One of the major objectives in the field of aerosol studies today is to understand the factors affecting the variability of aerosols and to characterize that variability. There is a dearth of information about many aspects of the atmospheric aerosol cycle; this is especially true for the marine atmosphere. The study of marine aerosols presents a particularly great challenge because of the vast areas involved, the wide range of sources that impact on the marine atmosphere, and the many types of physical environments involved.

2.1.3
The Global Distribution of Aerosols Over the Oceans

Satellites provide us with graphic evidence of aerosol distributions over the oceans. An example is AVHRR (Advanced Very High Resolution Radiometer), which measures the radiation backscattered to space by aerosols over the oceans; these data are converted to equivalent aerosol optical thickness (*EAOT*), which is a measure of the column-integrated aerosol loading (Husar et al. 1997). Figure 2.3 shows the global average distribution of oceanic *EAOT* for the four seasons. Clear patterns are evident. First of all, the highest values of *EAOT* (and hence, the greatest column concentrations of aerosols) are found in regions close to the continents. This distribution affirms the fact that in many ocean regions the aerosol character is defined to a large extent by the transport of materials by winds from the continents. A second feature is that there are large seasonal differences in aerosol concentrations as can be seen by comparing the December–January distributions with June–August in Fig. 2.3. These differences are due to a variety of factors, including the seasonal variability of source emissions and meteorology. Third, some continents emit more aerosols than others, which suggests large differences in source and transport conditions. Especially notable is that large "plume" of high *EAOT* values over the tropical Atlantic. The plume extends from the coast of Africa to South America (in January) and to the Caribbean (in July); this plume is attributed to the transport of African dust. The large region of high *EAOT* over the Arabian Sea in June–August is due to dust carried from Africa and the Middle East. A large plume is seen off the western coast of southern Africa in June–August and in September–November; this is due to intense biomass burning in this region. Substantial plumes are seen over the North Atlantic during all seasons but winter; these are attributable to the transport of pollutants from North America and Europe. Similarly, large regions with high *EAOT* values are seen along the coast of Asia during most seasons; these are largely due to pollution transport. The Asian plume is most prominent in the spring when large quantities of soil dust are carried out of Asia along with pollution. The attribution of these plumes to these dominant aerosol types is supported by evidence from field studies in the regions covered by the plumes (Husar et al. 1997).

The Total Ozone Mapping Spectrometer (TOMS) also provides information on the distribution of aerosols, especially absorbing species – predominantly mineral dust and smoke (Herman et al. 1997). The TOMS product is especially useful, because (in contrast to AVHRR) the system can detect aerosols over land as well as over water surfaces. Figure 2.4 shows the global frequency of occurrence (days per month) of moderate-to-high absorbing aerosol concentrations during January and July. Similarly to AVHRR, TOMS shows prominent plumes over the tropical North Atlantic (January and July) and the Arabian Sea (July) due to dust and the large plume in the South Atlantic in July due to smoke. In contrast to AVHRR, TOMS does not show large pollu-

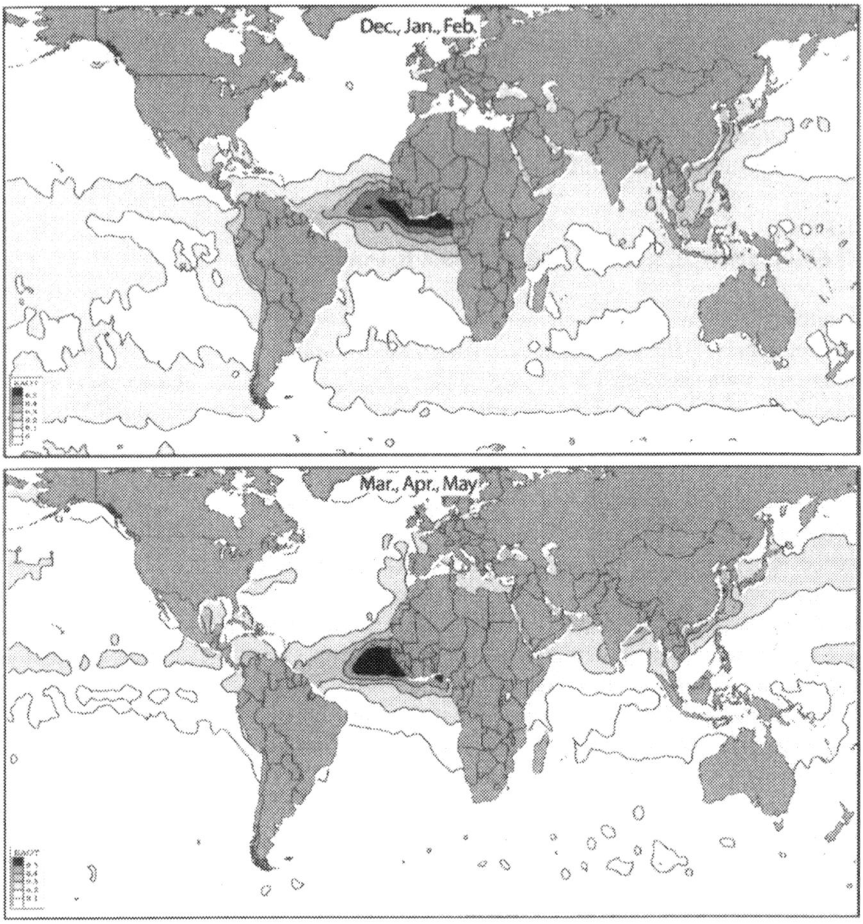

Fig. 2.3 a,b. Aerosol distributions over the oceans derived from the National Oceanic and Atmospheric Administration (NOAA) advanced very high resolution radiometer (AVHRR) satellite (Husar et al. 1997). Distributions are shown in units of equivalent aerosol optical thickness (*EAOT*), which is derived from measurements of visible-spectrum solar radiation reflected from aerosols back to space; *EAOT* values are roughly proportional to the integrated column concentration of aerosol. Distributions are shown for each of the four seasons; **a** December–February; **b** March–May (Figure from R. Husar, personal communication, based on Husar et al. 1997)

tion plumes emerging from Europe and Asia; this is because pollution plumes contain high concentrations of highly reflective aerosol (such as sulphates) that are detectable by AVHRR but only small amounts of absorbing aerosols (such as soot) that are detectable by TOMS.

Aerosol transport varies greatly from season to season. Consequently, any impacts that aerosol transport might have on ocean processes should also show a large seasonal and spatial variability. Especially notable is the large seasonal migration of the African dust plume. It reaches its northernmost position in July and August and its southernmost position in the low latitudes of the North Atlantic in the winter. The

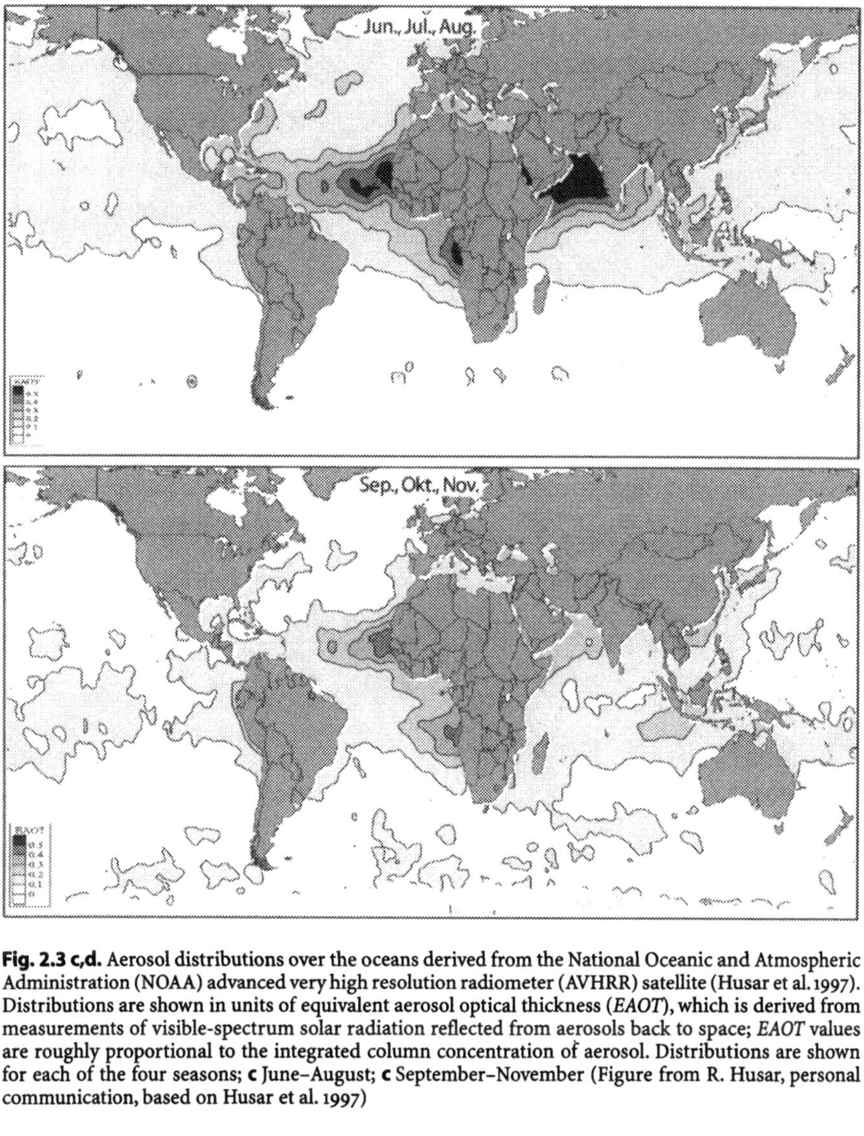

Fig. 2.3 c,d. Aerosol distributions over the oceans derived from the National Oceanic and Atmospheric Administration (NOAA) advanced very high resolution radiometer (AVHRR) satellite (Husar et al. 1997). Distributions are shown in units of equivalent aerosol optical thickness (*EAOT*), which is derived from measurements of visible-spectrum solar radiation reflected from aerosols back to space; *EAOT* values are roughly proportional to the integrated column concentration of aerosol. Distributions are shown for each of the four seasons; **c** June–August; **c** September–November (Figure from R. Husar, personal communication, based on Husar et al. 1997)

seasonal oscillation of the plume is consistent with the very large seasonal cycle of dust concentrations, as measured in South Florida, and Barbados, West Indies (Husar et al. 1997; Chiapello et al. 1999; Prospero 1999); indeed, during the summer months African dust has been measured over the eastern United States as far north as the New England states (Perry et al. 1997).

The TOMS aerosol distributions shown in Fig. 2.4 can be used to identify dust sources. In many cases, the distributions over land show a clearly defined geometry that can be matched to topographical features and geology. Note for example the isolated source "hot-spots" in North Africa and Australia in January and Central Asia in

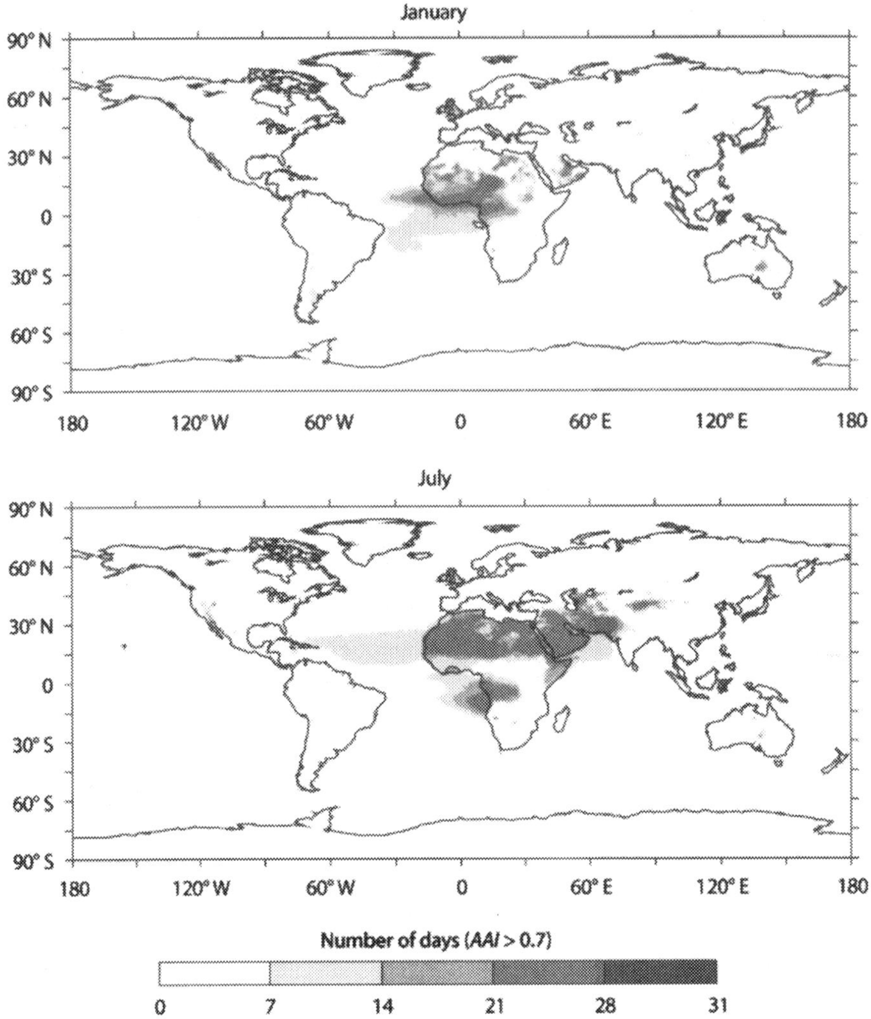

Fig. 2.4. The distribution of absorbing aerosols derived from the Total Ozone Mapping Spectrometer (TOMS) satellite. The absorbing aerosol index (*AAI*) provides a measure of the concentration of dust and black carbon in the atmosphere. The figure shows the number of days in January (*top*) and July (*bottom*) when the TOMS *AAI* exceeded 0.7, a value associated with moderately high dust and smoke concentrations (redrawn from Prospero et al. 2002)

July. In a later section, the environmental characteristics of the dust source regions will be discussed. TOMS also is sensitive to biomass burning aerosols because of the presence of highly-absorbing soot. In Fig. 2.4, the large plume emerging from southern Africa in July is almost entirely due to biomass burning; in January the plume that emerges from equatorial Africa and extends across the Atlantic along the equator almost to Brazil is largely attributable to biomass burning, while the more northerly portion of the plume is mostly due to dust.

Figures 2.3 and 2.4 clearly show that the major sources of aerosols, both pollution and mineral dust, lie in the Northern Hemisphere. The contrast between the Northern and Southern Hemispheres is important both from the standpoint of climate issues and also the impact of aerosol transport on ocean biogeochemical processes.

2.1.4 Aerosol Composition

The large fraction of the aerosol mass over the ocean is comprised of a relatively small number of species as shown in Table 2.1, which presents a selection of aerosol concentrations at various mostly remote ocean sites. The annual means are compiled from long-term (multi-year) measurements made at ocean network stations operated by the aerosol group at the University of Miami (Prospero et al. 1989; Savoie et al. 1989a; Arimoto et al. 1995, 1996; Savoie et al. 1993, 1994). The sea salt aerosol concentration is calculated from the measured concentration of Na^+ in aerosol based on the ratio of the total concentration of salts in sea water to that of Na^+ (i.e. sea salt aerosol = 3.256 × Na, by weight). The nss-SO_4^{2-} fraction is calculated by measuring the Na^+ concentration in the aerosol; on the assumption that the SO_4^{2-}/Na^+ ratio in the sea-spray droplets is the same as that in bulk sea water, the Na^+ value is converted to a sea-salt SO_4^{2-} by multiplying by 0.2517. This quantity is subsequently subtracted from total SO_4^{2-} yield of the nss-SO_4^{2-} concentration.

There are many other important marine-aerosol species that are not included in Table 2.1. In particular, organic materials can constitute a substantial fraction of the total aerosol mass (Jacobson et al. 2000). Most notable is soot ("black carbon" or "*BC*"), produced by fuel combustion and biomass burning. *BC* is found almost everywhere, even over the most remote oceans (Heintzenberg et al. 2000). In addition, as discussed below, over many ocean regions a large fraction of the aerosol mass remains uncharacterized (Quinn et al. 2000).

As one might expect, sea salt aerosol is a major constituent at most sites. (At some sites the high concentrations are attributable in part to local surf conditions, which are not representative of concentrations on larger scales.) Mineral dust shows an extremely wide range of concentrations. Note the very high concentrations in the tropical North Atlantic sites (Izaña, Barbados) and the sites in the western North Pacific (Cheju, Okinawa, Hong Kong). These reflect the impact of dust transport from North Africa and China, respectively. Substantial dust concentrations are also noted in higher latitude stations in the North Atlantic (Bermuda). Annual mean dust concentrations over the central North Pacific (Shemya, Midway, Oahu) are low, but as will be shown below, they are significant. Dust concentrations in the southern oceans tend to be extremely low (for example, see American Samoa) due to the dearth of strong dust sources on the southern continents and the very great transport distances to the central ocean regions.

The impact of pollutant sources is evident at many sites in the Northern Hemisphere. Extremely high nss-SO_4^{2-} and NO_3^- concentrations are seen at the sites near the coast of Asia in the western Pacific; these are attributable to the high levels of atmospheric pollution found over much of Asia due to the limited use of emission controls. Moderately high pollutant levels are seen over the North Atlantic as well (Bermuda, Mace Head) as a result of emissions from North America and Europe. In contrast, the concentrations of NO_3^- and nss-SO_4^{2-} at American Samoa and the Antarctic stations,

Table 2.1. Annual mean aerosol concentrations measured at remote ocean stations (Prospero et al. 1989; Savoie et al. 1989a; Arimoto et al. 1995, 1996; Savoie et al. 1993, 1994)

Station	Station Location[a]		Na	Sea salt[b]	NO_3	$nssSO_4$	MSA	NH_4	Al	Dust[c]	V	Sb	Total[d]
	Lat.	Long.	Mean	Mean	Mean	Mean	Mean	Mean	Mean	Mean	Mean	Mean	
	°N	°E	µg m^{-3}	µg m^{-3}	µg m^{-3}	µg m^{-3}	µg m^{-3}	µg m^{-3}	µg m^{-3}	Calc	ng m^{-3}	ng m^{-3}	µg m^{-3}
North Pacific													
Western Pacific													
Cheju, Korea	33.5	126.5	3.07	19.76	4.12	7.23	34.75	2.95	1.24	15.47	4.09	2.35	49.5
Okinawa	26.9	128.3	7.10	23.11	1.80	4.16	19.80	1.07	0.79	9.84	2.08	1.63	40.0
Taiwan	21.9	120.9	5.57	18.14	1.93	4.05	11.34	1.31	0.30	3.77	1.85	0.80	29.2
Hong Kong	22.6	114.3	3.01	9.79	3.03	7.41	23.11	2.79	1.05	13.13	6.20	2.05	36.2
Central Pacific													
Shemya	52.9	174.1	20.72	67.47	0.20	0.33	58.56	0.13	0.11	1.33	0.50		69.5
Midway	28.2	−177.4	4.23	13.77	0.27	0.53	20.32	0.07	0.06	0.72	0.19	0.07	15.4
Oahu	21.3	−157.7	4.65	15.14	0.35	0.51	18.93	0.03	0.05	0.66	0.26	0.03	16.7
North Atlantic													
Heimaey, Iceland	63.4	−20.3	8.79	28.62	0.23	0.65	38.37	0.38					29.9
Mace Head	53.3	−9.9	4.34	14.13	1.49	2.03	50.51	0.91	0.04	0.47	0.93	0.14	19.0
Izaña, Tenerife	28.3	−16.5	0.38	1.25	0.77	0.92	4.58	0.33	1.78	22.28	2.86	0.08	25.5
Bermuda	32.3	−64.9	4.19	13.65	1.06	2.19	35.12	0.31	0.45	5.59	1.28	0.09	22.8
Barbados	13.2	−59.4	5.08	16.53	0.53	0.78	19.70	0.11	1.16	14.55	1.85	0.03	32.5
South Pacific													
American Samoa	−14.3	−170.6	5.14	16.74	0.11	0.37	22.90		0.00	0.02	0.07	0.00	17.2
Antarctica													
Mawson	−67.6	62.5	0.10	0.33	0.03	0.09	22.90	0.00					0.5
Palmer Station	−64.8	−64.1	1.20	3.91	0.02	0.10	48.80	0.02					4.0

[a] Station location: negative latitudes = Southern Hemisphere; negative longitudes = Western Hemisphere.
[b] Sea salt calculated by multiplying Na concentration by 3.256, the sea salt/Na mass ratio in sea water.
[c] Dust is calculated from the Al concentration assuming that Al is 8% of total as in average crustal material (Taylor and McLennan 1985).
[d] Total aerosol is the sum of sea salt, soil dust, nss-SO_4, NO_3 and NH_4.

Mawson and Palmer, are extremely low; these levels are representative of conditions that one might expect under "natural" conditions over the oceans when there are no pollution impacts.

The presence of pollution aerosol is also reflected in the concentration of certain trace species. Table 2.1 shows the concentration of V and Sb as examples. Vanadium is emitted in a number of processes, and it is abundant in petroleum obtained from some sources. Antimony is present in the emissions from many types of industrial processes, especially smelters. The concentrations of V and Sb are very high in aerosols over the western North Pacific sites; in contrast, the concentrations at American Samoa are extremely low, in most samples near the detection limit (Savoie et al. 1994). The great differences between the hemispheres is also reflected in the latitudinal distribution of NO_3^- and nss-SO_4^{2-} in marine aerosols based on data compiled from measurements made on islands and aboard ships (Heintzenberg et al. 2000).

The aerosol data in Table 2.1 are consistent with the picture that we obtain from satellites (Fig. 2.3 and Fig. 2.4). They show that the highest levels of aerosol are found in the Northern Hemisphere, that the continents are the dominant sources of the aerosol that is most visible in satellite products, that dust is one of the most dominant aerosol species, and that pollution aerosols are a major factor in many regions. In contrast, aerosol concentrations in the Southern Hemisphere are often extremely low.

2.1.5
Temporal Variability of Marine Aerosols

The satellite data in Figs. 2.3 and 2.4 show that aerosol concentrations are highly variable on a seasonal basis. They are also variable on shorter time scales. Figure 2.5 shows daily aerosol concentrations of nss-SO_4^{2-}, NO_3^- and MSA measured over a four year

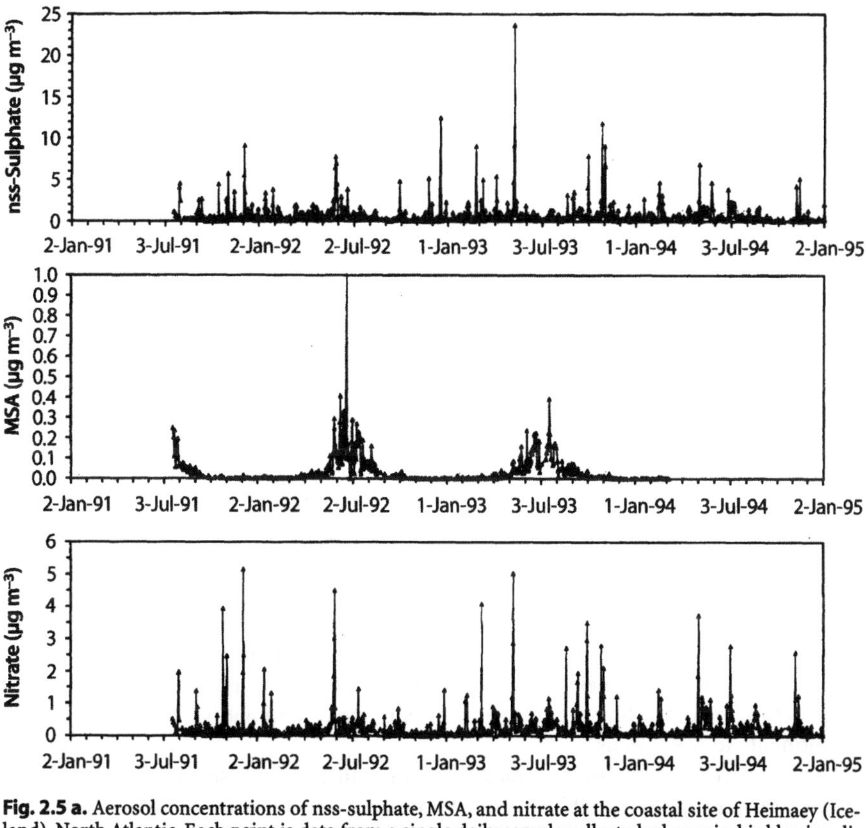

Fig. 2.5 a. Aerosol concentrations of nss-sulphate, MSA, and nitrate at the coastal site of Heimaey (Iceland), North Atlantic. Each point is data from a single daily sample collected when wind is blowing directly from the ocean to the sampling site (D. L. Savoie and J. M. Prospero, unpublished data)

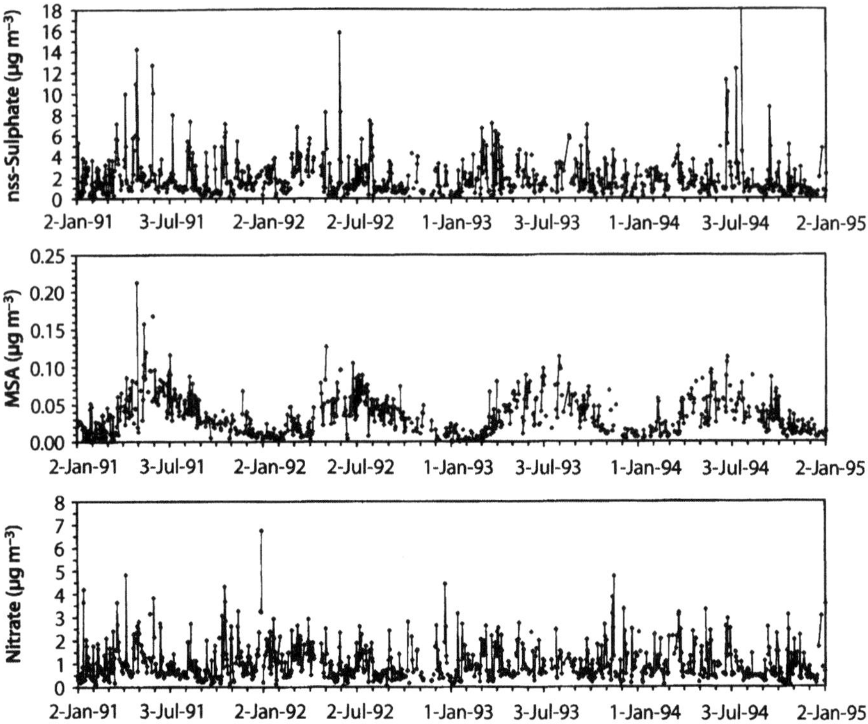

Fig. 2.5 b. Aerosol concentrations of nss-sulphate, MSA, and nitrate at the coastal site of Bermuda, North Atlantic. Each point is data from a single daily sample collected when wind is blowing directly from the ocean to the sampling site (D. L. Savoie and J. M. Prospero, unpublished data)

period (1991–1994) at three sites in the North Atlantic: Heimaey, an island that lies off the southern coast of Iceland; Bermuda; Barbados, and the West Indies. These figures clearly show the variability of concentrations on time scales ranging from annual to seasonal and day-to-day. Note that at Heimaey, the concentrations of nss-SO_4^{2-} and NO_3^- tend to be quite low most of the time, consistent with the relatively low annual mean concentrations shown in Table 2.1. However, there are occasional periods when concentrations increase sharply. These sharp "spikes", often lasting only a day or two, are associated with the transport of polluted air masses, usually from Europe (Prospero et al. 1995). Note that the large peaks in NO_3^- concentrations match in time those in nss-SO_4^{2-}. Compare the Heimaey time series of NO_3^- and nss-SO_4^{2-} with that at Bermuda and Barbados, both in terms of the peak concentrations, mean levels, and yearly variability. We might expect that the concentrations of other pollutants would also increase proportionately at such times. These data demonstrate the importance of meteorological factors in controlling the transport of pollutants (and natural materials) from the continents to the oceans.

The MSA aerosol concentration data from these three sites also show great differences. A large part of the variability is due to the seasonal character of DMS emissions from the ocean.

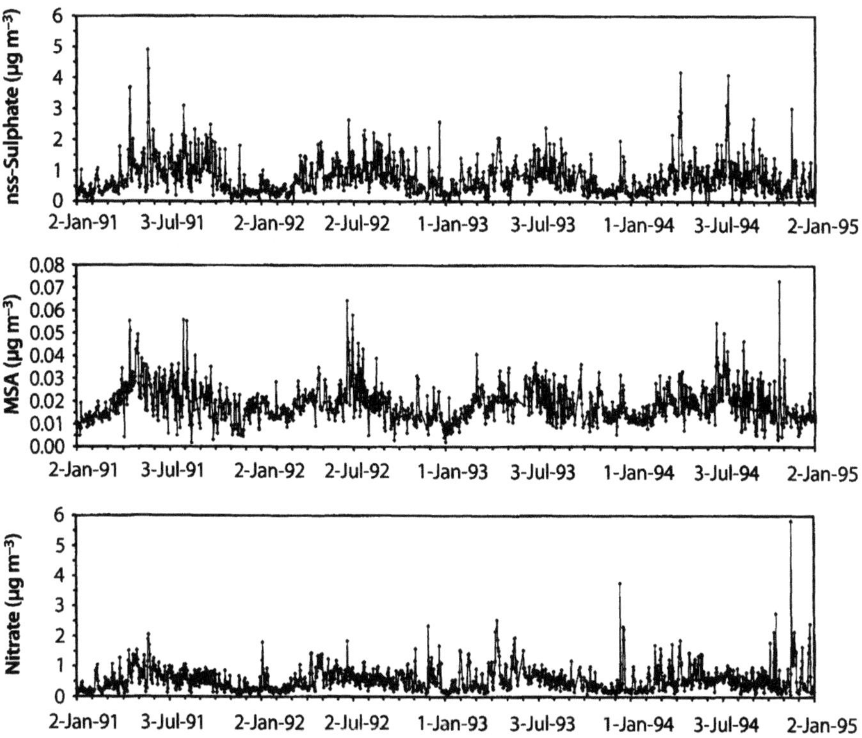

Fig. 2.5 c. Aerosol concentrations of nss-sulphate, MSA, and nitrate at the coastal site of Barbados, North Atlantic. Each point is data from a single daily sample collected when wind is blowing directly from the ocean to the sampling site (D. L. Savoie and J. M. Prospero, unpublished data)

In the following sections, we will examine the factors affecting the concentration distribution of these and other aerosol species in the marine atmosphere.

2.2 Sea Salt Aerosols

2.2.1 Sea Salt Production and Size Distribution

Sea salt aerosols can be generated by a wide range of physical processes. The most effective mechanism is by the bursting of air bubbles entrained in the ocean surface during whitecap formation. The production of sea salt aerosol and its size distribution is very sensitive to wind speed and the sea state, among other factors. This makes it very difficult to estimate production rates of sea salt. There are a number of models that estimate the temporal and spatial distribution of sea salt over the global oceans (Gong et al. 1997; Erickson and Duce 1988; Tegen et al. 1997). These yield estimates are in the range of 1 000–3 000 Tg yr^{-1}; the large range indicates the difficulty of the modelling task.

Bubble rupture produces aerosol droplets in a size range that extends from under about 0.01 μm diameter to more than 10 μm diameter (Fitzgerald 1991; O'Dowd et al. 1997). The concentration and size distribution of sea salt aerosols is dependent on a number of factors, especially the wind speed and the altitude above the sea surface (Fitzgerald 1991). Table 2.1 shows that on a mass basis, sea salt aerosol is dominant at all sites except Izaña, which is located on a mountain (altitude 2 360 m). As shown in Fig. 2.1, the size distribution of sea salt is such that the total surface area and volume distribution peaks in the supramicrometre size range (O'Dowd et al. 1997). Thus, sea salt aerosol offers a large surface area for chemical and physical reactions with other aerosol species and with gases (O'Dowd et al. 1997). Such large particles have relatively high settling velocities. For example, a 10 μm diameter sea salt aerosol particle has a settling velocity of 0.307 cm s^{-1} (0.265 km d^{-1}). Consequently, most of the sea salt mass has a relatively short residence time in the atmosphere, and much of the very large annual production rate (1 000–3 000 Tg yr^{-1}) is rapidly returned to the ocean. As we shall see in a later section, sea salt aerosol can serve as a very effective mechanism for removing many atmospheric species, both gaseous and aerosol, from the marine boundary layer (MBL).

2.2.2 The Contribution of Sea Salt to Submicrometre Aerosol

Although most of the sea salt aerosol mass is in the size fraction above 1 μm diameter, a small but significant fraction of the sea salt aerosol is in the submicrometre fraction (O'Dowd et al. 1999). Quinn et al. (2000) report on measurements of submicrometre aerosol made in the Pacific Ocean. They collected the aerosol fraction below 1 μm diameter and analyzed it for a wide range of soluble components including nss-SO_4^{2-} and sea salt.

Quinn et al. (2000) also weighed each of the filters so as to obtain a total aerosol mass below 1 μm. The total concentration of measured ions is always less than the total weighed mass; the difference is referred to as the "residual mass" which includes unanalyzed components such as mineral dust, organic material, soot, etc. Figure 2.6 shows the concentration of these three components (nss-SO_4^{2-}, sea salt, and residual mass) for various latitude bands in the Pacific Ocean. A number of features stand out in the figure. First of all, in most latitude bands the concentration of submicrometre sea salt is comparable to that of nss-SO_4^{2-}. The exceptions are in the mid-latitude North Pacific (20–40° N), where submicrometre sea salt is much lower and the high latitude (40–60° S) South Pacific, where it is much higher. The relative sea salt concentrations reflect the large differences in wind speeds in these latitude bands. In the case of nss-SO_4^{2-}, continental sources impact on the North Pacific, whereas over most of the South Pacific the only significant source of nss-SO_4^{2-} is the oxidation of DMS (see below). The other notable feature of this figure is that the residual mass is very large in most latitude bands, usually comparable to (and in some cases greater than) those of nss-SO_4^{2-} and sea salt. One might expect such a result for the North Pacific because of the impact of the transport of soil dust, soot, and organics from the continents, especially Asia as evident in Table 2.1. But the residual mass is also quite large in the South Pacific where there is relatively little land mass and the transport distances are very great.

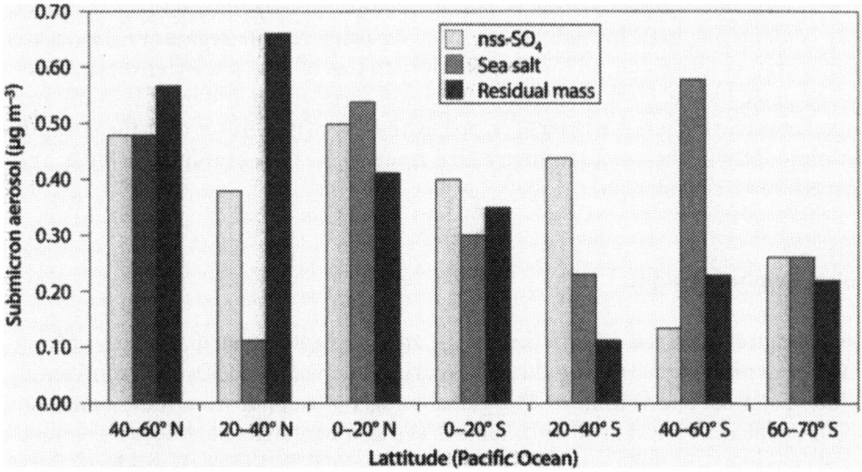

Fig. 2.6. Latitudinal distribution of submicrometre aerosol concentrations over the Pacific Ocean (redrawn from Table 3 in Quinn et al. 2000)

2.2.3
Sea Salt Aerosol and New Particle Production

In remote ocean regions, the optical properties of the MBL are largely controlled by the concentration of sea salt aerosol and by nss-SO_4^{2-} particles produced from the oxidation of DMS emitted from the ocean (see the following section). For example, Quinn et al. (1998) show that in some areas, sea salt aerosol is the dominant light scattering aerosol. Sea salt is also important as a source of cloud condensation nuclei (O'Dowd and Smith 1993). Because of the importance of sea salt in climate, there is great interest in developing models that can accurately reproduce sea salt aerosol distributions on a global scale. But the development of such models is hampered by the dearth of accurate and precise sea salt size and concentration data on the oceans.

Although it is clear that nss-SO_4^{2-} and sea salt aerosols are important, it is significant that a very large fraction of the radiatively important submicrometre aerosol over the oceans is largely uncharacterized as shown in Fig. 2.6. One of the major objectives in marine aerosol studies today is to focus on this important aerosol component.

2.3
The Oceanic Atmospheric Sulphur Cycle

There is great interest in the role of SO_4^{2-} in aerosols because of its role in climate-related processes. As stated earlier, most of the SO_4^{2-} aerosol mass produced from the oxidation of gaseous precursors (e.g. SO_2, DMS) is in the size range 0.1–1.0 µm diameter; particles in this size range are very efficient in the scattering and absorption of solar radiation. Also, because sulphate particles are hygroscopic, they can play an important role in cloud nucleation processes. Consequently, sulphates – both natural and anthropogenic – could have a large impact on climate.

Measurements of SO_4^{2-} over the oceans are complicated by that fact that sea water spray contains sulphate. Indeed, in aerosols over many remote regions of the ocean, the mass ratio of total SO_4^{2-} to Na^+ is quite close to that of bulk sea water, 0.2517. However, as stated before, most of the sea salt aerosol surface area and mass is in the size range between 1–10 μm diameter (Fig. 2.1), while nss-SO_4^{2-} aerosol over the ocean is predominantly in the submicrometre size fraction (see for example, Li-Jones and Prospero 1998; Li et al. 1996).

2.3.1
Global Sulphur Budgets

The concentration of nss-SO_4^{2-} in the marine environment is principally derived from two gaseous precursors: DMS produced by marine organisms and SO_2 from continental pollution sources and volcanoes. The global budget of sulphur from these sources is shown in Table 2.2 (Graf et al. 1997, Table 6; see also Benkovitz et al. 1996). The total global emissions of sulphur are 100 Tg S yr^{-1}. Of this total, two-thirds is produced from anthropogenic sources. Thus, anthropogenic sulphur has dramatically impacted the global atmospheric sulphur cycle. It is for this reason that so much attention is focused on the possible implications of pollution emissions on climate processes. Prior to the human era, the atmospheric sulphur cycle was dominated by DMS from the oceans. Volcanic inputs are also important, but they are sporadic. Also, in most cases, the impact of volcanic eruptions is limited to regional scales; this is especially true for volcanoes whose emissions are largely confined to the troposphere, where aerosol lifetimes are relatively short. In contrast, highly explosive volcanoes that inject material into the stratosphere can have long-range effects that can persist for years. In some cases, the effects can be global; for example, the eruption of Pinatubo in 1991 emitted 20 Tg SO_2, sharply increased the concentration of sulphate aerosol in the stratosphere and caused temperatures in the Northern Hemisphere to drop 0.2 °C (Hansen et al. 1992). The effects of Pinatubo aerosols could be detected for several years after the eruption. In contrast, oceanic DMS sources are widely distributed, and emissions are relatively steady from year to year (Kettle et al. 1999). Note in Table 2.2 that biomass burning is a minor source of sulphur emissions but it is a major source of nitrogen species emissions (see below). In the following sections, we briefly consider some of the more important aspects of the chemistry of SO_2 and DMS in the marine atmosphere.

2.3.2
SO_2 and nss-SO_4^{2-}

Anthropogenic sulphur is emitted primarily in the form of SO_2. The principal sources are fossil fuel combustion (especially coal) and the smelting of ores. In the atmosphere, SO_2 is oxidized in the gas phase by the OH radical to produce H_2SO_4. This reaction is rather slow; it yields an SO_2 atmospheric lifetime of 1–2 weeks. If the reaction with OH were the major controlling reaction for SO_2 in the atmosphere, then SO_2 would be transported over greater distances than is typically observed; indeed, under most conditions the concentration of pollutant-derived SO_2 over the oceans is generally quite

Table 2.2. Global annual mean sulphur budget (Graf et al. 1997)

Source	Sulfur emissions (Tg yr^{-1})
Anthropogenic	65.6
Biomass burning	2.5
DMS	18.2
Volcanoes	13.7
Total	100.0

low except close to continental sources. Under most conditions, the controlling reaction for SO_2 is that with H_2O_2 in cloud droplets; this reaction is very fast and results in the quantitative conversion of SO_2 to SO_4^{2-} as long as sufficient H_2O_2 is present (and in most environments, it usually is).

The atmospheric chemistry of SO_2 dramatically demonstrates the critical role that clouds can play in atmospheric chemistry in general and the impact that these cloud processes have on climate.

- In the absence of clouds, SO_2 has a lifetime of 1–2 weeks; within clouds, minutes.
- Other types of gases and particles are also incorporated into cloud droplets resulting in a complex composition.
- If the cloud evaporates (on a global average, the vast majority do), each droplet forms a complex particle containing SO_4^{2-}.
- If the cloud precipitates, the SO_4^{2-} is removed (along with other species) as precipitation; in regions affected by pollution, the precipitation is often very acidic because of the high concentrations of SO_4^{2-}.
- Many types of clouds (convective clouds, fronts) pump air (and aerosol) into the middle and upper troposphere where particles have relatively long lifetimes (weeks to months) and where winds can carry them over great distances.

Sulphate in aerosol particles is initially present as sulphuric acid, which rapidly picks up ammonia to form ammonium sulphate and various intermediate compounds, depending on the amount of gaseous ammonia available. The molar ratio of NH_4^+/SO_4^{2-} varies widely, but the global average tends to be about one (Adams et al. 1999) – that is, equivalent to a "pure" particle with a composition $(NH_4)HSO_4$. In reality, such pure particles are not normally found in the ambient atmosphere.

As a result of these processes, anthropogenic SO_2 does not directly play a prominent role in sulphur aerosol chemistry over the ocean. Most pollution SO_2 is deposited close to the sources on the continents. By the time polluted air masses reach the ocean, SO_2 has largely reacted to form sulphate. The primary processes that affect the distribution of anthropogenic SO_4^{2-} particles over the oceans are the transport meteorology (visible as plumes in satellite imagery such as shown in Fig. 2.3) and the removal processes to the ocean (primarily in rainfall). Thus, the production of new sulphate particles over the ocean must come largely from the oxidation of oceanic DMS.

2.3.3
DMS and the Atmospheric Sulphur Cycle

Charlson et al. (1987), in their classic paper, suggested that in the pre-human era, the radiative properties of the marine atmosphere were strongly modulated by nss-SO_4^{2-} that was derived from the oxidation of DMS emitted by various marine organisms. Even today, in remote ocean regions, the production of new ultrafine sulphate particles is linked to DMS production. The precursor of DMS is DMSP (dimethylsulphonium propionate), an osmolyte produced by many phytoplankton species, especially dinoflagellates, prymnesiophytes (including coccolithophores), and chrysophytes. Two species in particular, *Phaeocystis pouchetii* and *Emiliania huxleyi*, are known to be very strong producers of DMSP. DMSP is released by these organisms during senescence or when grazed; in water DMSP is enzymatically cleaved to produce a variety of compounds including DMS. The concentration of DMS in the ocean follows in a general way the seasonal cycle of oceanic primary productivity. Accordingly in the mid and high latitudes of the Northern Hemisphere, DMS concentrations increase in March or April and peak in May or June then decrease rapidly; in the Southern Hemisphere, the cycle is shifted by six months (Kettle et al. 1999). In contrast, in the tropics there is little evidence of a seasonal cycle. The concentration of MSA in the atmosphere mimics the seasonal cycle of DMS in the ocean. In Fig. 2.5, at Heimaey, Iceland, MSA shows a strong peak in June–July and very low concentrations during the remainder of the year. At Bermuda, the maximum MSA concentrations also occur during summer, but the peak is broader; winter concentrations are very low. At Barbados, the seasonal MSA cycle is even less evident; it is much broader and there is no well-defined winter minimum.

The cycle of DMS production and emission to the atmosphere is quite complex, and there is much that is not understood about the processes that affect DMS distributions in and over the oceans. Global surface-water DMS concentration data (over 10 000 measurements) were recently compiled and interpolated into a 1×1 monthly data set (Kettle et al. 1999), and the results compared to published fields of geophysical and biological parameters. Kettle et al. could not find any correlation between DMS and these parameters, and they could find no simple algorithm to create monthly fields of sea surface DMS concentrations based on these parameters. There clearly is much research to be done before we can understand the linkage between biological processes and the emission rate of DMS to the atmosphere.

The atmospheric chemistry of DMS is complex, and there are many unresolved issues (Berresheim et al. 1995; Ravishankara et al. 1997). Nonetheless, studies show that OH is the dominant oxidizing agent for DMS in unpolluted marine atmospheres where DMS has a lifetime of about 1 day. Reaction with OH can proceed by two dominant routes. Hydrogen can be abstracted by OH from the methyl group to form the radical group CH_3SCH_2, or OH can be added to the sulphur atom. The principal products are methanesulphonic acid (CH_3SO_3H, MSA) and SO_4^{2-}. Sulphate production is the higher energy route, and as a result, the relative yields of MSA and SO_4^{2-} would be expected to be temperature dependent, yielding higher ratios of nss-SO_4^{2-}/MSA in warmer climates.

2.3.4 MSA and nss-SO_4^{2-}

In remote ocean regions, the concentrations of nss-SO_4^{2-} and MSA are often correlated and yield a characteristic ratio that suggests a link between DMS emissions and nss-SO_4^{2-} production. Figure 2.7 shows scatter plots of nss-SO_4^{2-} against MSA from three ocean stations: America Samoa (Savoie et al. 1994), Mawson Station, Antarctica (Savoie et al. 1993) and Bermuda (Savoie and Prospero, unpublished data). At American Samoa and Mawson, remote sites where anthropogenic impacts are very small, the scatter plots yield relatively well-defined regression lines which suggest that the production of nss-SO_4^{2-} is linked to DMS emissions. In the low and mid-latitudes, the mass ratio of nss-SO_4^{2-}/MSA tends to be about 12–15; this is the case for the data from American Samoa (Fig. 2.7). In contrast, in the high latitudes it is about 3 as observed at Palmer Station (Fig. 2.7). The fact that the slopes of the regressions are markedly different reflects the different gas-to-particle conversion processes that apply in these two very different environments (i.e. tropical vs. the high latitudes). The fact that the ratio of nss-SO_4^{2-}/MSA is relatively low in the high latitudes compared to that in the low and mid-latitudes is often cited as verification of the temperature dependence of the reaction of DMS with OH, although there is a continuing debate about the role of temperature in causing the observed differences (Berresheim et al. 1995).

At ocean sites impacted by transport from pollution sources, there is no clear relationship between MSA and nss-SO_4^{2-}, because pollution-source SO_4^{2-} overwhelms the contribution form oceanic sources. The lack of correlation between MSA and nss-SO_4^{2-} at North Atlantic sites is clearly evident in Fig. 2.5; the nss-SO_4^{2-} time series at Heimaey appears very different from that of MSA and shows the dominance of pollution transport, which is evident as frequent "spikes" in the time series. The same situation is obtained at Bermuda (Fig. 2.5). Only at Barbados one can see at times a linkage between MSA and nss-SO_4^{2-} in the time series (Fig. 2.5). Nonetheless, it is possible to discern the impact of oceanic DMS-SO_4^{2-} at polluted sites such as Bermuda. Scatter plots usually show a well-defined lower boundary as seen in Fig. 2.7 for Bermuda. A line drawn along this boundary yields a nss-SO_4^{2-}/MSA slope of about 15, similar to that shown for American Samoa in Fig. 2.7. Thus, when pollution levels are low, the ocean DMS source is still significant and detectable at Bermuda. Data from other sites in the North Atlantic (and from other oceanic regions impacted by pollution sources) yield scatter plots of nss-SO_4^{2-}/MSA are similar to that shown for Bermuda in Fig. 2.7.

2.3.5 New-Particle Production from DMS over the Oceans

While it is now generally accepted that DMS plays a central role in the atmospheric sulphur cycle of the oceans, there are many uncertainties in the process and the ultimate impact on climate. The first problem has to do with the process of the biological processes in the water column that lead to the production of DMS and the physical processes that control the subsequent transfer from the ocean to the atmosphere. As

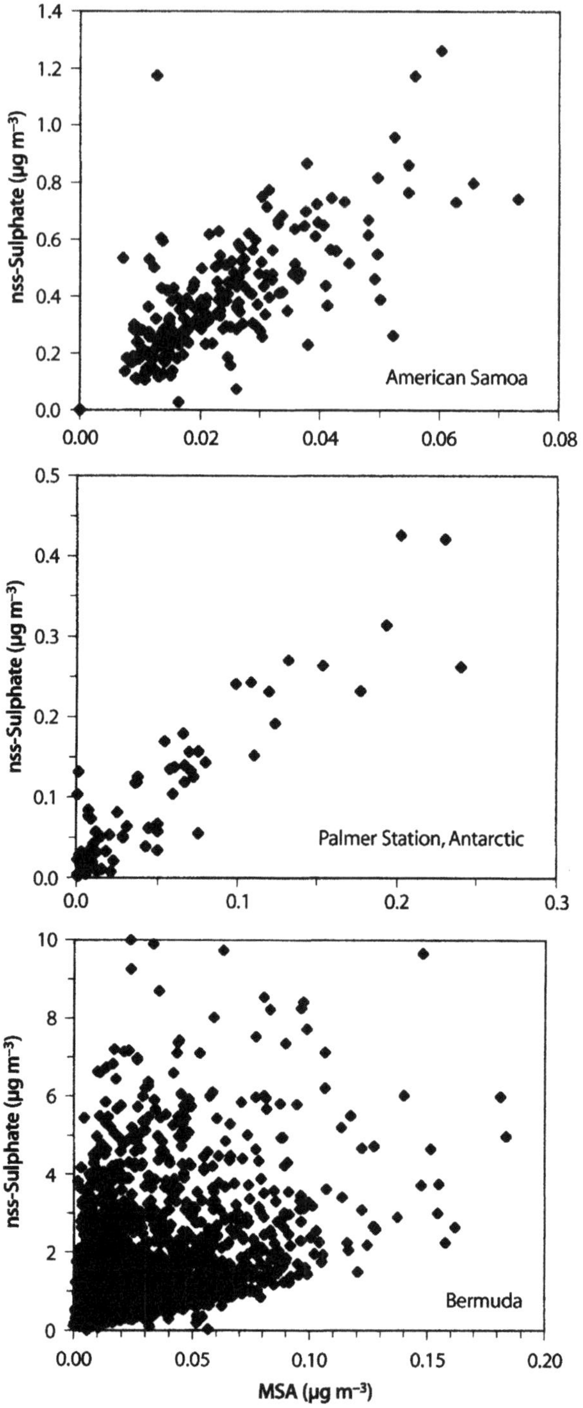

Fig. 2.7. Scatter plot of the daily concentrations of nss-SO_4^{2-} against MSA from aerosol samples obtained at coastal stations in American Samoa, Palmer Station (Antarctica), and Bermuda (D. L. Savoie and J. M. Prospero, unpublished data; see also Savoie et al. 1993, 1994)

stated above, Kettle et al. (1999) could not find any systematic relationships that could explain the observed distributions.

A second major uncertainty has to do with the chemical reactions of DMS in the atmosphere and the subsequent conversion of these products to the aerosol phase. DMS has a relatively short lifetime, about one day in the MBL under typical OH concentrations. Because of the high concentration (and large surface area) of sea salt aerosol in the MBL and the high mobility of the newly-formed ultrafine particles, DMS reaction products can rapidly diffuse to the existing aerosol phase as suggested in Fig. 2.1. Sievering et al. (1992) suggest that a large fraction of the DMS-SO_2 reacts directly with sea salt aerosols; because of the large settling velocity of these particles, this fraction of the DMS-SO_4^{2-} is rapidly recycled back to the ocean surface. From the standpoint of climate processes, the critical question is: How much of the DMS-SO_2-SO_4^{2-} goes into the formation of new aerosol particles in the size range of (roughly) 0.1–1.0 μm diameter – that is, particles that are efficient both as scatterers of light and as cloud-droplet nucleating particles? It is only through the production of new particles in this size range that DMS-SO_2-SO_4^{2-} can have a significant impact on radiation and on cloud nucleating processes. If the DMS reaction products (i.e. SO_2 and SO_4^{2-}) end up on the surfaces of existing particles, primarily sea salt particles, then the DMS-SO_4^{2-} source will have little impact on climate. A number of major field campaigns have attempted to address this issue (Bates et al. 1998; Raes et al. 2000). The general feeling at this time is that under most conditions, there is relatively little production of new particles in the MBL from DMS oxidation, because a large fraction of the DMS-SO_4^{2-} ends up on large sea-salt particles (Andreae et al. 1999, O'Dowd et al. 1997). However, recent work suggests that the direct reaction of SO_2 with sea salt is not as important as suggested by Sievering et al. (1992); van den Berg et al. (2000), using a sophisticated MBL chemical-physical model, found that the reaction of SO_2 with sea salt aerosol is quite complicated and that most SO_2 is oxidized in cloud droplets, not on sea salt aerosol.

The conclusion that new particle production is hindered in the MBL is generally consistent with recent studies that show that the concentration of nss-SO_4^{2-} over the oceans cannot be readily related to ocean water concentrations of DMS. Furthermore, it has not been possible to relate year-to-year variations in the concentrations of MSA and nss-SO_4^{2-} to observable climate changes. For example, measurements made over the equatorial Pacific throughout the 1980s and 1990s do not show any systematic change, despite the occurrence of major El Niño events during that time period (Bates and Quinn 1997).

Studies suggest that new particle production from DMS takes place primarily in the outflow from clouds that tap into the MBL (Perry and Hobbs 1994, 1995; Clarke et al. 1998). MBL air rising through the clouds is stripped of particles and reactive gases by cloud droplets. DMS, which is relatively insoluble and unreactive in the cloud environment, emerges from the top of the cloud into the middle and upper troposphere; here, because of the low particle concentrations and intense sunlight, the photochemical reaction products of DMS have a high probability of combining (condensing and coagulating) to form new particles. These newly-formed ultra-fine particles (smaller than about 0.01 μm diameter) are subsequently brought down into the MBL by subsiding air. In the MBL, the particles can grow and ultimately serve as cloud-nucleating particles that can contribute to the formation of new clouds, thereby continuing the cycle of particle production and transport. This process might well serve as the climate-controlling engine of the pre-human marine atmosphere and the present-day remote marine atmosphere.

2.3.6
Impact on Climate

A question that concerns us today is the degree to which humans have perturbed this natural aerosol production cycle over the world ocean and the impact on radiative processes. It has been clearly demonstrated that pollution aerosols modify cloud properties, often dramatically. This effect is readily visible as the appearance of "ship tracks" in stratus cloud over the ocean. These were first detected in satellite images (Coakley et al. 1987). It was noted that thin stratus cloud decks often were crossed by long thin lines of denser and brighter cloud. These tracks often appeared in regions where ship traffic was heavy. It was subsequently demonstrated that the exhaust from ships was entrained in clouds; the high concentration of exhaust aerosols resulted in a cloud that had higher concentrations of smaller droplets. This increased the scattering of sunlight and (hence) caused the clouds to appear "whiter". The increase in light scatter with increasing particle concentrations is known as the Twomey effect (Twomey 1991). This effect has been shown to be widespread, and it is central to the study of the role of anthropogenic aerosols in climate forcing.

There are still many unresolved issues about gas-to-particle conversion processes in the marine atmosphere. Although this discussion has focused on the DMS-SO_2-SO_4^{2-} aerosol system, it does demonstrate that there are many aspects of the aerosol formation process that are poorly characterized and that similar problems apply to other types of gas-aerosol systems.

The concern about the role of anthropogenic sulphate particles in climate has provided the impetus for a renewed interest in aerosol chemistry. The interest in marine SO_4^{2-} stimulated a strong research effort into oceanic sources, in particular the role of DMS. After more than a decade of intensive research, there are still many unanswered questions about the marine sulphur cycle and the relative importance of natural and anthropogenic sources in the atmospheric chemistry of sulphur over the oceans. Of particular interest is the linkage between marine primary productivity and the impact that it has on the atmospheric sulphur cycle. If climate is indeed changing as a result of human activities (or natural ones for that matter), then we might expect some affects on ocean productivity. Thus, the possible feedback to the sulphur cycle has important implications for future climate trends.

2.4
The Oceanic Atmospheric Cycle of Nitrates and Ammonium

Nitrate and ammonium are important to marine chemistry because of their possible impact on the oceanic nutrient cycle. The important precursor species of aerosol NO_3^- are NO and NO_2 (collectively referred to as NO_x). There are a number of organic nitrates that may play a significant role, in particular peroxyacetylnitrate (PAN). The entire ensemble of reactive gas phase N-species is generally referred to as NO_y, which is defined as the sum of NO_x, HNO_3, PAN and other related minor (mostly organic) species. Most NO_x is emitted into the atmosphere as NO, but there is rapid cycling in the atmosphere between NO and NO_2 on the time scale of a minute. Because of this rapid cycling, it is customary to think of the atmospheric chemistry of these species in terms of NO_x, although at night-time (i.e. in the absence of sunlight), NO_x is present entirely as NO_2. The main sink of NO_x is oxidation to HNO_3. In the daytime, the reaction is through NO_2 with OH; at night it is through NO_2 with O_3 to produce the NO_3^- radical, which reacts with NO_2 to form N_2O_5, which subsequently reacts with H_2O to from HNO_3. HNO_3 is extremely soluble in water and highly reactive. Consequently, it is rapidly removed from the atmosphere through precipitation and by direct deposition to surfaces. Over the ocean, HNO_3 reacts rapidly with sea-salt aerosol particles. It is because of these various factors, especially the latter, that the NO_x-NO_3^- system has relatively little impact on the radiative forcing of climate over the oceans, a subject discussed in greater detail below.

The dominant reduced nitrogen species in the atmosphere are NH_4^+ and NH_3 (which, as a pair, are commonly referred to as NH_x). Ammonia is important in atmospheric chemistry because it is by far the most important gaseous species available for the titration of acidic aerosol particles – primarily SO_4^{2-} particles. It is important to note that in the atmosphere, there is relatively little conversion or exchange between NO_x/NO_y species and the NH_x forms of nitrogen. In particular, NH_3 is quite resistant to oxidation under normal circumstances, although in tropical regions model results suggest that significant amounts of NH_3 may be oxidized due to high OH and low sulphate concentrations (Dentener and Crutzen 1994). As a result, the NH_x and NO_x cycles follow quite different routes and have completely different fates in the marine aerosol cycle as discussed below. Also in contrast to the NO_x cycle, where the ocean is invariably the sink (for NO_x as NO_3^-), the ocean can serve as either a source or a sink for NH_3, depending on the relative concentrations in the ocean and in the atmosphere (Quinn et al. 1996).

2.4.1 Global Budgets of NO_y and NH_x

The global budget of NO_y (see Prospero et al. 1996; Holland et al. 1999) is dominated by anthropogenic sources (Table 2.3). Energy production (combustion of coal, petroleum products, natural gas) produces about 20 Tg N yr^{-1} as NO_x. This source has been growing at a steady rate – from 1960 to 1986, at about 2.7% per year. The next largest source is biomass burning, which produces about 9 Tg N yr^{-1}. The primary natural sources of NO_x are biological fixation (about 6 Tg N yr^{-1}) and lightning (about 3 Tg N yr^{-1}). Note that in contrast to the sulphur cycle, which had a substantial oceanic source (i.e. DMS), there are no significant oceanic (water column) sources of NO_x. On the other hand, lightning and the stratosphere can be substantial sources of NO_x over the oceans. The present-day emissions of NO_y are roughly 40 Tg N yr^{-1}, while the preindustrial rate was 8 Tg N yr^{-1}. Thus, human activities have resulted in a five-fold increase in nitrogen emissions largely due to energy production and biomass burning.

Because of the highly reactive nature of NO_x species, they have a relatively short residence time in the atmosphere: about one day. Consequently, pollutant NO_x is not directly transported in large quantities over the oceans; the concentration of NO_x is

Table 2.3. Emissions of nitrogen species to the atmosphere (Prospero et al. 1996)

Source	Pre-industrial	Present day	Anthropogenic
NO_x (Tg N yr^{-1})			
Lightning	3	3	–
Soils (and crops)	3.6	5.5	1.9
Biomass burning	0.8	8.5	7.7
Stratospheric injection	0.6	0.6	–
Energy production	0	21.3	21.3
Aircraft	0	0.5	0.5
Total	8.0	39.4	31.4
NH_x (Tg N yr^{-1})			
Ocean	3–13	3–13	–
Soils (and fertilizer)	10	20	10
Biomass burning	0.5	5	4.5
Animal excretia	2.5	32	29.5
Total	16–26	60–70	44

low over the oceans except for those regions close to coastal urban complexes that emit large quantities of pollutants. However, pollutants do have a substantial impact on the NO_x chemistry of the remote ocean through PAN. PAN is quite stable at cold temperatures – it has a lifetime of months at 250 K – but it decomposes rapidly at typical ambient temperatures. PAN produced in polluted continental regions can be transported long distances through the upper troposphere; when it is brought down to the surface, it decomposes to produce NO_x; subsequent reactions produce the usual end product, HNO_3. PAN along with lightning and transport from the stratosphere are the major sources of NO_x over the ocean (see Table 2.3); in many ocean areas, PAN pollution sources completely dominate natural sources.

As shown in Table 2.3, human activities have had a great impact on the mobilization of NH_3 and NH_4^+ (Galloway et al. 1995; Bouwman et al. 1997; Benkovitz et al. 1996). The production of fertilizer converts about 80 Tg N yr^{-1} from N_2 to NH_3 with an annual rate of increase of 5.3% per year (Galloway et al. 1995); a substantial fraction of this NH_3 (about 10 Tg N yr^{-1}) is volatized directly from fertilized fields (Dentener and Crutzen 1994). The largest single source of ammonia (about 30 Tg N yr^{-1}) is from the excreta of domesticated animals; this source is estimated to be much greater today (about a factor of ten) than in preindustrial times because of the greatly increased world population and because of the increased consumption of meat in the diet (Dentener and Crutzen 1994; Prospero et al. 1996). Emissions of NH_3 from biomass burning have also greatly increased, although amounts are only modest in the overall budget (Dentener and Crutzen 1994). The present day emissions of NH_x (including both natural and anthropogenic sources) is about 60–70 Tg N yr^{-1} (Prospero et al. 1996). The total present-day emissions for both NO_x and NH_x is about 100–110 Tg N yr^{-1} (Table 2.3). Total preindustrial emission rates were about 24–34 Tg N yr^{-1}. Considering all continental sources of the NH_x budget, the emission rates today are about four to five times greater than the preindustrial rates, a factor similar to that found for NO_x emissions.

2.4.2
Concentrations of Nitrate and Ammonium in the Marine Atmosphere

The impact of pollution sources of NO_3^- and NH_4^+ is evident in Table 2.1. In remote ocean regions, nitrate aerosol concentrations are a few ten's of µg m^{-3} or less (see for example, American Samoa, Mawson, Palmer Station); in contrast, off the coast of Asia and in the North Atlantic, concentrations are ten to one hundred times greater. The same is true for NH_4^+ concentrations.

The NO_3^- time series in Fig. 2.5 show dramatic evidence of the impact of polluted air masses in the form of sharp "spikes" in the NO_3^- concentrations. These are most visible in the record from Heimaey where aerosol concentrations are quite low except on those occasions when polluted air is advected into the region, usually from Europe (Prospero et al. 1995). Note that the peaks in NO_3^- concentrations coincide with those of nss-SO_4^{2-}. The Bermuda aerosol time series also shows a lot of sharp peaks, much more than at Heimaey, because of transport from North American pollution sources; furthermore, there is little evidence of a seasonal cycle.

At Barbados (Fig. 2.5), the NO_3^- time series is relatively smooth, although some sharp peaks are evident. Barbados is affected by pollution sources in Europe and North Africa; Savoie et al. (1989b) estimate that approximately half of the NO_3^- and nss-SO_4^{2-} is natural and half is anthropogenic. There is a clear seasonal cycle in NO_3^- at Barbados; the spring maximum and sporadic peaks during winter are attributed to the transport of biomass burning products from Africa (Savoie et al. 1989b).

2.4.3
Nitrate and Ammonium Aerosol Properties

Aerosol nitrate has a more complex chemistry than sulphate, because under acid conditions NO_3^- can be volatilized as HNO_3, which can subsequently undergo further chemical reactions (including photochemistry). In contrast, once SO_4^{2-} enters the aerosol phase, it is essentially locked in the particle. Because of the volatility of NO_3^-, the size distribution of NO_3^- aerosol is very dependent on the chemical properties of the ambient aerosol. Under most conditions on the continents and almost invariably over the oceans, the submicrometre of aerosol is dominated by SO_4^{2-}, which is only partially neutralized by NH_4^+; as a consequence, NO_3^- is found almost exclusively in the supramicrometre size distribution, where there is little SO_4^{2-} and a relatively high concentration of basic material such as sea-salt and mineral dust. Over the ocean, the size distribution of NO_3^- typically follows the surface area distribution of sea salt (Li-Jones and Prospero 1998; Murphy and Thomson 1997; Gard et al. 1998). Because of the higher settling velocities of large sea salt particles, the marine NO_3^- aerosol has a shorter residence time than aerosol nss-SO_4^{2-}. Per unit mass, coarse particles are less efficient light scatterers than submicrometre particles; thus, in general, over the oceans the radiative impact of nitrate is insignificant relative to that of sulphate (Yang et al. 1994; Li-Jones and Prospero 1998).

2.4.4 Organic Nitrogen Aerosols

Organic nitrogen could play an important role as an N-nutrient source. Measurements of dissolved organic nitrogen (*DON*) in precipitation from various continental and marine locations including the North Atlantic (Cornell et al. 1995) show that concentrations are often comparable to that of dissolved inorganic nitrogen (*DIN*, mainly NO_3^- and NH_4^+). Organic nitrogen in aerosols and precipitation can result from reactions of gas phase species and by mechanical processes that release particles directly (e.g. plant fragments, sea spray droplets). The protein in pollens or oceanic surface films can degrade in the atmosphere to lower molecular weight organic species such as amino acids or primary amines. Because many different techniques have been used to study organic nitrogen, it is difficult to make direct comparisons of results. Most research has focused on the *DON* fraction.

More recently, Neff et al. (2001) reviewed the atmospheric organic nitrogen (*AON*) cycle, including both reduced and oxidized forms. They found that *AON* comprises about 1/3 of the total nitrogen deposited to the surface but that this fraction varies widely from region to region. The sources of *AON* do not appear to be dominated by pollution; there are no strong correlations between fluxes of nitrate and *AON* or ammonium and *AON*. They suggest that the reduced and bacterially-related forms of *AON* do not appear to play an important role in the overall flux of *AON* to the surface of the Earth; however, they suggest that both dust and organic nitrates such as PAN may be important. Neff et al. estimate a global *AON* flux between 10 and 50 Tg N yr^{-1}.

These studies show that there are large unresolved uncertainties in the organic nitrogen cycle. Nonetheless, it is clear that organic nitrogen is an important aspect of local and global atmospheric nitrogen budgets.

2.4.5 Trends in Nitrate and Ammonium in Pollution Aerosols

There are certain regions in the world where the emissions of NO_x and NH_3 are very high relative to those of SO_x. Examples are southern California, Western Europe and India. Under such conditions, the SO_4^{2-} aerosol is neutralized by NH_3, and NH_4NO_3 is found in the submicrometre size fraction. NH_4NO_3 concentrations can exceed those of $(NH_4)_2SO_4$ and thereby play an important role in radiative forcing on regional scales (ten Brink et al. 1996). This has important implications about future trends in air quality and radiative forcing. Air quality regulations have been most effective in reducing the emissions of SO_x than of NO_x. This is largely because the major sources of SO_x are industry and power plants that are easier to regulate than the principal sources of NO_x, usually cars and trucks. Thus, the ratio of NO_x/SO_x in emissions will continue to increase as will the importance of NH_4NO_3 aerosol. A substantial increase in the concentration of NH_4NO_3 has already been observed in the European plume emerging over the eastern North Atlantic (Andreae et al. 2000; Raes et al. 2000). On a global scale over the coming decades, the emissions of NO_x and NH_3 are expected to increase, while those of SO_x are expected to remain relatively stable; thus, NH_4NO_3 could play an increasingly important role in radiative processes. This factor could also be important with regard to the deposition of nutrients to the ocean, especially in coastal waters and regional seas.

2.4.6
Atmospheric Deposition and the Nitrogen-Nutrient Budget in the Oceans

The primary productivity in surface waters of the ocean depends on the availability of nutrients (see for example, Levitus et al. 1993), the most important being phosphate and nitrate. Historically, the major sources of nutrients in the photic zone were considered to be the fixation of nitrogen in surface waters; the upward transport of nutrients from deeper waters; and the recycling of nutrients in the surface waters. However, Duce (1986) showed that atmospheric inputs could be important to some ocean regions. Using a simple model, he made some preliminary estimates of the possible impact of the deposition of various aerosol species on the surface waters and compared these atmospheric inputs to ocean sources such as upwelling. Duce found that in some oligotrophic regions, the atmospheric input of NO_3^- and Fe could contribute a substantial and sometimes major fraction of these species to surface waters and thereby have an impact on biogeochemical processes. Of particular interest is the role of these species in primary productivity. In this section, we consider the possible significance of NO_x and NH_x species. In Section 2.5 we consider the role of Fe deposition on the ocean nutrient cycle.

The huge increase in anthropogenic emissions of NO_x and NH_x has resulted in major disruptions of the nitrogen-nutrient cycle over the continents, and there is concern about the impact of these inputs on coastal and ocean processes. It is known that pollutant nitrogen species such as NO_3^- and NH_4^+ (Paerl 1995) as well as various *DON* species (Seitzinger and Sanders 1999) can serve as a nutrients for microorganisms. The impact on coastal waters has been particularly notable (Paerl 1995; Howarth et al. 1996) but the importance to the larger ocean has been more difficult to characterize. Recently, an effort was made to assess the nitrogen budgets for the North Atlantic and the surrounding watersheds (Galloway et al. 1996). Prospero et al. (1996) estimated that the total present-day (natural and anthropogenic) deposition of NO_y to the North Atlantic Ocean (NAO) watershed, shelf, and open ocean is 670 Gmol yr^{-1}, 79 Gmol yr^{-1}, and 360 Gmol yr^{-1}, respectively; in addition, the estimated deposition of NH_x to these same regions is 390 Gmol yr^{-1}, 55 Gmol yr^{-1}, and 260 Gmol yr^{-1}. In contrast, Michaels et al. (1996) estimate that the source of NO_3^- within the main thermocline of the NAO due to N-fixation is in the range of 3 700 to 6 400 Gmol yr^{-1}. Thus, the total atmospheric deposition of nitrogen-nutrient species to the open ocean, about 600 Gmol yr^{-1}, constitutes a substantial fraction (roughly 10–15%) of the estimate production above the thermocline. If the deposition of *DON* is comparable to that of NO_3^- and NH_4^+ as suggested above, then the total nutrient-nitrogen deposition from the atmosphere is doubled and would constitute 20–30% of the open-ocean in-water source.

Thus, it appears that atmospheric transport and deposition has indeed had a major impact on the N-cycle in the NAO compared to pre-human times. Atmospheric transport has also impacted the N-cycle in other remote ocean regions. Figure 2.8 shows the deposition rate (mmol m^{-2} yr^{-1}) of NO_y and NH_x in latitude bands extending from the western North Pacific across North America and the Atlantic to Europe and Africa. Huge amounts of NO_x and NH_x are deposited on the continents. Also, the deposition rate to the NAO is substantially higher than to the Pacific. Note also the ratio of NO_x/NH_x. The ratio is much higher over North America compared to Europe. This reflects the very different character of the emissions from these two regions, a

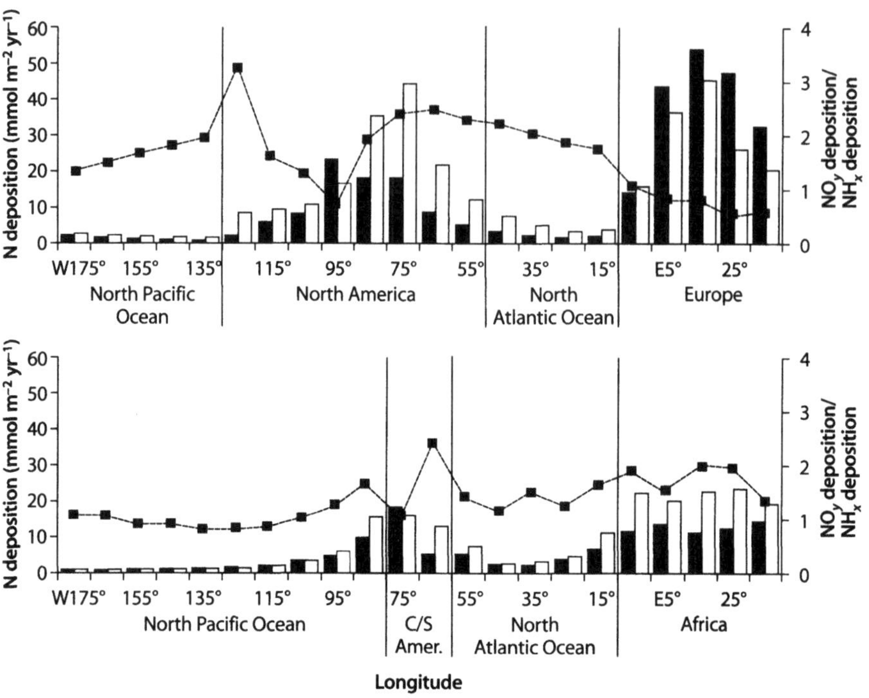

Fig. 2.8. Estimated deposition rates of NO_y (white bar, left axis) and NH_x (black bar, left axis) across the North Pacific, North America, the North Atlantic and Europe (redrawn from Galloway et al. 1996). The *square symbols* and *connecting line* show the ratio of NO_y to NH_x (right axis)

factor that could have implications regarding aerosol chemistry and the role of deposited N species to the respective watersheds, coastal waters, and adjacent ocean regions.

2.5 Mineral Dust in the Marine Atmosphere

2.5.1 Global Distribution of Dust

Large amounts of soil dust are mobilized by winds, mostly in arid regions, and substantial quantities can be carried great distances. Of all the aerosol species discussed in this paper, mineral dust is the one that most clearly leaves its signature in the oceans in the form of terrigenous minerals in deep-sea sediments. Indeed, the importance of aeolian transport to oceanic processes was first suggested by studies of the mineral distributions in pelagic ocean sediments. It was noted that the concentration patterns of certain minerals (e.g. quartz, kaolinite, illite) in sediments off the coasts of some continents (e.g. the western coast of North Africa and North America, the eastern coast of Asia) were not related to fluvial sources but rather to the pattern of large scale wind fields (Prospero 1981). Various aspects of dust transport and its effects have been extensively studied during the past two decades. For reviews of various areas of research relevant to the ocean, see Andreae 1995; Prospero 1981, 1996a,b; Pye 1987; Middleton et al. 1986; Duce et al. 1991; Duce 1995; Goudie 1983; Leinen and Sarnthein 1989; Golitsyn and Gillette 1993; Guerzoni and Chester 1996.

Satellite images provide the most graphic evidence of the widespread occurrence of dust. In Figs. 2.3 and 2.4, huge plumes of dust are seen to emerge from arid continental regions and extend over large ocean areas. Indeed, excluding clouds, dust is the most prominent aerosol feature over the oceans. Dust plumes are much more prominent than pollution plumes; they cover larger areas and are more persistent. Because of the prominence of dust over such large regions, dust has become a major focus of climate studies. Dust is a strong absorber and scatterer of solar and terrestrial radiation (Sokolik and Toon 1999). Thus, dust can be an agent of climate change. Conversely, the generation of dust is strongly dependent on climatic factors such as aridity and wind conditions. Thus, there is a strong possibility that dust can provide strong feedback in the climate forcing system.

Satellite images show that the distribution of dust over the oceans is highly variable. The most prominent dust plumes are found in the Northern Hemisphere. Satellite images show a dust belt that extends from the western coast of North Africa, through the Middle East, into Central Asia and reaching into China almost to the Pacific coast. This huge dust belt dominates the global dust budget. In this belt, the largest sources by far are found in North Africa. In contrast, there are almost no major dust sources in the Southern Hemisphere. This is surprising in light of the widespread arid and desert regions in southern Africa, South America, and Australia. The absence of major dust sources in Australia, a continent where about 80% of the area is arid, is especially notable.

2.5.2 Sources of Dust

Mineral dust is a primary aerosol product; it is lifted directly from soils in the source regions by winds. Soil particles are produced by the chemical weathering of rocks and by mechanical processes (e.g. grinding, fracturing, impaction, etc.). These processes mostly produce particles that are relatively large, tens to hundreds of micrometres in diameter, but there is a substantial yield of smaller particles as well. When winds lift soil particles into the air, the suspended mass initially consists largely of particles with sizes greater than 10 µm (e.g. silt and sand particles) (Duce 1995). Such large particles have a very short residence time in the atmosphere because of their high settling velocity. For example, a sand particle (density 2.6 gm/cm^3) with a diameter of 100 µm diameter has a Stokes settling velocity of 78.5 $cm\ s^{-1}$ (67.9 $km\ d^{-1}$). Because of the rapid fallout of large particles during transport, the peak in the suspended dust size distribution rapidly shifts to smaller particles; at distances of about 1 000 km or more from the source, the dust attains a relatively stable size distribution with a mass median diameter of several µm (Duce 1995), consistent with the size distributions shown schematically in Fig. 2.1. Dust particles in this size range can be carried great distances by winds; for example; a dust particle 2 µm diameter has a settling velocity of 0.034 $cm\ s^{-1}$ (0.029 $km\ d^{-1}$).

There is considerable uncertainty as to the specific sources of dust. It is known that certain types of soil environments readily yield large amounts of dust. Prospero et al.

(2002) used the TOMS satellite to identify the most active dust sources all over the Earth. They show that there is a clear geographical pattern to the frequency and intensity of dust activity. This relationship is apparent in Fig. 2.4, which shows the frequency of occurrence of high concentrations of dust (and smoke). Many of the source regions in Fig. 2.4 have a characteristic shape. These features (most clearly seen in the January distribution in Fig. 2.4) appear in TOMS year after year over a broad region of Africa, the Middle East and Asia. The pattern of activity in the most active dust sources can be closely matched to terrain contours and shows that the most active sites are located in topographical lows. Most of these regions had been inundated during pluvial periods in the Pleistocene/Holocene. Thus, these basins contain deep alluvial deposits, which in modern (arid) times are rapidly being deflated by winds. Finally, all major dust sources are located in regions where rainfall is less than about 200 mm yr^{-1}.

As stated earlier, the most active dust sources are concentrated in a band that extends from the western coast of North Africa almost to the North Pacific coast of China. In contrast, there are many deserts and arid regions in southern Africa, North America, South America and Australia that produce so little dust that they are essentially insignificant in the global dust budget. The absence of major dust sources in Australia is especially notable. About 80% of the area of Australia is arid and there are extensive deserts and arid regions. Yet there is very little dust activity in Australia. Nonetheless, it is significant that dust activity in Australia largely takes place over the Lake Eyre basin, a region that has the characteristics that Prospero et al. (2002) identify as important: it is a topographical low that was inundated in the Pleistocene; it has deep fluvial deposits; and it is now in an arid region with rainfall in the range of about 100–200 mm yr^{-1}.

Of all the dust regions identified in Prospero et al. (2002), North Africa is clearly the largest source of dust that can be transported great distances. Large amounts of dust are carried to the north across the Mediterranean to Europe (Guerzoni and Chester 1996) and to the west (Chiapello et al. 1995) across the North Atlantic to the Caribbean (Prospero and Nees 1986) and the eastern coast of North America (Prospero 1999; Perry et al. 1997). Figure 2.9 shows the monthly mean mineral dust concentration measured in the trade winds at Barbados starting in 1965. There is a clear seasonal cycle, which is linked to the seasonal shift in the large-scale wind patterns. Dust concentrations are relatively low during winter when dust from Africa is carried in the lower latitudes to South America (Prospero et al. 1981; Swap et al. 1992) as shown in the AVHRR (Fig. 2.3) and TOMS (Fig. 2.4) aerosol products.

Satellite imagery of aerosol optical thickness (*EAOT*) such as that in Fig. 2.3 shows that the highest values of *EAOT* and the largest areal coverage over the oceans is clearly related to dust sources (Husar et al. 1997). In contrast, the pollution plumes that emerge from the eastern coast of the United States and the western coast of Europe are relatively small and weak in comparison to the African dust plume. Furthermore, dense African dust plumes are highly visible all year long while the European and North American pollution plumes are prominent only during spring and summer.

Large amounts of dust are also transported out of Asia each spring (Prospero et al. 1989; Perry et al. 1999; Arimoto et al. 1996. Zhang et al. 1997). The dust is intermixed with substantial amounts of pollution aerosol as well (Savoie et al. 1989b; Arimoto et al. 1996). The dust and pollution plume is clearly visible in the AVHRR aerosol optical

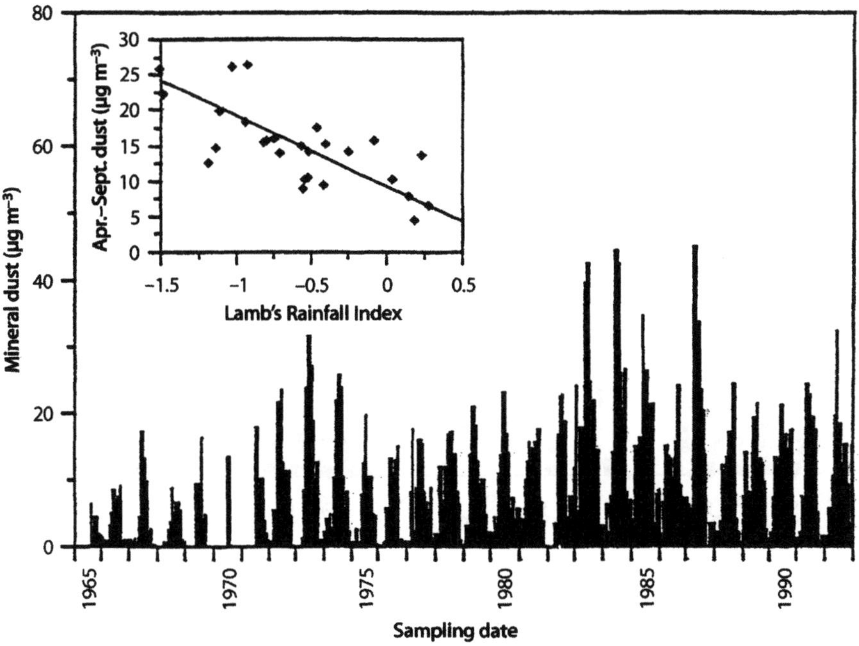

Fig. 2.9. Monthly mean mineral dust concentrations (µg m^{-3}) at Barbados, West Indies, 1965–1992. *Inset* is a scatter plot of mean annual August–September dust concentration at Barbados vs. the rainfall anomalies in the sub-Saharan region of North Africa (Prospero et al. 1996)

depth product (Husar et al. 1997) and in the TOMS absorbing aerosol product (Herman et al. 1997). Relatively high concentrations of mineral dust are measured at islands in the central North Pacific each spring (Prospero et al. 1989; Perry et al. 1999). Figure 2.10 shows the long term dust concentration record from Midway Island. Note the extremely high concentrations of dust in April 1998. This dust event was the largest ever observed over the North Pacific in the 20-year record of studies. Dust from this event was carried to the western and central United States; dust concentrations measured over the western coastal states was typically in the range 20–50 µg m^{-3} with local peaks over 100 µg m^{-3} (Husar et al. 2001).

The dust records at Barbados and Midway demonstrate that dust transport is extremely sensitive to a wide range of environmental factors; consequently, dust concentrations tend to be much more variable on a year to year basis than are the concentrations of other species, including pollutants. One might expect that this variability would be even greater over geological time scales because of the large extremes in climate.

2.5.3 Elemental Composition

The chemical and mineralogical composition of mineral dust over the oceans has been extensively studied and reviewed (see for example, Prospero 1981; Duce 1995; Guerzoni

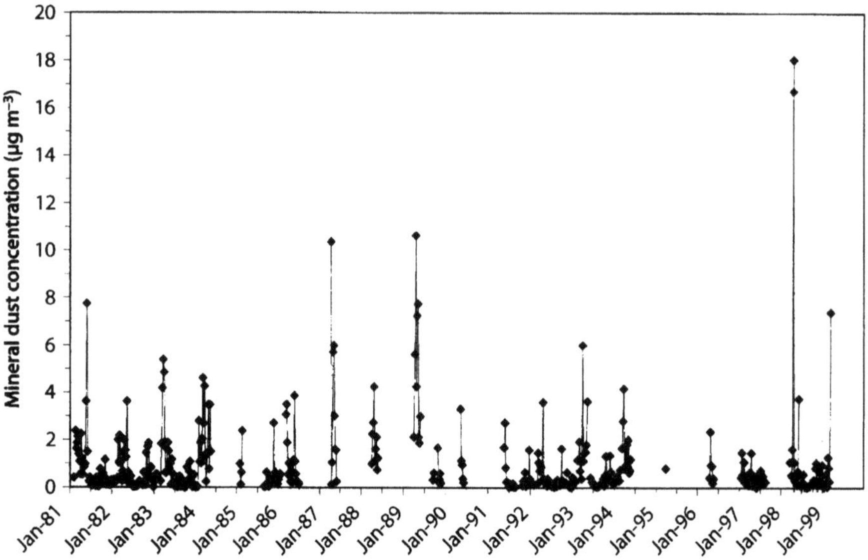

Fig. 2.10. Mineral dust concentrations at Midway Island in the central North Pacific. Data points are concentrations obtained from one-week-long samples collected at a coastal site on the island (Prospero et al. 1989; unpublished data, R. Arimoto and D. L. Savoie)

and Chester 1996; Pye 1987; Merrill et al. 1994; Leinen and Sarnthein 1989). The elemental composition of the dust is broadly similar to that of average crustal material (Zhuang et al. 1992; Uematsu et al. 1985; Arimoto et al. 1989, 1990, 1992; Prospero 1981; Prospero et al. 1987, 1989; Talbot et al. 1986; Graham and Duce 1979; Spokes et al. 1994). Indeed, there is a remarkable consistency in composition of dust from specific source regions. As an example, Fig. 2.11 presents data on the elemental composition of dust samples collected from eight different large African dust events that passed over Barbados, West Indies, over a two year period. The figure shows the mean ratio of the concentration of a specific element (Fe, Ti, Cd, Sb, etc.) measured in the dust to that of Al in the dust; these mean ratios are plotted against the same ratio in average crustal material (Taylor and McLennan 1985). Aluminum is used as the normalizing element because it is a major component in the Earth's crust and there are no major anthropogenic sources of Al-rich aerosol; but other major elements could have been (and are) used (e.g. Fe, Ti, and Si). As can be seen in Fig. 2.11, most elements fall very close to the 1:1 line, which suggests that the composition of the dust is very similar to that of average crustal material. Indeed, the composition of the individual samples shows a very similar pattern, suggesting that there is very little difference in composition from one dust event to another. When element ratios deviate markedly from the 1:1 line (for example, As, Sb, and Cd in Fig. 2.11), it suggests that other sources or processes could be influencing the sample, for example, pollution (Guieu and Thomas 1996; Chester et al. 1996; Gullu et al. 1996). In comparing aerosol composition to hypothesized sources, it is convenient to use enrichment factors (*EF*). The *EF* is based on the ratio

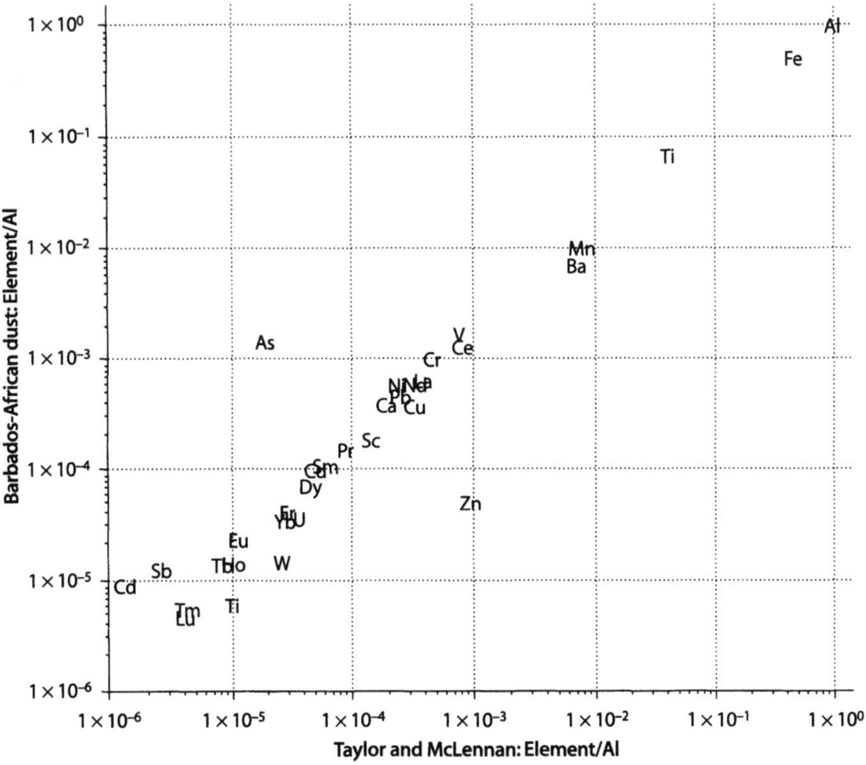

Fig. 2.11. Elemental concentrations in dust samples collected at Barbados, West Indies. Concentrations are normalized against Al and plotted against the ratio of these same elements in average crustal material as presented in Taylor and McLennan (1985) (Data courtesy of W. Landing, personal communication)

of the concentration of a specific element in the sample to that of a normalizing element (e.g. Al, Fe, Ti, and Si); the *EF* is calculated by dividing the sample ratio by the ratio in the hypothesized soil. That is, for soils, Al is used as the reference element.

$$EF = (X / Al)_{sample} / (X / Al)_{soil}$$

Figure 2.12 shows the *EF*s calculated for a large suite of elements in samples collected at sites in the Mediterranean (Gullu et al. 1996). The samples were of two types: ones that contained high concentrations of African dust and ones that contained little (if any) dust. In the dust samples, most elements yield *EF*s close to 1, but many elements yield ratios a factor of 10 or higher. Some elements yielded ratios over 100 times higher. These elements are commonly associated with pollution sources (e.g. Cd, As, Sb, Se, and Hg). These large enrichments are attributed to the mixing of European pollution with the African dust prior to (or during) the collection phase.

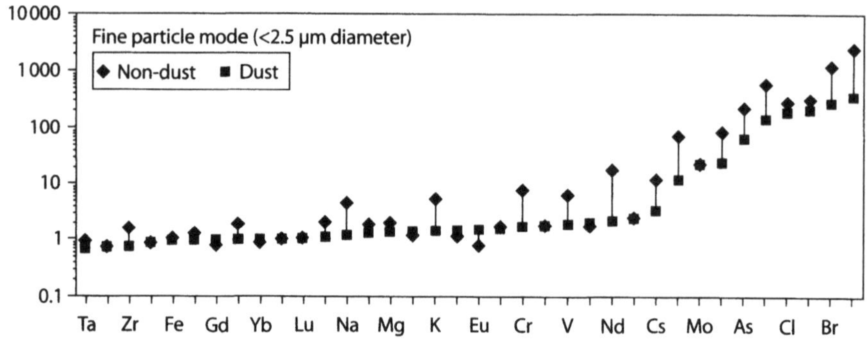

Fig. 2.12. Element enrichment factors calculated for the fine fraction (<2.5 µm diameter) of aerosol samples collected from Mediterranean air masses. Element concentrations are normalized to Al and compared to average crustal abundances. Data are plotted separately for African dust events and non-dust events. The larger enrichment factors obtained from non-dust events reflects the impact of pollutants from Europe, but even the dust events show the effects of the add-mixing of pollutant aerosols. (redrawn from Gullu et al. 1996, Fig. 5). The *square symbols* are data from samples dominated by dust. The *diamond symbols* are data from samples that contain substantial pollution aerosol from European sources

2.5.4 Mineralogical Composition

The mineralogy of dust has been widely studied. For reviews that touch on this subject, see Prospero 1981; Leinen and Sarnthein 1989; Guerzoni and Chester 1996; Pye 1987, 1989. Dusts typically contain minerals such as quartz, kaolinite, chlorite, illite, smectite, calcite, etc. The mineralogy of dust from specific source regions shows a consistency similar to that noted for elemental composition – for example, African dust carried to Barbados during the summer months (Glaccum and Prospero 1980). But various studies have shown regional differences in dust composition. Dusts originating from low-latitude dust sources (e.g. south of the Sahara in North Africa) tend to have more kaolinite than dusts from mid-latitude sources; conversely, mid-latitude sources tend to have more illite (Glaccum and Prospero 1980; Prospero 1981). The relative concentrations of these two minerals reflect climate differences that result in differences in mineral weathering processes. Indeed, differences in mineral composition can be used to broadly identify source regions, for example in North Africa (Herrmann et al. 1996; Molinaroli 1996). Similarly, large differences are noted for dusts transported over the North Pacific from sources in Asia and North America (Blank et al. 1985; Merrill et al. 1994; Leinen et al. 1994). The differences in mineralogy could conceivably be used to trace dust sources on a global scale, but this is difficult because the properties of potential soil sources are poorly characterized (Claquin et al. 1999).

2.5.5 Deposition of Dust to the Oceans

Estimates of the global dust budget vary widely, ranging from 1000 to 5000 Mt yr^{-1} (Duce 1995; Prospero 1996a). Because of the complexity of the generation and transport processes, it has been difficult to develop global dust models. A number of models have recently addressed this issue (see for example, Marticorena et al. 1997; Miller and Tegen 1998; Tegen and Miller 1998; Ginoux et al. 2001). Although these do replicate some features of the global dust cycle, there are major discrepancies. Most notable is that models tend to show very large amounts of dust transport in the Southern Hemisphere, whereas satellites show very little dust activity there (Prospero et al. 2002). For example, most models show very large plumes of dust emerging from Australia, when in fact Australia is a very weak dust source. This is apparent from satellite images such as AVHRR (Fig. 2.3) and TOMS (Fig. 2.4). The development of better models will hinge on a better understanding of the factors affecting dust mobilization and a more complete knowledge of the physical environments in the source regions. They will also require more and better data on the distribution of dust over the oceans.

There are major issues to be resolved with regard to the temporal and spatial variability of dust transport and deposition. Over the open ocean where the mass median diameter is only a few micrometres at most, the dominant deposition mechanism is generally by precipitation (i.e. "wet" removal). Removal by "dry" processes (principally by sedimentation) could be important in coastal regions close to sources (such as along the western coast of North Africa) where the dust size distribution is skewed towards large particles. There are very few measurements of dust deposition rates to the ocean; consequently, they must be estimated. Estimates of wet deposition rates are based on calculations using scavenging ratios (defined as the concentration of a substance in rain divided by the corresponding concentration in air). Scavenging ratios are empirically derived from measurements made with collocated precipitation and aerosol samplers. However, because of the dearth of long-term measurements of dust in precipitation and aerosols (Prospero 1996a,b), it has been necessary to extrapolate scavenging ratios to the world ocean. The most recent and most comprehensive estimate of dust deposition to the oceans is presented in Duce et al. 1991. For various reasons, Duce et al. 1991, used a scavenging ratio of 200 for the North Atlantic; for the remainder of the world ocean, they used a ratio of 1 000 (which will yield a deposition flux that is five times that obtained with the 200 value, all other things being equal). The Duce et al. estimates are shown in Table 2.4. To demonstrate the sensitivity of the estimated global fluxes to the scavenging ratio, Table 2.4 also shows wet deposition rates obtained using a global scavenging ratio of 200. The difference in the global flux is almost a factor of three. Despite the large differences in the flux estimates, it is clear that the highest basin rates are in the northern North Atlantic, the northern Indian Ocean, and the North Pacific.

Ultimately, the estimates of dust deposition to the ocean must be reconciled with the accumulation rates in the deep-sea sediments. (However, it should be noted that a major shortcoming of such comparisons is that sediment accumulation rates are in effect long-term averages, while the atmospheric deposition estimates are based on present day conditions.) Rea (1994) estimated dust deposition rates based on the analysis of aeolian materials in pelagic sediment cores. Rea's measured accumulation rates for aeolian materials in Holocene sediments in the central North Pacific Ocean seem to be in reasonable agreement with the estimates by Duce et al. (1991). Data on the Atlantic are sparse and concentrated in the eastern equatorial regions; but here too, the agreement seems acceptable. In contrast, Rea finds that the Duce et al. deposition rates to the southern oceans are too high by a factor of 5 to 10; however, Rea's data

Table 2.4. Dust deposition rates to various ocean regions (after Duce et al. 1991)[a]

Ocean	Mean basin flux ($g\ m^{-2}\ yr^{-1}$)[b]	Total basin deposition ($Tg\ yr^{-1}$)	
		Duce et al. (1991)[b]	*SR* = 200[c]
North Pacific	5.3	480	96
South Pacific	0.35	39	8
North Atlantic	4.0	220	220
South Atlantic	0.47	24	5
North Indian	7.1	100	20
South Indian	0.82	44	9
Global	2.5	910	358

[a] Prospero (1996a), Table 3.2.
[b] Duce et al. (1991). Scavenging Ratio (*SR*) = 1 000 except for North Atlantic, where *SR* = 200.
[c] Duce et al. (1991). Modified using *SR* = 200 globally.

density in these regions (especially in the Indian Ocean) is very sparse and does not warrant a strong conclusion in this regard.

2.5.6 Impact of Dust on Marine Biogeochemistry Cycles

Atmospheric dust could also supply other elements that might play a significant role in ocean biogeochemistry. For example, Measures and Vink (2000) surveyed data on the concentration of dissolved Al in surface waters from a wide range of ocean regions; they found that Al concentrations ranged over 3 orders of magnitude but were in good agreement with measurements (and estimates) of dust deposition to these ocean regions. Various studies (see Measures and Vink 2000) show that only a few percent of the Al in mineral aerosols is readily released in aqueous solutions. Nonetheless, dust deposition leaves a clearly discernible pattern in ocean surface waters. They suggest that Al could be used as a surrogate for dust deposition to the global ocean and thereby enable the study of the impact of other dust-borne species. For example, dust could be a significant source of rare-earth elements (REE) in ocean waters (Greaves et al. 1994; Sholkovitz et al. 1993). Approximately 1–3% of the REE in African dust is readily dissolved in surface waters in laboratory experiments (Greaves et al. 1994). Increased concentrations of REE were found in ocean surface waters in the western North Pacific, which yielded REE ratios similar to that of Asian soils (Greaves et al. 1999).

2.5.7 The Impact of African Deposition on the Nutrient Cycle

Much interest has focused on Fe in its role as an essential micronutrient. Pioneering work by J. H. Martin (Martin et al. 1994) showed that primary productivity could be significantly increased with the addition of soluble Fe to ocean surface waters. Martin suggested that the Fe carried by wind-borne dust could play a major role in controlling the primary productivity in vast ocean regions. Subsequent research has firmly established that Fe is an important micronutrient (Archer and Johnson 2000) and that Fe could play an important role in the Earth's carbon cycle, thereby mediating the ocean-atmosphere exchange of CO_2 (Behrenfeld et al. 1996; Coale et al. 1996a,b; Falkowski 1997; Falkowski et al. 1998; Fitzwater et al. 1996; Gordon et al. 1997). Furthermore, it is clear that except for coastal waters, the only significant source of Fe is that transported by winds from the continents (Johnson et al. 1997a,b).

The work cited above largely focuses on the results of laboratory experiments or on field experiments where Fe is added to ocean waters. Although there is considerable evidence to support the hypothesis that Fe plays an important role in the ocean primary productivity, it is difficult to estimate the importance of dust-Fe on the global ocean. Recently, Fung et al. (2000) presented an analysis of the iron budget in the upper ocean. They estimated the global distribution of annual iron assimilation by phytoplankton from satellite-derived estimates of oceanic primary production and on the measured ratios of Fe/C in cells. They took into account two sources of iron: that supplied by upwelling and mixing of ocean waters and that deposited from the atmosphere. On this basis, they computed the global assimilation rate of Fe by phytoplankton in the open ocean to be 12×10^9 mol Fe yr^{-1}. They estimated that the total deposition of Fe in dust was 96×10^9 mol Fe yr^{-1} in the open ocean. The soluble fraction of Fe depends on a variety of factors, especially the pH of the solution (Desboeufs et al. 1999); solubilities at pH's commonly found in precipitation and cloud water are typically in the range of 1 and 10% (Zhu et al. 1997; Spokes et al. 1994, Spokes and Jickells 1996). Thus, the estimated Fe-dust flux yields a total available-Fe deposition rate in the range of 1 to 10×10^9 mol Fe yr^{-1}. Thus, the estimated aeolian deposition rate of available Fe in dust could readily supply a substantial fraction of that required by phytoplankton. In comparison, the upwelling and entrainment supply of dissolved Fe to the upper ocean is relatively small: $\sim 0.7 \times 10^9$ mol Fe yr^{-1}. It should be noted that there are major uncertainties in the estimate of Fung et al, as large as a factor of 5 to 10. The largest limitations in the calculations are due to the dearth of information on the transport rates of dust to the oceans, the chemical form of the Fe in the dust, and the availability of dust-Fe to organisms in the water column. These are areas ripe for research.

In an earlier section, I showed that huge amounts of dust are carried from North Africa to the Atlantic (Table 2.4). Thus, there is considerable interest in the impact of this transport of biogeochemical processes in this region. The nutrient budgets of the North Atlantic have been extensively studied, although major uncertainties remain (Galloway et al. 1996). Michaels et al. (1996) compiled an N-nutrient budget for the region. They found that the pool of N-nutrients was substantially greater than could be accounted for by conventional oceanic sources. They concluded that though the anthropogenic inputs of NO_x and NH_x were significant as discussed in an earlier section, the greatest impact was through the input of Fe associated with African dust. Michaels et al. attribute this high rate of N-fixation to the colonial cyanobacterium, *Trichodesmium* sp. Nitrogen fixation by *Trichodesmium* sp. requires a great deal more Fe than that required by organisms that utilize nitrate. The ratio of N:Fe in *Trichodesmium* sp. is about 1 000 moles of N to 1 mole of Fe. Prospero et al. (1996) estimate the total deposition of dust to the latitude band 10° N to 40° N (excluding the areas near the coast of Africa) at about 44×10^{12} g yr^{-1}. The concentration of Fe in African dust is close to the average crustal abundance, 3.5% by weight. Thus, the annual deposition of Fe in dust in this latitude band (which includes the Sargasso Sea where enhanced fixation is most prominent) is 2.8×10^{10} mol Fe yr^{-1}. This amount of Fe could support an annual excess (i.e. "new") nitrate production of as much as 28×10^{12} mol N yr^{-1} if all the Fe were available to the *Trichodesmium*. Even if only a relatively small fraction of the Fe was available (about 10–25%), it could support a nitrate production comparable to the excess production observed in this region. This suggests that the excess nitrate in the North Atlantic may be largely controlled by dust transport and deposition.

If dust is playing a large role in nitrate nutrient production, then dust could cause large interannual changes in nitrate production (and primary productivity in general) in the North Atlantic. As shown in Fig. 2.9, there is a very large year-to-year variability in dust transport; Prospero and Nees (1986) showed that this variability was highly correlated to drought in sub-Saharan Africa (see also the inset in Fig. 2.9). Over the past three decades, dust concentrations in the Trade Winds has increased by a factor of three to four compared to dust transport during the mid-60s when rainfall rates were higher. Over longer time scales, the variability could be much larger. Palaeoclimate studies in North Africa show that this region has gone through extreme changes in climate. About 6–8 kyr B.P., the region experienced a pluvial phase which must have greatly reduced dust emissions. For example, the largest and most intense dust source on Earth today is found in northern Chad, in the Bodele Depression (Prospero et al. 2002). During the pluvial phase, the Bodele Depression was filled with water forming the palaeo Lake Chad, which covered a large area of the present-day country of Chad. Today Lake Chad covers a very small area far to the south of the depression. Further back in time, during glacial periods, dust activity must have been much greater than that today. Ice core studies in Greenland show that during the last glacial maximum, dust deposition rates were orders of magnitude greater than at present (Mahowald et al. 1999). Similarly, large changes in dust deposition have been observed in ice cores from the Antarctic; dust most likely was transported from sources in Australia and southern South America (Prospero et al. 2002).

2.6 Other Aerosol Species and the Impact of Continental Sources

This paper has focused on a limited number of species found in marine aerosols. There are many other species that could play an important role in the marine environment. Of particular interest are organic aerosols. Compared to the aerosol species discussed in this review, we know very little about the organic fraction. There are three major classes of primary organic aerosols: biological particles, carbonaceous material emitted by biomass burning, and particles emitted from anthropogenic sources. Secondary organic particles can be formed from the oxidation of volatile organic carbon (*VOC*) compounds that are emitted from biogenic and anthropogenic sources (Andreae and Crutzen 1997; Scholes and Andreae 2000). The chemical characterization or organic aerosols is extremely difficult because of the huge range of compounds present. Usually, only a small percentage of the total organic aerosol mass can be accounted for (Peltzer and Gagosian 1989). An alternative approach has been to broadly characterize the total organic carbon (*OC*) and that of an important subset of organics – *BC*. Both *OC* and *BC* can be measured relatively easily by a variety of techniques, although the data are subject to considerable uncertainty (Heintzenberg et al. 1997; Jacobson et al. 2000; Huebert and Charlson 2000). Liousse et al. (1996) provide a summary of global *BC* and total *OC* distributions. Typical concentrations at remote ocean sites are in the range of several hundred ng C m^{-3}, except for some locations impacted by proximate sources. *BC* concentrations were substantially lower, mostly in the range of 10's of ng C m^{-3} in the Northern Hemisphere and mostly below 10 ng C m^{-3} in the Southern Hemisphere.

There is great interest in *BC* because it is an extremely efficient absorber of solar radiation; consequently *BC* is believed to play major role in climate (Hansen et al. 2000). Although measurements are limited, studies show that *BC* and *OC* are important to climate related processes over the oceans. Extremely high concentrations of *OC* and *BC* (along with many other pollutant species) were measured in the Arabian Sea in air masses emerging from the Indian subcontinent; the radiative properties of the atmosphere were strongly affected by the organic component in the aerosol (Lelieveld et al. 2000; Satheesh et al. 1999). In haze layers, *BC* constituted as much as 17% of the fine particle mass (Novakov et al. 2000). A substantial fraction of the organic aerosol is water-soluble (Saxena and Hildemann 1996). Indeed, over the western North Atlantic, organic aerosols (and associated water) scattered more light than sulphate aerosol (Hegg et al. 1997).

The presence of *BC* and *OC* over the oceans is clearly associated with the transport of continental aerosols to the marine environment. Recently, Heintzenberg et al. (2000) presented a review of the size distribution and chemical composition of marine aerosols. Figure 2.13 (Heintzenberg et al. 2000, Fig. 7) shows the latitude distribution of four aerosol species over the oceans: nss-SO_4^{2-}, biogenic SO_4^{2-}, *BC*, and metals. The *top panel*

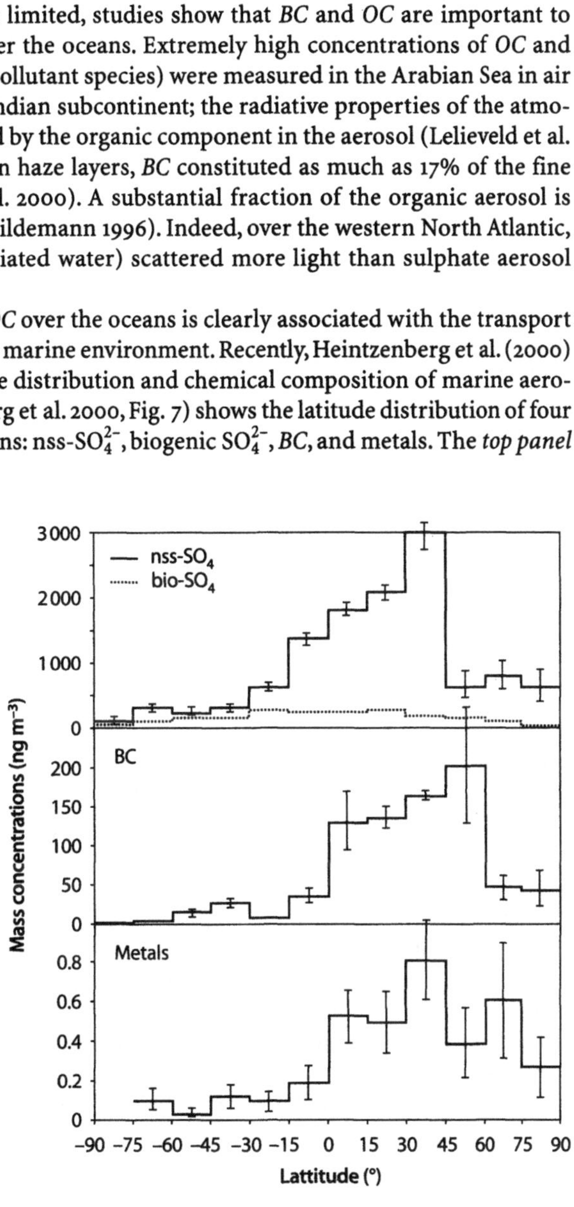

Fig. 2.13. Global annual average latitudinal distributions of nss-SO_4^{2-}, black carbon, and metals. Also shown in the top (nss-SO_4^{2-}) panel is the contribution from biological sources (i.e. DMS emissions) calculated from a model. The distribution of metals is given in units of the normalized (relative) latitudinal concentration of a suite of metals – see Heintzenberg et al. 2000 (modified after Heintzenberg et al. 2000, Fig. 7)

shows the measured distribution of nss-SO_4^{2-}; also shown is the computed distribution of biogenically-produced (from DMS) nss-SO_4^{2-}. This figure shows that the Northern Hemisphere is hugely impacted by continental anthropogenic sources and that the biogenic source is trivially small. The latitudinal distribution of *BC* (*middle panel*) is broadly similar to that of nss-SO_4^{2-}; one subtle difference is the presence of a small Southern Hemisphere maximum in the mid-latitudes (30–45° S), which is attributed to biomass burning. Finally, the *bottom panel* shows the relative latitudinal distribution of "metals"; this category includes V, Cr, Ni, Cu, Zn, Cd, and Pb – all species that are closely identified with pollution sources. The distribution of these metals also shows a large maximum in the Northern Hemisphere mid-latitudes.

Thus, there is clear evidence of the widespread impact of anthropogenic aerosols over large areas of the global ocean. This impact is evident for a wide range of materials, both inorganic and organic. The chemistry of these complex aerosol mixtures is expected to be very different from that of aerosols in remote ocean regions where pollution impacts are minimal.

2.7 Conclusions

Research shows that aerosols can play an important role in a number of important marine biogeochemical processes and in climate. However, at this time it is difficult to quantify these impacts. There are huge areas of the ocean for which we have little or no detailed knowledge of aerosol physical and chemical properties. We also lack detailed knowledge of the processes by which aerosols can affect climate. As for the impact of aerosol material on ocean biogeochemical processes, there are two problems that preclude a clear assessment. First of all, we have only a very rough idea of how much aerosol material is actually deposited; estimates are largely based on models that thus far have yielded widely varying results. Secondly, we have a very poor understanding of the ways in which deposited aerosol materials participate in sea water chemistry and biology. For example, we have very little information on the rates at which dust-borne metals are released to the water column, the speciation and chemical properties of the released materials, and their bioavailability. These are all ripe areas for research.

This review could only cover a relatively short list of species. While these species make up a substantial fraction of the aerosol mass, in many ocean regions there is a large fraction of the marine aerosol mass that has not been characterized. For example, the work of Quinn et al. (2000) suggests that 9–45% of the submicrometre aerosol mass over the remote Pacific Ocean consists of species other than nss-SO_4^{2-}, sea salt, NH_4^+, and MSA. We know that in some regions mineral dust and organic materials could make up much of the balance of the mass. Nonetheless, it is clear that there is much work yet to be done in this regard. In particular, there is considerable evidence that organic materials are ubiquitous over the oceans and that much of this material is transported from the continents. Unfortunately, much of the data on organic aerosols is semiquantitative. Thus, there is a great need to characterize this aerosol component. In the absence of such data it is not possible to assess the impact of organic aerosols on marine biogeochemistry.

References

Adams PJ, Seinfeld JH, Koch DM (1999) Global concentrations of tropospheric sulfate, nitrate, and ammonium aerosol simulated in a general circulation model. J Geophys Res 104:13791–13823

Andreae MO (1995) Climatic effects of changing atmospheric aerosol levels. In: Henderson-Sellers A (ed) World survey of climatology, future climates of the World. Elsevier, Amsterdam (vol XVI, pp 341–392

Andreae MO, Crutzen PJ (1997) Atmospheric aerosols: Biogeochemical sources and role in atmospheric chemistry. Science 276:1052–1056

Andreae MO, Elbert W, Cai Y, Andreae TW, Gras J (1999) Non-sea-salt sulfate, methanesulfonate, and nitrate aerosol concentrations and size distributions at Cape Grim, Tasmania. J Geophys Res 104(D17):21695–21706

Andreae MO, Elbert W, Gabriel R, Johnson DW, Osborne S, Wood R (2000) Soluble ion chemistry of the atmospheric aerosol and SO_2 concentrations over the eastern North Atlantic during ACE-2. Tellus 52B:1066–1087

Archer DE, Johnson K (2000) A model of the iron cycle in the ocean. Global Biogeochem Cycles 14:269–280

Arimoto R, Duce RA, Ray BJ (1989) Concentrations, sources, and air/sea exchange of trace elements in the atmosphere over the Pacific Ocean. In: Riley JP, Chester R, Duce RA (eds) Chemical oceanography. Academic Press, London (vol XX, pp 107–149)

Arimoto R, Ray BJ, Duce RA, Hewitt AD, Boldi R, Hudson A (1990) Concentrations, sources, and fluxes of trace elements in the atmosphere of New Zealand. J Geophys Res 95:22389–22405

Arimoto R, Duce RA, Savoie DL, Prospero JM (1992) Trace elements in aerosol particles from Bermuda and Barbados: Concentrations, sources, and relationships to aerosol sulfate. J Atmos Chem 14:439–457

Arimoto R, Duce RA, Ray BJ, Ellis WG Jr, Cullen JD, Merrill JT (1995) Trace elements in the atmosphere over the North Atlantic. J Geophys Res 100:1199–1214

Arimoto R, Duce RA, Savoie DL, Prospero JM, Talbot R, Cullen JD, Tomza U, Lewis NF, Ray BJ (1996) Relationships among aerosol constituents from Asia and the North Pacific during PEM-West A. J Geophys Res 101:2011–2024

Bates TS, Quinn PK (1997) Dimethylsulfide (DMS) in the equatorial Pacific Ocean (1982 to 1996): Evidence of a climate feedback? Geophys Res Lett 24(8):861–864

Bates TS, Huebert BJ, Gras JL, Griffiths FB, Durkee PA (1998) International Global Atmospheric Chemistry (IGAC) Project's First Aerosol Characterization Experiment (ACE 1): Overview. J Geophys Res 103(DD13):16297–16318

Behrenfeld MJ, Bale AJ, Kolber ZS, Aiken J, Falkowski PG (1996) Confirmation of iron limitation of phytoplankton photosynthesis in the equatorial Pacific Ocean. Nature 383:508–511

Benkovitz CM, Scholtz MT, Pacyna J, Tarrason L, Dignon J, Voldner EC, Spiro PA, Logan JA, Graedel TE (1996) Global gridded inventories of anthropogenic emissions of sulfur and nitrogen. J Geophys Res 101:29239–29253

Berg A van den, Dentener F, Lelieveld J (2000) Modeling the chemistry of the marine boundary layer: Sulphate formation and the role of sea-salt aerosol particles. J Geophys Res 105(D9):11671–11698

Berresheim H, Wine P, Davis D (1995) Sulfur in the atmosphere. In: Singh H (ed) Composition, chemistry, and climate of the atmosphere. Van Nostrand Reinhold, New York, pp 251–307

Blank M, Leinen M, Prospero JM (1985) Major Asian aeolian inputs indicated by the mineralogy of aerosols and sediments in the western North Pacific. Nature 314:84–86

Bouwman AF, Lee DS, Asman WAH, Dentener FJ, Van Der Hoek KW, Olivier JGJ (1997) A global high-resolution emission inventory for ammonia. Global Biochem Cycles 11:561–588

Brink HM ten, Veefkind JP, Waijers-Ijpelaan A, Hage JC van der (1996) Aerosol light-scattering in the Netherlands. Atmos Environ 30:4251–4261

Charlson RJ, Heintzenberg J (eds) (1995) Dahlem Workshop on Aerosol Forcing of Climate (1994: Berlin). John Wiley, Chichester

Charlson RJ, Lovelock JE, Andreae MO, Warren SG (1987) Oceanic phytoplankton, atmospheric sulphur, cloud albedo, and climate. Nature 326:655–661

Chester R, Nimmo M, Keyse S (1996) The influence of Saharan and Middle Eastern desert-derived dust on the trace metal composition of Mediterranean aerosols and rainwaters: An overview. In: Guerzoni S, Chester R (eds) The impact of desert dust across the Mediterranean. Kluwer Academic Publishers, Dordrecht, pp 253–273

Chiapello I, Bergametti G,Gomes L, Chatenet B, Dulac F, Pimenta J, Santos Soares E (1995) An additional low layer transport of Sahelian and Saharan dust over the northeastern Tropical Atlantic. Geophys Res Lett 22:3191–3194

Chiapello I, Prospero JM, Herman JR, Hsu NC (1999) Detection of mineral dust over the North Atlantic Ocean and Africa with the Nimbus 7 TOMS. J Geophys Res 104:9277–9292
Clarke AD, Varner JL, Eisele F, Mauldin RL, Tanner D, Litchy M (1998) Particle production in the remote marine atmosphere: Cloud outflow and subsidence during ACE 1. J Geophys Res 103(DD13):16397–16410
Claquin T, Schulz M, Balkanski YJ (1999) Modeling the mineralogy of atmospheric dust sources. J Geophys Res 104(D18):22243–22256
Coakley JA Jr, Bernstein RL, Durkee PA (1987) Effect of ship-stack effluents on cloud reflectivity. Science 237:1020–1022
Coale KH, Fitzwater SE, Gordon RM, Johnson KS, Barber RT (1996a) Control of community growth and export production by upwelled iron in the equatorial Pacific Ocean. Nature 379:621–624
Coale KH et al. (1996b) A massive phytoplankton bloom induced by an ecosystem-scale iron fertilization experiment in the equatorial Pacific Ocean. Nature 383:495–501
Cornell S, Rendell A, Jickells T (1995) Atmospheric inputs of organic nitrogen to the oceans. Nature 376:243–246
Dentener FJ, Crutzen PJ (1994) A three dimensional model of the global ammonia cycle. J Atmos Chem 19:331–369
Desboeufs KV, Losno R, Vimeux F, Cholbi S (1999) The pH-dependent dissolution of wind-transported Saharan dust. J Geophys Res 104(D17):21287–21300
Duce RA (1986) The impact of atmospheric nitrogen, phosphorus, and iron species on marine biological productivity. In: Buat-Manard P (ed) The Role of air-sea exchange in geochemical cycling. Reidel, Dordrecht, pp 497–528
Duce RA (1995) Sources, distributions, and fluxes of mineral aerosols and their relationship to climate. In: Charlson RJ, Heintzenberg J (eds) Dahlem Workshop on Aerosol Forcing of Climate. John Wiley, Chichester, pp 43–72
Duce RA, Liss PS, Merrill JT, Atlas EL, Buat-Menard P, Hicks BB, Miller JM, Prospero JM, Arimoto R, Church T, Ellis W, Galloway JN, Hansen L, Jickells TD, Knap AH, Reinhardt KH, Schneider B, Soudine A, Tokos JJ, Tsunogai S, Wollast R, Zhou M (1991) The atmospheric input of trace species to the world ocean. Global Biogeochem Cycles 5:193–259
Erickson DJ III, Duce RA (1988) On the global flux of atmospheric sea salt. J Geophys Res 93:14079–14088
Falkowski PG (1997) Evolution of the nitrogen cycle and its influence on the biological sequestration of Co_2 in the ocean. Nature 387:272–275
Falkowski PG, Barber RT, Smetacek V (1998) Biogeochemical controls and feedbacks on ocean primary production. Science 281:200–206
Finlayson-Pitts BJ, Pitts JN Jr (2000) Chemistry of the upper and lower atmosphere. Academic Press, San Diego
Fitzgerald JW (1991) Marine aerosols: A review. Atmos Environ 25A:533–545
Fitzwater SE, Coale KH, Gordon RM, Johnson KS, Ondrusek ME (1996) Iron deficiency and phytoplankton growth in the Equatorial Pacific. Deep-Sea Res II 43:995–1015
Fung I, Meyn S, Tegen I, Doney SC, John J, Bishop JKB (2000) Iron supply and demand in the upper ocean. Global Biogeochem Cycles 14:281–296
Galloway JN, Schlesinger WH, Levy H II, Michaels A, Schnoor JL (1995) Nitrogen fixation: Anthropogenic enhancement-environmental response. Global Biogeochem Cycles 9(2):235–252
Galloway JN, Howarth RW, Michaels AF, Nixon SW, Prospero JM, Dentener FJ (1996) Nitrogen and phosphorous budgets of the North Atlantic Ocean and its watershed. Biogeochemistry 35:3–25
Gard EE, Kleeman MJ, Gross DS, Hughes LS, Allen JO, Morrical BD, Fergenson DP, Dienes T, Galli, ME, Johnson RJ, Cass GR, Glen R, Prather KA (1998) Direct observation of heterogeneous chemistry in the atmosphere. Science 279:1184–1187
Ginoux P, Chin M, Tegen I, Herman J, Torres O, Holben B (2001) Global modeling of mineral dust with the Goddard transport model. J Geophys Res (to be published)
Glaccum RA, Prospero JM (1980) Saharan aerosols over the tropical North Atlantic – mineralogy. Mar Geol 37:295–321
Golitsyn G, Gillette DA (1993) Introduction: A joint Soviet-American experiment for the study of Asian desert dust and its impact on local meteorological conditions and climate. Atmos Environ 27A:2467–2470
Gong SL, Barrie LA, Prospero JM, Savoie DL, Ayers GP, Blanchet J-P, Spacek L (1997) Modeling sea-salt aerosols in the atmosphere. 2. Atmospheric concentrations and fluxes. J Geophys Res 102:3819–3830
Gordon RM, Coale KH, Johnson KS (1997) Iron distributions in the Equatorial Pacific: Implications for new production. Limnol Oceanogr 42:419–431
Goudie AS (1983) Dust storms in space and time. Prog Phys Geog 7:502–530
Graedel TE, Crutzen PJ (1993) Atmospheric change: An Earth system perspective. W. H. Freeman and Co., New York

Graf H-F, Feichter J, Langmann B (1997) Volcanic sulfur emissions: Estimates of source strength and its contribution to the global sulfate distribution. J Geophys Res 102:10727–10738
Graham WF, Duce RA (1979) Atmospheric pathways of the phosphorus cycle. Geochim Cosmoschim Acta 43:1195–1208
Greaves MJ, Statham PJ, Elderfield H (1994) Rare earth element mobilization from marine atmospheric dust into seawater. Mar Chem 46:255–260
Greaves MJ, Elderfield H, Sholkovitz ER(1999) Aeolian sources of rare earth elements to the Western Pacific Ocean. Mar Chem 68:31–38
Guerzoni S, Chester R (eds) (1996) The impact of desert dust across the Mediterranean. Kluwer Academic Publishers, Dordrecht
Guieu C, Thomas AJ (1996) Saharan aerosols: From the soil to the ocean. In: Guerzoni S, Chester R (eds) The impact of desert dust across the Mediterranean. Kluwer Academic Publishers, Dordrecht, pp 207–216
Gullu GH, Olmez I, Tuncel G (1996) Chemical concentrations and elements size distributions of aerosols in the eastern Mediterranean during strong dust storms. In: . In: Guerzoni S, Chester R (eds) The impact of desert dust across the Mediterranean. Kluwer Academic Publishers, Dordrecht, pp 339–347
Hansen JE, Lacis A, Ruedy R, Sato M (1992) Potential climate impact of Mount Pinatubo eruption. Geophys Res Lett 19:215–218
Hansen JE, Sato M, Ruedy R, Lacis A, Oinas V (2000) Global warming in the twenty-first century: An alternative scenario. Proc Nat Acad Sci 97:9875–9880
Hegg DA, Livingston J, Hobbs PV, Novakov T, Russell P (1997) Chemical apportionment of aerosol column optical depth off the mid-Atlantic coast of the United States J Geophys Res 102(D21):25293–25304
Heintzenberg J, Charlson RJ, Clarke AD, Liousse C, Ramaswamy V, Shine KP, Wendisch M, Helas G (1997) Measurement and modeling of aerosol singlescattering albedo: Progress, problems and prospects. Beitr Phys Atmosph 70:249–263
Heintzenberg J, Covert DC, Dingenen R Van (2000) Distribution and chemical composition of marine aerosols: A compilation and review. Tellus 52B:1104–1122
Herman JR, Bhartia PK, Torres O, Hsu C, Seftor C, Celarier E (1997) Global distribution of UV-absorbing aerosols from Nimbus-7/TOMS data. J Geophys Res 102:16911–16922
Herrmann L, Jahn R, Stahr K (1996) Identification and quantification of dust additions in peri-Saharan soils. In: . In: Guerzoni S, Chester R (eds) The impact of desert dust across the Mediterranean. Kluwer Academic Publishers, Dordrecht, pp 173–182
Hobbs PV (ed) (1993) Aerosol-cloud-climate interactions. Academic Press, San Diego
Hobbs PV (2000) Introduction to atmospheric chemistry. Cambridge University Press, Cambridge
Holland EA, Dentener FJ, Braswell BH, Sulzman JM (1999) Contemporary and pre-industrial global reactive nitrogen budgets. Biogeochemistry 46:7–43
Howarth RW, Billen G, Swaney D, Townsend A, Jaworski N, Lajtha K, Downing JA, Elmgren R, Caraco N, Jordan T, Berendse F, Freney J, Kudeyarov V, Murdoch P, Zhao-Liang Z (1996) Regional nitrogen budgets and riverine N and P fluxes for the drainages to the North Atlantic Ocean: Natural and human influences. Biogeochemistry 35:75–139
Huebert BJ, Charlson RJ (2000) Uncertainties in data on organic aerosols. Tellus 52B:1249–1255
Husar RB, Prospero JM, Stowe LL (1997) Characterization of tropospheric aerosols over the oceans with the NOAA advanced very high resolution radiometer optical thickness operational product. J Geophys Res 102:16889–16909
Husar RB, Tratt DM, Schichtel BA et al. (2001) The Asian dust events of April 1998. J Geophys Res 106(D16):18317–18330
IPCC (1996) Climate change 1995. In: Houghton JT, Meira Filho LG, Callander BA, Harris N, Kattenberg A, Maskell K (eds) The science of climate change. Cambridge University Press, Cambridge
Jacob DJ (1999) Introduction to atmospheric chemistry. Princeton Univ. Press, Princeton
Jacobson MC, Hansson H-C, Noone KJ, Charlson RJ (2000) Organic atmospheric aerosols: Review and state of the science. Rev Geophys 38:267–294
Johnson KS, Gordon RM, Coale KH (1997a) What controls dissolved iron in the world ocean? Mar Chem 57:137–161
Johnson KS, Gordon RM, Coale KH (1997b) What controls dissolved iron in the world ocean? Author's closing comments. Mar Chem 57:181–186
Kettle AJ, Andreae MO, Amouroux D, Andreae TW, Bates TS, Berrresheim V, Bingemer H, Boniforti R, Curran MAJ, DiTullio GR, Helas G, Jones GB, Keller MD, Kiene RP, Leck C, Levasseur M, Malin G, Maspero M, Matrai P, McTaggart AR, Mihalopoulos N, Nguyen NC, Novo A, Putaud JP, Rapsomanikis S, Roberts G, Schebeske G, Sharma S, Simo R, Staubes R, Turner S, Uher G (1999) A global database of sea surface dimethylsulfide (DMS) measurements and a procedure to predict sea surface DMS as a function of latitude, longitude and month. Global Biogeochem Cycles 13:399–444

Leinen M, Sarnthein M (eds) (1989) Paleoclimatology and paleometeorology: Modern and past patterns of global atmospheric transport. Kluwer Academic Pubs., Norwell, Mass.
Leinen M, Prospero JM, Arnold E, Blank M (1994) Mineralogy of aeolian dust reaching the North Pacific Ocean 1. Sampling and analysis. J Geophys Res 99:21017–21023
Lelieveld J, Crutzen PJ, Ramanathan V, Andreae MO, Brenninkmeijer CAM, Campos T, Cass GR, Dickerson RR, Fischer H, Gouw JA de, Hansel A, Jefferson A, Kley D, Laat ATJ de, Lal S, Lawrence MG, Lobert JM, Mayol-Bracero O, Mitra AP, Novakov T, Oltmans SJ, Prather KA, Reiner T, Rodhe H, Scheeren HA, Sikka D, Williams J (2001) The Indian Ocean experiment: Widespread air pollution from South and Southeast Asia. Science 291(5506):1031–1036
Levitus S, Conkright ME, Reid JL, Najjar RG, Mantyla A (1993) Distribution of nitrate, phosphate and silicate in the world oceans. Prog Oceanogr 31:245–273
Li X, Maring H, Savoie D, Voss K, Prospero JM (1996) Dominance of mineral dust in aerosol light scattering in the North Atlantic trade winds. Nature 380:416–419
Li-Jones X, Prospero JM (1998) Variations in the size distribution of non-sea-salt sulfate aerosol in the marine boundary layer at Barbados: Impact of African dust. J Geophys Res 103(DD13):16073–16084
Liousse C, Penner JE, Chuang C, Walton JJ, Eddleman H, Cachier H (1996) A global three-dimensional model study of carbonaceous aerosols. J Geophys Res 101(D14):19411–19432
Mahowald N, Kohfeld K, Hansson M, Balkanski Y, Harrison SP, Prentice IC, Schulz M, Rodhe H (1999) Dust sources and deposition during the last glacial maximum and current climate: A comparison of model results with paleodata from ice cores and marine sediments. J Geophys Res 104:15895–15916
Marticorena B, Bergametti G, Aumont B, Callot Y, N'Doume' C, Legrand M (1997) Modeling the atmospheric dust cycle. 2. Simulation of Saharan dust sources. J Geophys Res Atmos 102:4387–4404
Martin JH et al. (1994) Testing the iron hypothesis in ecosystems of the equatorial Pacific Ocean. Nature 371:123–129
Measures CI, Vink S (2000) On the use of dissolved aluminum in surface waters to estimate dust deposition to the ocean. Global Biogeochem Cycles 14(1):317–328
Merrill J, Arnold E, Leinen M, Weaver C (1994) Mineralogy of aeolian dust reaching the North Pacific Ocean. 2. Relationship of mineral assemblages to atmospheric transport patterns. J Geophys Res 99:21025–21032
Michaels AF, Olson D, Sarmineto JL, Ammerman W, Fanning K, Jahnke R, Knap AH, Lipschultz F, Prospero JM (1996) Inputs, losses and transformations of nitrogen and phosphorus in the pelagic North Atlantic Ocean. Biogeochem 35:181–226
Middleton NJ, Goudie AS, Wells GL (1986) The frequency and source areas of dust storms. In: Nickling WG (ed) Aeolian geomorphology. Allen and Unwin, N.Y., pp 237–259
Miller RL, Tegen I (1998) Climate response to soil dust aerosols. J Clim 11:3247–3267
Molinaroli E (1996) Mineralogical characteristics of Saharan dust with a view to its final destination in Mediterranean sediments. In: Guerzoni S, Chester R (eds) The impact of desert dust across the Mediterranean, Kluwer Academic Publishers, Dordrecht, pp 153–162
Murphy DM, Thomson DS (1997) Chemical composition of single aerosol particles at Idaho Hill: Negative ion measurements. J Geophys Res 102:6353–6368
Neff JC, Holland EA, Dentener FJ, McDowell WH, Russell KM (2001) Atmospheric organic nitrogen: Implications for the global N cycle. Biogeochem (to be published)
Novakov T, Andreae MO, Gabriel R, Kirchstetter TW, Mayol-Bracero OL, Ramanathan V (2000) Origin of carbonaceous aerosols over the tropical Indian Ocean: Biomass burning or fossil fuels? Geophys Res Lett 27:4061–4064
O'Dowd CD, Smith MH (1993) Physico-chemical properties of aerosols over the Northeast Atlantic: Evidence for wind-speed-related sub-micron sea-salt aerosol production. J Geophys Res 98:1137–1149
O'Dowd CD, Smith MH, Consterdine IE, Lowe JA (1997) Marine aerosol, sea-salt, and the marine sulfur cycle: A short review. Atmos Environ 31:73–80
O'Dowd CD, Lowe JA, Smith MH, Kaye AD (1999) The relative importance of Nss-sulphate and sea-salt aerosol to the marine CCN population: An improved multi-component aerosol-cloud droplet parameterization. Quart J Roy Meteorol Soc 125(B):1295–1314
Paerl HW (1995) Coastal eutrophication in relation to atmospheric nitrogen deposition: Current perspectives. Ophelia 41:237–259
Peltzer ET, Gagosian RB (1989) Organic geochemistry of aerosols over the Pacific Ocean. In: Riley JP, Chester R, Duce RA (eds) Chemical oceanography, vol X. Academic Press, San Diego, pp 283–338
Perry KD, Hobbs PV (1994) Further evidence for particle nucleation in clear air adjacent to marine cumulus clouds. J Geophys Res 99(D11):22803–22818
Perry KD, Hobbs PV (1995) Correction to "Further evidence for particle nucleation in clean air adjacent to marine cumulus clouds". J Geophys Res 100(D9):18929–18929
Perry KD, Cahill TA, Eldred RA, Dutcher DD, Gill TE (1997) Long-range transport of North African dust to the eastern United States. J Geophys Res 102:11225–11238

Perry KD, Cahill TA, Schnell RC, Harris JM (1999) Long-range transport of anthropogenic aerosols to the National Oceanic and Atmospheric Administration baseline station at Mauna Loa Observatory, Hawaii. J Geophys Res 104(D15):18521–18534
Prospero JM (1981) Aeolian transport to the World Ocean. In: Emiliani C (ed) The Sea, vol VII: The oceanic lithosphere. Wiley Interscience, New York, pp 801–974
Prospero JM (1996a) The atmospheric transport of particles to the ocean. In: Ittekkott V, Honjo S, Depetris PJ (eds) Particle flux in the ocean. John Wiley and Sons, New York (SCOPE Report 57, pp 19–52)
Prospero JM (1996b) Saharan dust transport over the north Atlantic Ocean and Mediterranean: An overview. In: Guerzoni S, Chester R (eds) The impact of desert dust across the Mediterranean. Kluwer Academic Pub., Dordrecht, pp 133–151
Prospero JM (1999) Long-term measurements of the transport of African mineral dust to the Southeastern United States: Implications for regional air quality. J Geophys Res 104:15917–15927
Prospero JM, Nees RT (1986) Impact of the North African drought and El Niño on mineral dust in the Barbados trade winds. Nature 320:735–738
Prospero JM, Glaccum RA, Nees RT (1981) Atmospheric transport of soil dust from Africa to South America. Nature 289:570–572
Prospero JM, Nees RT, Uematsu M (1987) Deposition rate of particulate and dissolved aluminum derived from Saharan dust in precipitation at Miami, Florida. J Geophys Res 92:14723–14731
Prospero JM, Uematsu M, Savoie DL (1989) Mineral aerosol transport to the Pacific Ocean. In: Riley JP, Chester R, Duce RA (eds) Chemical oceanography, vol X. Academic Press, San Diego, pp 188–218
Prospero JM, Savoie DL, Arimoto R, Olafsson H, Hjartarson H (1995) Sources of aerosol nitrate and non-sea-salt (nss) sulfate in the Iceland region. Sci Tot Environ 160/161:181–191
Prospero JM, Barrett K, Church T, Dentener F, Duce RA, Galloway JN, Levy H II, Moody J, Quinn P (1996) Atmospheric deposition of nutrients to the North Atlantic basin. Biogeochem 35:27–73
Prospero JM, Ginoux P, Torres O, Nicholson S (2002) Environmental characterization of global sources of atmospheric soil dust derived from the NIMBUS-7 TOMS absorbing aerosol product. Rev Geophys (in press)
Pye K (1987) Aeolian dust and dust deposits. Academic Press, London
Pye K (1989) Processes of fine particle formation, dust source regions, and climatic changes. In: Leinen M, Sarnthein M (eds) Paleoclimatology and paleometeorology: Modern and past patterns of global atmospheric transport. Kluwer Academic Pubs., Dordrecht, pp 3–30
Quinn PK, Barrett KJ, Dentener FJ, Lipshcultz F, Six KD (1996) Estimation of the air/sea exchange of ammonia for the North Atlantic Basin. Biogeochem 35:275–304
Quinn PK, Coffman DJ, Kapustin VN, Bates TS, Covert DS (1998) Aerosol optical properties in the marine boundary layer during the First Aerosol Characterization Experiment (ACE 1) and the underlying chemical and physical aerosol properties. J Geophys Res 103:16547–16563
Quinn PK, Bates TS, Miller TL, Coffman DJ, Johnson JE, Harris JM, Ogren JA, Forbes G, Anderson TL, Covert DS, Rood MJ (2000) Surface submicron aerosol chemical composition: What fraction is not sulfate? J Geophys Res 105(D5):6785–6806
Raes F, Bates T, McGovern F, Liedekerke M Van (2000) The second Aerosol Characterization Experiment (ACE-2): General overview and main results Tellus 52B:111–125
Ravishankara AR, Rudich Y, Talukdar R, Barone SB (1997) Oxidation of atmospheric reduced sulphur compounds; perspective from laboratory studies. Philos Trans R Soc London 352(B):171–182
Rea DK (1994) The paleoclimatic record provided by eolian deposition in the deep sea – The geologic history of wind. Rev Geophys 32:159–195
Satheesh SK, Ramanathan V, Li-Jones X, Lobert JM, Podgorny IA, Prospero JM, Holben BN, Loeb NG (1999) A model for the natural and anthropogenic aerosols over the tropical Indian Ocean derived from Indian Ocean Experiment data. J Geophys Res 104(D22):27421–27440
Savoie DL, Prospero JM, Saltzman ES (1989a) Nitrate, non-seasalt sulfate and methanesulfonate over the Pacific Ocean. In: Riley JP, Chester R, Duce RA (eds) Chemical oceanography, vol X. Academic Press, London, pp 219–250
Savoie DL, Prospero JM, Saltzman ES (1989b) Non-seasalt sulfate and nitrate in trade wind aerosols at Barbados: Evidence for long-range transport. J Geophys Res 94:5069–5080
Savoie DL, Prospero JM, Larsen RJ, Huang F, Izaguirre M, Huang T, Snowdon TH (1993) Nitrogen and sulfur species in Antarctic aerosols at Mawson, Palmer Station, and Marsh (King George Island). J Atmos Chem 17:95–122
Savoie DL, Prospero JM, Arimoto R, Duce RA (1994) Non-sea-salt sulfate and methanesulfonate at American Samoa. J Geophys Res 99(D2):3587–3596
Saxena P, Hildemann LM (1996) Water soluble organics in atmospheric particles: A critical review of the literature and application of thermodynamics to identify candidate compounds. J Atmos Chem 24:57–109
Scholes M, Andreae MO (2000) Biogenic and pyrogenic emissions from Africa and their impact on the global atmosphere. Ambio 29:23–29

Seinfeld JH, Pandis SN (1998) Atmosperhic chemistry and physics: From pollution to climate change. Wiley-Interscience, New York
Seitzinger SP, Sanders RW (1999) Atmospheric inputs of dissolved organic nitrogen stimulate estuarine bacteria and phytoplankton. Limnol Oceanogr 44:721–730
Sholkovitz ER, Church TM, Arimoto R (1993) Rare earth element composition of precipitation, precipitation particles, and aerosols. J Geophys Res 98:20587–20599
Sievering H, Boatman J, Gorman E, Kim Y, Anderson L, Ennis G, Luria M, Pandis S (1992) Removal of sulphur from the marine boundary layer by ozone oxidation in sea-salt aerosols. Nature 360:571–573
Sokolik IN, Toon OB (1999) Incorporation of mineralogical composition into models of the radiative properties of mineral aerosol from UV to IR wavelengths. J Geophys Res Atmos 104:9423–9444
Spokes LJ, Jickells TD (1996) Factors controlling the solubility of aerosol trace metals in the atmosphere and on mixing into seawater. Aquat Geochem 1:355–374
Spokes LJ, Jickells TD, Lim B (1994) Solubilisation of aerosol metals by cloud processing: A laboratory study. Geochim Cosmochim Acta 58(15):3281–3287
Swap R, Garstang M, Greco S, Talbot R, Kallberg P (1992) Sahara dust in the Amazon basin. Tellus 44B:133–149
Talbot RW, Harriss RC, Browell V, Gregory GL, Sebacher DI, Beck SM (1986) Distribution and geochemistry of aerosols in the tropical North Atlantic troposphere: Relationship to Saharan dust. J Geophys Res 91:5173–5182
Taylor SR, McLennan SM (1985) The continental crust: Its composition and evolution. Blackwells, Oxford
Tegen I, Miller R (1998) A general circulation model study of the inter-annual variability of soil dust aerosol. J Geophys Res Atmos 103:25975–25995
Tegen I, Hollrig P, Chin M, Fung I, Jacob D, Penner JE (1997) Contribution of different aerosol species to the global aerosol extinction optical thickness: Estimates from model results. J Geophys Res Atmos 102:23895–23915
Twomey S (1991) Aerosols, clouds, and radiation. Atmos Environ 25A:2435–2442
Uematsu M, Duce RA, Prospero JM (1985) Deposition of atmospheric mineral particles to the North Pacific Ocean. J Atmos Chem 3:123–138
Wang C, Prinn RG (2000) On the roles of deep convective clouds in the tropospheric chemistry. J Geophys Res 105:22269–22297
Yang Z, Young S, Kotamarthi V, Carmichael GR (1994) Photochemical oxidant processes in the presence of dust: An evaluation of the impact of dust on particulate nitrate and ozone formation. J Appl Meteorol 33:813–824
Zhang XY, Arimoto R, An Z, Zhi S (1997) Dust emission from Chinese desert sources linked to variations in atmospheric circulation. J Geophys Res 102:28041–28047
Zhu XR, Prospero JM, Millero FJ (1997) Diel variability of soluble Fe(II) and soluble total Fe in North African dust in the trade winds at Barbados. J Geophys Res 102(DD17):21297–21306
Zhuang G, Yi Z, Duce RA, Brown PR (1992) The chemistry of iron in marine aerosols. Global Biogeochem Cycles 6:153–171

Chapter 3

Photochemical Processes in the Euphotic Zone of Sea Water: Progress and Problems

E. Pelizzetti · P. Calza

3.1 Introduction

There has been an increasing interest in the photochemical processes that occur in the surface waters of the oceans and other natural waters. The sunlight aquatic environment, which includes the euphotic zone, aerosols, the surface microlayer and the sediment-water interface in shallow areas, is a likely site for photochemical transformations of dissolved and particulate non-living matter, both organic and inorganic (Zafiriou et al. 1984). The most obvious evidence of photoreaction in aquatic environments is the widespread presence of phytoplankton and other light-dependent underwater plants, and much effort has been directed towards the mathematical description of photosynthesis by freshwater as well as marine phytoplankton.

In addition to being naturally significant, the zone of surface waters is used as the receptacle for many liquid, solid and airborne wastes. This surface layer is also essential for recreation, transportation, material exchanges, and biology; these processes are more intense in this area than they are in deeper waters. A significant number of experimental studies recently have appeared that provide some efforts to understand the processes occurring, the substrates involved, their reaction rates, products and associated effects on the environment.

Considerably less attention has been paid to possible environmental significance of photoreactions of xenobiotics in water. As interest in transformation of xenobiotics in the environment has increased in recent years, so have efforts to quantitatively describe the dynamics of such transformations. The chemical structures of the chemical pollutants present in the water are often very different from those of natural substances and it may take therefore considerable periods of time to degrade these structures by biological pathways. Thus, even when part of the solar energy will be stored in biological systems present in aquatic environment, this energy will not be readily available for the breakdown of man-made pollutants. Although there are cases in which a clear distinction between biological breakdown and abiotic degradation cannot be made, there are cases where both these degradation types are complementary. Early studies (Leighton 1961; Howard 1975) paved the way for a quantitative understanding of photochemical smog formation and other atmospheric photoreactions.

Recent studies of the dynamics of photoreactions in water have demonstrated that adsorption of sunlight by xenobiotics can energize transformations that are surprisingly rapid. Direct photolysis half-lives of minutes or seconds have been observed in water for xenobiotics having a wide variety of molecular structures, such as nitrosamines, polycyclic aromatic compounds, various chlorinated organic compounds and metal complexes (Zepp 1980). Evidence has also begun to emerge that the natural components of aquatic environments such as suspended sediments, aquatic humus and algae also have an important effect on photoreactions in water (Zepp and Wolfe 1987; Pelizzetti et al. 1990). In some cases, xenobiotics that do not react on exposure to sunlight in distilled water have been shown to rapidly react in natural water samples.

In the present chapter, after an overview on the principal light-induced processes occurring in the water, the attention will be focused on the role of two important constituents of sea water: iron and chloride. Iron is an essential metal, involved in the main redox reactions and electron transport, while chloride is an abundant constituent of sea water, and both may strongly influence the photoinduced processes.

3.2 General Framework

The spectral irradiance at depth depends on three factors: atmosphere-transmitted irradiance, which is incident on the water surface, transmission through the air-water interface and optical properties of water and dissolved and suspended materials, which determine the spectral attenuation properties of the water. The role of these three factors will be discussed below.

3.2.1 Solar Flux

The flux of solar energy that enters in the sea and is converted into other forms of energy by interaction with molecules is enormous. The global average (as photons) is roughly equal to the input of water in the form of rain (as molecules). These photons have energies ($14.0–33.5 \times 10^3$ cm^{-1}) in excess of the activation energies for most chemical reactions (Lee et al. 1977).

Values for intensity of the radiation with wavelengths of 390–800 nm can be obtained by calculations through data available for extraterrestrial solar irradiance, atmospheric transmissivity in respect to scattering (Leighton 1961) and absorption coefficient for ozone.

3.2.2 Light Attenuation

The photoprocesses occurring in the euphotic zone may be schematically represented as shown in Fig. 3.1.

Solar light can be reflected at the sea surface or can penetrate in sea water; in the second case, absorption and scattering can occur. Absorption and scattering of radiation cause an attenuation of the intensity of solar radiation, and the extent of attenuation may vary from one water body to another. Attenuation coefficients are usually wavelength-dependent, and the attenuation of light intensity generally increases with decreasing wavelength in the visible and ultraviolet region.

The incident light is greatly absorbed by various chromophores, such as dissolved organic matter, living and dead inorganic and organic particulate matter, and water itself; however, the water itself is an important absorber only under highly transparent conditions. Inland surface waters contain dissolved organic matter, which prevents the sunlight from penetrating to large depths.

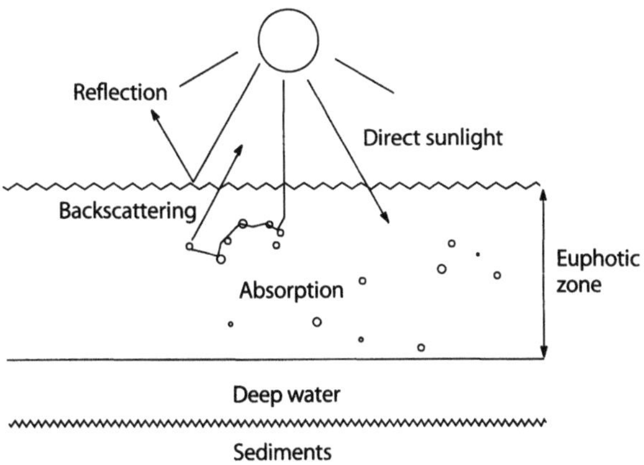

Fig. 3.1. Photoprocesses occurring at the water/air interface

In the oceans, the absorption of solar radiation is primarily due to the water itself. Water is most transparent for the blue region of the radiation spectrum and scattering is relatively wavelength-independent. As a result, solar radiation in clear ocean water assumes that blue hue at great depths (Roof 1982).

In the aquatic environment, scattering is usually less important than absorption as a contribution to attenuation, especially in the ultraviolet region. Scattered light in an aquatic environment is caused by the interaction of light with suspended matter in the water bodies. In coastal and inland water, the photic zone is much shallower, and light attenuation is almost completely attributable to the dissolved substances and suspended particles in the water. Efforts have been made to quantify the effect of these natural substances on the transmission of solar radiation into the aquatic environment. These efforts have involved relating observed attenuation coefficients to environmental properties such as chlorophyll or suspended sediment concentrations. The presence of light-absorbing materials in waters has a screening effect that makes the photolysis rate slower than that in pure water solutions, except for those cases in which the presence of natural sensitizers speeds up photoreaction.

The difficult parameter to evaluate is the wavelength-dependent underwater light field, which also depends on the highly variable absorption properties of water. The diffuse attenuation coefficient, $K_T(\lambda)$, which characterizes light penetration and mathematically defines absorption in a scattering medium, can be accounted for by the empirical model of Baker et al. (1982). This model agrees with marine observations within ±10% between 300 and 700 nm. It accounts for $K_T(\lambda)$ as a sum of contributions from water, chlorophyll, dissolved organic matter, and mineral matter. Models such as this one will prove useful for coastal and open ocean problems, although spatial-temporal variation in the amount and kinds of mineral matter and biopigments may limit their usefulness in estuaries and highly coloured waters.

In open ocean water the depth of 99% attenuation (the photic zone) ranges from about 30 m for middle ultraviolet light to over 100 m for near ultraviolet radiation. The light field is best understood (and is often best quantified) by means of models. Measuring the underwater light field is an obvious alternative to model it.

It is also possible to use light fields in the laboratory. Systems such as solar simulators can approximate surface sunlight at most wavelengths. Since the underwater light spectrum varies markedly with depth, broadband studies can accurately measure only the near-surface effects for optically thin samples or vertically integrated effects and absorption coefficients as a function of wavelength.

3.2.3 Factors Influencing Photoreactions

Photoreaction rates are determined by two factors:

1. The rate of absorption of light by the chromophores;
2. Quantum yield with which chemical change results after excitation of the chromophore.

These factors may all vary over an enormous range, and accordingly the transformations involved vary enormously in their capacity to act as sources or sinks for different compounds. Compounds with a sufficiently high absorption coefficient and a very efficient photodecomposition rate (high quantum yield) for instance will generally have little chance of being detected at the water surface or in the euphotic zone, because their steady state concentration will be kept very low as a result of the extensive photochemical sink.

Basically, any solution photoprocess involves two, or often three relatively distinct steps:

1. Absorption of a single photon by a single chromophore C with the generation of an electronically excited state, C*;
2. Parallel and consecutive processes that degrade the electronic energy to heat, emitted radiation and primary photochemical products;
3. Secondary reactions of these primary photoproducts into the water.

It is frequently desired to measure or calculate the excitation rate of a chromophore under various conditions such as in situ or in an experimental apparatus simulating environmental light intensities and spectral distributions.

Several cases are possible. An example can be that of an optically thin layer near the water surface in which the excitation rate is a function only of the spectrum and intensity of the incident solar light and of the concentration and absorption spectrum of the chromophores. Consideration of this system can be simplified by neglecting reflection, backscattering, and internal scattering, so that in-air insolation measurements are assumed to represent the underwater light intensity.

An optically thick water layer that has a homogeneous chromophore and scatterer distribution may represent another case. In addition to the parameters involved in the optically thin case, the excitation rate now depends also on the optical properties of the water. Since all photons are absorbed, the excitation of a given chromophore depends on its ability to compete with all other chromophores for photons. The depth of such an optically thick layer may extend a few centimetres into turbid and/or absorbing waters, or nearly 100 m into relatively transparent oceanic waters.

3.3 Main Photoprocesses Occurring in Water and Air

Solar radiation is fundamental to all the photochemical and photobiological processes of natural waters. The amount, spectral quality and spatial-temporal distribution of sunlight are key variables at the surface and within the water.

Various photoreactions occur in the environment; hydrogen peroxide forms in fresh and sea water, the free radical NO reaches detectable steady state levels in the equatorial Pacific, a photochemical rearrangement product is found in shallow-water encrusting corals (Look and Fenical 1984), and so on. On the one hand, solar radiation promotes photosynthesis processes and influences phytoplankton growth; on the other, light activated abiotic processes, through absorption by chromophores, can occur both by direct and indirect photoreactions.

3.3.1 Direct Photolysis

The direct photolysis of known dissolved molecules with known chromophores is the simplest type of photochemical reaction. The degree of spectral overlap between the electronic absorption spectrum of a chemical species and the emission spectrum of the light source is one of the primary determinant factors of direct photolysis rates. Compounds that strongly absorb at wavelengths greater than 320 nm have the potential to rapidly undergo direct photolysis in sunlight, especially if absorption extends into the visible region. On the other hand, if absorption is weak or non-existent at wavelengths greater than 300 nm, direct photolysis is usually negligible. Most xenobiotics absorb in the ultraviolet spectral region, but there are notable exceptions, such as dyes and certain nitroaniline herbicides, that absorb visible light. Electronic absorption spectra can be employed to compute direct photolysis rate constant in sunlight (Zepp 1978).

The simple inorganic components of natural waters are also generally transparent to sunlight. In the domain of particle-free water without dissolved organic chromophores, water absorbs nearly all the light. The rarity of direct photoreactions of inorganic chromophores is underscored by the fact that nitrite and nitrate are the only well-studied cases. Nitrogen oxides and OH radicals are photoproducts; nitrate is quite unreactive, while at the sea surface nitrite shows a net loss of about 10% per day. A few additional weak chromophores with some known photochemistry are iodate, uranyl ion, hydrogen peroxide and ferrous ions (Braterman et al. 1984).

It is not possible to derive much useful information about direct photolysis from theory, although the literature may indicate the most probable products and mechanisms. For these reasons, direct reactions are best studied experimentally in the medium of interest by monitoring starting material disappearance as well as product appearance. Simpler experiments may fail to reveal the true quantum yield because of unexpected products, or they may give unrealistic rates and product distributions ascribable to medium effects. The photolysis of complex organic chromophore molecules of unknown structures is particularly important in natural waters because simple natural chromophores are rare (Zika 1981; Zepp 1980; Zafiriou 1983). The most important chromophores seem to be diverse organic molecules of unknown structure and they may be partially colloidal. Harvey (1983) have extracted weakly coloured hydrophobic organic materials called marine humus; these materials have chemical properties that suggest they are formed in situ by auto-oxidative reactions of plankton-derived organic compounds.

3.3.2 Indirect Photoreactions

Indirect photoprocesses are common and important because they can alter molecules resistant to direct photolysis, such as transparent species chromophores. There is evidence that dissolved substances in natural waters can photosensitize a variety of reactions; these photosensitized reactions are sometimes the dominant pathway for photochemical transformation, especially when direct photolysis is negligible.

In indirect photolysis, a reaction is initiated through light absorption by a chromophore other than the substrate itself. If the chromophore is regenerated, it plays a photocatalytic role; this occurs in energy transfer processes and in some cyclic redox reactions. If the chromophore changes irreversibly, it has undergone direct photolysis itself, which simultaneously causes indirect photolysis of other substances presented.

Various studies (Miller and Zepp 1979) have demonstrated that the photolysis rate of xenobiotics can be altered by the dissolved and suspended matter in the aquatic environment. The photolysis rate is effected by both physical and photochemical factors. Physical effects include light attenuation or scattering by natural substances. Changes in photochemistry occur through photosensitization by dissolved humic substances and other natural substances. Although semiconductor powder like TiO_2 promotes rapid photocatalysed reactions of xenobiotics, some studies have shown that natural suspended sediments do not mediate such photoreactions. A few qualitative studies suggest that certain xenobiotics are more photolabile when absorbed to algae than when dissolved in distilled water (Zepp 1980).

In this chapter we will consider only the processes involving indirect photoreactions. In Fig. 3.2, the possible absorbers the photoreactant and the main factors that influence the photoprocesses are summarized. The most common light absorbers present in water are dissolved organic matter (*DOM*), semiconductors (SC) and inorganic anions, such as NO_3^-. These absorbers, through photo-absorption, originate active species (B); the more common are the generation of radicals, such as hydroxyl radical and peroxy radical, and of oxidants, such as hydrogen peroxide and singlet oxygen.

The sources and sinks of those species will be discussed below. These species, reacting with an organic compound, give origin to a photo-transformed compound. The characteristics and concentration of the dissolved organic matter, concentration of dissolved oxygen and the quantity of the activated species photogenerated influence the rate of this process.

The known indirect photoreactions involve initial excitation of chromophore followed by energy transfer, by transfer of electrons or hydrogen atoms to or from other system components. The energy transfer reaction is the first recognized photoprocess in natural waters (Joussot-Dubien and Kadiri 1970) and has been widely studied; it almost represents the largest quantum yield as well. A significant fraction of ultraviolet sunlight absorbed in natural waters excites organic chromophores, particularly at wavelengths below 450 nm. Even processes with low efficiencies originating from these chromophores may be important because of these high excitation rates. Electronic

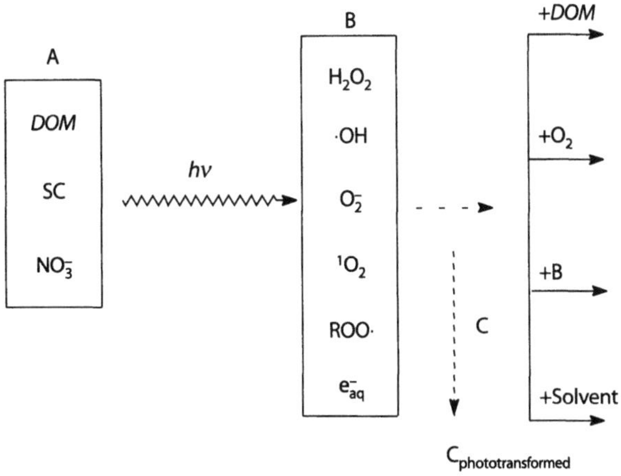

Fig. 3.2. Schematic representation of indirect photochemical processes

energy transfer, giving rise to newly excited molecules that may react without having absorbed a photon, deactivates a fraction of these chromophores. The likely donor state are organic singlets and triplets, but the lifetime of singlets in solution is so short ($<10^{-8}$ s) that only acceptors in the concentration range $>10^{-4}$ M level in natural waters have an energy gap smaller than even the highest available photon energy (Millero 1996); so, singlet-singlet energy transfer is inefficient in aquatic systems. However, this is not valid for longer-lived organic triplet states, which may form by intersystem crossing from organic singlets with a different efficiency, depending on structure and medium.

There is also evidence of the photoionization and electron transfer pathway. Flash photolysis studies of humic chromophores have shown the formation of a transient, which adsorbs at a long wavelength and is quenched by nitrous oxide (Newman 1984), suggesting that the hydrated electron is produced. Furthermore, indirect arguments implicate superoxide ion-radical in the widespread formation of hydrogen peroxide from irradiated organic matter. The O_2 may form from unknown photoreactive chromophores by electron ejection with simultaneous or sequential attachment to oxygen or by hydrogen transfer forming HO_2, followed by ionization to $\cdot O_2^-$.

3.3.2.1
Short-Lived Oxidants

Environmental photoreactions often produce free radicals (Zika 1981; Zepp 1980; Zafiriou et al. 1984) and other reaction products, particularly oxidants. Efficient oxidants found in water are singlet oxygen and hydrogen peroxide.

Singlet oxygen. In the environment, 1–2% of the UV-absorbing chromophores can generate long-lived triplet states with enough energy to interact with dissolved oxygen to form singlet oxygen. There is now evidence for singlet oxygen production in natural waters; the evidence seems to establish qualitatively the ubiquity of singlet oxygen formation, but quantitative aspects are uncertain. Lee et al. (1977) used 2,5-dimethylfuran as a singlet oxygen trap and found that singlet oxygen generation occurred in each of the water samples investigated. Although apparent rates of formation varied widely, the rate per unit of solution absorbency at 366 nm was less variable. The quantum yield for singlet oxygen formation ranged from 1.4 to 9.3% with an average of 4.3%. Despite uncertainties, if these experiments indicate the correct order of magnitude, singlet oxygen formation is very active in many coastal, estuarine and fresh waters.

The chemistry of singlet oxygen has been widely studied (Foote and Clennan 1995). A large percentage of it simply returns to the ground state oxygen in a fast interaction ($1.6 \times 10^{-5}\ s^{-1}$) with the vibrational overtone bands of water (Rodgers and Snowden 1982). This fast reaction is the dominant singlet oxygen sink and the major reason why, even with high formation rates, singlet oxygen concentrations are always expected to be very low. Singlet oxygen is predominantly quenched to ground state oxygen by water; its half-life is about 3 μs. Nevertheless, some reactions might compete with water quenching to give significant chemical changes. Joussot-Dubien and Kadiri (1970) suggested that singlet oxygen might be involved in chemical nitrification, as the oxidation of ammonia. They showed that singlet oxygen in water would oxidize ammonia, but that this process is unlikely to compete effectively with biological ammonia sinks in surface waters. Thus, in bulk solution the rates of energy-transfer reactions of even the most reactive materials now seem constrained, on the one hand by physical quenching of singlet oxygen, and on the other by the strong competition with direct transfer to organics by O_2. However, in oxygen-depleted or hydrophobic microenvironments, such reactions could proceed with higher rates and affect properties such as the surface chemistry and the composition of particulate material.

Hydrogen peroxide. The finding of hydrogen peroxide in surface sea waters has strongly supported the notion that photochemical processes occur in the oceans. The sources of H_2O_2 can result from three major processes (Millero 1996):

I. $$\mathrm{Org} + h\nu \rightarrow \mathrm{Org}^* \tag{3.1}$$

$$\mathrm{Org}^* \rightarrow \mathrm{Org}^+ + e^- \tag{3.2}$$

$$O_2 + e^- \rightarrow \cdot O_2^- \tag{3.3}$$

$$\cdot O_2^- + H^+ \rightarrow HO_2\cdot \tag{3.4}$$

$$HO_2\cdot + HO_2\cdot \rightarrow H_2O_2 + O_2 \tag{3.5}$$

II. $$\mathrm{Org}^* + O_2 \rightarrow \mathrm{Org}^+ + \cdot O_2^- \tag{3.6}$$

$$\mathrm{Org}^* + \mathrm{sub} \rightarrow \mathrm{Org}^- + \mathrm{sub}^+ \tag{3.7}$$

$$\mathrm{Org}^* + \mathrm{sub} \rightarrow \mathrm{Org}^+ + \mathrm{sub}^- \tag{3.8}$$

$$\mathrm{Org}^+ + \mathrm{sub} \rightarrow \mathrm{Org} + \mathrm{sub}^+ \tag{3.9}$$

$$\mathrm{Org}^- + O_2 \rightarrow \mathrm{Org} + \cdot O_2^- \tag{3.10}$$

$$\mathrm{sub}^- + O_2 \rightarrow \mathrm{sub} + \cdot O_2^- \tag{3.11}$$

III. $ML^{n+} + h\nu \rightarrow M^{(n+1)} + L^-$ (3.12)

$M^{(n-1)} + O_2 \rightarrow M^{\nu+} + \cdot O_2^-$ (3.13)

In all the considered pathways, the superoxide anion is a key intermediate in the formation of H_2O_2. Thus, all the reactions that lead to O_2^- will increase the concentration of H_2O_2.

Studies by Zika et al. (1982) indicate that the formation of H_2O_2 is higher in coastal waters with higher humic concentrations than in oligotrophic waters. The production rates also appear to be directly related to the concentration of the humic in the water as measured by absorbance at 300 nm. These results support the notion that abiotic photochemical processes are responsible for the hydrogen peroxide in the surface waters of the oceans. The biological and non-photochemical processes known to form hydrogen peroxide are generally insignificant except in oligotrophic waters. The portion of the sunlight responsible for most of the formation of H_2O_2 is below 400 nm.

Since steady-state levels of H_2O_2 are found at 10–200 nM in surface water, the decay of H_2O_2 must be slow. Studies also indicate that the lifetimes are much longer in deep waters; these results indicate that particles or biological processes may control the lifetime of the decay. The decay appears to be partly related to biological particles, both living and dead cells. Nevertheless, further work is necessary to elucidate the decay mechanism of H_2O_2 and the role that cells have on the formation and destruction of H_2O_2. The decomposition is not affected by light and occurs in a matter of days. Enzymes as well as particles may be important. Although abiotic decomposition processes are small, they may be important in the open oceans.

3.3.2.2
Free Radicals

The involvement of radicals of various origins in natural waters has been suggested repeatedly (Swallow 1969; Zafiriou 1983) but not on the basis of strong evidence. However, recent research continues to suggest a nearly ubiquitous formation of a variety of radicals in natural waters. Many radicals live long enough to show a chemistry that is independent of their immediate sources, and some generalizations can be made about their chemical behaviours that are often interconnected.

Sources: Some studies (Sehgal et al. 1980) show that radical formation occur when ultrasound energy interacts with gas-satured aqueous solutions. The environmental impact of these processes has not been investigated at all, but is probably minor. A second potential source of radicals in natural waters is supply from the atmosphere of OH, HO_2, CH_3O_2 or precursor species such as O_3, HOOH and peroxy acetyl nitrate. Although a lack of evidence makes it impossible to estimate the fluxes of such radicals with confidence, approximate upper limits can be set by taking the expected concentrations of these species either from atmospheric measurements or from models, and then assuming that they deposit to water surface rapidly because of their high reactivity and/or high water solubility.

Biological emission of free radicals and their precursors is a little-understood source. Free radicals play key metabolic roles in vivo and may be responsible for cellular malfunctions. It is unclear, however, whether their loss by active or passive processes is common for either healthy or senescent cells. A variety of thermal chemical processes may produce radicals. It has been suggested that thermal oxidation of phenols by MnO_2 may generate humic materials (Shindo and Huang 1982).

Radical fate: The principal fate of many of the more reactive radicals is their transformation to a new radical; generally the daughter radical will be less reactive and more selective than its parent. This type of transformation is quite common and occurs during the reactions of radicals with the major constituents of fresh and natural waters, because these components are always light-element even-electron structures. Harvey (1983) has noted that marine humic materials may be primarily lipid-derived autoxidation/condensation products, formed in part from the action of thermally and photochemically generated radicals as well as other reactants with biogenic polyunsaturated triglycerides.

Finally, to a limited extent, less reactive radicals may simply accumulate in certain water systems until their steady state levels are significant. The fate of radicals in the environment may therefore be summarized as follows: transformation, redox stabilization, radical formation and possibly simple accumulation.

The oxidants participate in important non-radical reactions and also frequently generate new radicals, thus serving as radical reservoirs. The efforts to understand the role of photochemically produced free radicals in natural waters are complicated by several factors. Free radicals react readily with one another, with transition metals, and more slowly with many organic compounds.

The ·OH radical: the ·OH radical is the most reactive photochemically produced free radical in the atmosphere, while its role in an aqueous system is less clearly understood. Flash photolysis studies have demonstrated that it is formed in sea water. The possible sources of the ·OH radicals in sea water are NO_2^-, NO_3^-, H_2O_2, Fe^{2+} and dissolved organic matter. The significant known sources and sinks, together with typical daytime steady-state concentration of hydroxyl radical in atmospheric and surface waters are reported in Table 3.1.

Peroxy radicals and superoxide radical anion. In addition to the ·OH radical formation rate, Hoignè (1990) has demonstrated the formation and estimated the concentrations in some natural waters of other major radical types, as organic peroxy radicals and superoxide radical anion. They have suggested that the fastest ROO· secondary reactions would give half-lives of a few days at the surface for reactive compound classes, such as phenols, aromatics, amines and hydroquinones.

The major fonts of both radicals, together with their fates, are reported in Table 3.2.

3.3.2.3
Heterogeneous Photochemistry

All natural waters contain non-living organic particles, mineral particles and living organisms, as well as colloidal materials. In addition to these core materials, binding to particles by ions and solutes (especially hydrophobic organic compounds) also exposes the colloidal materials to the different and often enhanced reactivity of the particulate phase.

Table 3.1. Sources and sinks for ·OH radicals

Natural Water	OH sources	OH sinks	Typical daytime steady-state aqueous concentrations
Continental clouds	$\cdot OH_{(g)} \leftrightarrow \cdot OH_{(aq)}$	*DOC*	1×10^{-14}–1×10^{-12} M
	$Fe(II) + H_2O_2 \rightarrow Fe(III) + \cdot OH_{(aq)} + \cdot OH^-$	HSO_3^-/HSO_2^-	
	$O_3(aq) + \cdot O_2^- \rightarrow \cdot OH_{(aq)} + 2O_2$	Cl^-/Br^-	
	$Fe(OH)^{2+} + h\nu \rightarrow Fe^{2+} + \cdot OH_{(aq)}$		
	$NO_3^- + h\nu \rightarrow NO_2 + \cdot OH_{(aq)}$		
Fresh surface waters	$NO_3^- + h\nu \rightarrow NO_2 + \cdot OH$	*DOC*	1×10^{-18} -1×10^{-15} M
	$NO_2^- + h\nu \rightarrow NO + \cdot OH$	HCO_3^-/CO_3^{2-}	
	$Fe(II) + H_2O_2 \rightarrow Fe(III) + \cdot OH + OH^-$		
Marine waters	$NO_3^- + h\nu \rightarrow NO_2 + \cdot OH$	Br^-/Cl^-	1×10^{-19} -1×10^{-17} M
	$NO_2^- + h\nu \rightarrow NO + \cdot OH$	*DOC*	
	DOC $+ h\nu \rightarrow \cdot OH$?	HCO_3^-/CO_3^{2-}	
	$Fe(II) + H_2O_2 \rightarrow Fe(III) + \cdot OH + OH^-$		

Table 3.2. Sources and sinks form ·ROO and $\cdot O_2^-$ radicals

Natural water	$\cdot ROO/\cdot O_2^-$ sources	$\cdot ROO/\cdot O_2^-$ sinks	Typical daytime steady-state aqueous concentrations
Continental clouds	$\cdot ROO(g) \rightarrow \cdot ROO(aq)$	$\cdot ROO(aq)/\cdot O_2^-$	1×10^{-9} -4×10^{-8} M
	$RC(O)R' + PhOH + h\nu(+O_2) \rightarrow \cdot O_2^- +$ products	Cu(II)/Cu(I), Fe(III)/Fe(II) *DOC*	
Fresh and marine waters	*DOC* $+ h\nu(+O_2) \rightarrow \cdot O_2^- +$ oxidized *DOC*	$\cdot ROO(aq)/\cdot O_2^-$	1×10^{-9} -1×10^{-8} M?
Surface waters		Cu(II)/Cu(I), DOC	

Fundamental chemical considerations suggest many ways in which particulate or absorbed molecules differ from their dissolved forms. Highly significant spectral shifts occur for absorbed molecules (Leermakers et al. 1966), while the quenching of excited states of absorbed pesticides is observed (Hautala 1978).

Studies of xenobiotics in the presence of particulate materials, such as desert soils and sand, have shown bathochromic spectral shifts of sorbed anthracene, polychlorinated aromatic compounds and pesticides (Leermakers et al. 1966) moving photochemically active strong bands into the sunlight spectral region. There have been several general and mechanism-oriented studies on the effect of particulate on photoprocesses (Miller and Zepp 1979). In particular, there has been a great deal of interest in the possible involvement of semiconductor-type mechanisms in natural water photochemistry (Calza et al. 2001).

The possible photolysis of abundant hydrophobic and largely particulate mono- and polyunsaturated fatty acids in the marine environment is of biogeochemical interest. Numerous pathways are known for such molecules once liberated from cells, and they are the suspected precursors of some humic and fulvic acid components.

Finally, there is evidence that the surfaces of algae (living and dead) are active in some photoprocesses. Heat-killed and ascorbate-washed cells also promote the reaction, so although the cells are living, the process does not appear to be metabolic. Similar effects have been seen for various other organophosphorus compounds.

3.3.2.4
Organic Compounds in Sea Water

Although the concentration of organic matter in the oceans is less than 0.01% of the total amount of salts, these compounds are important modifiers of many of the biological and chemical reactions that take place in the sea. They provide a nutritional and energy base for micro and macroorganisms, have a major impact on the speciation of many metals by such processes as complexation and adsorption, and are precursors of fossil fuels such as petroleum and oil shale. These organic compounds can also provide growth promoters and inhibitors that may control phytoplankton succession.

The source of most of the organic compounds in sea water is from primary production in the euphotic zone. Part of the organic matter from photosynthesis is partially metabolized in order to satisfy the energy requirements of phytoplankton. Much of the organic matter is consumed by organisms, but a significant amount is decomposed at the surface, while 1 to 12% of the annual production remains as a refractory fraction, quite resistant to biological and chemical transformations. The residence time of this residual dissolved organic matter in the sea has been estimated by Williams and Druffel (1987) to be 1310 yr for surface and 6240 yr for deep dissolved organic matter. Most of this residual organic carbon consists of humic substances. The other sources of input of organic matter to sea water, in addition to primary productivity by marine plankton, include terrestrial input by rivers and by the atmosphere, excretion by marine organisms and the resuspension of organic matter from marine sediments. The direct addition of organic matter due to oil spills is an additional source.

In aquatic systems, it is traditional to divide organic matter into two major categories: dissolved and particulate (*DOM* and *POM*). *POM* includes those species that will be retained by a glass fibre filter of 0.45 μm.

Dissolved organic carbon (*DOC*) is about 50% of the *DOM*, and likewise particulate organic carbon (*POC*) is about 50% of the value of *POM*. The concentration of dissolved organic carbon in sea water is typically significantly greater than that of *POC*. Seasonal variations of *DOC* and *POC* occur primarily in shallow waters and are similar to the variations in primary productivity. At depths below a few hundred metres, the average *DOC* is 40 μM, while the average *POC* is 0.8 μM (Millero 1996). The *POC* frequently goes through a maximum in the oxygen minimum zone. Some have suggested that this is a result of sinking particulate material having a density similar to water at a depth of 500 to 1000 m. The oxygen minimum is thus a result of bacteria oxidizing this material in this zone. The dissolved and particulate organic carbon, nitrogen and phosphorus levels in surface, deep, and coastal waters are compared to the levels at the oxygen minimum. The coastal levels are higher than those in surface waters of the oceans. The levels of dissolved and particulate concentrations of all the organic compounds are much lower in deep waters than in surface waters.

Dissolved marine organic matter is such an extremely complex and dilute mixture of compound that only 10 to 20% can be full characterized. Measurements of dissolved organic carbon, nitrogen and phosphorus compounds are standard measurements used to obtain an understanding of the major types of organic compounds present in the oceans. The dissolved organic matter in sea water consists largely of humic substances and more labile compounds from the major biochemically important compound classes such as carbohydrates, steroids, alcohol, amino acids, hydrocarbons and fatty acids.

Particulate organic matter consists of a mixture of living and dead phytoplankton and zooplankton, bacteria, their degradation products and macroscopic aggregates. The distribution and nature of particulate organic matter has been found to be quite variable both geographically, vertically, diurnally, and seasonally, and it is influenced by a complex set of equilibria between sources, sinks and circulation patterns. In the euphotic zone, the major portion of *POM* is due to phytoplankton, the chemical composition of which will vary with species as well as with environmental conditions, hence, leading to significant variations in the nature of the *POM*. The metabolic products of phytoplankton can vary with changes in temperature, light intensity and nutrient availability. In the euphotic zone, the majority (90%) of *POC* is due to living matter, while at depths of 2 400 m and more, less than 1% of the *POC* is due to living organisms.

3.4 Role of Iron and Chlorine

Iron is an essential trace metal. In fact, as redox reactions and electron transport are two major functions of iron, it affects the enzyme systems. Iron, along with molybdenum, is also essential for nitrogen fixation and nitrate reduction and thus for the assimilation of the major nutrient nitrogen.

It is hypothesized that iron availability limits specific rates of phytoplankton growth (Martin et al. 1990; Brand 1991). Most marine photosynthetic organisms can only take up iron in the dissolved form. Under sea water conditions, the solubility of iron is extremely low. Careful analytical studies in the oceans have shown that the concentrations of total dissolved iron range from 0.02 to 10 nM.

The role of chloride in the marine environment is also crucial. Several classes of mechanisms for dechlorination of sea salt aerosol have been hypothesized. More than one kind of mechanism is probably operating in the marine boundary layer, and the relative importance of different mechanisms probably varies over space and time. An improved understanding of inorganic Cl chemistry in the marine boundary layer is needed because of its potential importance for global chemistry (Keene 1995).

3.4.1 Inorganic Cl Formation in the Marine Environment

The possible mechanisms for formation of atomic chlorine in the marine environment have been widely investigated, and several possible sources have been identified.

Reactions of sodium chloride, the major component of sea salt particles, with nitrogen oxides generate chlorine atom precursors. Chlorine atoms formed from the reactions of sea salt particles can destroy ozone and greenhouse gas through direct reactions. Alternatively, Cl reacts rapidly with organic molecules, which can, in turn, lead to ozone formation in the presence of sufficient nitrogen oxides (Finlayson-Pitts and Pitts 1986).

The injection into the atmosphere of sea salt aerosol generated by breaking waves on the ocean surface is the major global source of tropospheric chlorine. Most of this material remains in the aerosol and is redeposited to the ocean surface, but important fractions ranging from averages of 3 to 35% are released from the aerosol as inorganic Cl vapour (Keene 1995). Another precursor of atomic Cl is HCl (Behnke et al. 1995). Although HNO_3 and H_2SO_4 displace HCl from sea salt particles, this is not a significant source of atomic Cl, because the subsequent reaction of HCl with OH is relatively slow (Singh and Kasting 1988).

Recent evidence indicates that other chemical processes may also dechlorinate sea salt aerosol. These processes generate highly reactive Cl gases that during the daytime undergo rapid photochemical conversion to HCl via Cl atoms. The aerosol eventually scavenges HCl. Processes involved in Cl cycling are very uncertain, however, partially because of the difficulties in reliable measures of principal reactant and product species.

3.4.2 Role of Iron in Surface Waters

In waters several reductants coming from biodegradation of the biomasses are present. Among these, it has to be reminded of iron(II) and H_2S. Even if those species are able to directly interact with the organic compounds, but usually through slow reactions, redox reactions in natural systems may occur more easily due to a microbial action and mediation through one or more electron transporter. The iron(III)/iron(II) system acts as a electron transporter. The oxido-reductive processes mediated by iron compounds are important, because in anaerobic conditions, such as sediments and deep waters, iron(III) oxides are the more abundant oxidant species. In various ecosystems the iron cycle depends on several physical, chemical and biological processes (Stumm 1992). The iron cycle is interdependent and often connected with the cycles of phosphorus, sulphur, heavy metals, oxygen and carbon; moreover, it is dependent on living matter and light intensity.

The importance of iron can be stressed by such considerations:

1. High quantities are present in rocks and it possesses a rapid rate of transformation. The ability of the species containing iron to oxidize or reduce and contemporary to precipitate or to be solubilized, links the iron cycle to the oxygen cycle (oxidant) and to the carbon cycle (reductant).
2. High surface areas of the iron oxides and their surface reactivity easily makes the adsorption of various solutes. This is one of the causes of interdependence between the iron cycle and the cycles of several other elements, such as heavy metals and phosphate.

3. It is able to act as semiconductor and to participate with the redox photoreactions. In those reactions, the electromagnetic radiation is utilized either to conduct reactions thermodynamically not favoured (heterogeneous photosynthesis) or to catalyse a reaction thermodynamically favoured (heterogeneous photocatalysis).

The iron cycle includes the reductive solubilization of iron(III) by organic ligands, a process that can also be photocatalysed in surface waters, and the oxidation of iron(II) by oxygen, a process catalysed by superficial interactions. During the iron(II) oxidation to iron(III), reactive compounds as heavy metals, phosphate ions and some organic compounds are linked to the precipitate surface. Those compounds are later released in solution during the reduction of iron oxides. The total balance of the cycle gives also the oxidation of organic matter by oxygen mediated by iron.

Fe(III) is also absorbed on external surfaces of humic materials, weakly bound octahedrally and easily complexed and reduced. When Fe(III) is added to the humic materials, most of the iron becomes bound to octahedral sites.

In Fig. 3.3 several important inorganic iron(III) species and their distribution as a function of pH are represented. While in acidic conditions, the there are several possible species (with Cl^-, SO_4^{2-} and F ligands), at the pH typical of sea water the only main species present at remarkable concentrations are the Fe-hydroxides. The hydrolysis of iron(III) is strongly dependent on pH. In particular, at strongly acidic pH, iron(III) is present in dissolved form and the prevalent species is $Fe(OH)^{2+}$. At pH levels typical of sea water (~8), iron is prevalently present in a precipitate form ($Fe(OH)_3$).

Interesting findings have recently been reported with regard to the speciation of dissolved iron in oceanic surface waters (Gledhill and van den Berg 1995, van den Berg 1995, Rue and Bruland 1995). Organic ligands present in the euphotic zone of marine

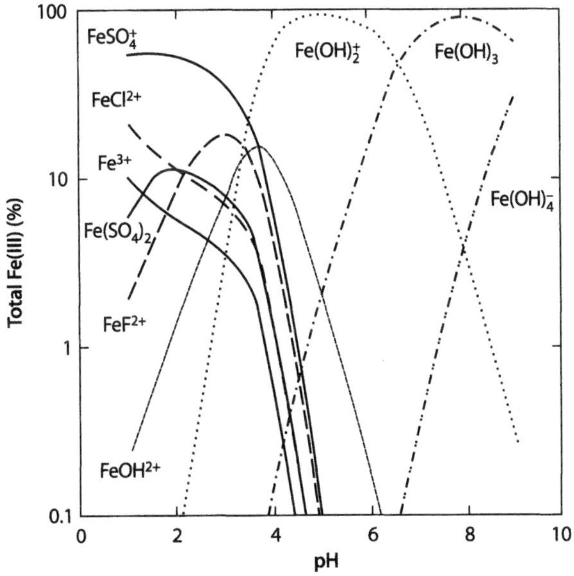

Fig. 3.3. Speciation of iron(III) in sea water as function of pH

waters form very stable complexes with dissolved Fe(III). Thus, it is most likely that the concentration of inorganic, dissolved Fe(III) species determines the upper limit for bioavailable Fe(III). The Fe(II) produced is likely to form less stable complexes with organic ligands present in the euphotic zone of sea water and hence is biologically more available than Fe(III). In the case of Fe(II), organic complexes may be rather weak, and hence ligands at the cell surface may form stronger complexes with Fe(II) than organic ligands in the medium (Anderson and Morel 1982).

3.4.2.1
Photoreactions Occurring on Iron

The speciation, and thus the biological availability, of redox-active metals are often strongly influenced by light.

In Fig. 3.4 the absorption spectra registered for several iron-complexes are reported. While Fe^{3+} gives no adsorption in the visible region, the absorption spectrum is extended in the visible zone in the presence of ligands, such as chloride and bromide. The extent of absorption is higher in the presence of bromine and the maximum lays in the visible light. So, light-induced reactions are favoured when iron is complexed, above all if it is linked to chlorine or bromine.

An important reaction taking place under light is the photo-reduction of Fe^{3+} to give ·OH; under the conditions here the most important photolabile form is the monohydroxy complex (Faust and Hoignè 1990) with a quantum yield of 0.14 at 313 nm and 0.017 at 360 nm:

$$FeOH^{2+} + h\nu \rightarrow Fe^{2+} + \cdot OH \qquad (3.14)$$

In the presence of hydrogen peroxide, the Fe^{2+} formed in Eq. 3.14 generates additional ·OH via the Fenton reaction.

Because light affects to a larger extent the reduction of Fe(III), the concentration of Fe(II) in the euphotic zone of surface water is expected to depend on the light intensity and thus on the time of day. A diurnal variation in the Fe(II) concentration has

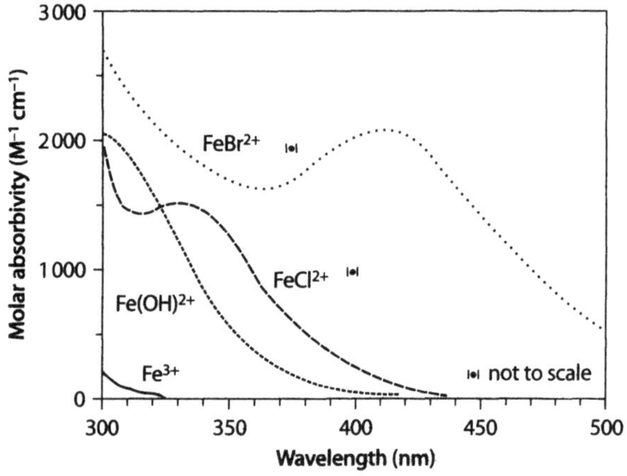

Fig. 3.4. Absorption spectra of Fe(III) species

been observed by several research groups, not only in acidic lakes and rivers (McKnight et al. 1988), but also in sea waters. The concentration of Fe(II) is much higher in acidic surface waters than in sea water because of the much slower oxidation of Fe(II) at low pH-values as compared to pH-values of sea water. Oxidation kinetics of Fe(II) by O_2 and H_2O_2 depends in fact strongly on the Fe(II) speciation.

Organically and inorganically complexed Fe(III) can be reduced under the influence of sunlight. Although the average steady-state concentration of Fe(II) may be very low in the euphotic zone of sea water, considerably higher Fe(II) concentrations can be expected upon illumination with high intensity solar radiation.

The cycling of iron between its reduced and oxidized forms and between particulate and dissolved species in natural waters has a number of environmental consequences beyond iron's role as a nutrient and adsorbent. Iron photo-redox cycling has been shown to catalyse the oxidation of dissolved organic matter (Miles and Brezonik 1981). This process could represent a significant sink of biologically refractory materials, such as humic substances, and is closely coupled to the cycling of reactive transient species such as H_2O_2, ·OH, and $HO_2/{\cdot}O_2^-$.(Faust and Zepp 1993).

Within the acidic pH range, the oxidation of dissolved Fe(II) by oxygen is expected to be very slow compared to other processes. However, the irradiation of humic substances in the presence of oxygen results in the formation of $HO_2/{\cdot}O_2^-$ (Blough and Zepp 1995; Hoignè et al. 1989). The mechanism of this reaction is not known but may occur via reduction of oxygen by aqueous electrons or triplet state, both formed when humic substances are photo-excited. The end product of $HO_2/{\cdot}O_2^-$ dismutation is hydrogen peroxide. This reaction is catalysed by the presence of dissolved iron by the reactions:

$$HO_2/{\cdot}O_2^- + Fe(II) \rightarrow Fe(III) + H_2O_2 \tag{3.15}$$

$$HO_2/{\cdot}O_2^- + Fe(III) \rightarrow Fe(II) + O_2 \tag{3.16}$$

The hydrogen peroxide formed from $HO_2/{\cdot}O_2^-$ is another potentially significant oxidant of Fe(II). It is clear that both iron and light have an accelerating effect on the oxidation of humic substances by oxygen. Voelker et al. (1997) have observed two photo-oxidation processes of fulvic acid. The first, which does not seem to require the presence of metals, results in the reduction of oxygen to $HO_2/{\cdot}O_2^-$.

The second process, photo induced ligands to metal charge transfer reaction of Fe(III)-fulvate complex, occurs both in solution and on the surface of the iron oxide and results in the reduction of Fe(III). Dissolved Fe(III) is continuously supplied via re-oxidation of Fe(II) by both $HO_2/{\cdot}O_2^-$ and H_2O_2. The OH radicals produced by the reaction of Fe(II) with H_2O_2 further oxidize fulvic acid.

3.4.3
Interactions Between Iron and Chloride

Recently chlorine was identified and measured in coastal areas at concentrations up to 150 parts per trillion. This is a much larger concentration than can be attributed to known reactions of sea salt particles, suggesting that there must be an unrecognized source producing Cl_2. Photocatalytic reactions involving metal oxides have been studied. Behnke et al. (1995) have produced, for example, chlorine during irradiation of most metal-oxide aerosol (TiO_2, Fe_2O_3 and SiO_2) in the presence of HCl and O_3. The authors postulate a photocatalytic mechanism involving electron transfer and free-radical reactions to explain these observations.

Recently, we have observed that photocatalytic transformation of trichloromethane on titanium dioxide was inhibited by chloride and that tetrachlorometane is formed during the process (Minero et al. 1997). The suggested mechanism involves the oxidation of chloride through reaction with the valence band holes, according to:

$$h^+_{VB} + Cl^- \rightarrow \cdot Cl \tag{3.17}$$

and subsequent reaction of $\cdot Cl$ with $\cdot CCl_3$.

Several oxides and chalcogenides that occur in the environment can be activated by absorption of ground-level solar light, such as widely diffuse iron oxides. Fe_2O_3 is a semiconductor widely diffuse in nature and the photocatalytic halogenation with Fe_2O_3 may represent an important natural source of halogenated compounds in the environment.

The influence of iron(III) on the photoinduced transformations of phenol in the presence of halides can be adopted as a model system. The photodecomposition of phenol in the presence of Fe_2O_3 and NaBr at pH 7, together with evolution of bromophenols, is reported in Fig. 3.5. In this case, the formation of 2- and 4-bromophenol is observed.

The E^0_{VB} potential for Fe_2O_3 is reported to be ca. 2.8 V (Gerisher 1979) and is able to guarantee the oxidation of Br-ion to the correspondent radical. This should allow the oxidation of bromide, according to Eq. 3.17, being E^0=2.0 V for $\cdot Br/Br^-$ and 1.6 V for Br_2^-/Br^-., so both species result active in the halogenation of phenol.

Similarly, addition of chloride ions causes the formation of chlorophenols (Calza et al. 2001). In fact, E^0=2.5 V for $Cl^-/\cdot Cl$ and 2.3 V for Cl_2^-/Cl^- (Wardman 1989) should allow the oxidation of chloride in agreement with Eq. 3.17.

Interaction between iron and chloride can also occur in a homogeneous phase. Experiments conducted in conditions that simulate solar irradiation have shown that

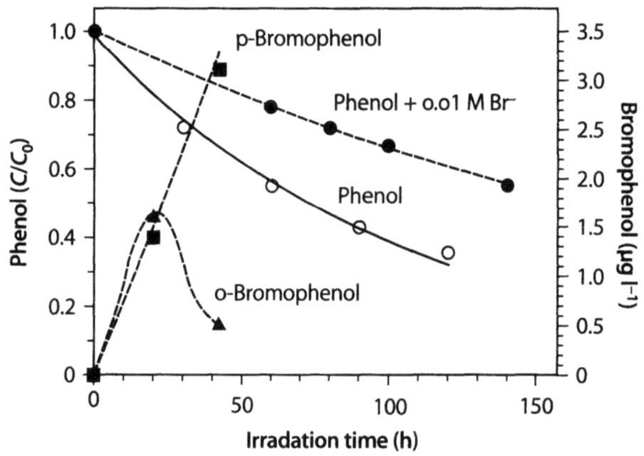

Fig. 3.5. Photodegradation of phenol (2×10^{-4} M) and time evolution of bromophenols in presence of Fe_2O_3 and NaBr 0.01 M at pH 7

phenol, in the presence of iron(III) and halide ions gives the formation of halogenated phenols.

Photodecomposition of phenol and evolution of chlorophenols in the presence of Fe(III) and NaCl at pH 2 is reported in Fig. 3.6. At pH 2, the iron(III) is prevalently present in dissolved form so the process occurs in a homogeneous phase and $Fe(OH)^{2+}$ is the predominant species (see Fig. 3.3). It is also a photoreactive species and can originate OH radicals through Eq. 3.14.

In such experimental conditions, quinone, cathecol and quinol have been identified as intermediates.

In the presence of halide X^- ions (chloride or bromide), also the following reactions take place:

$$Fe^{3+} + X^- \leftrightarrow FeX^{2+} \tag{3.18}$$

$$FeX^{2+} + h\nu \rightarrow Fe^{2+} + X\cdot \tag{3.19}$$

Atomic Cl reacts rapidly with many hydrocarbons and dimethylsulphide in an analogous manner to, but at different rates than, OH radicals. Phenol, 2- and 4-chlorophenol have been identified as shown in Fig. 3.6.

Interestingly, also at neutral and basic pH conditions, the halophenol formation is observed both in the presence of chloride and bromide (Calza et al. 2001).

From these results it emerges that the formation of chloro and bromophenols may occur under solar light in chloride and bromide rich environments in the presence of iron species. This process should be considered as a possible abiotic source of natural halogenation of organic matter at the air/sea water boundary. The widespread occurrence of chlorinated structures in humic substances provides another example of the difficulties in quantifying the role of different natural halogenation processes. However, the same type of reaction products can be expected in other chlorinating processes involving the formation of active chlorine species. This implies that at best, the environmental compartments responsible for the large-scale natural production of organohalogens can be identified, whereas much of the present uncertainty regarding the predominant reaction mechanism is likely to persist for a long time.

Fig. 3.6. Photodegradation of phenol (2×10^{-4} M) and time evolution of chlorophenols in the presence of Fe(III) and NaCl 0.01 M at pH 2

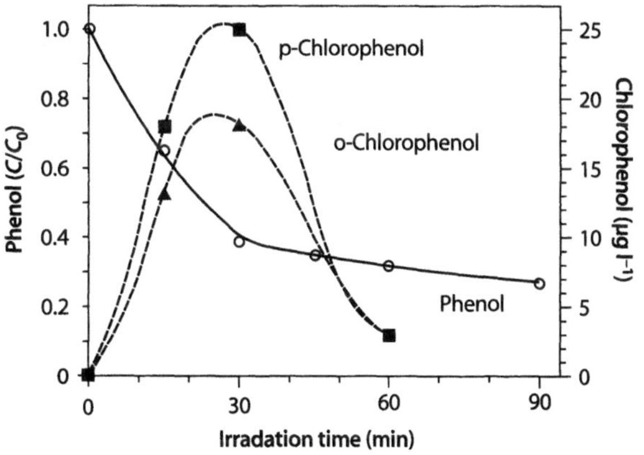

References

Anderson MA, Morel FM (1982) The influence of aqueous iron chemistry on the uptake of iron by the coastal diatom *Thalassiosira weissflogii*. Limnol Oceanogr 27:789–813

Baker KS, Smith RC, Green AES (1982) Middle ultraviolet irradiance at the ocean surface: Measurements and models. LimnolOceanogr 27:500–509

Behnke W, Scheer V, Zetzsch C (1995) Production of photolytic precursor of atomic Cl from aerosols and Cl^- in the presence of O_3. In: Grimvall A., Leer EWB de (eds) Naturally produced organohalogens. Kluwer Academic Publishers, Dordrecht, pp 375–384

Berg CMG Van den (1995) Evidence for organic complexation of iron in seawater. Mar Chem 50:139–157

Blough NV, Zepp RG (1995) Reactive oyxgen species in natural waters. In: Foote CS, Valentine JS (eds) Active oxygen in chemistry. Chapman and Hall, New York, pp 280–333

Brand (1991) Minimum iron requirements of marine phytoplankton and the implications for biogeochemical control of new production. Limnol Oceanogr 36:1756–1771

Braterman PS, Cairns-Smith AG, Sloper RW (1984) Photo-oxidation of iron(II) in water between pH 7.5 and 4.0. J Chem Soc Dalton Trans 1441–1445

Calza, P, Maurino V, Minero C, Pelizzetti E, Sega M, Vincenti M (2001) Chloro and bromophenols formation from phenol and halides by simulated solar light irradiation in the presence of iron ions, iron oxides and cadmium sulfide. (to be published)

Faust BC, Hoignè J (1990) Photolysis of Fe(III)-hydroxy complexes as sources of OH radicals in clouds, fog and rain. Atmos Environ 23:235–240

Faust BC, Zepp RG (1993) Photochemistry of aqueous iron(III)-polycarboxylate complexes: roles in the chemistry of atmospheric and surface waters. Environ Sci Technol 27:2517–2522

Finlayson-Pitts BJ, Pitts JN Jr (1986) Atmospheric chemistry: Fundamentals and experimental techniques. Wiley, New York

Foote ST, Clennan EL (1995) Properties and reactions of synglet dioxygen. In: Foote CS, Valentine JS, Greenberg A, Liebman JF (eds) Active oxygen in chemistry, vol II. Blackie Academic and Professional, Glasgow, UK, pp 105–140

Gerisher H (1979) Solar photoelectrolysis with semiconductor electrodes. Appl Physics 31:115–172

Gledhill M, Berg CMG Van den (1995) Measurement of the redox speciation of iron in seawater by catalytic cathodic voltammetry. Mar Chem 50:51–61

Harvey GR (1983), Dissolved carbohydrates in the New York bight and the variability of marine organic matter. Mar Chem 12:333–339

Hautala R (1978) U.S. Environmental Protection Agency. U.S. National Technical Information Service, Springfield, VA. (EPA-600/3-78-060 Report PB-285)

Hoignè J (1990) Formulation and calibration of environmental reaction kinetics; oxidation by aqueous photooxidants as an example in aquatic chemical kinetics. In: Stumm W (ed) Reaction rates of processes in natural waters. Wiley Interscience, New York, pp 43–70

Hoignè J, Faust BC, Haag WR, Scully FE, Zepp RG (1989) Aquatic humic substances as sources and sinks of photochemically produced transient reactants. In: Suffert EM, Mac Carthy P (eds) Aquatic substances: Influence on fate and treatment of pollutants. American Chemical Society, Washington D.C., pp 363–381

Howard PH (1975) EPA report No. EPA-560/5-75-006. Washington, DC

Joussot-Dubien J, Kadiri A (1970) Photosensitized oxidation of ammonia by singlet oxygen in aqueous solution and in seawater. Nature 227:700–701

Keene WC (1995) Inorganic Cl cycling in the marine boundary layer: A review. In: Grimvall A, Leer WB de (eds) Naturally-produced organohalogens. Kluwer Academic Publishers, Dordrecht, pp 363–371

Lee WN, Zepp, RG, Gordon JA, Baughman GL, Cline DM (1977) Kinetics of chemical degradation of malathion in water. Environ Sci Technol 11(1):88–93

Leermakers PA, Thomas HT, Weis LD, James FC (1966) Spectra and photochemistry of molecules adsorbed on silica gel IV. J Am Chem Soc 88:5075–5083

Leighton PA (1961) Photochemistry of air pollution. Academic Press, New York, pp 6–41

Look SA, Fenical W (1984) Erythrolides: Unique marine diterpenoids interrelated by a naturally occurring di-ð-methane rearrangement. J Am Chem Soc 106:5026–5027

Martin JH, Fitzwater SE, Gordon RM (1990) Iron deficiency limits phytoplankton growth in Antartic waters. Global Biogeochem Cycles 4:5–12

McKnight DM, Kimball BA, Bencala KE (1988) Iron photoreduction and oxidation in an acidic mountain stream. Science 240:637–640

Miles CJ, Brezonik PL (1981) Oxygen consumption in humic colored waters by a photochemical ferrous-ferric catalytic cycle. Environ Sci Technol 15:1089–1095

Miller GC, Zepp RG (1979) Effects of suspended sediments on photolysis rates of dissolved pollutants. Wat Res 13:453–485
Millero FJ (1996) Chemical oceanography, 2nd edn. CRC Press, Boca Raton
Minero C, Maurino V, Calza P, Pelizzetti E (1997) Photocatalytic formation of tetrachloromethane from chloroform and chloride ions. New J Chem 21:841–842
Newman L (ed) (1984) Gas-liquid chemistry of natural waters, vols I and II. Brookhaven National Laboratory, Upton, N.Y. (BNL 51757)
Pelizzetti E, Minero C, Maurino V (1990) The role of colloidal particles in the photodegradation of organic compounds of environmental concern in aquatic systems. Adv Colloid Interface Sci 32:271–316
Rodgers MAT, Snowden PT (1982) Lifetime of O_2 ($^1\Delta_g$) in liquid water as determined by time-resolved infrared luminescence measurements. J Amer Soc 104:5541–5561
Roof AAM (1982) Aquatic photochemistry. In: Hutzinger O (ed) The handbook of environmental chemistry, vol IIB: Reactions and processes. Springer-Verlag, Berlin, pp 43–65
Rue EL, Bruland KW (1995) Complexation of iron(III) by natural organic ligands in the Central North Pacific as determined by a new competitive ligand equilibration/adsorptive cathodic stripping voltammetric method. Mar Chem 50:117–138
Sehgal C, Sutherland RG, Verrall RE (1980) Optical spectra of sonoluminescence from transient and stable cavitation in water satured with various gases. J Phys Chem 84:388
Shindo H, Huang PM (1982) Role of manganese IV oxide in abiotic formation of humic substances in the environment. Nature 298:363
Singh HB, Kasting JF (1988) Chlorine-hydrocarbon photochemistry in the marine troposphere and lower stratosphere. J Atmos Chem 7:262–285
Stumm W (1992) Chemistry of the solid-water interface. John Wiley and Sons, New York
Swallow JC (1969) Hydrated electrons in seawater. Nature 222:369–370
Voelker BM, Morel FMM, Sulzberger B (1997) Iron redox cycling in surface waters: Effects of humic substances and light. Environ Sci Technol 31(4):1004–1011
Wardman P (1989) Reduction potentials of one-electron couples involving free-radicals in aqueous solution. J Phys Chem Ref Data 18:1637–1755
Williams PM, Druffel ERM (1987) Radiocarbon in dissolved organic matter in the Central North Pacific Ocean. Nature 330:246–248
Zafiriou OC (1983) Naturally water photochemistry. In: Riley JP, Chester R (eds) Chemical oceanography, vol VIII. Academic Press, New York, pp 339–379
Zafiriou OC, Joussot-Dubien J, Zepp RG, Zika R (1984) Photochemistry of natural waters. Environ Sci Technol 18(12):358–371
Zepp RG (1978) Quantum yields for reaction of pollutants in dilute aqueous solution. Environ Sci Technol 12(3):327–329
Zepp RG (1980) Assessing the photochemistry of organic pollutants in aquatic environments. In: Haque R (ed) Dynamics, exposure and hazard assessment of toxic chemicals. Ann Arbor Science, Ann Arbor, Mich., pp 69–110
Zepp RG, Wolfe NL (1987) Abiotic transformation of organic chemicals at the particle-water interface. In: Stumm W (ed) Aquatic surface chemistry. Wiley, New York, pp 423–455
Zika R (1981) Marine organic photochemistry. In: Duursma EK, Dawson R (eds) Marine organic chemistry. Elsevier, Amsterdam, The Netherlands, pp 299–326
Zika R, Salzman E, Chameides WL, Davis DD (1982) Hydrogen peroxide levels in rainwater collected in south Florida and the Bahamas Islands. J Geophys Res 87:5015–5017

Chapter 4

Sedimentary Organic Matter Preservation and Atmospheric O_2 Regulation

J. I. Hedges

4.1 Introduction

The global cycles of organic carbon (*OC*) and molecular oxygen (O_2) are inextricably linked by the fact that both substances are uniquely produced and destroyed in equimolar amounts by photosynthesis and respiration. These opposing processes have been balanced on the geochemical equivalent of a knife's edge for at least the last 600 million years, over which O_2 dependent metazoans are continuously represented in the geologic record. Throughout this period, O_2 release from the preservation of organic matter (~50 wt% *OC*) in marine sediments has been closely compensated for by simultaneous uptake of O_2 during weathering of organic matter (and reduced inorganic minerals) in continental rocks. In this way, the atmospheric reservoir of O_2 has been maintained within the relatively narrow concentration range (±50%), above which runaway vegetation fires occur and below which anoxia becomes deadly to multicellular life (Berner 1989). Given that the mean residence time of atmospheric O_2 is four million years with respect to contemporary sedimentary *OC* burial, over the last 600 million years the Earth's global O_2 control system has had roughly 300 chances to fail on either the side of conflagration and anoxia – and has not (Walker 1974; Watson et al. 1978; Garrels et al. 1976; Jones and Chaloner 1991). Such an extended planetary winning streak, at least from a human perspective, suggests the existence of an effective control system for global-scale cycling of bioactive elements for at least the most recent eighth of Earth history (Van Valen 1971; Petsch and Berner 1998).

This chapter lays forth the hypothesis that a tectonically-driven "mineral conveyer belt," modulated by a shifting "organic carbon compensation depth" along continental margins, has provided the basis for sensitive atmospheric O_2 control over Phanerozoic time (0–0.6 billion years B.P.). In this scenario, flow of mineral debris from weathering mountains to depositing marine sediments physically links these geographically separate sinks and sources of O_2 on a geologically short time scale. In an enmeshed cycle, "excess" organic matter depositing in marine sediments is oxidized by downward mixing of surface ocean waters, whose initial oxygen content is directly proportional to the current atmospheric O_2 concentration. A small fraction of this depositing organic matter, however, escapes remineralization to be preserved in sediments, thereby releasing just enough O_2 to compensate globally for weathering of continental rocks. This combination of mineral mass balance modulated by negative-feedback control during sedimentary organic matter preservation along continental margins appears to exhibit the capacity, sensitivity, and response time required to provide the Earth's hidden O_2 safety net.

4.2
Global Cycles of Carbon and Oxygen

Photosynthesis, in its simplest chemical representation (excluding nutrients), can be thought of as the solar-powered conversion of carbon dioxide and water into organic matter (CH_2O) and molecular oxygen (O_2). The net result of this highly endothermic reaction is the production of one mole each of both the strongest naturally occurring reducing agent (organic matter) and oxidizing agent (molecular oxygen) on Earth.

$$CO_2 + H_2O \rightarrow CH_2O + O_2 \qquad (4.1)$$

In this greatly simplified formulation, CH_2O represents organic matter of all types and forms. Among all the reactants and products of photosynthesis, organic matter is unique in that it is predominantly: (*a*) nonvolatile, (*b*) solid, and (*c*) macroscopic. One outcome of this combination of characteristics is that in natural environments, particulate organic matter tends to sink and collect within a thin layer of solids deposited at the base of the atmosphere (in soils and peats) and ocean (in sediments). In contrast, oxidizing power represented by gaseous O_2 primarily "floats" in the Earth's atmosphere (>99% of all O_2) or as a minor dissolved component (~350 μM at saturation) in the ocean, where ~0.5% of all O_2 resides (Fig. 4.1). One result of this separation of products is that oxic environments supporting multicellular life are restricted largely to a thin shell comprising the atmosphere, land surface and the surface

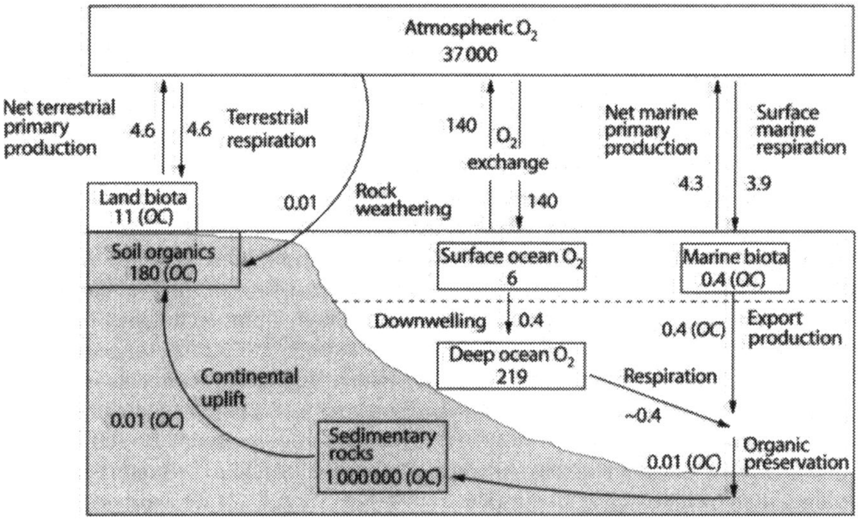

Fig. 4.1. The global cycle of molecular oxygen and other redox active elements (adapted from Keeling et al. 1993). All reservoirs and fluxes (per year) are in units of 10^{15} moles O_2. In this figure, (*OC*) indicates that the molar O_2 equivalent occurs in the form of organic carbon, which inverts flux directions from their O_2 counterpart. Molecular O_2 is the only oxidizing agent whose major reservoir is the atmosphere

millimetres to a few metres of marine sediments. The rest of the internal planet is anoxic, just as all the Earth was prior to photosynthesis (Des Marais 1997).

Respiration, the reverse of photosynthesis, releases solar energy stored in organic matter by reaction with molecular oxygen (Eq. 4.1 reversed). In contrast to photosynthesis, aerobic respiration can occur anywhere molecular oxygen and organic matter coexist in the presence of life. Respiration has thus allowed heterotrophic life to pervade deeply into soil profiles, sedimentary deposits and dark subterranean waters. Deep respiration is similarly constrained by O_2 input from the lighted surface of the planet, although the delivery mechanisms for this gas depend largely on advection (or diffusion) in air and water. Once available O_2 is consumed by respiration, organic matter degradation is carried out exclusively by single-cell organisms that often specialize both in the biochemical substrates and oxidizing agents they process. The general pattern for anaerobic microbial degradation is to employ electron acceptors in the order of their maximal free energy yield (e.g. $NO_3^- > Fe^{3+} > MnO_2 > SO_4^{2-} > CO_2$). An attending pattern is toward decreasing flexibility in substrate utilization at lower free energy yield, such that sulphate and carbon dioxide reducers are heavily reliant on fermenters and other microorganisms to generate the simple substrates (e.g. acetate, formate, oxylate) they metabolize. A consequence of these combined trends is for structurally complex, hydrolysis-resistant organic substances such as lignin and algaenans (De Leeuw and Largeau 1993) to become increasingly difficult to metabolize under more reducing conditions.

Because preservation of organic carbon in marine sediments is essentially the only source for release of a molar equivalent of photosynthetically produced O_2, it follows that "An organic carbon paved is an O_2 saved." The net global production rate of "saved" O_2 can thus be estimated by multiplying the global average sediment accumulation rate by the average *OC* content of those deposits (Berner 1989; Hedges and Keil 1995). The resulting flux is on the order of 0.15×10^{15} g *OC* yr^{-1}, which is equivalent to $\sim 0.01 \times 10^{15}$ moles *OC* and O_2 per year (Fig. 4.1). To maintain a steady reservoir of atmospheric O_2 (38×10^{18} moles; Van Cappellen and Ingall 1996), O_2 must be taken up at the same average rate by the weathering of continental rocks and oxidation of volcanic gases. The latter sink, however, presently accounts for only ~1% of O_2 uptake by rock weathering (Holland 1973). Among rocks, over 90% of O_2 uptake is by sedimentary forms (largely shales), vs. granites and basalts (Walker 1974). Oxidation of organic matter, ferrous iron and sulphides accounts for most of the total oxygen uptake by sedimentary rocks (Broecker 1970). Although both climate and vegetation affect the weathering rate of sedimentary rocks, the long-term control must be the rate of continental uplift (Walker 1974; Berner and Canfield 1989; Des Marais 1997).

This close coupling presents a major challenge, because in the absence of appreciable O_2 removal by volcanic gases, it is difficult to imagine how O_2 concentrations might be linked to tectonics in a negative feedback loop that could explain the long-term stability of the atmospheric pool. An additional challenge is that preservation rates of organic carbon appear to have varied appreciably over the Phanerozoic, with an especially pronounced high 300 million years B.P. during the formation of Permo-Carboniferous deposits (Berner 1987; Berner and Canfield 1989). Global uplift and weathering rates of sedimentary rocks are also likely to have varied widely over the last 600 million years. Any mechanism that might moderate atmospheric O_2 fluctuations must therefore respond to substantially changing sources and sinks on a relatively short time scale (millions of years). A weathering sink whose magnitude varies with O_2 concentration could help minimize atmospheric O_2 fluctuations, but there is a paucity of evidence for fossil carbon in modern marine sediments (Emerson et al. 1987), or for a build-up of sedimentary *OC* over the Phanerozoic (Broecker 1970). Thus, it appears that oxidative weathering of sedimentary rocks is essentially complete in today's world, and hence has limited potential for counteracting upswings in O_2 concentration (Walker 1974). All indications, therefore, are that the control valve for the concentration of atmospheric O_2 must operate on the source side during preservation of sedimentary organic matter. Nevertheless, a responsive link to fluctuating sinks on relatively remote continental surfaces appears necessary. In fact, research in the last five years (Hedges and Keil 1995) points toward an effective communication link between the continental sinks and sea floor sources of O_2, as well as an O_2-based negative feedback control mechanism on sedimentary organic matter preservation (Hedges et al. 1999).

4.3 Organic Matter Preservation and Sediment Texture

Preservation in marine sediments is a rare fate for organic matter. Only one organic carbon in ~1 000 ultimately escapes respiration and recycling back to CO_2. Although the pelagic ocean accounts for over 80% of the surface area and primary production of the global ocean (Fig. 4.2), over 90% of contemporary *OC* preservation occurs within a relatively narrow band of shallow sediments flanking continental margins (Berner 1989; Hedges and Keil 1995). The corresponding depth interval (~0–2 000 m water depth) includes deltas, continental shelves, and upper continental slopes, which together are generally characterized by the relatively rapid accumulation of organic-rich sediments beneath oxygenated bottom waters. The observation that approximately twice as much total organic matter is presently delivered from land to the ocean than is preserved in all marine sediments (Smith and Mackenzie 1987) indicates that severe recycling mechanisms occur in the sea. Whatever processes are involved, they must allow extensive remineralization of intrinsically resistant terrigenous organic matter at sea and subsequent preservation of seemingly more labile marine-derived counterparts in marine sediments (Keil et al. 1997; Mayer et al. 1998). Thus, the key to understanding atmospheric O_2 stabilization is to identify the mechanisms that modulate the minuscule leak of organic matter preserved in near-shore marine sediments over geologic time.

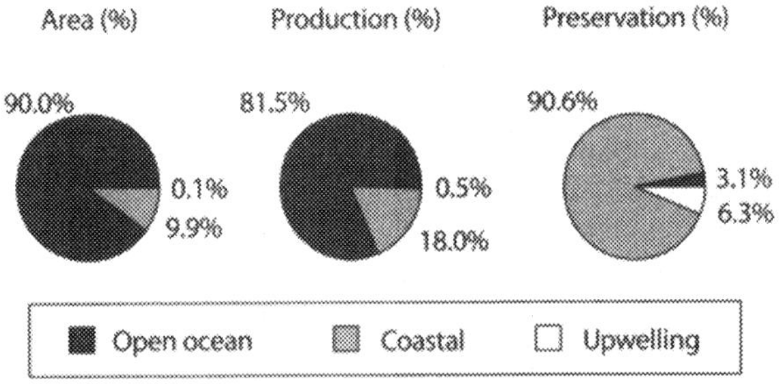

Fig. 4.2. The percentage distribution of area, productivity and organic carbon preservation among anoxic, coastal and open ocean marine waters. Although the coastal ocean accounts for only a small fraction of the ocean's area and primary productivity, it is the predominant site of sedimentary organic carbon preservation

One of the most widely observed patterns in the composition of marine sediments is that their organic carbon content increases with decreasing mean particle size within deposits of a given depositional regime (Bordovskiy 1965; Premuzic et al. 1982). This relationship was once thought to occur because particles of organic matter and finer mineral grains are similarly sorted during transport and because the supply rate of oxidizing agents to organic debris is more restricted in clay-rich deposits. This earlier notion was fundamentally changed by the finding that organic matter and coexisting sedimentary minerals could not be physically separated hydrodynamically or with heavy liquids (e.g. Mayer et al. 1993; Keil et al. 1994a). Recognition that most sedimentary organic matter is directly associated with mineral grains has been one of the most formative conceptual breakthroughs in the study of organic matter distribution and preservation.

Mayer (1994a,b) demonstrated that the weight percentage of organic carbon (wt% *OC*) is directly correlated to the specific surface area (*SA*) of bulk sediments and soils, rather than simply to texture. Moreover, he observed that coastal marine sediments depositing outside deltas typically exhibit *OC*/*SA* ratios in the range of 0.5–1.0 mg *OC* m^{-2}. This range corresponds to an organic carbon "loading" similar to that expected for average protein molecules adsorbed to mineral surfaces in a single layer, and was thus referred to as a "monolayer-equivalent" concentration. At the time, Mayer (1994a,b) stressed that this term refers to an equivalent concentration and not a mechanism, because the type and uniformity of organic molecule association with sedimentary minerals was unknown. Subsequent characterizations of mineral/organic associations by microscopic techniques (e.g. Ransom et al. 1998), gas adsorption energetics (Mayer 2000) and electron spectroscopy (Furukawa 2000) indicate that most sedimentary organic matter associated with minerals occurs in clumps rather than being distributed evenly in complete monolayers. This finding confirms intimate organic-mineral associations down to the nanometre scale (Furukawa 2000), but indicates that the molecules involved are not simply sorbed and thus may not have necessarily once been dissolved. Parallel studies in soils indicate that organic association is selective, dependent upon the surface characteristics of the mineral and often involves multivalent cation bridging (Kaiser and Guggenberger 2000; Baldock and Skjemstad 2000). The observation that organic matter separated from the mineral matrix of soils (Nelson et al. 1994) and sediments (Keil et al. 1994b) becomes orders of magnitude more reactive, also points toward a physical protection mechanism that would explain observed patterns in long-term preservation.

The wide range of *OC*/*SA* ratios observed in marine sediments from contrasting depositional regimes (Fig. 4.3), however, indicates that surface area is not the only determinant of sedimentary *OC* preservation rates (Hedges and Keil 1995). In comparison to typical continental margin deposits, sediments accumulating beneath anoxic bottom waters usually have higher *OC*/*SA* loadings (1–5 mg *OC* m^{-2}), whereas those deposited under the open ocean (depth > 2 000 m) almost always have lower ratios (<0.5 mg *OC* m^{-2}). Singling out the reasons for this nearly universal pattern is difficult, because many complex processes and properties covary in modern depositional settings. Primary production rates, water column depth, accumulation rates and varying degrees of biological mixing (bioturbation) and pumping (irrigation) are among frequently suggested causative factors. Particularly strong evidence has been presented that wt% *OC* increases with sediment accumulation rate (Heath et al. 1977;

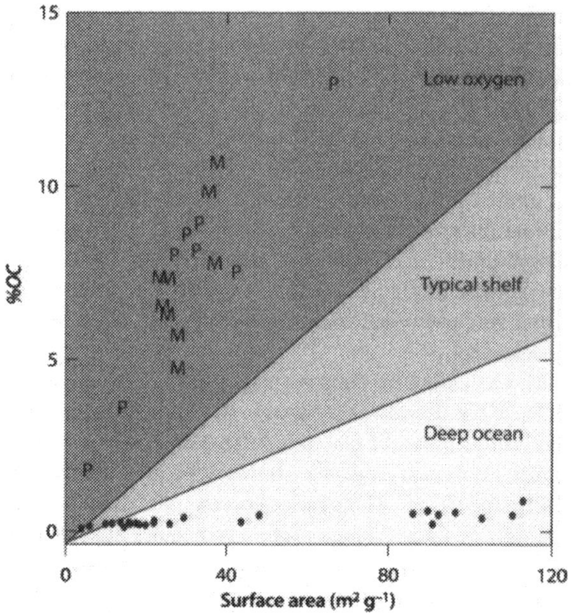

Fig. 4.3. Weight percent organic carbon (%*OC*) vs. surface area for marine sediments from a range of depositional regimes (adapted from Hedges and Keil 1995). *M* and *P* represent data for samples from the Mexican and Peruvian margins, respectively. In general, organic carbon content decreases with increasing availability of O_2, but varies directly with the surface area within different sedimentary regimes

Müller and Suess 1979) to the point where dilution by sand is observed (Doyle and Garrels 1985). This effect may result because high sedimentation rates decrease the time during which sediments are subjected to extensive biodegradation and strong oxidizing agents near the sediment-water interface (Henrichs 1992). In sharp contrast to the patterns in Fig. 4.3, and the parallel trends (Demaison and Moore 1980) observed for ancient rocks, Betts and Holland (1991) reported no significant correlation between *OC* burial efficiency and the O_2 content of bottom waters in contemporary marine depositional environments. As discussed in the next section, this apparent contradiction may result from the fact that bottom water O_2 contents do not directly indicate how long sedimentary particles are exposed to oxic degradation.

4.4 Oxygen Effects on Sedimentary Preservation

One of the strongest indications that molecular oxygen affects sedimentary organic matter preservation comes from unusual circumstances where *OC*-rich coastal sediments are transported within turbidity flows and redeposited at off-shore deep-ocean sites. The most studied example of such a phenomenon (Prahl et al. 1989, 1997) is the relict f-turbidite from the Madeira Abyssal Plain (MAP). This fine-grained deposit was emplaced approximately 140 000 years B.P., when an organic-rich deposit slumped off the continental slope of NW Africa and flowed down to spread as a 3-m thick deposit on the floor of the MAP region at ~5 500 m water depth (Thomson et al. 1993). After essentially instantaneous deposition, dissolved O_2 diffused into the pore water at the surface sediment and reacted slowly across a sharp interface with organic matter and reducing minerals. After approximately 10 000 years, the reaction interface "burned down" to a depth of ~half a metre, at which time diffusive impedance through the accumulating surface sediment covering returned the entire deposit to anoxic conditions (Buckley and Cranston 1988). Mineralogical compositions above and below the preserved redox interface are essentially identical in two cored sections of the f-turbidite recovered ~100 km apart in the MAP (Thomson et al. 1993), indicating that any differences in this initially uniform sequence should be attributable solely to their contrasting redox histories.

The pronounced compositional contrasts across the redox boundaries of the two MAP f-turbidite sequences give testament that long-term exposure to oxic conditions is sufficient to remineralize most forms of organic matter (Cowie et al. 1995). For example (Fig. 4.4), 10 000 years exposure to oxic pore water resulted in a loss of 80% of the *OC* and 100% of the physically recognizable pollen grains from the upper (oxidized) segment of this initially homogeneous deposit. Clearly, some aspect of prolonged exposure to oxic conditions decreased both components from concentrations characteristic of upper continental margin deposits to the low levels characteristic of the open ocean. To produce similar losses of *OC* and pollen in the laboratory, approximately 15 sequential treatments with heated H_2O_2 are necessary (Cowie 1990). The complete loss of pollen (Keil et al. 1994c) is particularly telling, because the covering of pollen grains, sporopollenin, is among the most resistant of all biopolymers to biodegrada-

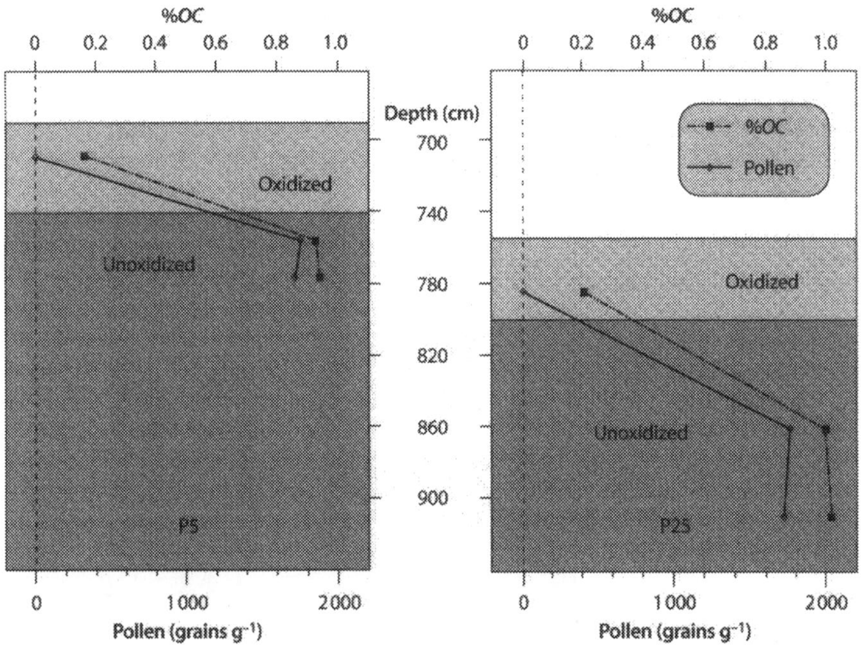

Fig. 4.4. Profiles of the weight percent organic carbon (%*OC*) contents and pollen abundances down two sequences of the f-turbidite from the Madeira Abyssal Plain (adapted from Keil et al. 1994; Cowie et al. 1995). Long-term exposure to oxic conditions on the surface of these deposits decreases organic carbon and pollen concentrations from values typical of continental margin deposits to low levels characteristic of the deep open ocean sediments

tion and typically persists for millions of years in anoxic deposits. These discrete particles (10–50 μm) cannot be lost by diffusion and therefore must have been destroyed as a result of severe in situ degradation. Extensive degradation is substantiated by a pronounced increase in the percentage of nonprotein amino acids (β-alanine plus γ-aminobutyric acid) across the oxidation front. These two diagenetic products increase from low relative concentrations (<10 mole%) typical of coastal sediments in the deeper turbidite to elevated sums (>30 mole%) in the oxidized surface horizon that are found only in deep ocean deposits (Cowie and Hedges 1994). Although the MAP turbidite demonstrates that conditions of long-term exposure to oxic conditions are sufficient to produce the low concentrations of highly degraded organic matter typical of open ocean sediments, they do not indicate how O_2 exposure might affect the organic compositions of incrementally depositing sediments along continental margins, where essentially all *OC* preservation presently occurs.

To quantitatively assess the potential importance of oxicity on *OC* preservation in modern deposits, it is useful to estimate the average time period that particulate material at the sediment surface is exposed to oxic conditions before accumulating to a depth below local O_2 penetration (Reimers 1989; Hartnett et al. 1998). This "oxygen exposure time" (*OET*) can be estimated (Fig. 4.5) for any given benthic site as the depth of O_2 penetration divided by the average rate of sediment accumulation (Hedges and

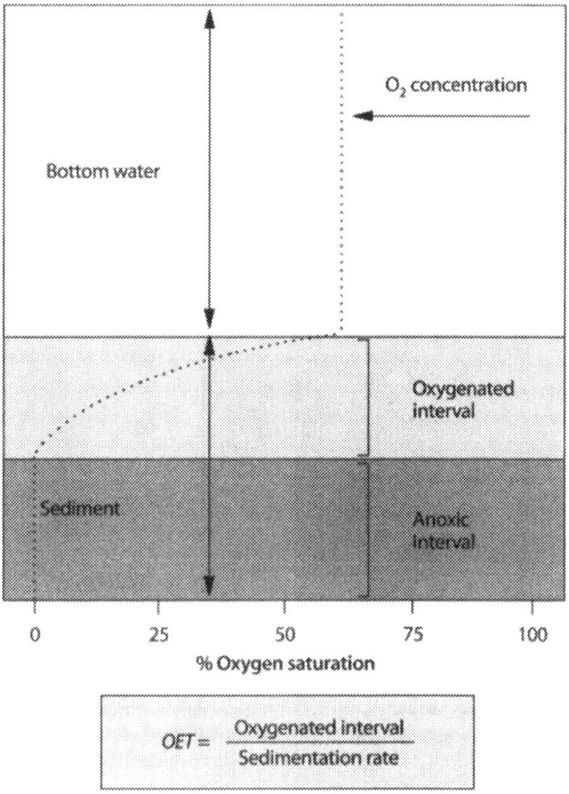

Fig. 4.5. An illustration of the oxygen exposure time (*OET*) concept. *OET* combines both sediment accumulation rate and the presence of O_2 into one measurable parameter

Keil 1995). Bioturbation sparingly affects *OET*, because particles should be mixed into and out of the thin oxic layer of continental margin sediments with comparable probabilities. *OET* incorporates the sediment accumulation rate as a prime determinant and is a direct indicator of oxic exposure within a sediment, as opposed to a remote proxy such as bottom water O_2 concentration (Betts and Holland 1991). The first published test of this hypothesis was the demonstration by Hartnett et al. (1998) that the efficiency of organic carbon burial in surface sediments along the western Pacific margin is indirectly related to the log of *OET*, as would be expected for an oxic remineralization mechanism exhibiting first-order kinetics. This study indicated that burial efficiency (preservation rate divided by delivery rate to the sea floor) appears to be sensitive to oxic conditions on time scales of weeks to months, and thus that oxic degradation is important on short as well as long time scales.

A more detailed study of the role of O_2 in organic matter preservation was later carried out for sediments depositing along a transect off the Washington State coast, USA (Hedges et al. 1999). In this study, %*OC*, surface area, C/N, biochemical (individual amino acids, and carbohydrates) compositions and O_2 penetration depths were measured in 16 sediment cores. For six of these cores, $\delta^{13}C$ compositions and ^{14}C-based deposition rates were also determined. The measured elemental and stable carbon isotope compositions both indicate that all these sediments are predominantly marine derived and compositionally uniform with depth. Due to off-shore increases in O_2 penetration depth and attending decreases in sediment accumulation rates, *OET*s increase exponentially off-shore in the six ^{14}C-dated cores (Fig. 4.6a). Exposure periods ranged from decades on the continental shelf and upper continental slope, to hundreds of years on the lower continental slope, to approximately 1 000 years at the most off-shore sampling site. A corresponding plot of *OC*/*SA* vs. log *OET* gives a fit to a straight line with a slope corresponding to an approximate half-life of 150 years (Fig. 4.7). Although this result is largely constrained by the intrinsic time scale represented by the studied cores (Middelburg et al. 1993), a major fraction of the organic matter in these sediments appears to be remineralized under oxic conditions with a time constant on the order of 100 years.

To more extensively test the inference of progressive off-shore oxic degradation, the freshness of the organic matter in these deposits was tested by two different biochemical indicators and by an assessment of the percentage of total pollen grains showing physical evidence of degradation. Both molecular indicators of organic matter "freshness" (%(BALA + GABA), and the percent of glucose within the aldose suite) indicated that the remnant organic matter in farther off-shore sediments is more degraded. In contrast, progressive *OC* depletion and degradation were not observed downcore at individual sites. Organic matter degradation was therefore effectively complete below the shallow (<3 cm) oxic surface horizons of these deposits. These observations indicate that most of the sedimentary organic matter remaining along the Washington continental margin is susceptible to remineralization under oxic conditions, but is degraded slowly, if at all, in the absence of molecular O_2. This finding, and parallel results from the MAP turbidites, point towards a consistent oxic effect over a range of time scales. Such "O_2 sensitivity" does not necessarily hold for all organic matter, most of which (e.g. polysaccharides and proteins) is easily fermented and rapidly remineralized regardless of redox conditions (e.g. Lee 1992; Canfield 1989, 1994). Oxygen sensitive organic matter appears to concentrate in the latter stages of

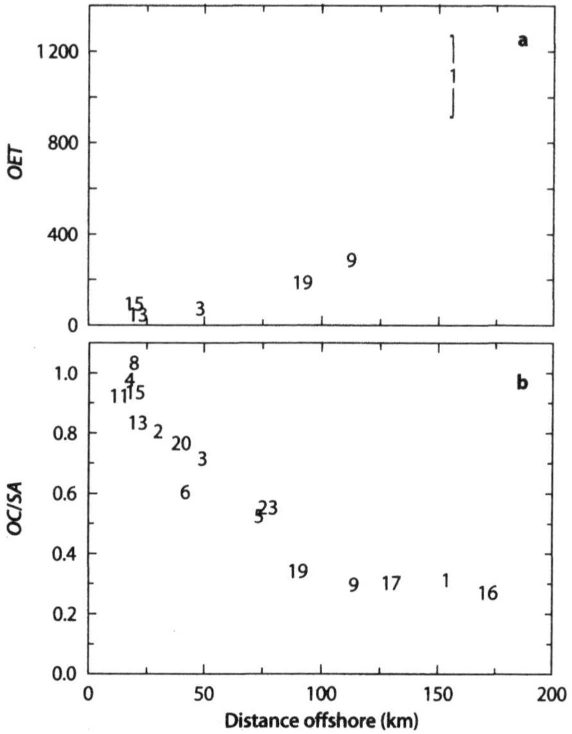

Fig. 4.6. Plots of; **a** oxygen exposure time (*OET*); **b** average organic carbon/surface area (*OC/SA*) vs. distance off-shore from the Washington State coast (adapted from Hedges et al. 1999). *OET* increases systematically off-shore in combined response to deeper O_2 penetration depths and slower sediment accumulation rates, causing a pronounced decrease in the amount of organic matter that is preserved in association with mineral surfaces

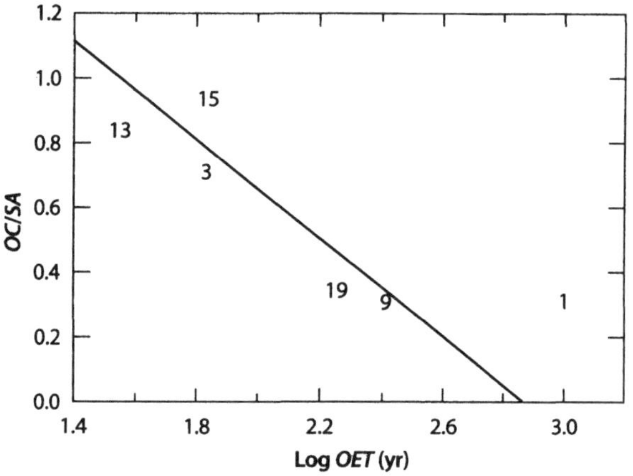

Fig. 4.7. Plot of organic carbon/surface area (*OC/SA*) vs. the log of oxygen exposure time (*OET*) for the same Washington margin sediments in Fig. 4.6 (adapted from Hedges et al. 1999). The numbers correspond to core collection sites described in the original paper. Organic carbon loading decreases systematically in response to increasing *OET*, with an average half-life on the order of 10^2 yr.

degradation where preservation extents, rather than degradation kinetics, are the key characteristics of preservation potential (Fig. 4.8).

Overall, the potential of modern continental margin sediments to preserve organic matter varies directly with the average surface area of the deposit (Fig. 4.3) and is modified greatly by O_2 availability at the deposition site (Fig. 4.9). Given that mineral

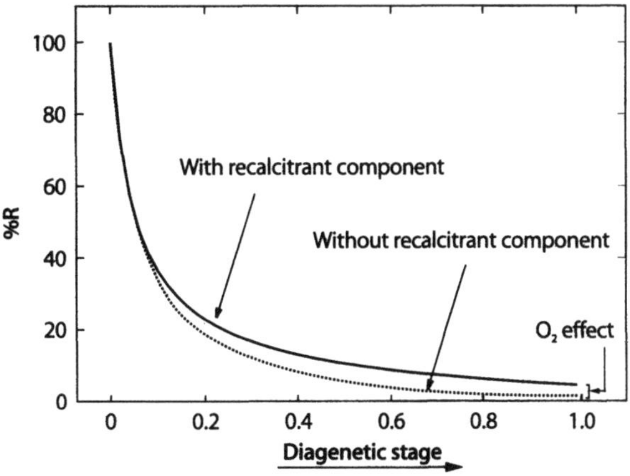

Fig. 4.8. Two hypothetical curves of percent remaining material (%R) vs. an arbitrary increase in diagenetic stage. This comparison illustrates the effect of a small resistant component on the overall extent, vs. the rate of organic matter degradation (adapted from Cowie et al. 1995). These two lines were generated for organic mixtures containing equal initial amounts of five components that are lost exponentially with first-order rate constants, which vary from each other by a factor of two. The only difference between the two curves is that the most recalcitrant component in the mixture indicated by the *solid line* is twice as stable as the most recalcitrant component in the mixture corresponding to the *dotted line*. As in sedimentary mixtures, such differences are seen only at later stages of degradation, and are most evident as a contrast in reaction extent, as opposed to reaction kinetics

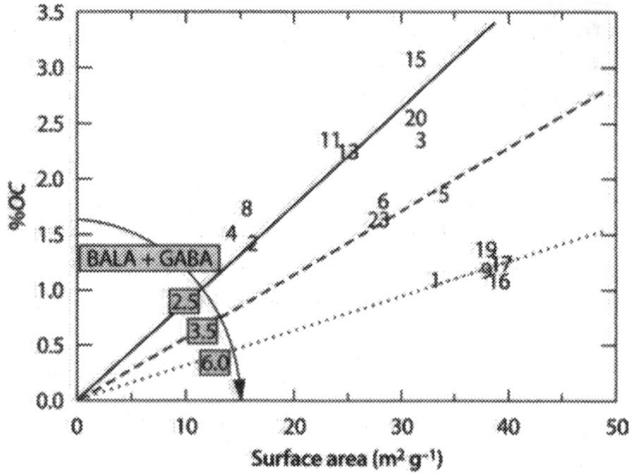

Fig. 4.9. A "degradation fan" plot of the average weight percentage of organic carbon (%*OC*) vs. surface area for the above suite of Washington coast sediments (adapted from Hedges et al. 1999) The *arrow* indicates increasing average mole percentages of β-alanine plus γ-aminobutyric acid, (BALA + GABA) in the same Washington margin sediments. The potential for organic matter preservation in these sediments increases in proportion to their average surface area, but decreases in response to increasing in situ oxic degradation, as indicated by increasing (BALA + GABA)

surface area and *OET* can both fluctuate by two orders of magnitude or more in nature (Hedges and Keil 1995; Hedges et al. 1999), the combined effect of these two variables can have a dynamic range in excess of 10^4 for a given mass of sediment. Because of this range of response along continental margins, where essentially all *OC* burial presently occurs, mineral surface area and O_2 availability have the combined potential to modulate the rate of O_2 generation on a global scale. Because surface area delivery to coastal zones is ultimately controlled by continental uplift and O_2 availability to the ocean floor is directly dependent upon the contemporary atmospheric concentration, a tectonic transmission and geochemical governor for a global O_2 control apparatus appear to be in place.

4.5 Maintaining Atmospheric O_2 within Safe Bounds

The geochemical challenge of maintaining relatively constant (within a factor or two) atmospheric concentrations of O_2 (mean residence time ~4 million years) over the last 600 million years has been recognized for several decades (Broecker 1970; Van Valen 1971). Twenty-five years ago, Walker (1974) pointed out that due to nearly complete remineralization of reduced elements during sedimentary rock weathering, the inferred global O_2 control mechanism must operate on the source side within depositing marine sediments. The first comprehensive attempts to quantitatively model reservoir sizes and exchange rates among major redox-active elements in the Earth's surface (exogenic) cycle came in the mid 1970s (e.g. Holland 1973; Garrels and Perry 1974; Garrels et al. 1976). The latter authors demonstrated mathematically that the network of biogeochemical exchanges among the major reduced and oxidized forms of carbon (*OC* and $CO_2 + CaCO_3$) and sulphur (SO_4^{2-} and FeS_2) at the Earth's surface appears to embody "effective feedback mechanisms" for controlling atmospheric CO_2 and O_2. For example, they pointed out that an increase in continental erosion rates would decrease the partial pressure of atmospheric O_2, which would lead (through gas exchange) to lower O_2 concentrations in sea water and (presumably) to more efficient sedimentary organic matter preservation (Fig. 4.1). The eventual system response would be a net release of more O_2 to partially offset the initial atmospheric decrease. The functional relationships used in this model to determine the exchange fluxes of O_2 and the various C and S forms, however, were largely empirical or presumed, with weak and incomplete mechanistic underpinnings.

It was also recognized about this time that large excursions in O_2 and CO_2 could be avoided if the Earth's exogenic redox cycles were confined largely among a select set of solid reactants and products. The overall global redox reaction can be represented as the reduction of carbon in carbonate and dolomite with sulphur in pyrite, to produce organic matter, iron oxide and sulphate (Garrels and Perry 1974).

$$4\,FeS_2 + CaCO_3 + 7\,CaMg(CO_3)_2 + 7\,SiO_2 + 15\,H_2O$$
$$\leftrightarrow 15\,CH_2O + 8\,CaSO_4 + 2\,Fe_2O_3 + 7\,MgSiO_3 \qquad (4.2)$$

This formulation bypasses the intermediary role of O_2 and CO_2 in photosynthesis/respiration and weathering and has the budgetary advantage that all the major electron exchangers occur predominantly in the rock reservoir. The masses of these redox-sensitive minerals in sedimentary rocks are so huge that any imbalance of the reaction network could rapidly change the comparably small amounts of sulphate and bicarbonate dissolved in the ocean, as well as of O_2 and CO_2 in the atmosphere. An illustration of the geological processes linking the key redox forms in Eq. 4.2 is given in

Fig. 4.10. A schematic illustration of the long-term global cycles of carbon and sulphur (from Berner 1999). If stoichiometric fluxes are perfectly balanced between the four principal rock reservoirs, the relatively small amounts of bicarbonate and sulphate dissolved in the ocean, and of CO_2 and O_2 in the atmosphere, will remain unchanged. The major compensation mechanism is to bury compensating masses of C and S in oxidized vs. reduced form

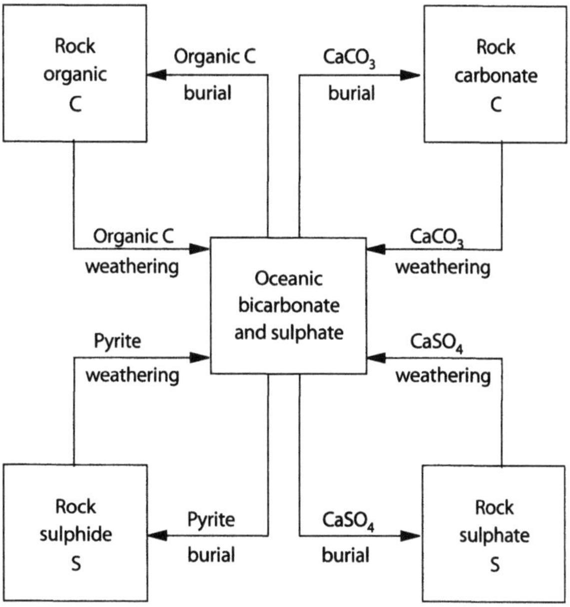

Fig. 4.10. This reaction network exemplifies how massive fluxes of redox-active elements can cycle through the surface of the Earth without causing catastrophic changes in ocean and atmospheric chemistry, but it does not rule out such fluctuations.

Another advantage of the global redox cycle expressed in Eq. 4.2 is that the stable isotopic compositions of sedimentary carbon ($\delta^{13}C$) and sulphur ($\delta^{34}S$) can be incorporated into mass balances (Garrels and Lerman 1981, 1984) to constrain the amounts of these elements in the major reservoirs and the fluxes between them (Fig. 4.10) over geologic time. The stable isotopic records of sea water sulphate (in gypsum) and carbonate (in marine limestones) have been particularly useful indicators of major shifts over the Phanerozoic among the four major rock reservoirs, whose original sizes are often poorly known. Berner (1987) later modified this isotopic budgetary approach by splitting each of the four rock reservoirs into younger (faster cycling) and older (slower cycling) compartments. An accompanying subroutine to this "rapid recycling" model was then used to estimate atmospheric O_2 trends over geologic time. A major rationale for subdividing the rock reservoirs was that conventional model results for the then available $\delta^{13}C$ data for marine limestones indicated what would have been catastrophic fluctuations in atmospheric O_2 concentrations. These calculations were incompatible with other geologic evidence that atmospheric O_2 concentrations were maintained within 50% of present-day levels throughout the Phanerozoic (Berner 1987). Even with subdivided rock reservoirs, it was necessary in Berner's model to introduce an empirical feedback control on atmospheric O_2 that operated on the sink (weathering) side (see Kump and Garrels 1986), which was at odds with the earlier inference of Garrels et al. (1976) that control must ultimately reside in the marine source region.

Subsequent generations of isotope/mass balance models have been formulated that account for additional factors such as redistribution of sediment type (Berner and

Canfield 1989), temporal trends in stable isotope composition (Lasaga 1989), and phosphorus limitation on marine primary production (Van Cappellen and Ingall 1996; Petsch and Berner 1998). The latter constraint is based on the observation that burial of the limiting nutrient P is less efficient when ocean bottom waters are low in oxygen. The driving mechanism appears to be that a more oxidizing ocean leads to globally increased sequestration of phosphate into sedimentary iron oxyhydroxides (or bacteria). The net result is that an increase in atmospheric O_2 leads, via increased downwelling of O_2 enriched surface sea water, to a more oxidizing ocean floor, larger P uptake by sediments and a reduced reservoir of dissolved phosphate to support marine photosynthesis. Models incorporating this feedback mechanism (without a weathering counterpart) are able to generate numeric results that match in amplitude and duration the history of atmospheric O_2 change inferred from the sedimentary record (Petsch and Berner 1998). Recognition of phosphate as a major constraint on the intermeshed cycles of carbon and oxygen is critical, because without this basic constraint, marine primary productivity has the potential to generate large amounts

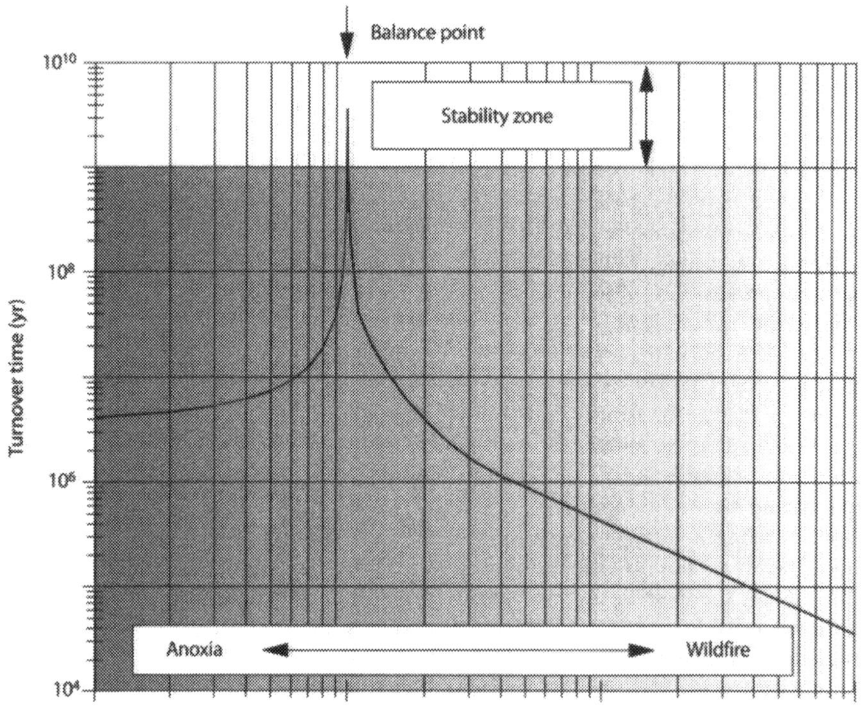

Fig. 4.11. Turnover time τ (τ) of atmospheric O_2 as a function of the magnitude of imbalances vs. the present-day rate of *OC* preservation and O_2 production (~0.012×10^{15} moles yr^{-1}). Without other constraints (see text), the atmosphere of a photosynthetically dead Earth would lose its O_2 in roughly 4 million years, whereas photosynthesis in the absence of respiration could (without P-control) double present atmospheric O_2 levels in a matter of a tens of thousands of years

of O_2 in a short period of time (Fig. 4.11). Because the only long term source of P is rock weathering, and the ocean reservoir is relatively small and can be depleted in a matter of thousands of years (Broecker and Peng 1982), phosphate serves as a sensitive tectonic throttle for bioactive element cycling.

Another important constraint on O_2 overproduction is that sulphate and carbon dioxide reduction convert insoluble sedimentary organic matter to dissolved sulphide and methane. In many coastal zones, these two reduction products diffuse upward into oxic zones where they remove the same amount of O_2 that was generated when the parent organic matter was photosynthesized (Berner 1982). At extremely rapid sediment accumulation rates, this mechanism breaks down because sulphide and methane are buried faster than they can diffuse upward. Such conditions are rare, however, and at sediment accumulation rates less than approximately 1 g cm^{-2} yr^{-1} greater than 95% of the generated sulphide is diffusively lost and oxidized at the ultimate expense of atmospheric O_2 (Morse and Berner 1995).

A recent modelling effort involving a detailed sulphur isotope record for Cenozoic sea water sulphate preserved in sedimentary barite (Payton and Arrigo 2000) demonstrated that the stable sulphur and carbon isotopic records are not consistent with atmospheric O_2 control solely by burial of pyrite S and organic matter (Fig. 4.10). The difficulty with stable-isotope based models in general may be traced to their extreme sensitivity to the isotopic compositions of the C and S rock reservoirs (Fig. 4.10). The almost universal assumption in these models that isotopic fractionations have remained constant over the Phanerozoic as C and S are transferred among the major geologic reservoirs may simply not be realistic (Payton and Arrigo 2000). Additional mechanisms for constraining atmospheric O_2 content over geologic time may well exist, and in fact be necessary.

4.6 The Mineral Conveyer Belt and Sedimentary Afterburner

The "mineral conveyer belt" model for control of atmospheric O_2 (Fig. 4.12) combines the concept of mass balance in the weathering/deposition cycle (e.g. Van Capellen and Ingall 1996) with a variant of Broecker's (1970) early concept for O_2-controlled marine sedimentary preservation. The assumption that mineral transport from weathering rocks to coastal marine sediments occurs with little net change in organic matter loading is based on the observation that most marine sediments deposited on upper continental margins exhibit a relatively uniform surface area "loading" of 0.5–1.0 mg *OC* m^{-2}. Because the weight percentages of *OC* in ancient shales and nearshore fine-grained marine sediments are similar (Hunt 1996), this concentration factor has not changed greatly over the Phanerozoic. A key assumption of the conveyer belt model is that sedimentary rocks weather to primary particles that exhibit a surface area comparable to the mineral grains from which the rocks were originally formed. This assumption is critical only for the clay and silt fractions of sediments (primarily shales), which account for most of the total buried surface area (Keil et al. 1994a; Bergamaschi et al. 1997). The assumption that mineral surface area controls the maximal burial potential of sediments at a given stage of oxic degradation is consistent with field observations (e.g. Figs. 4.3 and 4.9) and does not conflict with previous inferences that phosphate may be co-limiting.

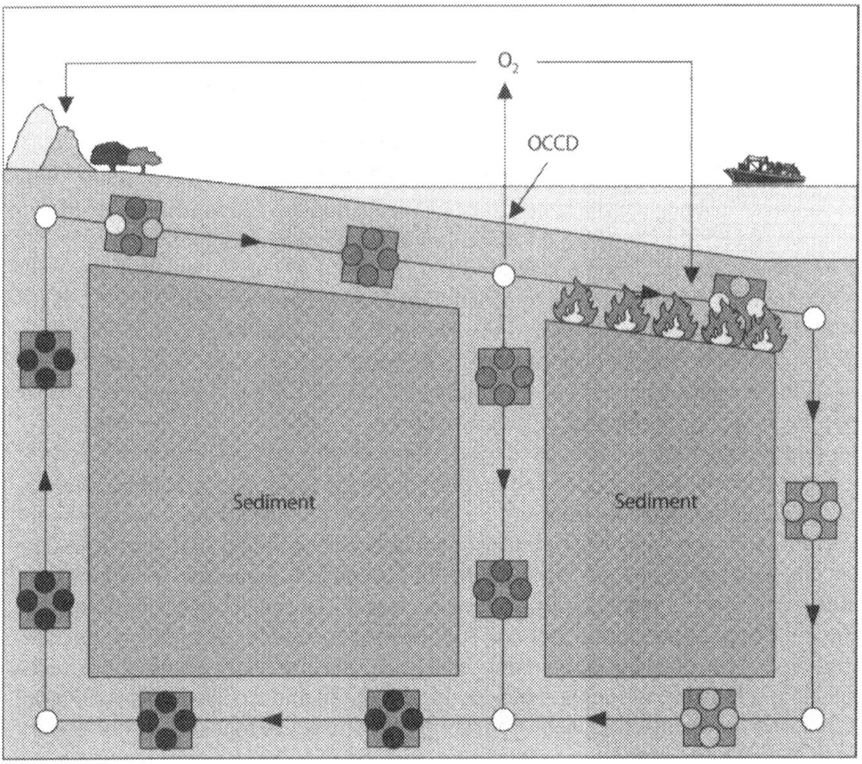

Fig. 4.12. A schematic representation of the hypothetical "mineral conveyer belt." The two cycles represent the transport of mineral surface area (*squares with holes*) with a typical organic matter loading (*ball in holes*) through the tectonic (anoxic) and weathering (oxic) cycles. The outer cycle represents deposition of organic-depleted sediments off-shore of the marine *OC* compensation depth (OCCD), where long term exposure to pore water O_2 is sufficient to oxidize most of the mineral-associated organic matter. Surface area conservation during weathering and deposition modulated by a negative feedback mechanism involving oxic degradation could provide a sensitive, quantitatively ample and intrinsically stable control system for atmospheric O_2

Although the idea that global marine sedimentary preservation is tied in a negative feedback loop to atmospheric O_2 concentration is not new (Broecker 1970), evidence in support of this inference has increased substantially in the last few years (e.g. Hartnett et al. 1998; Hedges et al. 1999). As previously discussed, several observations point specifically toward the period of oxic exposure (*OET*) during sediment accumulation as the key determinant of preservation efficiency. Geochemists continue to debate whether *OC* preservation rates are controlled by primary production in the surface ocean (e.g. Calvert and Pedersen 1992) vs. conditions (such as *OET*) prevailing near the sea floor (Hedges and Keil 1995). This chicken-or-egg argument is circuitous to the extent that the delivery rate of organic reducing power to the sea floor ultimately is influenced by primary production, which must therefore be important. Although it has been speculated that more extensive remineralization under oxic conditions may result from the use of oxygen-specific enzymes (oxygenases and peroxidases) to degrade recalcitrant, carbon-rich substances such as lignin (Emerson and Hedges 1988), the specific mechanisms involved are presently unknown. Thus, the inferred "oxic effect" should be regarded as a general phenomenon that characteristically occurs in the presence of molecular O_2, but does not necessarily involve this chemical species directly.

What is abundantly clear, however, is that deep open ocean sediments contain (Fig. 4.3) and bury (Fig. 4.2) very little organic matter (Premuzic et al. 1982). Where extensive off-shore gradients in sedimentary *OC* have been analyzed in detail, sharp decreases have been observed along lower (>2 000 m) continental margins (Hedges and Keil 1995; Van der Weijden et al. 1999; Graneshram et al. 1999). These trends, viewed within the context of an oxygen exposure model, suggest the existence of an "organic carbon compensation depth" (OCCD), much in parallel to the carbonate compensation depth (CCD) that has been recognized for decades on sea floor promontories (Broecker and Peng 1982). In both instances, slow external erosion appears to compete with sediment burial to determine preservation patterns that exhibit relatively sharp boundaries determined by bottom water conditions. In the case of the OCCD, however, the sharpness of the boundary and hence the sensitivity of the feedback mechanism may be enhanced by the sharp increase in *OC* preservation per unit surface area that typically occurs (Fig. 4.3) when bottom water O_2 falls to low values as are now seen along the Peruvian and NW Mexican margins. Although no attempt is made here to numerically model the combined effects on atmospheric O_2 concentrations that might result from a mineral conveyer belt operating in conjunction with a marine sedimentary "afterburner" (Fig. 4.12), such quantitative simulations should be possible. To understand the past expressions and the future functions of this proposed addition to the global safety net, it will be necessary to learn more about how organic matter becomes associated with sedimentary minerals and by what mechanism oxic degradation proceeds on land and in the ocean.

Acknowledgements

Cindy Lee and Kenia Whitehead made helpful comments on drafts of this chapter.

References

Baldock JA, Skjemstad JO (2000) Role of the soil matrix and minerals in protecting natural organic materials against biological attack. Org Geochem 31:697–710

Bergamaschi BA, Tsamakis E, Keil RG, Eglinton TI, Montluçon DB, Hedges JI (1997) The effect of grain size and surface area on organic matter, lignin and carbohydrate concentrations and molecular compositions in Peru Margin sediments. Geochim Cosmochim Acta 61:1247–1260

Berner RA (1982) Models for carbon and sulfur cycles and atmospheric oxygen: Application to Paleozoic geologic history. Am J Sci 287:177–196

Berner RA (1987) Burial of organic carbon and pyrite sulfur in the modern ocean: Its geochemical and environmental significance. Am J Sci 282:451–473

Berner RA (1989) Biogeochemical cycles of carbon and sulfur and their effect on atmospheric oxygen over Phanerozoic time. Palaeogeogr Palaeoclimatol Palaeoecol 73:97–122

Berner RA (1999) Atmospheric oxygen over Phanerozoic time. Proc Natl Acad Sci 96:10955–10957

Berner RA, Canfield DE (1989) A new model for atmospheric oxygen over Phanerozoic time. Am J Sci 289:333–361

Betts JN, Holland HD (1991) The oxygen content of ocean bottom waters, the burial efficiency of organic carbon, and the regulation of atmospheric oxygen. Palaeogeogr Palaeoclimatol Palaeoecol 97:5–18

Bordovskiy KO (1965) Accumulation and transformation of organic substances in marine sediments. Part I and II. Mar Geol 3:3–34
Broecker WS (1970) A boundary condition on the evolution of atmospheric oxygen. J Geophys Res 75:3553–3557
Broecker WS, Peng T-H (1982) Tracers in the Sea. Lamont-Doherty Geological Observatory, Palisades, New York
Buckley DE, Cranston RE (1988) Early diagenesis in deep-sea turbidities: The imprint of paleo-oxidation zones. Geochim Cosmochim Acta 52:2925–2939
Calvert SE, Pedersen TF (1992) Organic carbon accumulation and preservation in marine sediments: How important is anoxia? In: Whelan J, Farrington JW (eds) Organic matter. University Press, New York, pp 231–263
Canfield DE (1989) Sulfate reduction and oxic respiration in marine sediments: implications for organic carbon preservation in euxinic environments. Deep-Sea Res 6:121–138
Canfield DE (1994) Factors influencing organic carbon preservation in marine sediments. Chem Geol 114:315–329
Capellen P Van, Ingall ED (1996) Redox stabilization of the atmosphere and oceans by phosphorus-limited marine productivity. Science 271:493–496
Cowie GL (1990) Marine organic diagenesis: A comparative study of amino acids, neutral sugars and lignin. PhD dissertation, University of Washington
Cowie GL, Hedges JI (1994) Biochemical indicators of diagenetic alteration in natural organic matter mixtures. Nature 369:304–307
Cowie GL, Hedges JI, Prahl FG, de Lange GJ (1995) Elemental and major biochemical changes across an oxidation front in a relict turbidite: An oxygen effect. Geochim Cosmochim Acta 59:33–46
Demaison GJ, Moore GT (1980) Anoxic environments and oil source bed genesis. Am Assoc Petrol Geol Bull 64:1179–1209
Des Marais DJ (1997) Isotopic evolution of the biogeochemical carbon cycle during the Proterozoic Eon. Org Geochem 27:185–193
Doyle LJ, Garrels RM (1985) What does percent organic carbon in sediments measure? Geo-Mar Let 5:51–53
Emerson S, Hedges JI (1988) Processes controlling the organic carbon content of open ocean sediments. Paleoceanography 3:621–634
Emerson S, Stump C, Grootes PM, Stuiver M, Farwell GW, Schmidt FH (1987) Estimates of degradable organic carbon in deep-sea sediments from ^{14}C concentrations. Nature 329:51–53
Furukawa Y (2000) Energy-filtering transmission electron microscopy (EFTEM) and electron energy-loss spectroscopy (EELS) investigation of clay-organic matter aggregates in aquatic sediments. Org Geochem 31:735–744
Garrels RM, Lerman A (1981) Phanerozoic cycles of sedimentary carbon and sulfur. Proc Natl Acad Sci 78:4652–4656
Garrels RM, Lerman A (1984) Coupling the sedimentary sulfur and carbon cycles-an improved mode. Am J Sci 284:989–1007
Garrels RM, Perry AE (1974) Cycling of carbon, sulfur and oxygen through geologic time. In: Goldberg ED (ed) The Sea, vol V. Wiley, New York, pp 303–316
Garrels RM, Lerman A, Mackenzie FT (1976) Controls on atmospheric O_2 and CO_2: Past, present, and future. Amer Sci 64:306–315
Graneshram RS, Calvert SE, Pedersen TF, Cowie GL (1999) Factors controlling the burial of organic carbon in laminated and bioturbated sediments of NW Mexico: Implications for hydrocarbon preservation. Geochim Cosmochim Acta 63:1723–1734
Hartnett HE, Keil RG, Hedges JI, Devol AH (1998) Influence of oxygen exposure time on organic carbon preservation in continental margin sediments. Nature 391:572–574
Heath GR, Moore TC, Dauphin JP (1977) Organic carbon in deep-sea sediments. In: Andersen NR Malahoff A (eds) The fate of fossil fuel CO_2 in the Oceans. Plenum, New York, pp 605–625
Hedges JI, Keil RG (1995) Sedimentary organic matter preservation: An assessment and speculative synthesis. Mar Chem 49:81–115
Hedges JI, Hu FS, Devol AH, Hartnett HE, Keil RG (1999) Sedimentary organic matter preservation: A test for selective oxic degradation. Am J Sci 299:529–555
Henrichs SM (1992) The early diagenesis of organic matter in marine sediments: Progress and perplexity. Mar Chem 39:119–149
Holland HD (1973) Systematics of the isotope composition of sulfur in the oceans during the Phanerozoic and its implications for atmospheric oxygen. Geochim Cosmochim Acta 37:2605–2616
Hunt JM (1996) Petroleum geochemistry and geology. Freeman, New York
Jones TP, Chaloner WG (1991) Fossil charcoal, its recognition and paleoatmospheric significance. Global Planet Change 97:39–50
Kaiser K, Guggenberger G (2000) The role of DOM sorption to mineral surfaces in the preservation of organic matter in soils. Org Geochem 31:711–725

Keeling RF, Najjar RP, Mender ML, Tans PP (1993) What atmospheric oxygen measurements can tell us about the global carbon cycle. Global Biogeochem Cycles 7:37–67

Keil RG, Tsamakis E, Fuh CB, Giddings JC, Hedges JI (1994a) Mineralogical and textural controls on organic composition of coastal marine sediments: Hydrodynamic separation using SPLITT fractionation. Geochim Cosmochim Acta 57:879–893

Keil RG, Montluçon DB, Prahl FG, Hedges JI (1994b) Sorptive preservation of labile organic matter in marine sediments. Nature 370:549–552

Keil RG, Hu FS, Tsamakis E, Hedges JI (1994c) Pollen grains deposited in marine sediments are degraded only under oxic conditions. Nature 369:639–641

Keil RG, Mayer LM, Quay PD, Richey JE, Hedges JI (1997) Loss of organic matter from riverine particles in deltas. Geochim Cosmochim Acta 61:1507–1511

Kump LR, Garrels RM (1986) Modeling atmospheric O_2 in the global sedimentary redox cycle. Am J Sci 286:337–360

Lasaga AC (1989) A new approach to isotopic modeling of the variation of atmospheric oxygen through the Phanerozoic. Am J Sci 289: 411–435

Lee C (1992) Controls on organic carbon preservation: The use of stratified water bodies to compare intrinsic rates of decomposition in oxic and anoxic systems. Geochim Cosmochim Acta 56:3323–3335

Leeuw JW de, Largeau C (1993) A review of macromolecular organic compounds that comprise living organisms and their role in kerogen, coal, and petroleum formation. In: Engle MH, Macko SA (eds) Organic geochemistry. New York, Plenum, pp 23–72

Mayer LM (1994a) Surface area control of organic carbon accumulation in continental shelf sediments. Geochim Cosmochim Acta 58:1271–1284

Mayer LM (1994b) Relationships between mineral surfaces and organic carbon concentrations in soils and sediments. Chem Geol 114:347–363

Mayer LM, Keil RG, Macko SA, Joye SB, Ruttenberg KC, Aller RC (1998) Importance of suspended particulates in riverine delivery of bioavailable nitrogen to coastal zones. Global Biogeochem Cycles 12:573–579

Mayer LM (2000) Extent of coverage of mineral surfaces by organic matter in marine sediments. Geochim Cosmochim Acta 63:207–215

Mayer LM, Jumars PJ, Taghon GL, Macko SA (1993) Low-density particles as potential nitrogenous food for benthos. J Mar Res 51:373–389

Middelburg JJ, Vlug T, Nat FJWA Van der (1993) Organic matter mineralization in marine systems. Global Planet Change 8:47–58

Morse JW, Berner RA (1995) What determines sedimentary C/S ratios? Geochim Cosmochim Acta 59:1073–1077

Müller PJ, Suess E (1979) Productivity, sedimentation rate, and sedimentary organic matter in the oceans. I. Organic carbon preservation. Deep-Sea Res 26:1347–1362

Nelson PN, Dictor M-C, Soulsa G (1994) Availability of organic carbon in soluble and particle-size fractions from a soil profile. Soil Biol Biochem 26:1549–1555

Payton A, Arrigo KR (2000) The sulfur-isotopic composition of Cenozoic seawater sulfate: Implications for pyrite burial and atmospheric oxygen. Internat Geol Rev 42:1–8

Petsch ST, Berner RA (1998) Coupling the geochemical cycles of C, P, Fe, and S: The effect of atmospheric O_2 and the isotopic records of carbon and sulfur. Amer J Sci 298:246–262

Prahl FG, Lange GJ de, Lyle M, Sparrow MA (1989) Post-depositional stability of long-chain alkenones under contrasting redox conditions. Nature 341:434–437

Prahl FG, Lange GJ de, Scholten S, Cowie GL (1997) A case of post-depositional aerobic degradation of terrestrial organic matter in turbidite deposits from the Madeira Abyssal Plain. Org Geochem 27:141–152

Premuzic ET, Benkovitz CM, Gaffney JS, Walsh JJ (1982) The nature and distribution of organic matter in surface sediments of world oceans and seas. Org Geochem 4:63–77

Ransom B, Kim D, Kastner M, Wainwright S (1998) Organic matter preservation on continental slopes: Importance of mineralogy and surface area. Geochim Cosmochim Acta 62:1329–1345

Reimers CE (1989) Control of benthic fluxes by particulate supply. In: Berger WH, Smetacek S, Wefer G (eds) Productivity of the ocean: Present and past. John Wiley and Sons, New York, pp 217–233

Smith SV, Mackenzie FT (1987) The ocean as a net heterotrophic system: Implications for the carbon biogeochemical cycle. Global Biogeochem Cycles 1:187–198

Thomson J, Higgs NC, Croudace IW, Colly S, Hydes DJ (1993) Redox zonation of elements at an oxic/post-oxic boundary in deep-sea sediments. Geochim Cosmochim Acta 57:579–595

Valen L Van (1971) A history and stability of atmospheric oxygen. Science 171:439–443

Walker JCG (1974) Stability of atmospheric oxygen. Amer J Sci 274:193–214

Watson A, Lovelock JE, Margulis L (1978) Methanogenesis, fires and regulation of atmospheric oxygen. Biosystems 19:293–298

Weijden CH Van der, Reichart GJ, Hendrik JV (1999) Enhanced preservation of organic matter in sediments deposited within the oxygen minimum zone in the northeastern Arabian Sea. Deep-Sea Res 46:807–830

Chapter 5

Particulate Organic Matter Composition and Fluxes in the Sea

C. Lee

5.1 Introduction

For at least 40 years, the rain of particulate organic matter falling through the ocean has been known to exist and to constitute a source of food for organisms on the sea floor (Honjo 1990, 1996). Sinking of particulate material from the surface to deeper waters and the sea floor is one of the major pathways for the transport of carbon and other bioelements within the ocean, and a variety of biological, physical and chemical processes alter the organic and inorganic composition of particles as they sink. In the past four decades, marine geochemists have learned much about sinking particulate organic matter in the sea, yet many questions remain. This chapter reviews what is known about the quantity and quality of sinking particulate organic matter in the sea, describes how it varies in time and space, and considers several larger questions yet to be answered.

Organic compounds are synthesized from inorganic carbon in the sunlit surface layer of the world ocean, where primary production by phytoplankton is clearly the largest source of organic carbon. Heterotrophic consumption (or respiration) by zooplankton and bacteria removes most of the organic matter in the surface waters before it is exported below the euphotic zone. On a global average, only a small fraction (5–10%) of the total primary production sinks below the euphotic zone, and a small fraction of that (1–10%) survives transport to the sea floor to be preserved in sediments. Thus, the majority of exported organic matter is remineralized (returned to inorganic form) on its way to the sea floor. Despite this great loss, the surface productivity signal can extend to the deep-sea floor and into the sediments, and the composition of organic matter in the deep ocean reflects its phytoplankton source. We can use organic biomarkers to investigate the sources of organic mater and specific diagenetic indicator compounds to ascertain the extent of its degradation. Complicating this picture, however, is the fact that the fraction of chemically uncharacterizable organic matter increases in importance with depth and makes up most of the bulk carbon in sediments. Explaining the fates of different classes of organic compounds, including the relationship between molecularly characterizable and uncharacterizable fractions of organic matter, is a central problem that will require the development and application of new indicators of source and degradation as well as new analytical tools.

A second question concerns the interaction between organic matter and minerals. Material exported from the euphotic zone leaves as large, fast-sinking particles (McCave 1975). Sinking particles include the biogenic mineral phases of planktonic diatoms, radiolaria, foraminifera, coccolithophorids, and pteropods and lithogenic minerals (e.g. from dust) (Honjo 1996); these minerals may serve as dense ballast to allow the particles to sink. Sinking speed in turn influences the profile of organic matter remineralization with depth and the effectiveness of the deep ocean as a carbon sink. The deeper remineralization occurs, the longer recycled carbon is kept from contact with surface waters and the atmosphere. However, the extent to which dense mineral ballast determines how fast particles sink is currently unknown, and this is one of the critical, outstanding problems in the study of particulate matter in the sea.

5.2 Relation of Carbon Flux with Primary Production

Sediment traps capture large particles that sink in the water column and can be used to estimate the flux of these particles and their constituents (Honjo 1996). Sediment trap studies have demonstrated a direct correlation between primary production and the downward flux of bulk particulate organic carbon in regions of different average productivity and over time at individual sites with seasonally variable primary production. Below, details of these relations are investigated.

5.2.1 Spatial Relation

Early sediment trap studies suggested that carbon, and frequently mass fluxes, were dependent on total primary production. However, large variations in the nature of the relationship between flux and productivity were observed, depending on location (Fig. 5.1). It was quickly realized that carbon export from the euphotic zone is more closely related to "new production" (Eppley and Peterson 1979), and later models re-

Fig. 5.1. Sediment trap studies have shown a spatial correlation between the flux of bulk carbon (or mass) and primary production. Three different curves from the literature are shown here (after Suess 1980, Betzer et al. 1984; Pace et al. 1987)

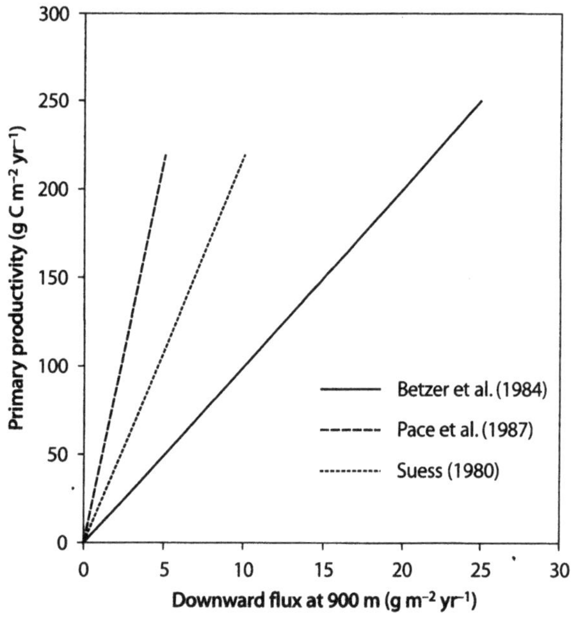

lated flux at depth to euphotic zone export (or new production) rather than to primary production (Martin et al. 1987; Pace et al. 1987). However, due to the effects of seasonality and complex food chains on flux, these models are not universally applicable.

Vertical fluxes of individual classes of biochemicals out of the surface ocean also directly reflect local primary production (Lee and Cronin 1984; Ittekkot et al. 1984a,b; Wakeham and Lee 1989, 1993). These relationships typically are much more variable than for organic carbon alone, reflecting their greater sensitivity to food-web dynamics, source and other factors.

Relationships between primary production and the fluxes of various classes of organic compounds and primary production are illustrated in Fig. 5.2. This comparison shows how different organic compound classes behave differently than total carbon. This difference could have several explanations. The flux of material from the euphotic zone reflects its production by phytoplankton; plankton growing in different areas could have varying relative amounts of certain biochemicals, for example, higher storage lipid contents in colder waters (Sargent 1976). Export from the euphotic zone is also dependent on the community of consumers present, because heterotrophs may preferentially degrade certain compounds over others. For example, bacteria may selectively degrade certain compounds compared to zooplankton. These factors also influence the variability of the flux-productivity relationship.

5.2.2 Temporal Relation

In addition to the spatial correlation of primary productivity with particle flux, there exists a well-defined temporal relationship. Close temporal coupling between phytoplankton blooms and particle flux maxima in the underlying water column clearly indicates rapid downward transport of biogenic debris (Deuser et al. 1981, Ittekkot et al. 1984a,b, Deuser 1986). This close coupling can be seen in data showing carbon fluxes to a 3 200-m sediment trap in the Sargasso Sea (Fig. 5.3). Fluxes peak each spring about a month after the pigment maxima produced during the spring bloom, suggesting particle sinking rates of about a hundred metres per day. This seven-year record illustrates the importance of long-term sampling for detecting changes that occur from year to year. Clearly an unusual event in 1981 triggered higher fluxes than those normally observed. Observations at this site continue today (http://www.bbsr.edu/cintoo/s202/s202_element/s202_element.html#flux) and have provided a wealth of data on seasonal and interannual variability in the North Atlantic. More recent work during the U.S. Global Ocean Flux Study (U.S. JGOFS) using time-series sediment traps has shown a similar general relation between flux and productivity over time and space in the Arabian Sea (Lee et al. 1998). However, temporal coupling between flux maxima and productivity events in the Arabian Sea is most pronounced when productivity is dominated by diatoms, indicating that biological community structure and/or composition strongly influence productivity-flux relations.

Specific organic compounds also show temporal productivity-flux relations. Amino acid and carbohydrate fluxes in the Sargasso Sea (Fig. 5.4) show the same close coupling with the spring peak in pigment concentrations observed for total organic carbon seen in the satellite data in Fig. 5.3. These relationships again will be affected by

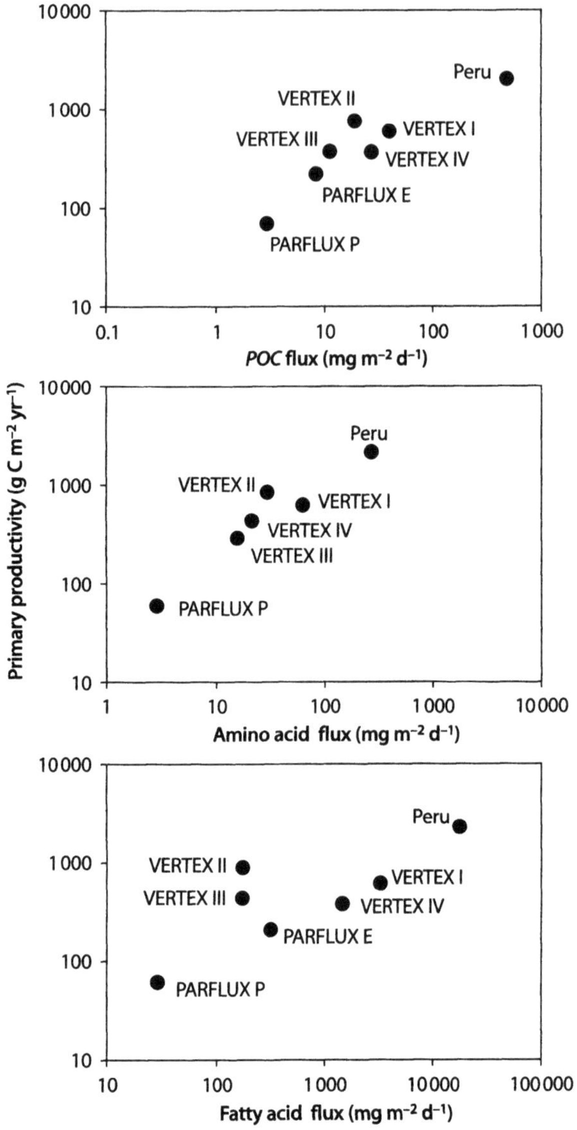

Fig. 5.2. Fluxes of *POC*, amino acids and fatty acids out of the euphotic zone (or the shallowest depth measured) in relation to primary production at various locations (reproduced from Wakeham and Lee 1993)

seasonal changes in biological community structure and in the production by phytoplankton of certain compounds relative to others.

Fluxes of specific organic compounds vary not only seasonally, but also on much shorter time scales. Diel patterns have been commonly observed for a wide variety of lipid compounds and amino acids as well as *POC* in the Peru upwelling area (Lee and Cronin 1982; Wakeham et al. 1983). Fluxes are usually much larger at night, and the particles collected then are dominated by faecal pellets, suggesting that these diel changes are related to zooplankton and anchovy vertical migration.

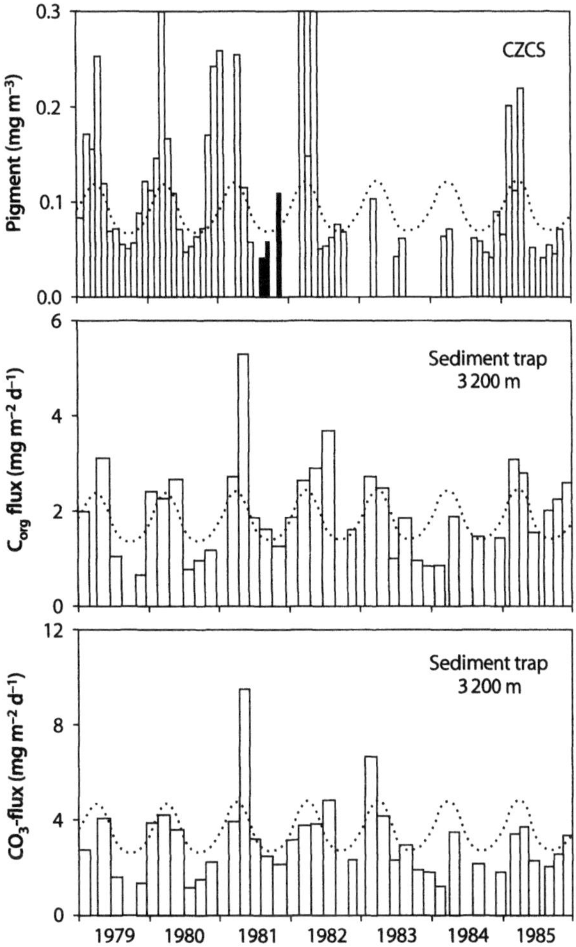

Fig. 5.3. Variation with time in fluxes of organic and inorganic carbon measured in a sediment trap deployed at 3200 m in the Sargasso Sea, compared to satellite (CZCS)-derived pigment concentrations for the same time period (reproduced from Deuser et al. 1990). *Dotted line* signifies average timing of high and low particle mass flux at 3200 m

Because the relation between productivity and flux is dependent on the production of organic matter by phytoplankton, one might not expect to see any specific relation between flux of non-biogenic inorganic materials and productivity. In fact, this is not the case. The major source of aluminum in particles is from terrestrial sources via atmospheric transport or direct land runoff. Yet, in the Sargasso Sea, which is roughly 1800 km from North America and 5000 km from Africa, Al clearly shows the same flux peak each spring as in organic carbon and the biogenic inorganic elements, Ca, Sr, Mg, I, and Ba (Fig. 5.5). These biogenic elements are constituents of skeletal material or organic matter of both phytoplankton and zooplankton, so their relationship with productivity over time is not surprising. Although fluxes of all these elements exhibit a seasonal variation, concentrations of the elements of terrigenous origin (K, Ti, La, V, and Co) do not correlate with *POC*; they are present at close to crustal abundance relative to Al. Elements that have fluxes that vary with productivity over time but with concentrations that do not vary with *POC* must have sinking mechanisms in

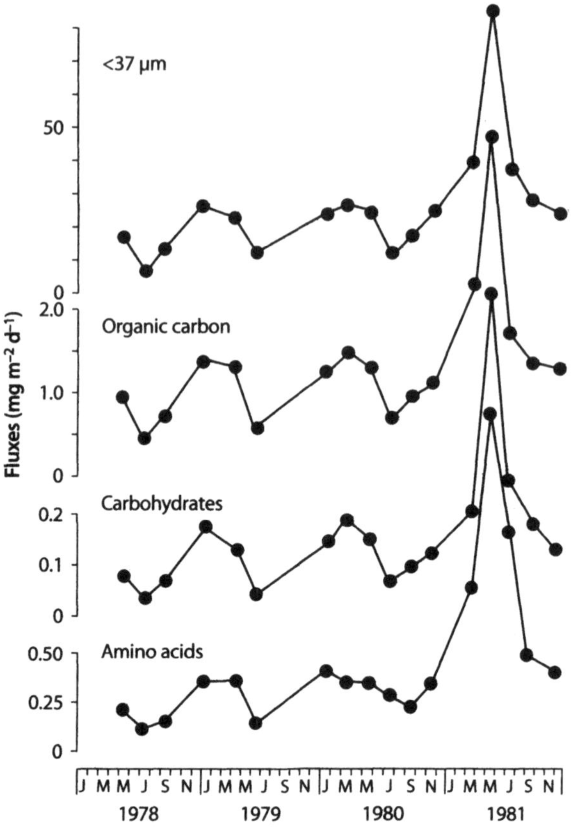

Fig. 5.4. Fluxes of mass, organic C and individual compound classes to a 3200 m sediment trap in the Sargasso Sea (data from Ittekkot et al. 1984a, reproduced from Wakeham and Lee 1993)

common but different sources. Their most likely source is from atmospheric transport of lithogenic material. As expected, concentrations of biogenic elements (Ca, Sr, Mg, I, and Ba) correlate with C and hence deviate from their crustal abundances. The strong relation between inorganic element flux and the spring bloom likely occurs because clays entering the ocean from terrestrial sources sink with marine snow and zooplankton faecal pellets that form after the bloom. Marine snow is a conglomeration of particulate matter onto a sticky amorphous matrix that may be derived, for example, from senescing diatoms, gelatinous zooplankton or from bacterial exoenzymes (Silver and Alldredge 1981; Honjo 1996). Lithogenic material that is blown into the open ocean can then adhere to this sticky material and sink with it. Clay particles can also be incorporated into faecal pellets by direct ingestion of the clay or material that the clay is sticking to.

5.3 Relation of Carbon Flux with Depth

In addition to varying spatially and temporally with primary productivity, the distribution of particulate organic matter varies with depth. Because of dissolution, disag-

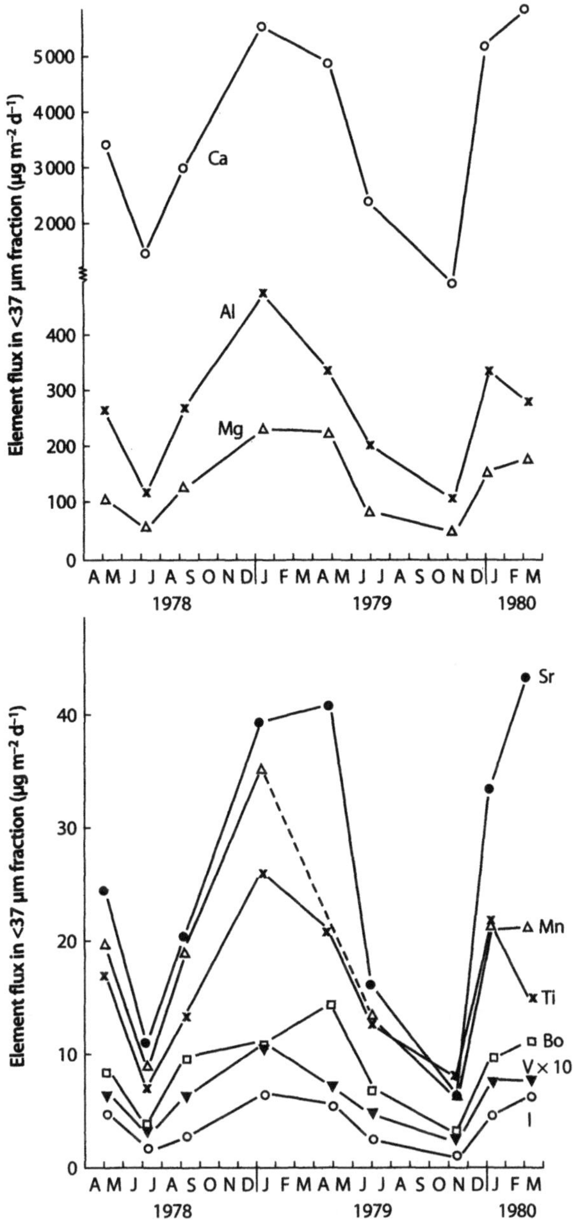

Fig. 5.5. Changes over time in the fluxes of inorganic elements in particles sinking to 3200 m in the Sargasso Sea (reproduced from Deuser et al. 1981)

gregation and heterotrophic consumption by zooplankton and bacteria, this variation is usually a decrease in flux with depth. As mentioned above, comparisons of particle rain rates at various depths over the entire marine water column clearly demonstrate that typically <1% of the organic matter produced by phytoplankton in the surface ocean reaches the sea floor in the open ocean (e.g. Suess 1980; Martin et al. 1987).

Measurements of individual compounds indicate that flux attenuation factors in the upper water column are consistent with the biochemical stability of different compound classes. Information on relative reactivities is useful both for judging the quality of organic matter at different stages of degradation and for more accurately applying biomarkers as source indicators. For example, at a site in the equatorial North Atlantic, *POC* flux estimated from sediment trap collections decreased by a factor of 5 between 400 and 5 000 m. In contrast, hydrocarbon fluxes decreased by a factor of 14, amino acids by 25, sterols by 45, and fatty acids by 60 (Lee and Cronin 1982; Wakeham 1982; Gagosian et al. 1982; de Baar et al. 1983; Wakeham et al. 1984; Wakeham and Lee 1993). These individual compounds are all more labile than bulk carbon and thus degrade faster. A large fraction of the *POC* cannot be readily identified at the molecular level by standard analytical techniques, and the proportion of this uncharacterized fraction in *POC* increases with depth (Wakeham et al. 1997).

The weight percent of organic matter in particles collected in sediment traps decreases with depth. At the same time, the fraction of carbonate and silicate increases (e.g. Table 5.1), as organic matter decomposes faster than these biogenic mineral phases dissolve. This leads to much denser particles with depth, as the mineral phases are denser than organic matter. Even though both %C and %N decrease with depth in the water column, the C/N ratio increases, because C is typically remineralized faster than N. This results from the comparatively greater lability of organic nitrogen compounds like amino acids relative to bulk carbon and more carbon-rich compounds. This varies between areas of different primary productivity. For example, elemental cycling results in an enrichment of C relative to N and P in open ocean areas compared to more eutrophic coastal waters (Table 5.2). Both organic nitrogen and organic phosphorus compounds are generally more labile than compounds without these heteroatoms. Even though fractions of both C and N generally decrease with depth, C/N and C/P are higher for particles exiting the euphotic zone and increase with depth in the open ocean more than in coastal areas. This results from more efficient open-ocean recycling, both within the euphotic zone and deeper in the water column. Thus material sinking out of the euphotic zone in open oceans is more degraded than that in coastal waters.

An example of the decrease with depth in *POC*, fatty acid and amino acid fluxes is shown in Fig. 5.6 with results of many studies by a number of investigators throughout the world oceans. Other less extensive flux measurements of carbohydrates, amino sugars, pigments, wax esters, triacylglycerols, and sterols show comparable trends with

Table 5.1. Composition of North Pacific Gyre sediment trap samples (from Honjo 1980)

Depth (m)	Carbonate (% of total mass)	Silicate	Organic	C_{org} (% of organic fraction)	H	N
378	35.1	5.2	59.5	52.3	7.8	6.8
978	72.1	11.7	16.2	45.1	5.7	5.7
2778	68.4	17.7	14.0	45.4	5.8	4.9
4280	71.6	17.6	10.7	48.9	5.8	5.3
5582	61.4	25.0	13.5	44.3	6.0	5.4

Table 5.2. Atomic ratios of C, N and P relative to P = 1 observed in sinking particles from the northeast Pacific Ocean (from Knauer et al. 1979)

Depth (m)	C_{org}	N	P	C_{org}/N
Coastal upwelling				
50	190	22	1	8.8
250	160	14	1	12
700	180	17	1	11
Coastal non-upwelling				
50	260	27	1	9.9
250	330	33	1	10
700	370	25	1	15
Open ocean				
75	410	29	1	14
575	460	34	1	13
1 050	910	31	1	29

depth (Ittekkot 1984a,b; Repeta and Gagosian 1984; Wakeham 1982; Gagosian et al. 1982). The *lines* shown in Fig. 5.6 are exponential regressions for flux data below 400 m. In analogy to half-lives, the $z_{1/2}$ *values* shown are calculated half-depths, the depth range over which half of the measured material has been lost. Compounds with smaller half-depths are more rapidly degraded. Thus, the amino acids with $z_{1/2}$ of 600 m are lost more quickly than fatty acids with $z_{1/2}$ of 1 400 m, and both are lost more rapidly than bulk carbon with $z_{1/2}$ of 1 700 m, as might be expected from their biochemical lability. This calculation of $z_{1/2}$ assumes first-order decomposition kinetics. If C is the flux of carbon, a compound class or an individual compound, then its loss with depth, z, is described as $-d[C] / dz = k[C]$. Thus $\ln [C] = -kz$, and $C = C_0 e^{-kz}$, where C_0 is the initial flux. Then the half depth, $z_{1/2}$, occurs when $C / C_0 = 1 / 2$. This sort of calculation makes many assumptions. First, we assume a constant sinking rate. In reality, different sized particles will sink at different rates. If we could assume that particle sinking rates, dz/dt, were uniform with depth, we could estimate a decomposition rate, $-d[C]/dt$. However, particles originate from various sources and include different amounts of ballast materials (usually minerals), so that their sinking rates will vary greatly with location and even with depth as their ratio of organic matter to mineral content changes during decomposition and dissolution (see discussion above). Estimates of sinking rates for open ocean particles average 100–200 m d^{-1} (Honjo 1996). Other assumptions are that all flux is vertically downward. There is some evidence that particles rich in lipid materials can float, and if the contribution of upward to total flux were substantial, estimates of half-depths would be biased (Smith et al. 1989; Wakeham et al. 2000). Although virtually all primary production is limited to the euphotic zone, chemosynthesis can occur in anoxic parts of the water column to form new organic matter from CO_2 (Karl et al. 1984). Individual compounds can also be synthesized from other compounds through alteration reactions (discussed below), resulting in local maxima or minima that are not easily modelled by first-order kinetics. A problem inherent to sediment trap methodology is that samples must be poisoned to prevent organic matter

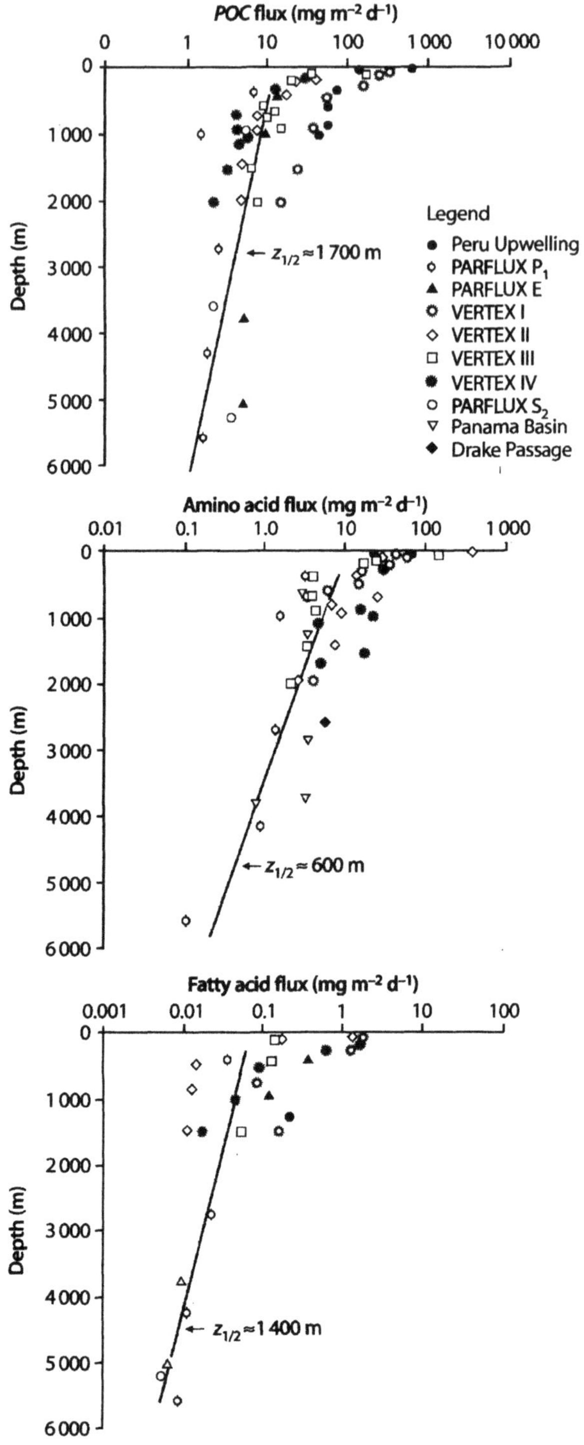

Fig. 5.6. Fluxes of *POC*, amino acids and fatty acids for many locations (reproduced from Wakeham and Lee 1993). PARFLUX P and VERTEX IV are in the central Pacific Gyre, PARFLUX E the equatorial North Atlantic, PARFLUX S the Sargasso Sea, VERTEX I the California Current, and VERTEX II and III eastern tropical North Pacific. *Lines* and $z_{1/2}$ are explained in text

decomposition before they are recovered. This treatment can result in an overestimated flux due to the collection of "swimmers," that is, zooplankton or fish that swim into the trap and die from contact with the poison (Lee et al. 1988). In shallow traps placed in coastal areas, much of the material collected can be swimmers. Technological changes in sediment traps address this problem (Peterson et al. 1993; Buesseler et al. 2000).

The preferential removal of certain components, both organic and inorganic, from sinking particles leads to major changes in their composition. Decomposition and transformation rates vary depending on the molecular structure of individual compounds and their availability as substrates for heterotrophic metabolism.

5.4
Compositional Changes During Degradation

5.4.1
Initial Composition

Just as primary productivity affects the amount of material sinking in the ocean, compounds produced by plankton influence the composition of organic matter in sinking particles. Some of the organic matter produced in the euphotic zone does sink deep in the water column, eventually reaching the sediment, and the composition of this particulate organic matter can vary significantly depending on it original source. The fact that some of the original composition is preserved is the basis of using "molecular biomarkers" as indicators of the source of sedimentary organic matter. Variation in the complex biochemical composition of living organisms provides a variety of natural biomarkers. The great number and chemical diversity of biomarkers among the biochemicals typical of living organisms provides a "fingerprint" that can be used to determine sources of organic matter (Lee and Wakeham 1989; Wakeham and Lee 1989). Such fingerprints are intrinsic characteristics of organic matter; as concentrations relative to total carbon or as individual compound concentration ratios, they can be applied without many of the problems inherent in estimates of sediment trap fluxes. For example, just the presence of certain fatty acids and sterols in particles from such biologically diverse oceanic regimes as the highly-productive Peru upwelling area and the oligotrophic North Central Pacific reflects the different plankton communities found in these areas (Wakeham and Lee 1989). In addition, specific lipids can clearly distinguish marine from terrestrial sources; marine hydrocarbons and fatty acids typically have shorter carbon chain-lengths than their terrestrial counterparts. Compositional differences among carbohydrates and amino acids can also reflect biological sources, although these compound classes are usually not as source-specific as lipids and pigments. Carbohydrates and amino acids also make up a substantial amount of the organic matter in marine organisms. Some organisms like siliceous diatoms are reportedly characterized by elevated percentages of glycine and fucose, while organisms with carbonate tests like coccolithophores often exhibit unusually high concentrations of aspartic acid and arabinose (Mitterer 1968; Hecky et al. 1973; Ittekkot et al. 1984a,b). Although both plankton and bacteria are characterized by high relative abundances of ribose and fucose (Cowie and Hedges 1984), O-methyl sugars and uronic acids may provide a means of discriminating between these two important sources of organic matter (Mopper and Larsson 1978). In addition, muramic acid and certain branched-chain fatty acids can serve as specific markers of bacteria (Lee et al. 1983; de Baar et al. 1983; Wakeham and Canuel 1988).

5.4.2 Diagenetic Indicators

Although phytoplankton producers initially dominate the organic composition of sinking particles, consumers can substantially alter this composition through degradation and alteration reactions. Not only can organic compounds be consumed, but waste products that are a structural portion of the original compound can be left behind. Or, bacteria colonizing a particle can synthesize new bacterial biomass either from the organic matter within the particle or by consuming *DOC*. A number of individual compounds indicate the freshness or diagenetic state of organic matter. For example, certain labile phytoplankton constituents, such as polyunsaturated fatty acids, are readily degraded in the environment and/or herbivore guts, and thus are depleted in more degraded particles (de Baar et al. 1983; Wakeham and Canuel 1988). Preferential loss of labile algal fatty acids resulting in the enrichment of more stable components in the products of heterotrophic metabolism has been observed both in field studies and laboratory feeding experiments (Prahl et al. 1985; Wakeham and Canuel 1988; Harvey et al. 1987). Similarly, high relative percentages of nitrogen and carbon in the form of chromatographically-measured amino acids (Whelan 1977; Lee and Cronin 1984; Cowie and Hedges 1992) and carbohydrates (Cowie et al. 1992; Cowie and Hedges 1992) are also indicative of relatively undegraded organic remains. Conversely, high relative concentrations of characteristic diagenetic products, such as the non-protein amino acids ornithine and β-alanine can indicate the presence of bacterially degraded material (Lee and Cronin 1982, 1984; Ittekkot et al. 1984a,b). Use of multiple compounds that indicate the stage of diagenesis provides a sensitive and consistent means of comparing the quality of organic matter in aquatic environments. For example, using a combination of only sixteen lipid, amino acid, pigment and carbohydrate compounds or groups of compounds, Wakeham et al. (1997) divided organic matter collected in the water column and sediment of the equatorial Pacific into four clearly-defined diagenetic classes that could be traced from sea surface to sediment (Fig. 5.7).

Freshness of organic matter can also be influenced by an intimately associated mineral or organic matrix. Resistant organic matrices may protect otherwise labile lipids. For example, waxy coatings common to land plants apparently protect higher plant alkanes, fatty acids, and fatty alcohols from degradation. Thus, terrestrial biomarkers are abundant in abyssal sediments, even though they are minor components of the particles produced in the overlying waters (Wakeham et al. 1984; Volkman et al. 1983; Gagosian et al. 1983). Marine-derived compounds seem to lack such a protective matrix and thus are preferentially degraded in the water column.

Several studies have reported that the adsorption of organic compounds serves to protect them from microbial degradation (Christensen and Blackburn 1980; Marshman and Marshall 1981; Gordon and Millero 1985). Mayer (1994) has recently postulated that adsorption of organic matter into micropores of inorganic sedimentary material might physically remove it from the action of hydrolyzing enzymes, which cannot function within the tiny pores. This mechanism may be responsible for the almost universal correlation between organic carbon content and mineral grain size observed in marine sedi-

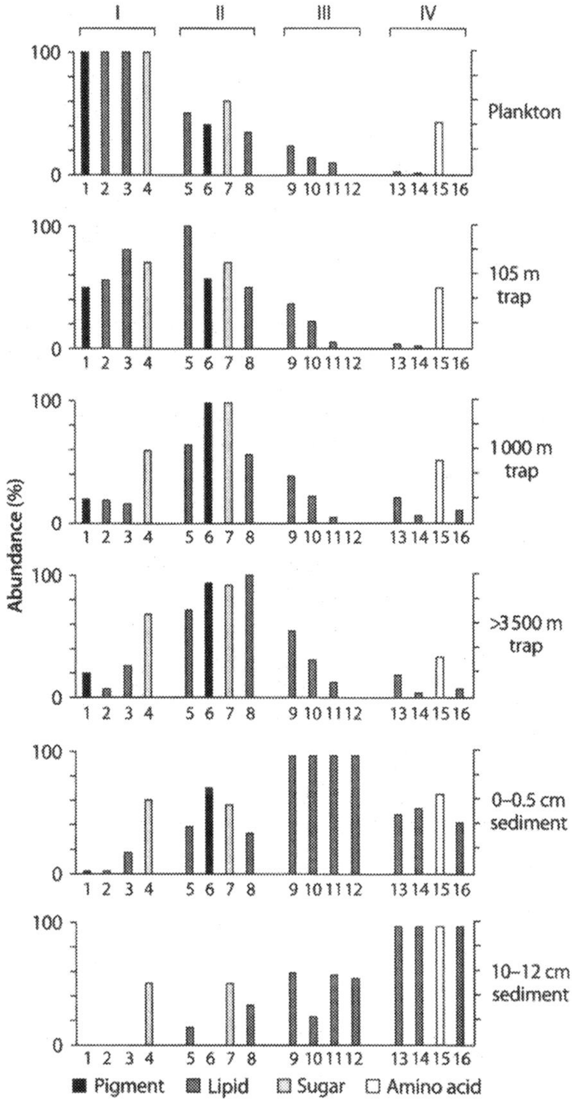

Fig. 5.7. Relative amounts of diagenetic indicators in plankton, sediment traps and sediment from the equatorial Pacific (Wakeham et al. 1997). Groups *I–IV* were categorized initially solely on the basis of their behaviour in the sample set, but clearly represent compounds of similar diagenetic fate. Compounds are: *1:* chlorophyll *a*; *2:* 22:6ω3 polyunsaturated fatty acid; *3:* 20:5ω3 polyunsaturated fatty acid; *4:* glucose; *5:* 24-methylcholesta-5,24(28)-dien-3β-ol; *6:* pheophorbide a; *7:* galactose; *8:* oleic acid; *9: cis*-vaccenic acid; *10: anteiso*-15:0 fatty acid; *11:* 24-ethyl-choles-5-en-3β-ol; *12:* bisnorhopane; *13:* summed $C_{24}+C_{26}+C_{28}+C_{30}$ *n*-fatty acids; *14:* summed $C_{37}+C_{38}$ methyl and ethyl, di- and tri-unsaturated alkenones; *15:* glycine; *16:* squalene

ments. Hedges and Keil (1995) developed this idea further and suggested that organic matter in association with mineral material beyond that equivalent to a monolayer coating was at least partially protected from decomposition, possibly due to protection from oxygen.

Preservation of certain amino acids and sugars occurs in both siliceous and carbonate tests, as mentioned above. Organic matter is incorporated into silicate and carbonate tests during biological deposition of these minerals. For example, diatom cell walls contain a protein-silica complex (Hecky et al. 1973), whose preservation is thought to account for the increase in glycine and serine commonly observed with depth in marine particles (Siezen and Mague 1978; Lee and Cronin 1984; Müller et al. 1986).

Carbonates are often enriched in acidic amino acids, which are thought to participate in the carbonate precipitation mechanism (Lowenstam and Weiner 1989). These acidic amino acids and glycine are the dominant compounds found in deep equatorial Pacific sediment traps and sediment (Lee et al. 2000). This material does not appear to be decomposed until it is released by dissolution of the mineral phase. The mechanism of association of organic matter to mineral grains is not well-understood and is probably multi-faceted. Simple adsorption undoubtedly plays a major role. If the mineral is terrestrially derived, adsorption could occur before deposition in the ocean; the organic coating could remain with the particle when the mineral grain enters the water, could be replaced by marine *DOC*, or further adsorption of marine *DOC* could occur. Recent work by Hedges and Keil (1999) and Keil et al. (1997) demonstrate this process. Berner (1995) suggested that benthic animals might coat fine-grained minerals as they pass through animal guts; this process could also occur in the water column as particles pass through zooplankton guts. And, as mentioned above, organic matter could be contained within the inorganic skeletal matrices of organisms.

Individual changes with depth also occur in inorganic components of sinking particles, with material of marine and terrestrial sources behaving differently. In an example from the deep North Atlantic Ocean (Table 5.3), biogenic elements like Ca, I and Ba in particles decreased in concentration with depth. Concentrations of other biogenic elements increased slightly (Sr and Mg), or considerably (Si), but still decreased relative to Al. Al and the other terrestrial elements (K, Ti, La, V, and Co) all increased in weight percent with depth. This trend reflects the stability of lithogenic material to dissolution; this component has already been subjected to various leaching and dis-

Table 5.3. Elemental composition of sediment trap samples collected in the North Atlantic Ocean (from Brewer et al. 1980)

Element	389 m	988 m	3755 m	5086 m
Si (%)	8.6	11.0	13.6	14.9
Al (%)	0.8	1.3	2.9	3.6
Ca (%)	21.7	20.5	20.9	18.3
Mg (%)	0.6	0.6	0.8	0.8
Fe (%)	0.3	0.8	1.6	1.9
K (%)	0.2	0.4	0.7	0.8
Mn (ppm)	56	78	330	460
Ba (ppm)		670	658	511
Ti (ppm)	537	927	614	2176
Sr (ppm)	1354	1514	1581	1250
Cu (ppm)		29	80	129
V (ppm)	11	36	52	66
I (ppm)	219	169	132	135
La (ppm)	6.8	14.8	17.8	27.9
Sc (ppm)	0.9	2.5	4.9	6.6
Zn (ppm)	600	440	340	470
Co (ppm)		2.5	7.9	11.2

solution processes on land and during transport. Most biogenic material, on the other hand, is formed in situ and readily begins to dissolve once the organism that formed it dies.

5.4.3 Heterotrophic Alteration

Degradation of organic matter in the sea occurs primarily through the action of bacteria and zooplankton. Biomarkers and diagenetic indicators can be used to distinguish between the presence and/or activity of each type of heterotroph.

5.4.3.1 *Bacteria*

Several organic compounds specific to bacteria such as muramic acid or *anteiso*-fatty acids can serve as specific source indicators of the presence of these organisms, as mentioned above. In the example from the equatorial Pacific shown in Fig. 5.7, several bacterial biomarkers were used to show the later stages (III and IV) of diagenesis, particularly *cis*-vaccenic acid and *iso*- and *anteiso*-15:0 fatty acids, which are known to have bacterial sources (Wakeham et al. 1997). These compounds were found in sinking particles throughout the water column but generally increased in relative abundance with depth. Bisnorhopane is also a bacterial biomarker, but is found only in sediments (Fig. 5.7). Bacterial biomarkers like muramic acid have been used to quantify bacterial biomass in particulate matter and sediments (King and White 1977; Moriarty 1977).

Other compounds, especially the organic nitrogen compounds, can be particularly useful as markers of bacterial alteration rather than indicators of bacterial biomass per se. For example, aspartic and glutamic acid, two commonly found amino acids in phytoplankton, decarboxylate as shown below to form non-protein amino acids that are not commonly found in phytoplankton and zooplankton. The ratios of protein to non-protein (*) amino acids (Fig. 5.8) have been used as indicators of bacterial activity in the Peru upwelling region (Lee and Cronin 1982) and the Sargasso Sea (Ittekkot et al. 1984a). These ratios can increase with depth in both water column particles and in sediments as bacteria degrade the original compounds. Another non-protein amino acid, ornithine, is a decomposition product of the protein amino acid arginine. Particulate ornithine fluxes increase with depth in the strong oxygen minimum off the coast of Mexico (Lee and Cronin 1984).

$$\mathrm{HOOC-CH_2-CH(NH_2)-COOH} \xrightarrow{-CO_2} \mathrm{H_2N-CH_2-CH_2-COOH}$$

Aspartic acid → β-alanine*

$$\mathrm{HOOC-CH_2-CH_2-CH(NH_2)-COOH} \xrightarrow{-CO_2} \mathrm{H_2N-CH_2-CH_2-CH_2-COOH}$$

Glutamic acid → γ-aminobutyric acid*

Fig. 5.8. Decarboxylation of the protein amino acids, aspartic acid and glutamic acid to the non-protein amino acids, β-alanine and γ-aminobutyric acid

5.4.3.2
Zooplankton

Consumption by zooplankton can lead to marked changes in composition between the phytoplankton food and excreted faecal pellets. Less than 5% of the carbon ingested by zooplankton ends up in faecal pellets (Copping and Lorenzen 1980), but pellets and other amorphous faecal material can be a significant part of the vertical flux, in some areas occasionally accounting for all of it (Honjo 1996). Some compounds are not altered extensively in composition during this process. Laboratory studies have shown, for example, that the amino acids in zooplankton faecal pellets are not very different from their algal food in composition (Cowey and Corner 1966; Tanoue et al. 1982). Lipid compounds are much better indicators of zooplankton feeding processes. Zooplankton edit the composition of their food by selectively removing some compounds while adding others. For example, in studies using calanoid copepods feeding on a dinoflagellate, Harvey et al. (1987) found that most fatty acids in the food were assimilated, but the polyunsaturated components were preferentially metabolized. Sterols were removed less than the fatty acids, but unsaturated sterols were again removed preferentially. In similar laboratory feeding experiments, Prahl et al. (1984) also found unsaturated fatty acids and sterols to be preferentially removed from a green alga eaten by a calanoid copepod. Zooplankton faecal pellets can also be enriched in individual lipids over those present in the algal food. For example, the sterol cholesterol and wax esters, which are for the most part absent in phytoplankton, are found in faecal pellets produced by herbivores (Volkman et al. 1980; Prahl et al. 1984). These compounds are even more enriched in carnivores over their food source.

Perhaps the most well-known alteration reaction mediated by zooplankton is the conversion of chlorophyll to its degradation products, particularly pheophorbide (Shuman and Lorenzen 1975; Welschmeyer and Lorenzen 1985). Similar alteration reactions during zooplankton grazing hydrolyze the algal carotenoid esters, fucoxanthin and peridinin, to the corresponding alcohols fucoxanthinol and peridininol, as shown in Fig. 5.9 (Repeta and Gagosian 1984). These four compounds are the major carotenoids transported to the sea floor in sinking particles.

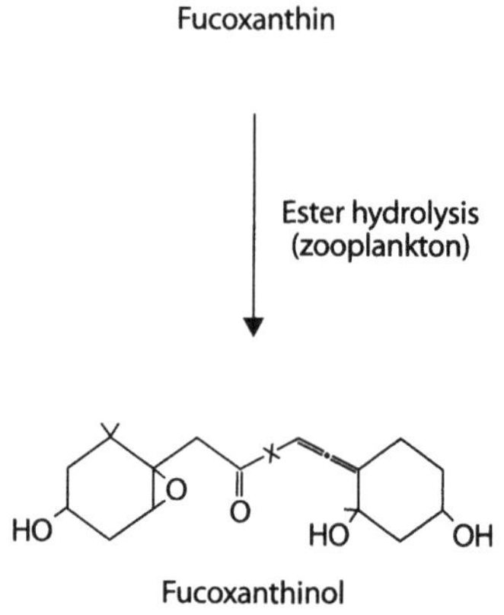

Fig. 5.9. Hydrolysis of the carotenoid pigment fucoxanthin to fucoxanthinol

5.4.4 Uncharacterized Material

In phytoplankton, most of the organic matter is made up of amino acids, carbohydrates and lipids that can be efficiently characterized at the molecular level with conventional chemical hydrolysis and chromatographic methods. Particles exported from the euphotic zone of the equatorial Pacific Ocean contain only a slightly lower proportion of characterizable material. However, the percentage of total organic matter in sinking particles that can be molecularly-characterized by conventional analysis decreases rapidly down the water column to become a minor component at depth (Fig. 5.10; Wakeham et al. 1997). Several hypotheses exist for this decrease in molecularly-resolvable organic matter during decomposition (Hedges et al. 2001). Labile intermediates released during microbial degradation of biomacromolecules (e.g. polysaccharides and proteins) may spontaneously recombine to form chemically complex "geopolymers" (de Leeuw and Largeau 1993). These geopolymers may be structurally too complex to analyze chemically, and they may increase preferentially with depth, because they do not degrade enzymatically. Alternatively, the biomacromolecules themselves may be in-

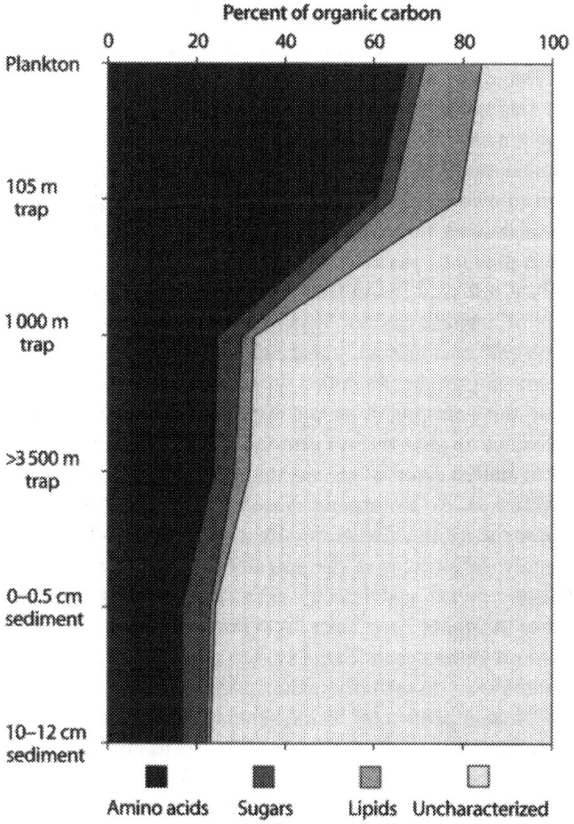

Fig. 5.10. Percent organic carbon in various biochemical classes within sinking particles in the equatorial Pacific Ocean (reproduced from Wakeham et al. 1997). Note the increase in the uncharacterized fraction

trinsically resistant to biodegradation and thus may also be selectively preserved. It is also possible that refractory organic or inorganic matrices may protect intrinsically labile organic substances (Knicker et al. 1996). A key difference among these protection mechanisms is that humification and selective preservation should result in pronounced changes in organic structure, whereas protection does not require major compositional changes.

Hedges et al. (2001) used solid-state ^{13}C NMR to investigate the composition of the bulk organic carbon used in the analyses shown in Figs. 5.7 and 5.10. They found that the major biochemical composition for bulk organic carbon was similar to that obtained by chromatographic analysis. This result fits more with the hypothesis of physical protection rather than selective preservation as described above. Several mechanisms could account for physical protection of organic matter by minerals. For example, as mentioned earlier, some organic matter is incorporated into silicate and carbonate tests during biological deposition of these minerals (Lowenstam and Weiner 1989). This material is entombed within the mineral matrix and will not be exposed to decomposing enzymes until it is released by dissolution of the mineral phase. Organic matter may also be protected on the surface of mineral particles if adsorption of the organic matter into. Mayer (1994) postulated that adsorption of organic matter into micropores of inorganic sedimentary material removes it physically from the action of large hydrolyzing enzymes. Inorganic matter makes up most of the total mass of sinking particles, so it would not be surprising if physical protection of the organic fraction by mineral components occurred.

The idea that minerals preserve associated organic matter and the fact that they are responsible for adding the ballast to particles that allows them to sink has suggested new concepts about organic matter fluxes in the ocean (Armstrong et al. 2002). Typical mixes of organic carbon compounds have densities 1.1 times that of sea water, while the density of typical inorganic mineral ballast is about 2.5 times that of sea water. Since sinking velocities are proportional to the excess in density over that of the fluid through which particles sink, a particle that is half organic matter and half inorganic ballast will sink many times faster than a particle of comparable size composed totally of organic matter. Wind-blown dust particles, opaline silica (produced by diatoms and radiolarians), and calcium carbonate (produced by coccolithophorids and foraminifera) are the major types of mineral ballast in the world ocean (Honjo 1996). Dust, for example, does not dissolve appreciably with depth, so that organic matter adsorbed to dust will be protected during its transit to the sea floor. In contrast, organic matter external or internal to opal and carbonate tests will be subject to decomposition as the biomineral dissolves. The mineral phase with which organic material is associated may therefore affect organic carbon flux in two ways: through mineral-specific differences in the amount of organic carbon that can be protected per unit ballast mineral, and through differential dissolution of the ballast minerals themselves. These thoughts have been incorporated into a new model of organic matter decomposition in the ocean (Fig. 5.11; Armstrong et al. 2002). Considering mineral ballast and protection, a quantitative description of *POC* remineralization must account both for *POC* that is "protected" by its association with ballast and for *POC* that is "unprotected" from degradation. Both types of *POC* are assumed to be associated with the same sinking aggregates (flocs and/or faecal pellets); the same ballast would then provide the excess density needed for both types of carbon to sink. Only further research on organic-inorganic associations will allow progress in making quantitative and predic-

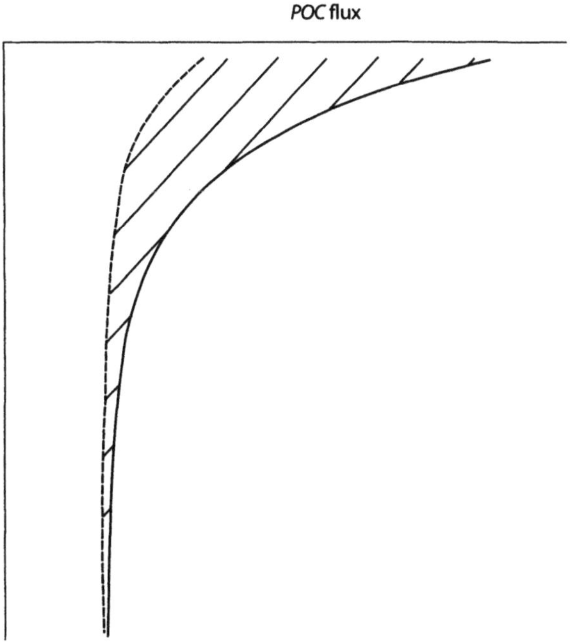

Fig. 5.11. Depth profiles of total *POC* flux (*solid line*) and protected organic carbon flux (*dashed line*) are illustrated in a schematic representation. Protected flux is assumed to be proportional to ballast flux. The *hatched area* between the two curves is the flux of unprotected *POC*. Neither protected nor unprotected *POC* flux are measured directly; their magnitudes are inferred by fitting measured values to a model (Armstrong et al. 2002)

tive models of organic matter flux that are useful over broad regions of the ocean and over seasonal and interannual time scales.

Acknowledgements

The author wishes to thank the Oceanographic Division of the U.S. National Science Foundation for supporting decades of study of particle fluxes, and Stuart Wakeham and John Hedges for their collaboration during these studies, and for constructive comments on this chapter.

References

Armstrong RA, Lee C, Hedges JI, Honjo S, Wakeham SG (2002) A new model for deep-ocean remineralization of organic carbon and mineral ballasts. Deep-Sea Res II 49:219–236

Baar HJW de, Farrington JW, Wakeham SG (1983) Vertical flux of fatty acids in the North Atlantic Ocean. J Mar Res 41:19–41

Berner RA (1995) Sedimentary organic matter preservation: an assessment and speculative synthesis – a comment. Mar Chem 49:121–122

Betzer PR, Showers WJ, Laws EA, Winn CD, DiTullio GR, Kroopnik PM (1984) Primary productivity and particle fluxes on a transect of the equator at 153° W in the Pacific Ocean. Deep-Sea Res 31:1–11

Brewer P, Nozaki Y, Spencer DW, Fleer AP (1980) Sediment trap experiments in the deep North Atlantic: Isotopic and elemental fluxes. J Mar Res 38:703–728

Buesseler KO, Steinberg DK et al. (2000) A comparison of the quantity and composition of material caught in a neutrally buoyant versus surface-tethered sediment trap. Deep-Sea Res 47:277–294

Christensen D, Blackburn TH (1980) Turnover of tracer (C-14, H-3 labeled) alanine in inshore marine sediments. Mar Biol 58:97–103

Copping AE, Lorenzen CJ (1980) Carbon budget of a marine phytoplankton-herbivore system with carbon-14 as a tracer. Limnol Oceanogr 25:873–882

Cowey CB, Corner EDS (1966) The amino acid composition of certain unicellular algae, and of the faecal pellets produced by *Calanus finmarchicus* when feeding on them. In: Barnes H (ed) Some contemporary studies in marine science. Allen and Unwin, London, pp 225–231
Cowie GL, Hedges JI (1984) Carbohydrate sources in a coastal marine environment. Geochim Cosmochim Acta 48:2075–2087
Cowie GL, Hedges JI (1992) Sources and reactivities of amino acids in a coastal marine environment. Limnol Oceanogr 37:703–724
Cowie GL, Hedges JI, Calvert SE (1992) Sources and relative reactivities of amino acids, neutral sugars, and lignin in an intermittently anoxic marine environment. Geochim Cosmochim Acta 56:1963–1978
Deuser WG (1986) Seasonal and interannual variations in deep-water particle fluxes in the Sargasso Sea and their relation to surface hydrography. Deep-Sea Res 33:225–246
Deuser WG, Ross EH, Anderson FR (1981) Seasonality in supply of sediment to the deep Sargasso Sea and implications for the rapid transfer of matter to the deep ocean. Deep-Sea Res 28:495–505
Deuser WG, Muller-Karger FE, Evans RH, Brown OB, Esias WE, Feldman GC (1990) Surface-ocean color and deep-ocean carbon flux: How close a connection? Deep-Sea Res 37:1331–343
Eppley RW, Peterson BJ (1979) Particulate organic matter flux and planktonic new production in the deep ocean. Nature 282:677–680
Gagosian RB, Smith SO, Nigrelli GE (1982) Vertical transport of steroid alcohols and ketones measured in a sediment trap experiment in the equatorial Atlantic Ocean. Geochim Cosmochim Acta 46:1163–1172
Gagosian RB, Volkman J, Nigrelli GE (1983) The use of sediment traps to determine sterol sources in coastal sediments off Peru. In: Bjøroy M et al. (eds) Advances in organic geochemistry 1981. John Wiley and Sons, New York, pp 369–379
Gordon AS, Millero FJ (1985) Adsorption mediated decrease in the biodegradation rate of organic compounds. Microb Ecol 11:289–298
Harvey JR, Eglinton G, O'Hara SCM, Corner EDS (1987) Biotransformation and assimilation of dietary lipids by *Calanus* feeding on a dinoflagellate. Geochim Cosmochim Acta 51:3031–3040
Hecky RE, Mopper K, Kilham P, Degens ET (1973) The amino acid and sugar composition of diatom cell-walls. Mar Biol 19:323–331
Hedges JI, Keil RG (1995) Sedimentary organic matter preservation: an assessment and speculative synthesis. Mar Chem 49:81–115
Hedges JI, Keil RG (1999) Organic geochemical perspectives on estuarine processes: Sorption reactions and consequences. Mar Chem 65:55–65
Hedges JI, Baldock JA, Gèlinas Y, Lee C, Peterson M, Wakeham SG (2001) Non-selective preservation of organic matter in sinking marine particles. Nature 409:801–804
Honjo S (1980) Material fluxes and modes of sedimentation in the mesopelagic and bathypelagic zones. J Mar Res 38:53–97
Honjo S (1990) Particle fluxes and modern sedimentation in the polar oceans. In: Smith WO (ed) Polar oceanography. Academic Press, New York, pp 687–739
Honjo S (1996) Fluxes of particles to the interior of the open oceans. In: Ittekkot V, Schaefer P, Honjo S, Depetris PJ (eds) Particle flux in the ocean. John Wiley and Sons, New York, pp 91–154
Ittekkot V, Deuser WG, Degens ET (1984a). Seasonality in the fluxes of sugars, amino acids, and amino sugars to the deep ocean: Sargasso Sea. Deep-Sea Res 31:1057–1069
Ittekkot V, Degens ET, Honjo S (1984b) Seasonality in the fluxes of sugars, amino acids, and amino sugars to the deep ocean: Panama Basin. Deep-Sea Res 31:1071–1083
Karl DM, Knauer GA, Martin JH, Ward BB (1984) Bacterial chemolithotropy in the ocean is associated with sinking particles. Nature 309:54–56
Keil RG, Mayer LM, Quay PD, Richey JE, Hedges JI (1997) Loss of organic matter from riverine particles in deltas. Geochim Cosmochim Acta 61:1507–1511
King JD, White DC (1977) Muramic acid as a measure of microbial biomass in estuarine and marine samples. Appl Environ Microbiol 33:777–783
Knauer GA, Martin JH, Bruland KW (1979) Fluxes of particulate carbon, nitrogen and phosphorus in the upper water column of the northeast Pacific. Deep-Sea Res 26:97–108
Knicker H, Scaroni AW, Hatcher PG (1996) ^{13}C and ^{15}N NMR spectroscopic investigation on the formation of fossil algal residues. Org Geochem 24:661–669
Lee C, Cronin C (1982) The vertical flux of particulate organic nitrogen in the sea: decomposition of amino acids in the Peru upwelling area and the equatorial Atlantic. J Mar Res 40:227–251
Lee C, Cronin C (1984) Particulate amino acids in the sea: Effects of primary productivity and biological decomposition. J Mar Res 42:1075–1097
Lee C, Wakeham SG (1989) Organic matter in sea water: Biogeochemical processes. In: Riley JP (ed) Chemical oceanography, vol IX. Academic Press, New York, pp 1–51
Lee C, Wakeham SG, Farrington JW (1983) Variations in the composition of particulate organic matter in a time-series sediment trap. Mar Chem 13:181–194

Lee C, Wakeham SG, Hedges JI (1988) The measurement of oceanic particle flux – Are swimmers a problem? Oceanography 1:34–36
Lee C, Wakeham SG, Hedges JI (2000) Composition and flux of particulate amino acids and chloropigments in equatorial Pacific seawater and sediments. Deep-Sea Res I 47:1535–1568
Lee C et al. (1998) Particulate organic carbon fluxes: Compilation of results from the 1995 US JGOFS Arabian Sea Process Study. Deep-Sea Res 45:2489–2501
Leeuw JW de, Largeau C (1993) A review of macromolecular organic compounds that comprise living organisms an their role in kerogen, coal, and petroleum formation. In: Engel MH, Macko SA (eds) Organic geochemistry. Plenum Press, New York, pp 23–72
Lowenstam HA, Weiner S (1989) On biomineralization. Oxford University Press, New York
Marshman NA, Marshall KC (1981) Bacterial growth on proteins in the presence of clay minerals. Soil Biol Biochem 13:127–134
Martin JW, Knauer GA, Karl DM, Broenkow WW (1987) VERTEX: Carbon cycling in the northeast Pacific. Deep-Sea Res 34: 267–285
Mayer LM (1994) Surface area control of organic carbon accumulation in continental shelf sediment. Geochim Cosmochim Acta 58:1271–1284
McCave IN (1975) Vertical fluxes of particles in the ocean. Deep-Sea Res 22:491–502
Mitterer RM (1968) Amino acid composition of organic matrix in calcareous oolites. Science 162:1498–1499
Mopper K, Larsson K (1978) Uronic and other acids in Baltic Sea and Black Sea sediments. Geochim Cosmochim Acta 42:153–163
Moriarty DJW (1977) Improved method using muramic acid to estimate biomass of bacteria in sediments. Oecologia 26:317–323.
Müller PJ, Suess E, Ungerer CA (1986) Amino acids and amino sugars of surface particulate and sediment trap material from waters of the Scotia Sea. Deep-Sea Res 33:819–838
Pace ML, Knauer GA, Karl DM, Martin JH (1987) Primary production, new production and vertical flux in the eastern Pacific Ocean Nature 325:803–804
Peterson ML, Hernes PJ, Thoreson DS, Hedges JI, Lee C, Wakeham SG (1993) Field evaluation of a valved sediment trap. Limnol Oceanogr 38:1741–1761
Prahl FG, Eglinton G, Corner EDS, O'Hara SCM, Forsberg TEV (1984) Changes in plant lipids during passage through the gut of *Calanus*. J Mar Biol Assoc UK 64:317–334
Prahl FG, Eglinton G, Corner, EDS, O'Hara SCM (1985) Faecal lipids released by fish feeding on zooplankton. J Mar Biol Assoc UK 65:547–560
Repeta DJ, Gagosian RB (1984) Transformation reactions and recycling of carotenoids and chlorins in the Peru upwelling region (15° S, 75° W). Geochim Cosmochim Acta 48:1265–1277
Sargent JR (1976) The structure, function and metabolism of lipids in marine organisms. In: Malins DC, Sargent JR (eds) Biochemical and biophysical perspectives in marine biology, vol III. Academic Press, New York, pp 149–212
Shuman FR, Lorenzen CJ (1975) Quantitative degradation of chlorophyll by a marine herbivore. Limnol Oceanogr 20:580–586
Siezen RJ, Mague TH (1978) Amino acids in suspended particulate matter from oceanic and coastal waters of the Pacific. Mar Chem 6:215–231
Silver MW, Alldrege AL (1981) Bathypelagic marine snow: Deep-sea algal and detrital community. J Mar Res 39:501–530
Smith KL, Williams PM, Druffel ERM (1989) Upward fluxes of particulate organic-matter in the deep north Pacific. Nature 337:724–726
Suess E (1980) Particulate organic carbon flux in the oceans – surface productivity and oxygen utilization. Nature 288:260–263
Tanoue E, Handa N, Sakugawa H (1982) Difference of the chemical composition of organic matter between fecal pellet of *Euphausia superba* and its feed. Trans Tokyo Univ Fish 5:189–196
Volkman JK, Corner EDS, Eglinton G (1980) Transformations of biolipids in the marine food web and in underlying bottom sediments. In: CNRS (ed) Colloques internationaux du CNRS No 293. Editions CNRS, Paris, pp 185–197
Volkman JK, Farrington JW, Gagosian RB, Wakeham SG (1983) Lipid composition of coastal marine sediments from the Peru upwelling region. In: Bjøroy M et al. (eds) Advances in organic geochemistry 1981. John Wiley and Sons, New York, pp 228–240
Wakeham SG (1982) Organic matter from a sediment trap experiment in the equatorial North Atlantic: Wax esters, steryl esters, triacylglycerols and alkyldiacylglycerols. Geochem Cosmochim Acta 46:2239–2257
Wakeham SG, Canuel EA (1988) Organic geochemistry of particulate matter in the eastern tropical North Pacific Ocean: Implications for particle dynamics. J Mar Res 46:183–213
Wakeham SG, Lee C (1989) Organic geochemistry of particulate matter in the ocean: The role of particles in oceanic sedimentary cycles. Organic Geochem 14:83–96

Wakeham SG, Lee C (1993) Production, transport, and alteration of particulate organic matter in the marine water column. In: Engel MH, Macko SA (eds) Organic geochemistry. Plenum Press, New York, pp 145–169
Wakeham SG, Farrington JW, Volkman JK (1983) Fatty acids, wax esters, triacylglycerols, and alkyldiacylglycerols associated with particles collected in sediment traps in the Peru upwelling. In: Bjøroy M et al. (eds) Advances in organic geochemistry 1981. John Wiley and Sons, New York, pp 185–197
Wakeham SG, Lee C, Farrington JW, Gagosian RB (1984) Biogeochemistry of particulate organic matter in the oceans: Results from sediment trap experiments. Deep-Sea Res 31:509–528
Wakeham SG, Lee C, Hedges JI, Hernes PJ, Peterson ML (1997) Molecular indicators of diagenetic status in marine organic matter. Geochim Cosmochim Acta 61:5363–5369
Wakeham SG, Cowen JP, Burd BJ, Thomsen RE (2000) Lipid-rich ascending particles form the hydrothermal plume at Endeavour Segment, Juan de Fuca Ridge. Geochem Cosmochim Acta 65:923–939
Welschmeyer NA, Lorenzen CJ (1985) Chlorophyll budgets: Zooplankton grazing and phytoplankton growth in a temperate fjord and the Central Pacific Gyres. Limnol Oceanogr 30:1–21
Whelan JK (1977) Amino acids in a surface sediment core of the Atlantic abyssal plain. Geochim Cosmochim Acta 41:803–810

Chapter 6

Diagenesis of Organic Matter at the Water-Sediment Interface

S. Wakeham

6.1 Introduction

The water-sediment interface is an intense heterotrophic reactor through which organic matter must pass if it is to be preserved in sediments. The term "water-sediment interface" is not a simple descriptor. Strictly speaking, it implies a geometric surface between the water column and the sediments, but to a biogeochemist it is the "zone in which organic matter is first accumulated from the water column and is initially metabolized by the sediment heterotrophic community" (Mayer 1993). The depth of this zone may range from millimetres to a metre or more, depending on the perspective of the biogeochemist and the processes involved.

The physical character of the interfacial zone is dictated by a combination of physical and chemical factors (Rhodes 1974). It is characterized by high water content, various forms of biological aggregations of mineral grains (faecal pellets, burrows etc.), horizontal and vertical mixing, resuspension, and elevated biological activity. Compared to underlying sediments, the interface is usually enriched in labile organic matter that has recently been delivered from the water column but has not yet been degraded. It is a very patchy environment on both macro- and microscales, and both chemically and biologically. Resuspension and bioturbation alter the topography and sedimentological character on a variety of time scales by mixing and redistributing particles. Living organisms are present at higher concentrations (on a volume basis) than in any other marine zone (Mayer 1993). Bacteria are present at 10^3–10^4 times greater numbers and activities than in the overlying water column (Deming and Baross 1993). Most organisms in marine sediments are heterotrophs that rely on the settling of particulate organic matter from the water column and its accumulation at the interface. Feeding by macrofauna physically mixes sediments, and the degradation of organic matter by bacteria using various electron acceptors produces the marked biogeochemical zonation and gradients (redox, organic carbon content, chemical composition etc.) commonly observed in surface sediment layers. Biogeochemical processes at the water-sediment interface consume >90% of the organic carbon (*OC*) that rains down upon it, leaving <0.1% of the *OC* produced in the water column to be preserved in sediments. Chemical gradients are steep in the interfacial zone. The composition of the preserved material can be markedly different from that which is delivered.

Most changes in the concentration and composition of sedimentary organic matter take place in this interfacial zone, and it is the combined processes affecting these changes that is termed "diagenesis". Most remineralization and transformation reactions that accompany diagenesis occur because the processes yield energy for benthic organisms. Organic matter diagenesis is rapid on the geological scale, with most decomposition occurring in the upper ten's of centimetres of the sediment column over time scales of months to centuries. The study of rates and mechanisms of organic matter alteration have broad implications related to energy transfer in the benthic community and organic carbon preservation and control of global atmospheric oxygen (Berner 1989).

This chapter describes some of the processes involved in diagenesis, using examples of alterations of organic matter that occur at the water-sediment interface and the types of information that these changes impart to organic biogeochemists.

6.2 Controls on Organic Matter Diagenesis

Organic matter remineralization in marine sediments (reviewed by Henrichs 1993) correlates closely with carbon rain rate, or the flux of *OC* to the sediments (Fig. 6.1). Rates of remineralization are higher in sediments receiving higher amounts of *OC*, but the efficiency of remineralization is actually lower in rapidly accumulating sediments than in slowly depositing sediments. This inefficiency allows greater amounts of organic matter to pass through the diagenetic zone and become buried, helping to explain why many organic-rich sediments are deposited under highly productive surface waters. Further observations have found that *OC*-rich sediments often are found under suboxic and anoxic waters. These findings have been interpreted as suggesting

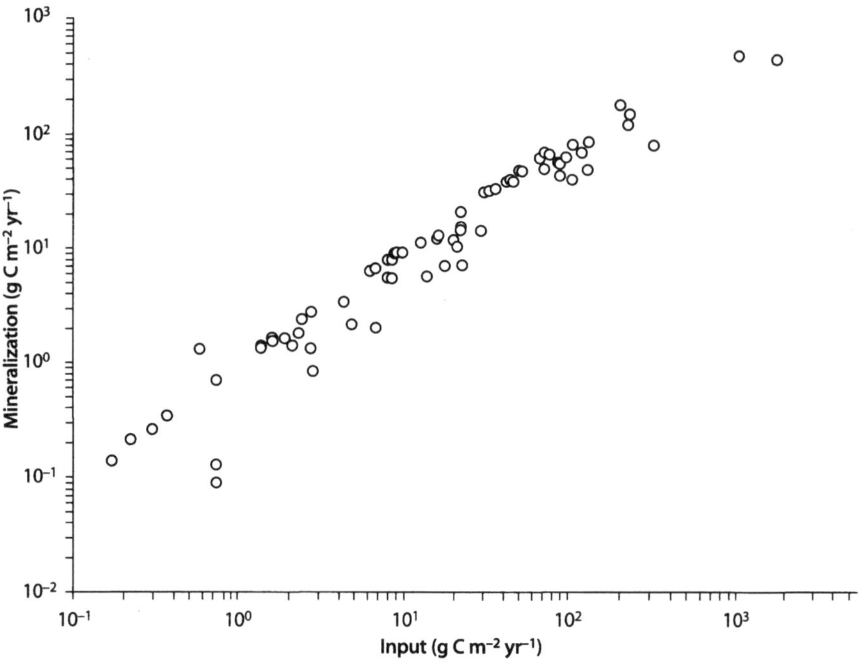

Fig. 6.1. Relationship between flux of organic carbon to the sea floor and organic carbon remineralization rate. Data are a compilation of measurements made globally (redrawn from Henrichs 1993)

that low bottom water oxygen concentrations retard decomposition, which then leads to accumulation of *OC* in sediments. The relative importance of these two factors – the supply of *OC* to the sea floor vs. bottom water oxygen concentration – has been hotly debated (e.g. Emerson and Hedges 1988; Calvert and Pedersen 1992; Lee 1992; Paropkari et al. 1992), and the debate has yet to be resolved (e.g. Keil and Cowie 1999; Cowie et al. 1999; Granesham et al. 1999). Nonetheless, there is reason to expect a coupling between the *OC* rain rate and water column oxygen, since higher availability of *OC* that can be remineralized leads to greater consumption of oxygen that produces suboxia and anoxia. An additional factor may be involved. Tegelaar et al. (1989) proposed that refractory macromolecules comprise the bulk of preserved organic matter. By this hypothesis, it is this recalcitrant macromolecular material that dominates the sedimentary *OC* that cannot be characterized at the molecular level (Wakeham et al. 1997; Hedges et al. 2000). In support of this concept is increasing evidence for a variety of biomacromolecules in algae and vascular plants that are resistant both to biodegradation and chemical analyses (de Leeuw and Largeau 1993). Thus, there may be molecular-structural factors that influence the intrinsic reactivity of organic substances.

Bacteria are major primary agents of early diagenesis (Deming and Baross 1993). They contain unique and versatile enzymes that can act on labile and refractory organic matter to satisfy their nutritional requirements. Acting in concert with each other and higher organisms, they produce powerful degradative systems beyond those of individual organisms. The classic description of diagenesis involves a stratification of sediments according to available oxidants that bacteria utilize in degrading organic matter to produce energy (Fig. 6.2). In reality, the situation is complicated by the fact that physiologically different types of bacteria degrade different classes of organic matter using specific oxidants, and there is considerable overlap and interaction of the zonation of the various respiratory functions, especially on the microscale. Because bacteria are unable to assimilate molecules larger than about 600 daltons, they must first hydrolyze complex molecules into simpler ones that can cross cellular membranes. End products from one functional group of bacteria are often substrates to be used by other groups in complex symbiotic relationships.

Recent research has expanded the number of mechanisms that may be involved in organic matter diagenesis to help explain changes in biochemical composition that occur during decomposition and preservation (reviewed by Hedges and Keil 1995). A sorption preservation hypothesis has been developed in which intimate association of organic matter with mineral grains protects *OC* from degradation (Mayer 1994a,b; Keil et al. 1994a, and earlier papers cited therein). Sorption of organic matter to mineral surfaces occurs in nature and is widely used to explain both the behaviour of hydrophobic xenobiotics in nature and why otherwise labile compounds are not always rapidly degraded in sediments. The new findings show that there is covariation of %*OC* and sediment surface area (*SA*) for surface sediments from a wide variety of continental margin environments (Fig. 6.3), suggesting stabilization of intrinsically labile organic matter if it is intimately associated with mineral grains. Mayer (1994a,b) originally hypothesized that organic matter coats mineral grains in continental margin sediments in a uniform "monolayer equivalent" coating of 0.5–1.0 mg $OC\,m^{-2}$. Sediments from non-continental margin areas have different %*OC*/*SA* relationships. Some fraction of the organic matter might be found in mesopores that protect it from degradation, because the mesopores are smaller than the hydrolytic exoenzymes released

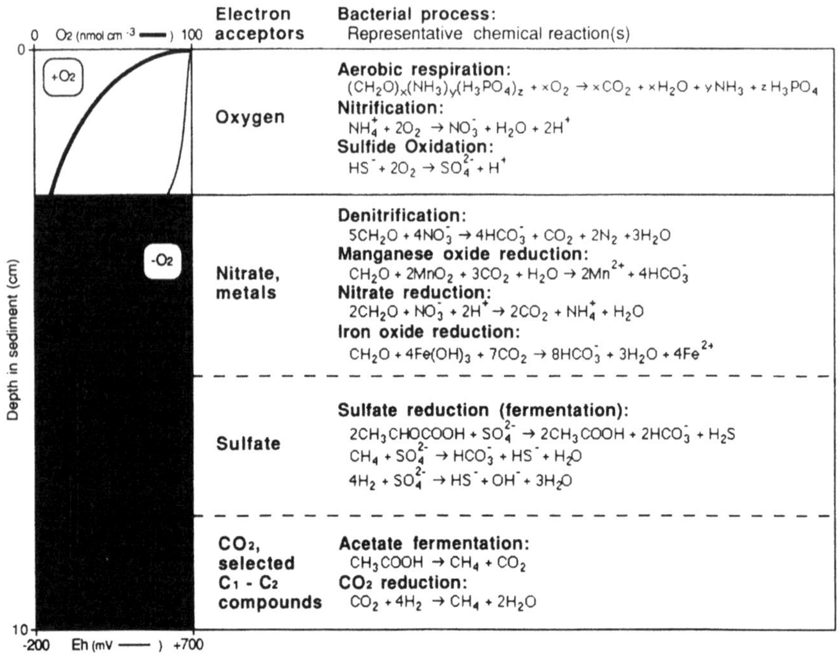

Fig. 6.2. A schematic of zones of organic matter degradation and respiratory functions of bacteria that occur during diagenesis of organic matter in sediments (from Deming and Baross 1993)

by bacteria. Recently, the mesopore hypothesis has given way to more discrete organic aggregates on mineral surfaces (Ransom et al. 1997, 1998; Mayer 1999). In either case, there appears to be a threshold below which sorbed organic matter cannot be attacked by exoenzymes. On the other hand, a surprisingly high proportion of organic matter sorbed to particles is reversibly bound. Experiments have shown that once desorbed, this organic matter is remarkably susceptible to microbial degradation (Wang and Lee 1993; Keil et al. 1994a). In addition, there are clear compositional variations as a function of particle size (Keil et al. 1994b, 1998; Bergamaschi et al. 1997) which may affect *OC* behaviour. For example, vascular plant-derived lignin phenols are enriched in the larger grain sizes consistent with plant debris being relatively large, while the finer sizes are characterized by higher ratios of vanillic acid/vanillin, indicating that the finer particles are more highly degraded. Hydrodynamic sorting of particulate matter on continental shelves results in selective deposition of vascular plant material in mid-shelf sediments, but little is transported further off-shore. Thus, the physical association of organic matter with mineral particles plays an important role in the degradation, preservation, and transport of organic carbon.

Of the bacterially-mediated chemical reactions depicted in Fig. 6.2, the most energetically favourable for bacteria are those in which oxygen is the electron acceptor. It follows that the extent of *OC* degradation (and preservation) in sediments is strongly controlled by the average time that *OC*-containing particles are exposed to pore water oxygen or the oxygen exposure time (*OET*) (Hartnett et al. 1998; Hedges et al. 1999).

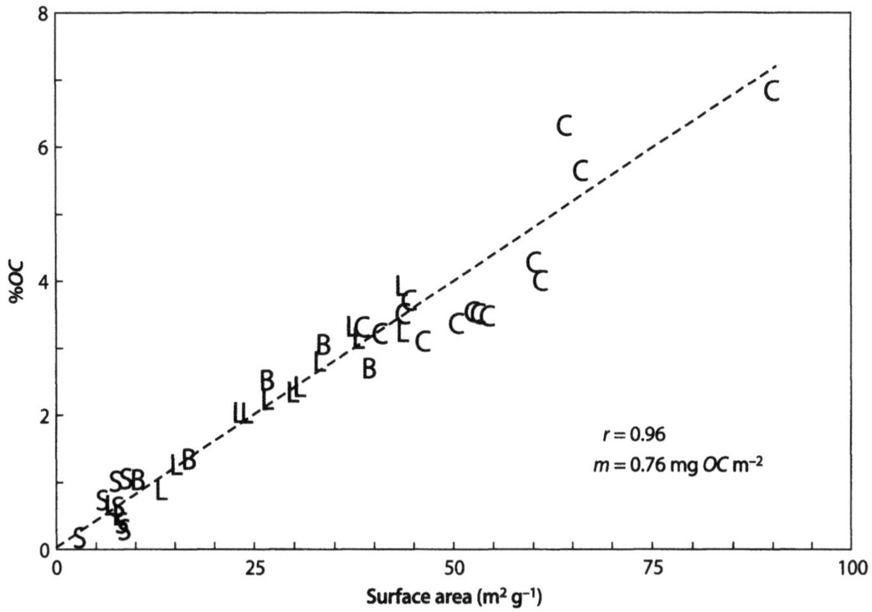

Fig. 6.3. Weight percent of organic carbon vs. mineral surface area for bulk (*B*), sand (*S*) silt (*L*), and clay (*C*) fractions of suspended sediment from the Columbia River estuary and the Washington continental margin (redrawn from Keil et al. 1994b; Hedges and Keil 1995)

Several lines of evidence support the *OET* hypothesis. Studies of a deep-sea turbidite characterized by an oxidation front (the Madeira Abyssal Plain (MAP) f-turbidite) clearly implicate molecular oxygen as a key agent in organic matter degradation (Keil et al. 1994c; Cowie et al. 1995). As the MAP f-turbidite was laid down some 140 000 years ago, the slumping sediment was thoroughly mixed mineralogically and presumably chemically. Following deposition, the upper half-metre of the turbidite was exposed to bottom water oxygen for about 10 000 years before being "capped" by the next deposit that returned the turbidite to a suboxic state. There has been marked degradation of organic matter in the oxidized zone compared to the unoxidized zone. Remineralization across the oxidation front of the MAP f-turbidite has removed >75% of the organic carbon initially deposited, producing sharp organic gradients across the redox front. The role of O_2 in organic matter degradation has been strengthened by further study on recent continental margin sediments (Hedges and Keil 1995; Hartnett et al. 1998; Hedges et al. 1999). In a transect across the Washington continental margin, slope and adjacent abyssal plain, measurements of penetration of O_2 into surface sediment along with sediment accumulation rates allowed calculation of *OET* (Hedges et al. 1999). There was a marked increase in *OET* in the farther off-shore sediments (Fig. 6.4) that corresponded with decreased *OC* concentrations and organic carbon/surface area ratios. Combined with molecular analyses, these results clearly indicate that exposure of *OC* to oxygen and hence organic matter degradation increased off-shore. Although the implication is that molecular O_2 ins the main oxidant, other electron acceptors (e.g. Fe and Mn) might also be involved and produce similar trends.

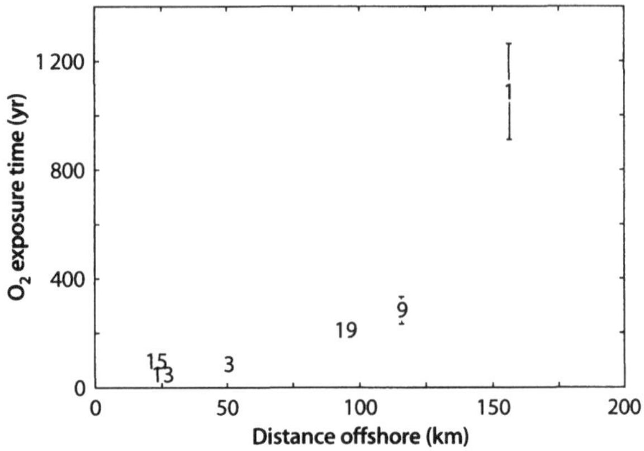

Fig. 6.4. Oxygen exposure times (*OET*) calculated for surface sediments off the coast of Washington State plotted vs. distance off-shore. Numbers refer to station numbers (redrawn from Hedges et al. 1999)

The relative reactivity or recalcitrance of organic matter seems to depend on intrinsic molecular structure. Unsaturated compounds tend to be more rapidly degraded than saturated ones; straight-chained molecules are more reactive than branched and cyclic compounds; the presence of heteroatoms (N, S, and O) sometimes inhibits decomposition. To better explain these trends, Hedges and Keil (1995) have described an oxygen-sensitive organic matter (OSOM) model in which there are (at least) three forms of organic matter in sediments. Materials such as charcoal are totally refractory. An oxygen sensitive fraction, lignin for example, degrades slowly in the presence of oxygen but not at all under anoxic conditions. Hydrolysable molecules such as proteins and polysaccharides are readily degraded regardless of conditions. Some combination of remineralization of these three types of *OC*, albeit on very different time scales, is additive in proportion to their abundance, generating the *OC* profiles observed in sediments (Fig. 6.5). The OSOM model is an extension of earlier work (Skopintsev 1981) that had suggested that algal organic matter did not degrade as a single pool but rather as a composite of multiple rates among the various classes of organic components. Subsequent work by Westrich and Berner (1984) led to a multi-first order model (the multi G model) that described decomposition of multiple pools of carbon. Experimental evidence has strengthened the multi G model (e.g. Hedges and Keil 1995; Sun and Wakeham 1994) and it is now widely used by biogeochemists. Examples of diagenetic trends in molecular indicators are discussed below. Note that the OSOM model relies on molecular properties as a key determinant of organic degradation rates, whereas the sorption model is largely based on sediment surface area, although the two are certainly not mutually exclusive.

6.3 Compositional Changes Resulting from Organic Matter Diagenesis

Diagenetic alteration of organic matter can be tracked using bulk elemental parameters or by the behaviour of organic biomarkers in the sediments. Elemental compositions provide information on behaviour of bulk organic matter but offer relatively little insight into the fate of different biochemical fractions of the bulk material. Organic biomarkers offer the advantage of often being source-specific (de Leeuw and

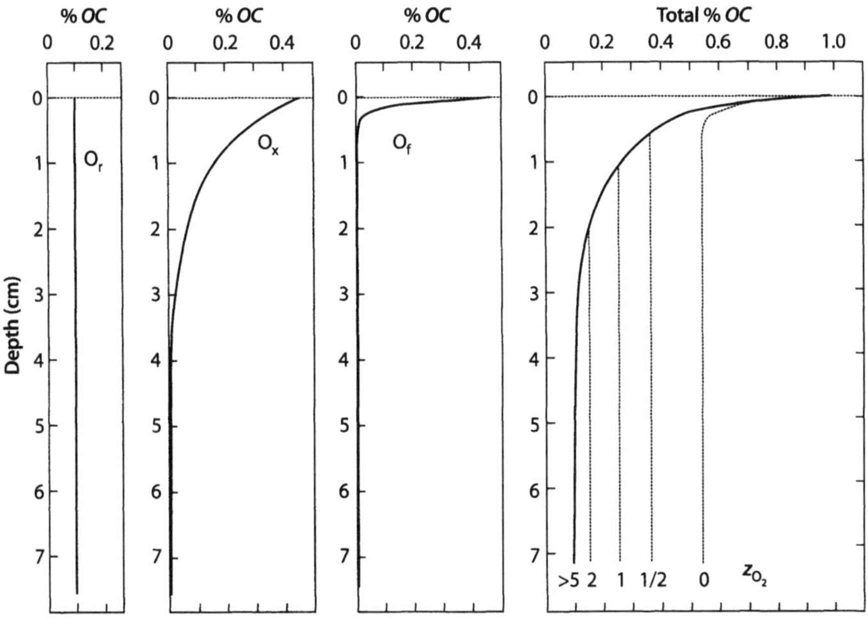

Fig. 6.5. Model-derived profiles of percent organic carbon for three different organic carbon fractions, including refractory (O_r), oxygen sensitive (O_x), and fermentable (O_f) fractions (redrawn from Hedges and Keil 1995)

Largeau 1993), and because analyses are made at the molecular level, it is often possible to follow the fate of specific compounds as they are either being degraded, produced, or structurally-altered during diagenesis. On the other hand, biomarkers often represent a very small fraction of the overall organic pool so that the picture obtained may be narrow in scope.

6.3.1 Elemental Compositions

Biogeochemists have long recognized that bulk organic carbon (*OC*) and total nitrogen (*TN*) concentrations and their atomic ratio [(C/N)a] could be used as indicators of diagenesis (e.g. Müller and Suess 1979). Typically *OC* and *TN* weight percentages decrease with increasing depth of burial as organic matter is remineralized. The atomic ratio C/N(a) increases as nitrogen-containing species are preferentially used relative to carbon.

An example of trends in *OC* and *TN* attending diagenesis in surface sediments and the potential role of oxygen availability in degradation can be found in a series of cores collected on the continental margin of the Black Sea (Cowie and Hedges 1991). Five box cores were collected along a 10 km transect that crossed the chemocline (redox boundary between oxic surface waters and anoxic bottom waters), thus giving a range of bottom water oxygen concentrations (Fig. 6.6). The assumption in this comparison, based on the proximity of the set of cores to each other, is that the organic matter

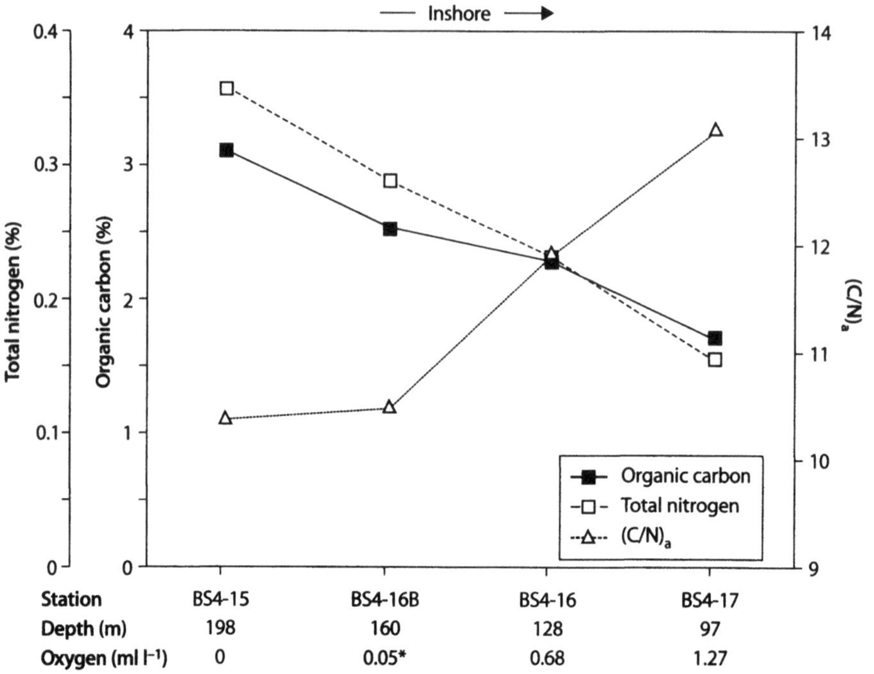

Fig. 6.6. Organic carbon and total nitrogen concentrations and atomic C/N ratios for fluff samples from sediment cores spanning the oxic/anoxic interface in the Black Sea. Also shown are the water depths and bottom water dissolved oxygen concentrations overlying the cores (redrawn from Cowie and Hedges 1991)

delivered to the sediments is uniform across the transect and differences in organic composition reflect different processes as a function of oxygen content. Decreasing *OC* and *TN* concentrations and increasing C/N(a) of interfacial "fluff" samples from the deepest, anoxic (BS4-15) to the shallowest, most oxygenated (BS4-17) coring stations are consistent with typical diagenetic trends. Accordingly, the extent of degradation in the fluff material varied with oxygen availability. *OC* and *TN* concentrations were also determined as a function of depth in the five cores (Fig. 6.7). *OC* and *TN* generally decreased with depth in the core and with bottom water oxygen present. For all five cores, there were substantial decreases in wt% *OC* and *TN* between the fluff layer and the underlying 0–0.5 cm sediment, confirming that most degradation takes place near the water-sediment interface.

Laboratory simulations have been useful for evaluating changes in elemental composition during organic matter diagenesis and possible effects of oxic and anoxic conditions on decomposition, although results regarding oxygen have been equivocal (Henrichs 1993). In a recent study, Harvey et al. (1995) examined the decay of two marine phytoplankton, the diatom *Thallasiosira weissflogii*, and the cyanobacterium *Synechococcus* sp., by measuring particulate organic carbon (*POC*), particulate organic nitrogen (*PON*) and biochemical compositions (discussed below) in oxic and anoxic treatments over the 3-month time-course of the experiment. First order decay con-

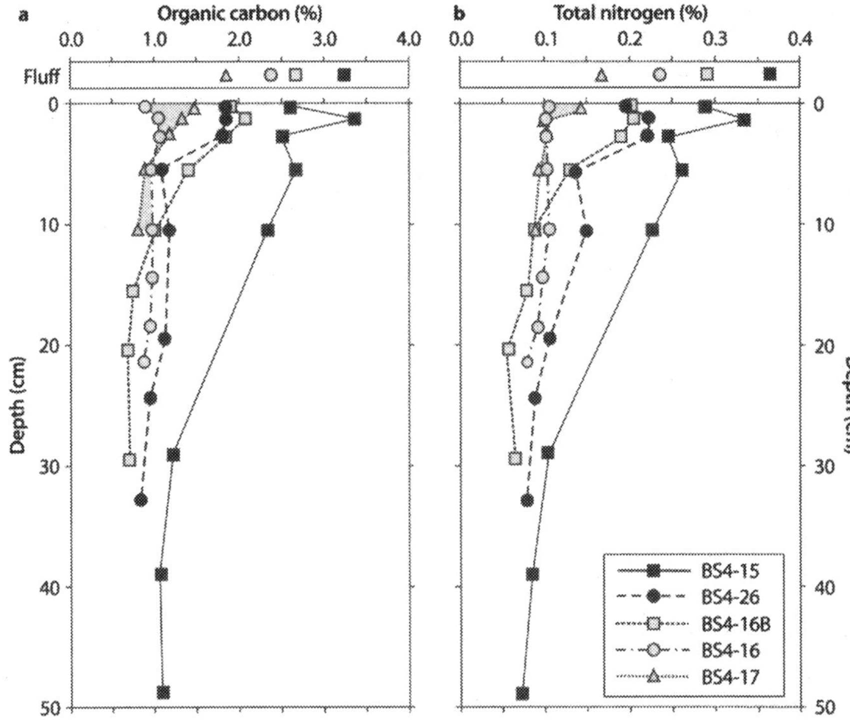

Fig. 6.7. Elemental compositions for Black Sea cores along a transect crossing the chemocline; **a** %*OC*; **b** %*TN* (redrawn from Cowie and Hedges 1991)

stants were significantly higher for both *POC* and *PON* under oxic conditions, and loss of nitrogenous material occurred preferentially to loss of carbon (Table 6.1). The absence of oxygen clearly resulted in slower decay for the phytoplankton and incomplete decomposition of the organic carbon pool, despite the observation that both bacterial abundances and metabolism were similar in the two treatments.

A very different approach was used by Lee (1992). Radiolabelled tracer compounds were used to compare aerobic and anaerobic microbial metabolism of organic substances. Differences in observed intrinsic rates of decomposition between the systems were small, but different compounds had very different labilities, suggesting a molecular structural effect. A significant observation was that high decomposition rates could occur when rate constants are low in situations when substrate concentrations are high. Thus, it is necessary to know both substrate concentrations and composition when decomposition rates are being calculated. Differences in apparent organic matter degradation between oxic and anoxic systems were postulated to be related to the presence of protozoa and meiofauna that graze bacteria (i.e. the microbial loop in sediments) in oxic systems and thus help increase the efficiency of remineralization. The absence of such grazers in anoxic systems may allow for greater sequestration of organic matter as bacterial biomass, thus leading to greater carbon preservation.

Table 6.1. Experimentally determined first order decay constants (k, yr^{-1}) and regression coefficients for particulate organic carbon, nitrogen and major biochemical fractions during oxic and anoxic phytoplankton decay. Data from Harvey et al. (1995)

Biochemical fraction	Oxic		Anoxic		
	k (yr^{-1})	r^2	k (yr^{-1})	r^2	Ratio O/A
T. weissflogii					
POC	12.8	0.90	2.9	0.77	4.4
	(19.5)[a]	0.97	(5.3)[a]	0.88	3.7
PON	17.3	0.95	3.5	0.88	4.9
	(24.0)[a]	0.98	(5.4)[a]	0.93	4.4
Lipid	8.3	0.88	2.7	0.60	3.1
	(13.1)[a]	0.96	(5.5)[a]	0.80	2.5
Protein	21.1	0.97	15.3	0.83	1.4
Carbohydrate	33.7	0.88	7.2	0.88	4.7
Synechococcus sp.					
POC	15.3	0.91	2.5	0.85	6.2
PON	14.7	0.74	3.0	0.90	4.9
Lipid	8.2	0.82	2.3	0.79	3.6
Protein	22.0	0.85	6.2	0.90	3.5
Carbohydrate	34.2	0.84	4.1	0.65	8.3

[a] Initial decay constants observed over first 41 d for oxic decay and first 92 d of anoxic incubations.

6.3.2 Biomarkers

Compound-specific biochemical compositions provide further evidence of the effects and degree of diagenetic alteration of organic matter. Like %*OC* and *TN*, concentrations of amino acids, sugars and lipids generally decrease over time in laboratory experiments or with depth in the sediment. Normalizing amino acid, sugar and lipid concentrations to %*OC* provides a means of gauging the relative reactivities of these biochemical classes compared to *OC*. Doing so clearly shows that the classes are usually more labile than bulk *OC* – implying some unidentified fraction that is more stable in the bulk *OC* – and that they exhibit differential reactivities relative to one another (Cowie 1990; Cowie and Hedges 1991; Lee 1992; Harvey et al. 1995). For example, in the Black Sea study of Cowie (1990), amino acids were preferentially degraded relative to *OC* in the fluff samples, but sugars were less affected. Amino acid and sugar concentrations were distinctly higher in fluff layers than in underlying sediments and concentrations gradually decreased with depth and under oxygenated bottom waters, providing further evidence of in situ sedimentary diagenesis. In the Harvey et al. (1995) laboratory study, carbohydrates were decomposed most rapidly under oxic conditions followed by protein and then lipids (Table 6.1). Under anoxic conditions, proteins were most labile. The ratio of carbohydrate:protein metabolism decreased from 15 during aerobic decomposition to 3 during anaerobic decomposition. These trends were interpreted as resulting from a metabolic shift in microbial community under anoxic conditions in favour of protein metabolism over that of carbohydrates.

In a follow-up of the Harvey et al. (1995) study described above, Harvey and Macko (1997) reported on the kinetics of lipid decay at the molecular level. Aerobic degradation had a significantly greater effect on total lipids and individual compounds than anaerobic decomposition (Table 6.2). At the individual compound level, there were significant differences in behaviour that appear related in part to molecular structure. Unsaturated lipids degraded faster than saturated counterparts, and in the case of polyunsaturated fatty acids, removal was complete under both oxic and anoxic conditions. Unsaturation is certainly an important control that regulates the lability of many

Table 6.2. Experimentally determined first order decay constants (k, yr^{-1}) and regression coefficients for particulate organic carbon, total lipid and major lipid classes during oxic and anoxic decay of diatom and cyanobacterial cells* (Data from Harvey and Macko 1997)

Biochemical fraction	Oxic		Anoxic		Ratio O/A
	k (yr^{-1})	r^2	k (yr^{-1})	r^2	
T. weissflogii					
POC	12.8	0.90	2.9	0.77	4.4
	(19.5)[a]	0.97	(5.3)[a]	0.88	3.7
Total lipid	8.3	0.88	2.7	0.60	3.1
	(13.1)[a]	0.96	(5.5)[a]	0.80	2.5
Algal fatty acids	30.1*	0.83	9.0*	0.84	3.4
Saturated	25.7	0.86	8.2	0.88	3.1
Monounsaturated	37.4	0.68	9.3	0.80	4.0
Polyunsaturated	34.3	0.82	13.6	0.82	2.5
Sterols	33.4	0.93	2.6	0.81	12.8
Phytol	31.9	0.88	3.0	0.64	10.6
Alkenes (21:6 – 21:5)	41.6	0.73	10.5	0.88	4.0
Synechococcus sp.					
POC	15.3	0.91	2.5	0.85	6.2
Total lipid	8.2	0.82	2.3	0.79	3.6
Algal fatty acids	23.0	0.84	4.2	0.88	5.5
Saturated	23.1	0.82	3.3	0.82	7
Monounsaturated	24.3	0.59	9.3	0.78	2.6
Phytol	15.2	0.82	3.4	0.75	4.5
Heptadecene	7.4	0.32	6.4	0.82	1.2

Regression calculations are based on cell death at day 5 and day 9 for *T. weissflogii* and *Synechococcus* sp., respectively.

[a] Rate observed over the first 41 d for oxic and 92 d for anoxic incubations.

lipids. Sterols were generally more stable than fatty acids and displayed the largest reduction in degradation rate when oxygen was absent. These trends could help account for the fact that unsaturated lipids are particularly vulnerable to loss, and ratios of fatty acid:sterol decrease during diagenesis as observed in field studies (e.g. Kawamura et al. 1980; Sun and Wakeham 1994; Canuel and Martens 1996). Remarkably, degradation patterns for lipids common to the two phytoplankton used by Harvey and Macko (1997) (*Thallasiosira weissflogii* and *Synechococcus* sp.) were not always similar, suggesting that factors other than molecular structure might affect the degradation rate. Furthermore, the proportion of total lipids that could be identified and quantified as individual compounds decreased with the degree of degradation, with the result that only a small fraction (18–24%) of lipids could be identified after the majority of *POC* and lipids were degraded. This loss of identifiable biochemicals has been described previously (e.g. Wakeham et al. 1997; Hedges et al. 2000), to the extent that the origin, reactivity and fate of a large amount of *OC* remains obscure.

Diagenesis not only results in absolute loss of organic compounds via degradation, but it is accompanied by changes in ratios of compounds as a function of their relative lability or due to in-growth of diagenetic products at the expense of precursors. Organic matter degradation is accompanied, for example, by decreases in weight percentages of glucose among total aldoses, but increases in percentages of the deoxy sugars, rhamnose and fucose (Hamilton and Hedges 1988; Cowie et al. 1992). Increases in mole-percentages of the non-protein amino acids β-alanine plus γ-aminobutyric acid (Henrichs et al. 1984; Cowie et al. 1995) also occur. In a detailed study of amino acids in sediments from the Peru upwelling region, Henrichs et al. (1984) detected high concentrations of β-aminoglutaric acid, a nonprotein isomer of glutamic acid. Although a specific source was not identified, increasing ratios of β-aminoglutaric acid:glutamic acid with depth in sediment cores (Fig. 6.8) strongly suggested an in situ bacterial source.

Among lipids, ratios of unsaturated to saturated fatty acids typically decrease down core (Fig. 6.9), because as noted above, unsaturated fatty acids that are derived from phytoplankton are more susceptible to degradation in sediments than are saturated acids (Kawamura et al. 1980; Sun and Wakeham 1994; Canuel and Martens 1996). From profiles such as these, it is uncertain whether the unsaturated fatty acids are oxidized partially and removed from the fatty acid "analytical window," remineralized completely to CO_2, or hydrogenated in situ to their unsaturated analogs. Branched-chain fatty acids (e.g. *iso*- and *anteiso*-C_{15} and C_{17}) that are common in bacterial cell membranes (Kaneda 1991) are often found to increase in relative abundance in sediments as diagenesis progresses. Enrichments in branched-chain fatty acids are widely used as indicators of bacterial alteration of organic matter (Perry et al. 1979). However, the reliance on branched-chain fatty acids as bacterial indicators must be tempered by the fact that many aerobic bacteria do not produce these compounds, so in situations where aerobic microbial decomposition is intense, branched-chain fatty acids may not be particularly abundant (Wakeham 1995). Branched-chain fatty acids at best may be semiquantitative indicators of bacterial biomass but are poor indicators of bacterial activity.

Sterols are popular biomarkers, because as a group they represent a wide variety of molecular structures that are remarkably useful as source indicators and because alterations to the sterol skeleton can be readily followed with the analytical tools avail-

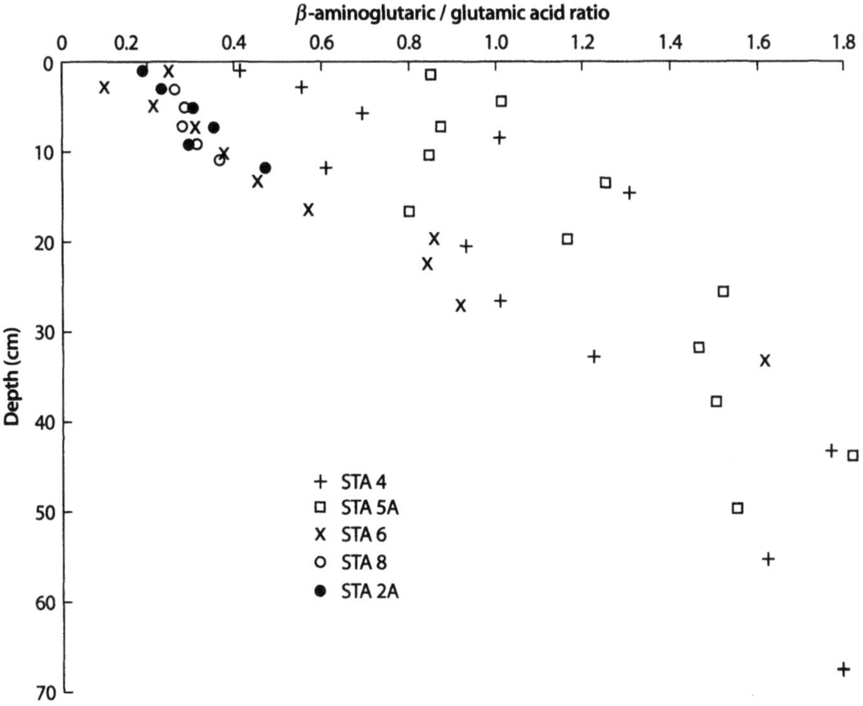

Fig. 6.8. Variation in the relative abundance of dissolved free glutamic acid and β-aminoglutaric acid in sediment cores from the Peru upwelling region (redrawn from Henrichs et al. 1984)

Fig. 6.9. Ratios of unsaturated fatty acids to their saturated counterparts in sediments of Lake Haruna, Japan (adapted from Kawamura et al. 1980)

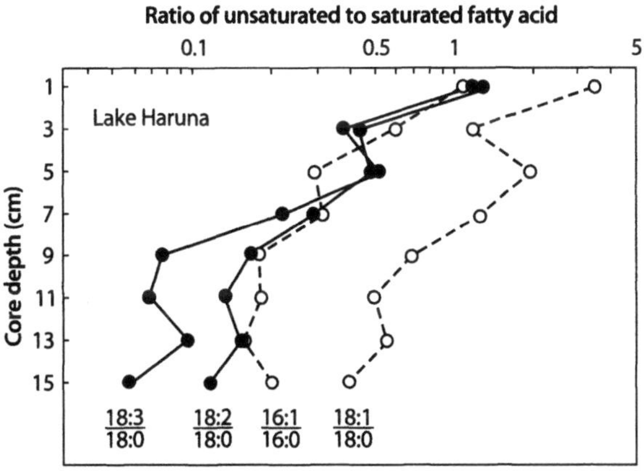

able to organic geochemists. Sterols are more reactive than bulk *OC*, such that sterol concentrations decrease during diagenesis relative to total organic carbon. However, sterols are generally less reactive than fatty acids and alcohols. Sterol diagenesis is often accompanied by bacterial conversion of stenols, the unsaturated biological form of most sterols, into stanols, the hydrogenated analogue (Gaskell and Eglinton 1975; Nishimura 1978), and subsequently by dehydration of stanols to sterenes, the steroidal hydrocarbons (Wakeham et al. 1984). Stenol:stanol ratios that decrease with depth in sediments (Fig. 6.10) are generally interpreted in terms of production of the hydrogenated stanol product at the expense of the unsaturated precursor stenol. This transformation appears to occur more readily under reducing conditions than in oxidizing environments, and is often used as evidence that the sediments are anoxic. However, since low amounts of stanols are also present in phytoplankton, the same sediment profile might just as convincingly be interpreted as resulting from selective preservation of the more stable saturated stanols (Nishimura and Koyama 1977).

One strength of the biomarker approach is in the ability to determine detailed compound chemical structures that can be informative for tracing diagenetic alteration processes at the molecular level. However, relatively few examples of biogenic precursor-diagenetic product relationships have been reported. The sedimentary conversion of fucoxanthin to loliolide (Repeta 1989) is one such case. Fucoxanthin (Fig. 6.11) is a major diatom carotenoid and a dominant pigment in diatomaceous sediments of the Peru upwelling region. Fucoxanthin concentrations decrease rapidly in upper layers of sediment cores. At first glance, the depth profile could be interpreted as being con-

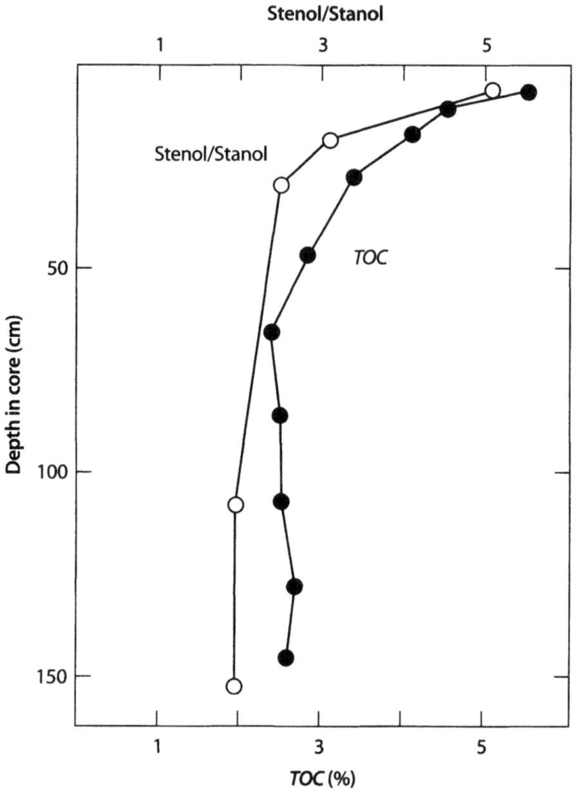

Fig. 6.10. Down core changes in total organic carbon (*TOC*) concentration and the ratio of stenols to stanols in sediments of Lake Suwa, Japan (adapted from Nishi-mura and Koyama 1977)

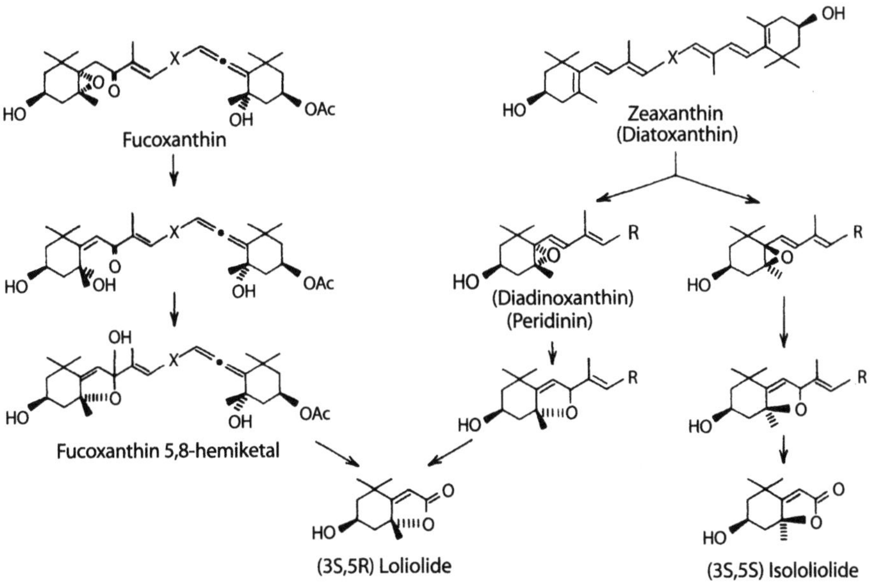

Fig. 6.11. Conversion of fucoxanthin to loliolide (after Repeta 1989)

sistent with rapid degradation of a highly unsaturated pigment. Repeta (1989) measured both fucoxanthin and loliolide in the same core (Fig. 6.12), building on previous work by Klok et al. (1984) that suggested that carotenoids might serve as biogenic precursors of loliolide. The distribution of loliolide mirrored that of fucoxanthin. For each mole of fucoxanthin degraded in Peru sediments, about 1 mole of loliolide was produced, demonstrating that fucoxanthin could be a unique source for loliolide. The results also cast doubt on the link between unsaturation in carotenoids and diagenetic lability. Repeta concluded that the rapid degradation of carotenoids was probably not due to their high degree of unsaturation, but rather their ability to form unstable epoxide intermediates without affecting the unsaturated carotenoid skeleton. Subsequent degradation of the epoxides would yield loliolides. Indeed, there was no evidence for saturated carotanes that would have been produced directly by hydrogenation of the unsaturated carotenoid precursors in the studied sediment. Thus, as is often the case during diagenesis, complex, high molecular weight biomolecules are degraded to simpler, low molecular weight products that would be available to heterotrophic bacteria.

6.4 Overview

The effects of diagenesis in near-surface sediments have a profound effect on the energy balance for the benthic community and on the potential preservation of organic matter in the sediment column. The case studies described here present but a brief introduction to diagenetic processes and should serve as a starting point for further investigation. Because the organic matter may be subject to intense alterations in its

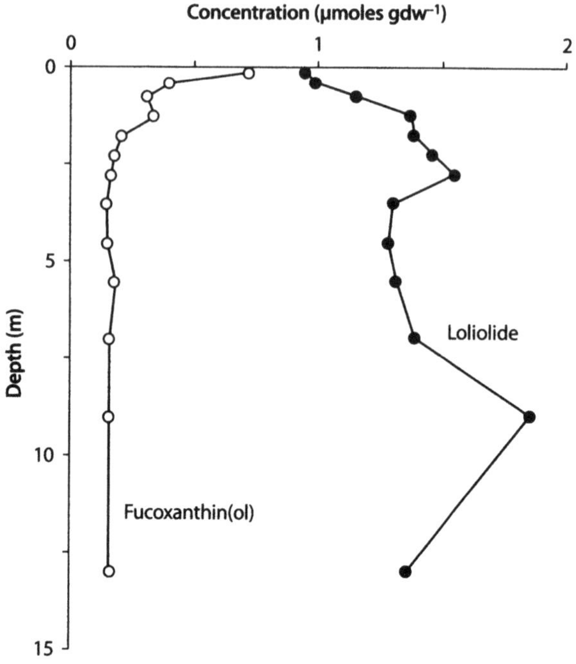

Fig. 6.12. Distributions of fucoxanthin and loliolide in a sediment core from the Peru upwell-ing region (redrawn from Repeta 1989)

composition, it is critical to evaluate these changes and the environmental factors involved both in the context of contemporary biogeochemical cycles and as a means of coupling the aquatic water column with the underlying sediment record in order to obtain realistic palaeoenvironmental reconstructions. Vital information can be obtained both at the bulk elemental and at the molecular levels, although the perspectives they present can be clearly different. Bulk compositions provide a view of the behaviour of a large fraction of the organic matter but are incapable of describing what happens to specific biochemical fractions that often are highly variable in reactivity. Analyses of biomarkers point out how specific biochemical fractions behave, but because they are only a small part of the bulk material, biomarkers may provide a very narrow perspective to organic matter diagenesis. Clearly, however, both approaches are needed.

Acknowledgements

Cindy Lee provided valuable comments that improved this paper. Support by the National Science Foundation is acknowledged during the preparation of this review.

References

Bergamaschi BA, Tsamakis E, Keil RG, Eglinton T, Montluçon DB, Hedges JI (1997) The effect of grain size and surface area on organic matter, lignin and carbohydrate concentrations and molecular compositions in Peru Margin sediments. Geochim Cosmochim Acta 61:1247–1260

Berner RA (1989) Biogeochemical cycles of carbon and sulfur and their effect on atmospheric oxygen over Phanerozoic time. Paleoceanogr Paleoclimatol Paleoecol 73:97–122

Billen G (1982) Modelling the processes of organic matter degradation and nutrients in sedimentary environments. In: Nedwell DB, Brown CM (eds) Sediment microbiology. Academic Press, New York, pp 15–52

Canuel EA, Martens CS (1996) Reactivity of recently deposited organic matter: Degradation of lipid compounds near the sediment water interface. Geochim Cosmochim Acta 60:1793–1806

Calvert SE, Pedersen TF (1992) Organic carbon accumulation and preservation in marine sediments: How important is anoxia? In: Whelan J, Farrington JW (eds) Organic matter. Univ. Press, New York, pp 231–263

Cowie GL (1990) Marine organic diagenesis: A comparative study of amino acids, neutral sugars, and lignin. PhD dissertation, University of Washington, Seattle

Cowie GL, Hedges JI (1991) Organic carbon and nitrogen geochemistry of Black Sea surface sediments from stations spanning the oxic/anoxic boundary. In: Izdar E, Murray JW (eds) Black Sea oceanography. Kluwer Academic Publishers, Dordrecht, pp 343–359

Cowie GL, Hedges JI, Calvert SE (1992) Sources and relative reactivities of amino acids, neutral sugars, and lignin in an intermittently anoxic sediment. Geochim Cosmochim Acta 56:1963–1978

Cowie GL, Hedges JI, Prahl FG, Lange GJ de (1995) Elemental and major biochemical changes across an oxidation from in a relict turbidite: A clear-cut oxygen effect. Geochim Cosmochim Acta 59:33–46

Cowie GL, Calvert SE, Pedersen TF, Schulz H, Rad U von (1999) Organic content and preservational controls in surficial shelf and slope sediments from the Arabian Sea (Pakistan margin). Mar Geol 161:23–38

Deming JW, Barros JA (1993) The early diagenesis of organic mater: Bacterial activity. In: Engel MH, Macko SA (eds) Organic geochemistry – principles and applications. Plenum Press, New York, pp 119–144

Emerson S, Hedges JI (1988) Processes controlling the organic carbon content of open ocean sediments. Paleoceanogr 3:621–634

Gaskell SJ, Eglinton G (1975) Rapid hydrogenation of sterols in a contemporary lacustrine sediment. Nature 254:209–211

Graneshram RS, Calvert SE, Pederson TF, Cowie GL (1999) Factors controlling the burial of organic carbon in laminated and bioturbated sediments off NW Mexico: Implications for hydrocarbon preservation. Geochim Cosmochim Acta 63:1723–1734

Hamilton SE, Hedges JI (1988) The comparative geochemistries of lignins and carbohydrates in an anoxic fjord. Geochim Cosmochim Acta 52:129–142

Hartnett HE, Keil RG, Hedges JI, Devol AH (1998) Influence of oxygen exposure time on organic carbon preservation in continental margin sediments. Nature 391:572–574

Harvey HR, Macko SA (1997) Kinetics of phytoplankton decay during simulated sedimentation: changes in lipids under oxic and anoxic conditions. Org Geochem 27:129–140

Harvey HR, Tuttle JH, Bell JT (1995) Kinetics of phytoplankton decay during simulated sedimentation: Changes in biochemical composition and microbial activity under oxic and anoxic conditions. Geochim Cosmochim Acta 59:3367–3377

Hedges JI, Keil RG (1995) Sedimentary organic matter preservation: An assessment and speculative synthesis. Mar Chem 49:81–115

Hedges JI, Hu FS, Devol AH, Hartnett HE, Tsamakis E, Keil RG (1999) Sedimentary organic matter preservation: A test for selective degradation under oxic conditions. Am J Sci 299:529–555

Hedges JI, Eglinton G., Hatcher PG et al. (2000) The molecularly-uncharacterized component of nonliving organic matter in natural environments. Org Geochem 31:945–958

Henrichs SM (1993) Early diagenesis of organic matter: The dynamics (rates) of cycling of organic compounds. In: Engel MH, Macko SA (eds) Organic geochemistry – principles and applications. Plenum Press, New York, pp 101–115

Henrichs SM, Farrington JW, Lee C (1984) Peru upwelling region sediments near 15° S. 2. Dissolved free and total hydrolyzable amino acids. Limnol Oceanogr 29:20–34

Kaneda T (1991) *Iso* and *anteiso*-fatty acids in bacteria: Biosynthesis, function, and taxonomic significance. Microbiol Rev 55:288–302

Kawamura K, Ishiwatari R, Yamazaki M (1980) Identification of polyunsaturated fatty acids in surface lacustrine sediments. Chem Geol 28:31–39

Keil RG, Cowie GL (1999) Organic matter preservation through the oxygen deficient zone of the NE Arabian Sea as discerned by organic carbon:mineral surface area ratios. Mar Geol 161:12–22

Keil RG, Montluçon DB, Prahl FG, Hedges JI (1994a) Sorptive preservation of labile organic matter in marine sediments. Nature 370:549–552

Keil RG, Tsamakis E, Fuh CB, Giddings JC, Hedges JI (1994b) Mineralogical and textural controls on organic composition of coastal marine sediments: Hydrodynamic separation using SPLITT fractionation. Geochim Cosmochim Acta 57:879–893

Keil RG, Hu FS, Tsamakis E, Hedges JI (1994c) Pollen degradation in marine sediments as an indicator of oxidation of organic matter. Nature 369:639–641

Keil RG, Tsamakis E, Giddings JC, Hedges JI (1998) Biochemical distribution (amino acids, neutral sugars and lignin phenols) among size classes of modern marine sediments from the Washington coast. Geochim Cosmochim Acta 62:1347–1364
Klok J, Baas M, Cox HC, Leeuw JW de, Rijpstra WIC, Schenck PA (1984) Loliolides and dihydroactinidiolides in a recent marine sediment probably indicate a major transformation pathway of carotenoids. Tetra Lett 1984:5577–5580
Lee C (1992) Controls on organic carbon preservation: The use of stratified water bodies to compare intrinsic rates of decomposition in oxic and anoxic systems. Geochim Cosmochim Acta 56:3323–3335
Leeuw JW de, Largeau C (1993) A review of macromolecular organic compounds that comprise living organisms and their role in kerogen, coal, and petroleum formation. In: Engel MH, Macko SA (eds) Organic geochemistry – principles and applications. Plenum Press, New York, pp 23–72
Mayer L M (1993) Organic matter at the sediment-water interface. In: Engel MH, Macko SA (eds) Organic geochemistry – principles and applications. Plenum Press, New York, pp 171–184
Mayer LM (1994a) Surface area control of organic carbon accumulation in continental shelf sediments. Geochim Cosmochim Acta 58:1271–1284
Mayer LM (1994b) Relationships between mineral surface and organic carbon concentrations in soils and sediments. Chem Geol 114:347–363
Mayer LM (1999) Extent of coverage of mineral surface by organic matter in marine sediments. Geochim Cosmochim Acta 63:207–215
Müller PJ, Suess E (1979) Productivity, sedimentation rate, and sedimentary organic matter in the oceans I. Organic carbon preservation. Deep-Sea Res 26:1347–1362
Nishimura M (1978) Geochemical characteristics of the high reduction zone of stenols in Suwa sediments and the environmental factors controlling the conversion of stenols into stanols. Geochim Cosmochim Acta 42:349–357
Nishimura M, Koyama T (1977) The occurrence of stanols in various living organisms and the behaviour of sterols in contemporary sediments. Geochim Cosmochim Acta 41:379–385
Paropkari AL, Prakash Babu C, Mascarenhas A (1992) A critical evaluation of depositional parameters controlling the variability of organic matter in Arabian Sea sediments. Mar Geol 107:213–226
Perry GA, Volkman JK, Johns RB (1979) Fatty acids of bacterial origin in contemporary marine sediments. Geochim Cosmochim Acta 43:1715–1725
Ransom B, Bennett RH, Baerwald R, Shea K (1997) TEM study of in situ organic matter on continental margins: occurrence and the "monolayer" hypothesis. Mar Geol 138:1–9
Ransom B, Kim D, Kastner M, Wainwright S (1998) Organic matter preservation on continental slopes: Importance of mineralogy and surface area. Geochim Cosmochim Acta 62:1329–1345
Repeta DJ (1989) Carotenoid diagenesis in recent marine sediments: II. Degradation of fucoxanthin to loliolide. Geochim Cosmochim Acta 53:699–707
Rhodes DC (1974) Organism-sediment relations on the muddy sea floor. In: Barnes H (ed) Annual reviews of oceanography and marine biology. George Allen and Unwin, London, pp 263–300
Skopintsev BA (1981) Decomposition of organic matter of plankton, humification and hydrolysis. In: Duursma EK, Dawson RA (eds) Marine organic chemistry. Elsevier, New York, pp 125–177
Sun M-Y, Wakeham SG (1994) Molecular evidence for degradation and preservation of organic matter in the anoxic Black Sea basin. Geochim Cosmochim Acta 58:395–3406
Tegelaar EW, Leeuw JW de, Derenne S, Largeau C (1989) A reappraisal of kerogen formation. Geochim Cosmochim Acta 53:3103–3106
Wakeham SG (1995) Lipid biomarkers for heterotrophic alteration of suspended particulate organic matter in oxygenated and anoxic water columns of the ocean. Deep-Sea Res 42:1749–1771
Wakeham SG, Gagosian RB, Farrington JW, Canuel EA (1984) Sterenes in suspended particulate matter in the eastern tropical North Pacific. Nature 308:840–843
Wakeham SG, Lee C, Hedges JI, Hernes PJ, Peterson ML (1997) Molecular indicators of diagenetic status in marine organic matter. Geochim Cosmochim Acta 61:5363–5369
Wang X-C, Lee C (1993) Adsorption and desorption of aliphatic amines, amino acids, and acetate by clay minerals and marine sediments. Mar Chem 44:1–23
Westrich JT, Berner RA (1984) The role of sedimentary organic matter in bacterial sulfate reduction: The G model tested. Limnol Oceanogr 29:236–249

Chapter 7

Sedimentary Geochemistry of the Carbonate and Sulphide Systems and their Potential Influence on Toxic Metal Bioavailability

J. W. Morse

7.1 Introduction

Two of the most biogeochemically dynamic and quantitatively important components of anoxic marine sediments are the carbonate and sulphide systems. They are in many ways remarkably similar (Fig. 7.1), as they are comprised of gases, are multiple species that are similarly dissolved, they form a variety of stable and metastable minerals, and they have widely used stable and radio isotopes. Both carbon and sulphur can occur in different redox states (Table 7.1), and because the dissolved systems include diprotic acids, they also exert a major influence on pH. Their dissolved concentrations can reach values of several millimoles. Carbonate and sulphide are thus generally regarded as the "master" components for controlling Eh-pH conditions in anoxic pore waters of marine sediments.

Many of the most important reactions in these systems are directly driven by organisms which may both oxidize and reduce components, adding further to their complexities. Sulphide is toxic to many organisms, but is actively consumed by others, while yet other organisms have developed various chemical "strategies" to protect themselves

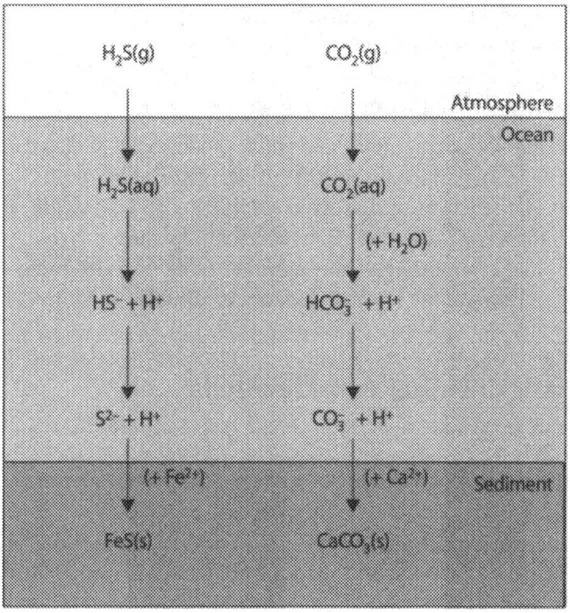

Fig. 7.1. Major components of the CO_2 and H_2S systems in natural waters

Table 7.1. Examples of common carbon and sulphur compounds in different valence states

Valence	Carbon	Sulphur
6		SO_4^{2-} (sulphate)
5		*sulphane S in thiosulphate
4	CO_2 (carbon dioxide) HCO_3^- (bicarbonate)	SO_3^{2-} (sulphite)
2	CO (carbon monoxide)	$S_2O_3^{2-}$ (thiosulphate average charge)
0	C^0 (graphite) CH_2O (much organic matter)	S^0 (elemental S) HS_n^- (polysulphides-mixed w –2)
–1		FeS_2 (pyrite average charge) *sulphonate S in thiosulphate
–2		H_2S (hydrogen sulphide)
–4	CH_4 (methane)	

from it. Therefore, sulphide plays a major role in benthic ecology. Carbonate and sulphide also have a major influence on the availability of toxic metals to benthic organisms by forming strong dissolved complexes, because of precipitation of low solubility toxic metal carbonate and sulphide minerals, and by coprecipitation and adsorption on authigenic calcium carbonate and iron sulphide minerals. In recent years, the study of these reactions and application of results to contaminated sediments has become one of the central themes in trying to relate metal concentrations to their effects on ecosystems.

The major elements of the sedimentary carbon-sulphur system that will be the focus of this chapter are shown schematically in Fig. 7.2. Although the relationships appear to be rather complex, they are in fact a substantial simplification of this dynamic biogeochemical system. The major elements, which will be discussed in more detail later can be divided into major interacting subsystems are:

1. The oxidation of metabolizable organic matter in which dissolved sulphate is used as the electron acceptor and which produces dissolved sulphide, bicarbonate and nutrients
2. The oxidation of sulphide to elemental sulphur, thiosulphate and sulphate
3. The reaction of dissolved sulphide with iron oxide minerals and dissolved iron to form iron sulphide minerals and sulphate
4. The interaction of the sulphate reduction products with calcium and/or calcium carbonate to either precipitate or dissolve calcium carbonate.

7.2 Basic Chemical Considerations

7.2.1 The Carbonic Acid and Hydrogen Sulphide Systems

The relations depicted in Fig. 7.1 for the CO_2 and H_2S systems can be described by a series of reactions and associated thermodynamic equilibrium constants (K; a is activity, f is fugacity). These are given below for 25 °C and an activity of water equal to

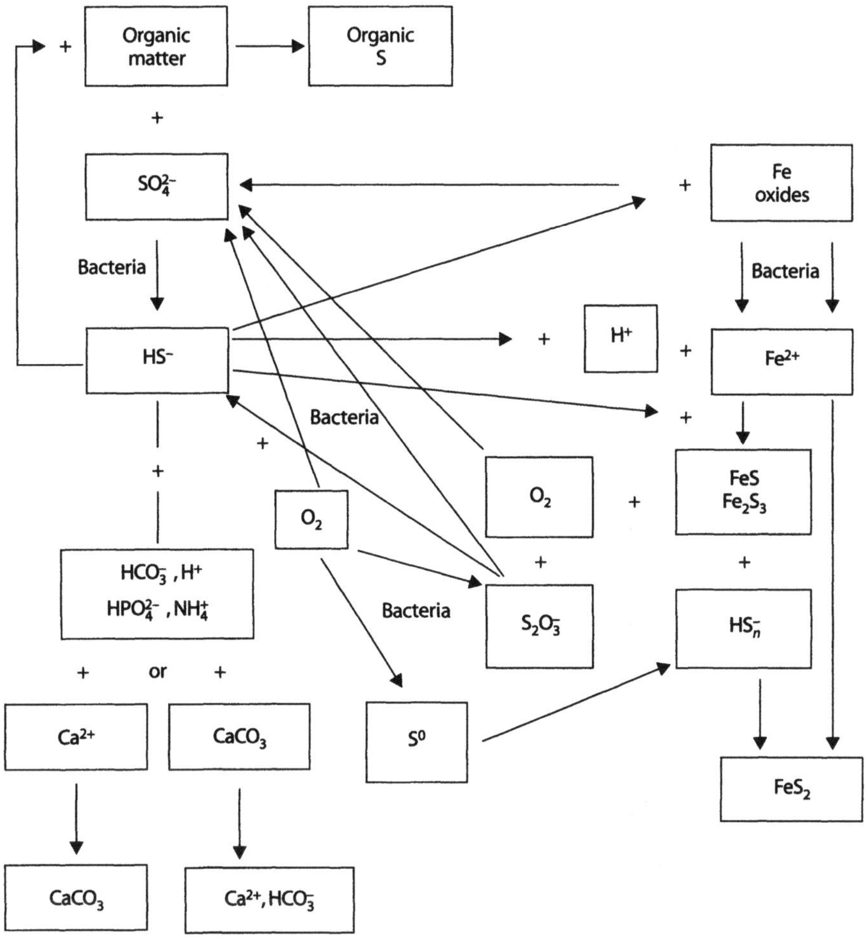

Fig. 7.2. A schematic representation of the sedimentary sulphur-carbon system

unity (see Goldhaber and Kaplan 1974; Morse et al. 1987; Morse and Mackenzie 1990; and Rickard et al. 1995 for literature reviews and discussions on this topic).

Gas-Liquid Equilibrium:

$$CO_{2(g)} \leftrightarrow CO_{2(aq)} \qquad K_{H,CO_2} = a_{CO_2(aq)} / f_{CO_2} = 3.39 \times 10^{-2} \tag{7.1}$$

$$H_2S_{(g)} \leftrightarrow H_2S_{(aq)} \qquad K_{H,H_2S} = a_{H_2S(aq)} / f_{H_2S} = 10.2 \times 10^{-2} \tag{7.2}$$

First Acid Dissociation Constant:

$$CO_{2(aq)} + H_2O \leftrightarrow HCO_3^- + H^+ \qquad K_{1,CO_2} = (a_{HCO_3^-} a_{H^+}) / a_{CO_2(aq)} = 4.47 \times 10^{-7} \tag{7.3}$$

$$H_2S \leftrightarrow HS^- + H^+ \qquad K_{1,H_2S} = (a_{HS^-} a_{H^+}) / a_{H_2S(aq)} = 1.05 \times 10^{-7} \tag{7.4}$$

Second Acid Dissociation Constant:

$$HCO_3^- \leftrightarrow CO_3^{2-} + H^+ \qquad K_{2,CO_2} = (a_{CO_3^{2-}} a_{H^+}) / a_{HCO_3^-} = 4.68 \times 10^{-11} \qquad (7.5)$$

$$HS^- \leftrightarrow S^{2-} + H^+ \qquad K_{2,H_2S} = (a_{S^{2-}} a_{H^+}) / a_{HS^-} < 1 \times 10^{-17} \qquad (7.6)$$

In the above relationships, it is noteworthy that, while the solubilities of the gases and first acid dissociation constants are similar for both systems, the second dissociation constant for H_2S is much smaller than for carbonic acid. It is in fact rather poorly known and may be orders of magnitude less than 10^{-17} (Rickard et al. 1995). This presents major problems for understanding solid phase solubilities and the importance of metal-S ligand complexes. Over the pH ranges commonly encountered in marine sediments, bicarbonate is the dominant form of the carbon dioxide system, as is bisulphite for the hydrogen sulphide system in alkaline waters. However, as near neutral pH values are approached, H_2S becomes an increasingly important component of the hydrogen sulphide system. It should also be kept in mind that the apparent constants for these systems in sea water are strongly influenced by temperature and salinity, and in the deep sea, by pressure.

7.2.2 Redox Reactions

Redox (oxidation-reduction) reactions are at the very heart of sedimentary carbon and sulphur systems and their interrelationships. Complicating things or making them interesting depending on your viewpoint, is the fact that many of these reactions are mediated by consortiums of bacteria, which in many cases are just beginning to be understood. In this section many of the basic redox reactions of general importance will be presented. It should be carefully noted that often these are only very simple approximations of what is occurring and should only be taken as "schematic" representations of processes. It will be useful to refer to Fig. 7.2 to keep the "big picture" in mind. Some of these redox processes will be treated in considerably more detail later in this chapter.

7.2.2.1 *Important General C-O-S Redox Reactions*

The simple redox reaction (Eq. 7.7) for the C/S system that is most often used as an example is the reaction of two organic-Cs with sulphate to produce hydrogen sulphide and carbon dioxide system species. At temperatures typical of recent sediments, this reaction takes place only via the

$$2\,CH_2O + SO_4^{2-} \rightarrow HS^- + 2\,HCO_3^- + H^+ \qquad (7.7)$$

activity of sulphate reducing bacteria such as *Desulfovibrio* (It is important to note that these organisms can only utilize relatively small organic compounds such as acetate that are produced in sediments by the breakdown of more complex organic matter by fermenters). Later, this important general reaction will be examined in a more complex form in the context of how it influences carbonate mineral dissolution and precipitation, nutrient regeneration and pore water pH.

Although organic-C of zero valence is generally assumed in sulphate reduction reactions, other forms are possible as well. For example, one of the more studied and still controversial reactions is where methane is used as the electron donor. This reaction can be simply written as in Eq. 7.8. However, it likely does not occur directly, and a consortium of bacteria, probably

$$CH_4 + SO_4^{2-} \rightarrow HS^- + HCO_3^- + H_2O \tag{7.8}$$

causes the above net reaction to occur via two steps as in Eqs. 7.9 and 7.10 (Hoehler et al. 1994).

$$CH_4 + 3H_2O \rightarrow HCO_3^{2-} + 4H_2 + H^+ \tag{7.9}$$

$$4H_2 + SO_4^{2-} + H^+ \rightarrow HS^- + 4H_2O \tag{7.10}$$

The sulphide that is produced by these reactions can meet a number of different fates. Generally, only a few percent of the produced sulphide survives to ultimately be buried, dominantly as pyrite-S. The two primary ways in which the sulphide is lost both involve oxidation. This oxidation can be accomplished by S oxidizing bacteria, with entire ecosystems using sulphide as an energy source, such as in hydrothermal ridge vents and in cold "seep" vents. Oxidation can also result from reaction of hydrogen sulphide with metal oxides, with iron and manganese oxides generally being most important.

Some examples of important redox reactions largely involving sulphur and oxygen follow. Many of these reactions are largely driven by bacteria in sediments. Jorgensen (1990) has found the disproportionation reaction (Eq. 7.19) for thiosulphate to be of particular importance in anoxic marine sediments.

$$2H_2S + O_2 \rightarrow 2S^0 + 2H_2O \tag{7.11}$$

$$2H_2S + 2O_2 \rightarrow S_2O_3^{2-} + H_2O^- + 2H^+ \tag{7.12}$$

$$2H_2S + 3O_2 \rightarrow 2SO_3^{2-} + 4H^+ \tag{7.13}$$

$$H_2S + 2O_2 \rightarrow SO_4^{2-} + 2H^+ \tag{7.14}$$

$$H_2S + (n-1)S^0 \rightarrow S_n^{-2-} + 2H^+ \tag{7.15}$$

$$2S^0 + O_2 + H_2O \rightarrow S_2O_3^{2-} + 2H^+ \tag{7.16}$$

$$S^0 + O_2 + H_2O \rightarrow SO_3^{-2-} + 2H^+ \tag{7.17}$$

$$SO_3^{2-} + S^0 \rightarrow S_2O_3^2 \tag{7.18}$$

$$S_2O_3^{2-} + H_2O \rightarrow SO_4^{2-} + H_2S \tag{7.19}$$

$$2S_2O_3^{2-} + O_2 \rightarrow 2SO_4^{2-} + 2S^0 \tag{7.20}$$

7.2.2.2
Reactions Involving Metals

Although it is possible to also write a large number of potential redox reactions between iron and manganese oxide minerals and assorted sulphur species (e.g. Aller and Rude 1988), the major reactions in sediments are almost certainly far more complex than these simple reactions would indicate. This is well-illustrated by the work of Pyzik and Summer (1981) on the sulphidization of goethite (FeOOH). They pointed out it could occur by four possible reactions (Eqs. 7.21–7.24):

$$2\,FeOOH + HS^- + H_2O \rightarrow S^0 + 2\,FeOH^+ + 3\,OH^- \qquad (7.21)$$

$$6\,FeOOH + 4\,HS^- + 2\,H_2O \rightarrow S_4^{2-} + 6\,FeOH^+ + 8\,OH^- \qquad (7.22)$$

$$8\,FeOOH + 5\,HS^- + 3\,H_2O \rightarrow S_5{}^{2-} + 8\,FeOH^+ + 11\,OH^- \qquad (7.23)$$

$$8\,FeOOH + 2\,HS^- + 3\,H_2O \rightarrow S_2O_3^{2-} + 8\,FeOH^+ + 8\,OH^- \qquad (7.24)$$

Mechanistically, there is an initial surface exchange of dissolved bisulphide for a surface hydroxyl group, followed by Eq. 7.21. Then there is a protonation of the surface hydroxide layer and dissolution of ferrous hydroxide to solution.

Other examples of types of redox reactions involving metals of concern to the sedimentary carbonate and sulphide systems include oxidation of iron sulphides, which produces acid that can result in carbonate mineral dissolution (Eq. 7.25) and use of metals oxides as electron acceptors for the oxidation of organic matter by bacteria resulting in the formation of metal carbonates (Eq. 7.26, siderite; Eq. 7.27 rhodochrosite).

$$4\,FeS_2 + 15\,O_2 + 14\,H_2O + 16\,CaCO_3 \rightarrow 4\,Fe(OH)_3 + 8\,SO_4^{2-} + 16\,HCO_3^- + 16\,Ca^{2+} \qquad (7.25)$$

$$2\,Fe_2O_3 + CH_2O + 6\,H^+ \rightarrow FeCO_3 + 3\,Fe^{2+} + 4\,H_2O \qquad (7.26)$$

$$2\,MnO_2 + CH_2O + 2\,H^+ \rightarrow MnCO_3 + Mn^{2+} + 2\,H_2O \qquad (7.27)$$

7.2.2.3
Influence of Closed System Sulphate Reduction on the Carbonic Acid System

A more complete representation of the oxidation of organic matter via sulphate reduction given in Eq. 7.7 would include the other major components of organic matter N and P. In marine plankton, these occur in a close to constant ratio of C:N:P of 106:16:1 known as the Redfield ratio (Redfield et al. 1963). When organic matter of this composition is oxidized via sulphate reduction, the nutrients phosphate and ammonia are also products (Eq. 7.28).

$$1/53\,(CH_2O)_{106}(NH_3)_{16}H_3PO_4 + SO_4^{2-}$$

$$\rightarrow CO_2 + HCO_3^- + HS^- + 16/53\,NH_3 + 1/53\,H_3PO_4 + H_2O \qquad (7.28)$$

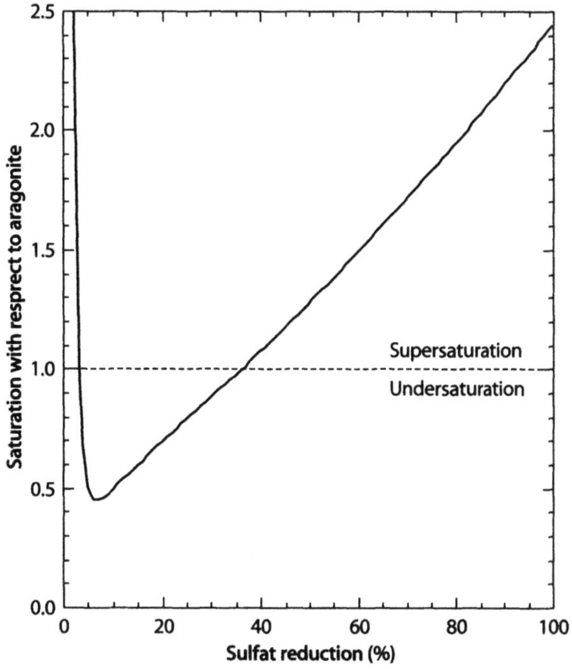

Fig. 7.3. The change in saturation state of sea water with respect to aragonite as a function of sulphate reduction as calculated by Morse and Mackenzie (1990)

Because they are produced in set ratios, fixed pH can be established after a moderate degree of sulphate reduction in a closed system (Ben-Yaakov 1973). During these early stages of reaction in sea water the pH drops to about 6.9 and remains constant. This can result in pore waters becoming undersaturated and carbonate dissolution occurring. Further sulphate reduction causes the alkalinity to rise, while the pH remains constant. At about 35% sulphate reduction, the pore waters regain supersaturation with respect to calcium carbonate, and further sulphate reduction can result in calcium carbonate precipitation (Fig. 7.3). Incorporation of iron oxides in this reaction results in pore waters being buffer at higher pH values. As noted by many investigators of these processes (e.g. Berner 1971), this reaction is of tremendous importance during the early diagenesis of marine sediments. Examples of this will be given later.

7.2.3 Carbonate and Sulphide Minerals

7.2.3.1 *Carbonate Minerals*

Carbonate minerals comprise about 20% of Phanerozoic sedimentary rocks. Ancient carbonate minerals are dominated by calcite and dolomite, in accordance with thermodynamic predictions. The fraction of carbonate minerals as dolomite increases with increasing geologic age. However, dolomite is rare in modern sediments where biogenic carbonates strongly dominate. It is convenient to divide modern sedimentary carbonates into those found in relatively shallow waters and those being deposited on deep ocean basins. Sources, mineralogy and diagenetic processes are generally quite different in these environments. In this chapter, we will not be dealing with deep ocean sediments. In shallow water marine sediments, aragonite is usually the most common carbonate mineral, followed by high-magnesian calcite (e.g. $Ca_{0.87}Mg_{0.13}CO_3$) and minor amounts of low-Mg calcite.

The pK (= $-\log K$) values for calcite and aragonite solubility products (Eq. 7.29) are respectively at 25 °C, 8.48 and 8.30. The solubility of high magnesian calcites is a complex and controversial topic (e.g. abiotic and biotic high magnesian calcites of the same composition appear to have different solubilities; see Morse and Mackenzie 1990 for extensive discussion of this topic).

$$CaCO3 \rightarrow Ca^{2+} + CO_3^{2-} \qquad K_{sp} = a_{Ca^{2+}} a_{CO_3^{2-}} \qquad (7.29)$$

Observations of the composition of pore waters in carbonate-rich shallow water sediments have demonstrated that no single value for the calcium carbonate ion activity product widely occurs. This results in a moderate range of saturation states relative to aragonite (Fig. 7.4; Morse et al. 1985) that have been demonstrated to represent conditions of "dynamic equilibrium" between the pore waters and carbonates (Bernstein and Morse 1985). For equilibrium with aragonite in a $S = 35$ pore water at 25 °C with normal Ca^{2+} concentration ($a_{Ca^{2+}} \cong 2 \times 10^{-3}$), the following approximate carbonic acid system parameter values at a pH of 7.2 (fairly typical of anoxic sediments) are total alkalinity about 2 × that of sea water (5 meq kg^{-1}), and p_{CO_2} about 10 matm. $a_{CO_3^{2-}}$ would then be about 2.5×10^{-6}.

7.2.3.2
Iron Sulphide Minerals

Just as with the carbonates, there are several sulphide minerals that are found in modern sediments. However, unlike the sedimentary carbonate minerals, they are dominantly of authigenic rather than biogenic origin. Sedimentary sulphides are usually divided into acid volatile sulphide (AVS) and pyrite (FeS_2), which is the thermodynamically stable phase.

AVS comprises an "operationally defined" group of what are generally believed to be metastable iron sulphide minerals plus dissolved H_2S species. Often, but with major exceptions, AVS is confined to a relatively small portion (<10%) of total sedimentary sulphides. It can be ephemeral. Spatially, AVS often appears with a maximum in the top few cm of anoxic sediments and disappears rapidly with depth to close to undetectable concentrations within the top 20 cm or less. Temporally, it can exhibit major changes on a seasonal basis in the upper few cm of sediments. AVS has received considerable attention, because it is one of the most chemically reactive components of anoxic sediments during early diagenesis. AVS is readily oxidized within hours or less when anoxic sediments are exposed to oxic waters. The iron sulphide minerals amorphous-FeS, mackinawite ($Fe_{0.9}S$) and griegite (Fe_3S_4) that are generally purported to be the major components of AVS have also traditionally (e.g. Berner 1984) been held to be precursor phases necessary for the formation of the dominant and thermody-

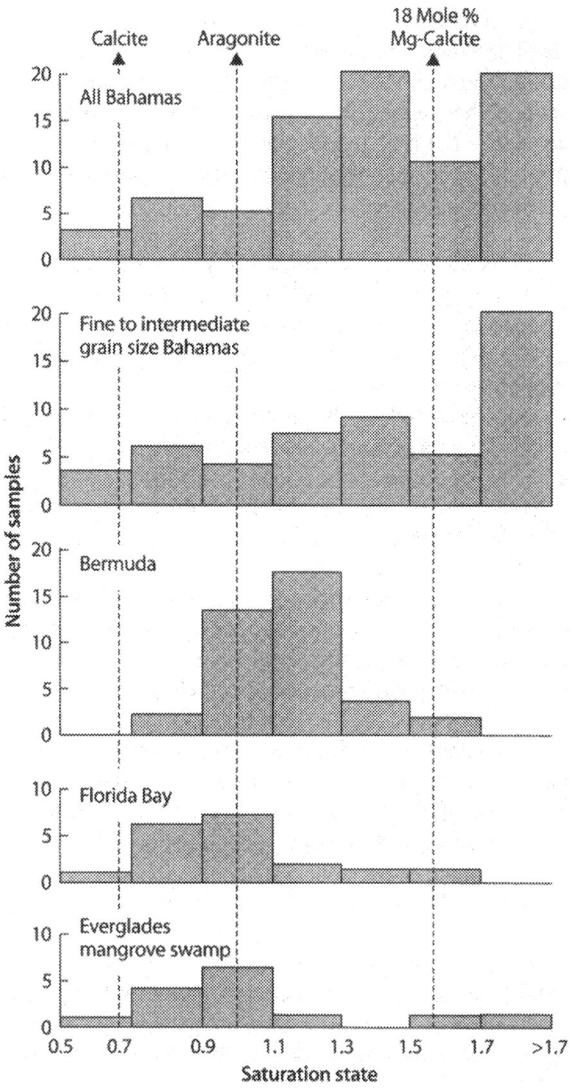

Fig. 7.4. Histograms for the num-ber of samples found in various concentration ranges relative to aragonite saturation in different carbonate-rich environments (redrawn from Morse et al. 1985)

namically stable sedimentary sulphide mineral pyrite (FeS_2). AVS minerals are, with rare exceptions, not directly observable by traditional techniques such as X-ray diffraction analysis and scanning electron microscopy (e.g. Morse and Cornwell 1987). Their presence is inferred by chemical leaching techniques and by comparing ion activity products to the equilibrium solubility products of the minerals. AVS minerals generally have not been the focus of studies of sulphidic sediments. However, they have been commonly measured.

The process by which pyrite forms in sediments has been among the most discussed and controversial topics in geochemistry for decades. The question still remains largely open, and no attempt will made to review the massive literature on the topic here. Instead, the major points on this topic in the review by Rickard et al. (1995) are briefly summarized and the reader should refer to this article for further details and referencing. The major problem in pyrite formation is how to produce the disulphide ligand and cause the Fe(II) to change from high spin in FeS to the low spin state found in pyrite. Different pathways have been proposed and demonstrated in laboratory experiments. These are:

1. *The FeS oxidation pathway.* Amorphous-FeS ages to mackinawite. Then under slightly oxidizing conditions, mackinawite converts to griegite and finally pyrite. This is the most often cited mechanism in the literature.
2. *The polysulphide pathway.* FeS (or $FeSH^+$) reacts with polysulphides forming a complex that then breaks down producing FeS_2. This reaction is relatively slow and does not necessarily involve a solid phase. It is a more realistic pathway than the classically cited schematic solid-solid reaction $FeS + S^0 \rightarrow FeS_2$.
3. *The H_2S pathway.* FeS reacts with hydrogen sulphide producing hydrogen gas ($FeS + H_2S \rightarrow FeS_2 + H_2$). Because this reaction involves H_2S, it is favoured at lower pH values and can be relatively rapid compared to the other pathways.

Pyrite occurs with different morphologies in sediments. Framboidal pyrite is most common and generally is assumed to be a relatively rapidly-formed early diagenetic product, whereas euhedral pyrite is usually presumed to form more slowly over long time periods during later stages of diagenesis.

Because of the large uncertainties in the value of the second dissociation constant for hydrogen sulphide, a special approach to representing metal sulphide solubility has been developed. This approach incorporates H^+ in the reaction, resulting in the formation of bisulphide (Eq. 7.30) and is therefore a pH dependent solubility constant ($^*K_s$, see Stumm and Morgan 1996 for discussion and examples).

$$MeS + H^+ \rightarrow Me^{2+} + HS^- \qquad K_s = (a_{Me^{2+}} a_{HS^-}) / a_{H^+} \tag{7.30}$$

The pK_s values for amorphous-FeS, mackinawite, griegite and pyrite are (using S^0 in the griegite and pyrite reactions), respectively, 2.95, 3.6, 4.4, 16.4 (Davison 1991). Many of the preceding concepts have been incorporated into constructing Fig. 7.5. In this figure, the activity of HS^- has been plotted vs. pH for a pore water of $S = 35$ at 25 °C containing 20 μM total H_2S over a pH range typical of most anoxic marine sediments of 6.5 to 8. This was accomplished using Eq. 7.31. Then the activity of Fe^{2+} was calculated assuming

$$a_{HS^-} = \gamma_{HS^-} \Sigma H_2S \left[\left(\frac{\gamma_{HS^-}}{\gamma_{H_2S}} \right) \left(\frac{a_{H^+}}{K_1} \right) + 1 \right]^{-1} \tag{7.31}$$

equilibrium with mackinawite according to Eq. 7.32 and also plotted against pH.

$$a_{Fe^{2+}} = {}^*K_s (a_{H^+} / a_{HS^-}) \tag{7.32}$$

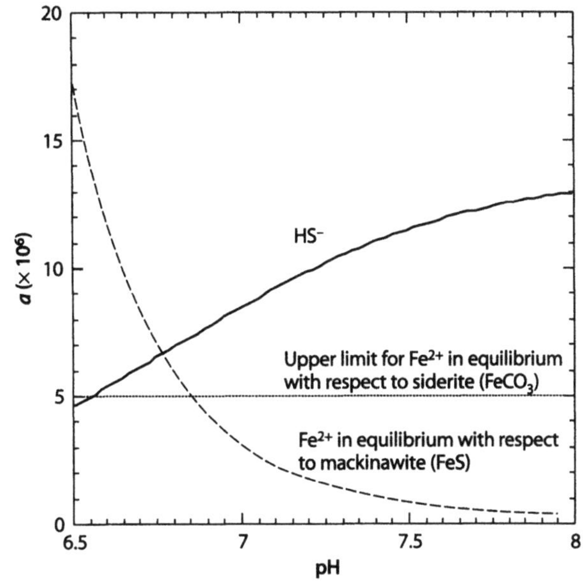

Fig. 7.5. A plot of the activities of bisulphite (*solid line*) and ferrous iron (*dashed line*) vs. pH for sea water $S = 35$, 25 °C, $\Sigma H_2S = 20\ \mu M$ and equilibrium with mackina-wite (FeS). *Dotted line* is upper limit of ferrous iron concentration when limited by the solubility of siderite ($FeCO_3$) when $a_{CO_3^=} = 2.5 \times 10^{-6}$ (see text)

Finally, the limiting activity of ferrous iron was calculated for a carbonate ion activity of 2.5×10^{-6} (see above). Even at this relatively low hydrogen sulphide concentration, low concentrations of ferrous iron are required over much of the pH range of anoxic marine sediments. Carbonate ion concentrations for equilibrium with aragonite also severely limit ferrous iron solubility.

7.2.3.3
Sulphide Minerals of Selected Toxic Metals

The association of trace metals enrichments with sulphidic sediments has long been known (e.g. Krauskopf 1956; Manheim 1961). Interest in these relationships has been sparked by studies that indicate trace metal-sulphide interactions, which may have a profound influence on the bioavailability of toxic metals in sediments (Di Toro et al. 1990, 1992; Morse 1994). Direct evidence for the interactions of trace metals with sulphides in anoxic sediments from numerous locations is now available, and clear patterns of behaviour for different trace metals have emerged (e.g. Huerta-Diaz and Morse 1992; Morse et al. 1993). These will be discussed near the end of this chapter along with possible explanations.

It is worthwhile at this point to consider some basic relationships and concepts. In Table 7.2, p^*K_s for simple metal sulphides and pK_{sp} values for simple metal carbonates are presented. From these values, at 25 °C and pH = 7.2, the activity of HS^- has been calculated for the metals in simultaneous equilibrium with their carbonate and sulphide minerals at a carbonate ion activity of 2.5×10^{-6} (aragonite equilibrium, see previous discussion). These results indicate that for most of the toxic metals (except Fe) considered, only extremely low bisulphite concentrations could occur under these conditions. Also calculated is the metal ion activity where carbonate is not con-

Table 7.2. Solubility relationships for different metal sulphides and carbonates (based on compilations by Morse and Mackenzie 1990 and Stumm and Morgan 1996)

Metal	p^*K_s	pK_{sp}	a_{Me}-CO_3	a_{HS^-}	a_{Fe}
Cd	14.4	13.7	7.3×10^{-9}	3.8×10^{-14}	2.8×10^{-17}
Co	7.4	9.7	8.4×10^{-5}	2.7×10^{-11}	2.3×10^{-10}
Fe	3.6	10.9	4.9×10^{-6}	3.2×10^{-6}	1.6×10^{-6}
Ni	5.6	6.9	5.4×10^{-2}	2.9×10^{-12}	1.6×10^{-8}
Pb	14.0	12.2	2.8×10^{-7}	2.4×10^{-15}	6.8×10^{-17}
Zn	10.9	10.0	4.0×10^{-5}	1.9×10^{-14}	7.4×10^{-14}

sidered and the concentration of total sulphide is set at 20 μM. Here again, with the exception of Fe, very low metal ion activities are permitted for equilibrium. A fairly regular relationship is shown in Fig. 7.6 between the values of p^*K_s for metal sulphides and pK_{sp} values for metal carbonates.

Unfortunately, the behaviour of trace metals in the real world is not so simple. Several different types of chemistry can occur (see Morse and Luther 1999 for discussion). These include formation of multiple different minerals (e.g. see Table 7.3 for Cu), coprecipitation reactions (Table 7.4) and largely irreversible adsorption. Often changes in metal valence occur (e.g. Cu(II) to Cu(I)), further complicating the situation. In this chapter we shall not delve into the complex literature on this topic, but rather later give some specific examples from natural sediments along with possible interpretations of the chemistry involved.

7.2.4 Isotopes

Another major similarity between the sedimentary carbon and sulphur systems is the use of stable isotope ratios ($^{13}C/^{12}C$ and $^{34}S/^{32}S$) and radioisotopes (^{14}C and ^{35}S). Measurements of these isotopes in various components of sediments and injection of radioisotopes as tracers of chemical pathways and rates are major elements of the study of carbon and sulphur in sediments.

Variations in stable isotope ratios are usually reported as "del" (δ) values (Eq. 7.41) which are in parts per mil (‰). The PDB fossil limestone is the primary C standard, and troilite from the Canyon Diablo meteorite is the generally accepted standard for S. The primary use of

$$\delta^{34}S = \left[\frac{\left(^{34}S/^{32}S\right)_{sample}}{\left(^{34}S/^{32}S\right)_{standard}} - 1\right]\times 1000 \qquad (7.41)$$

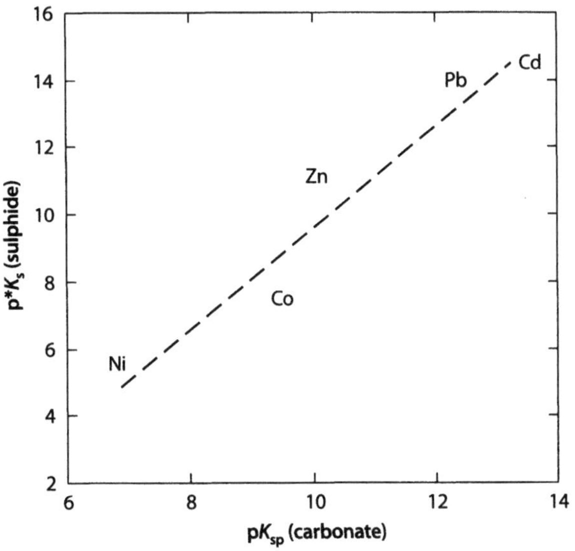

Fig. 7.6. A plot of p^*K_s for sulphide minerals vs. pK_{sp} for carbonate minerals of the same metals

Table 7.3. Copper sulphide and iron-sulphide phases (based on Vaughan and Craig 1978; from Morse and Luther 1999)

Copper sulfide minerals		Copper-iron sulfide minerals	
Chalcocite	Cu_2S	Chalcopyrite	$CuFeS_2$
Analite	Cu_7S_4	Bornite	Cu_5FeS_4
Digenite	Cu_9S_5	Fukuchilite	Cu_3FeS_8
Djurleite	$Cu_{1.97}S$	Talnakite	$Cu_9Fe_8S_{16}$
Geerite	$Cu_{1.6}S$	Mooiheckite	$Cu_9Fe_9S_{16}$
Spionkopite	$Cu_{1.39}S$	Haycockite	$Cu_4Fe_5S_8$
Yarrowite	$Cu_{1.12}S$	Cubanite	$CuFe_2S_3$
Covellite	CuS	Idaite	$Cu_{5.5}FeS_{6.5}$
Cubic	CuS_2		

carbon stable isotopes in the study of carbon in sediments is to identify the relative importance of different carbon sources. For example, $\delta^{13}C$ values for biogenic carbonates are usually close to 0, for marine organic-C −22, for terrestrial organic-C −28 and often very negative , −40 or less for thermogenic methane. Later in this chapter the use of carbon stable isotopes to solve carbon-sulphur diagenetic processes in seagrass root zones (Eldridge and Morse 2000) will be described.

The use of sulphur stable isotopes in studying sedimentary sulphides has proven more challenging. This is because the extent of fractionation depends to a significant extent on the rates of bacterial sulphate reduction. The cycling of sulphur between different oxidation states further complicates matters, as does the extent to which sulphate is reduced under conditions approximating to different extents open and close systems (Fig. 7.7). (See Thode 1991 for extensive overview and discussion.)

Table 7.4. Possible reactions for the incorporation of metals into FeS and FeS_2 phases. FeS_2 reactions are given for those where the kinetics are well-described (Rickard 1975; Luther 1991; Rickard 1997; Rickard and Luther 1997; from Morse and Luther 1999)

Eq.	FeS	
7.33	$Fe^{2+} + HS^{-} \rightarrow FeS + H^{+}$	(FeS formation)
7.34	$FeS + Me^{2+} \rightarrow Fe\text{-}S\text{–}Me^{2+}$	(Metal adsorption onto FeS)
7.35	$Fe\text{-}S\text{–}Me^{2+} \rightarrow Fe(Me)S + Fe^{2+}$	(Metal inclusion into FeS)
7.36	$FeS + Me^{2+} \rightarrow MeS + Fe^{2+}$	(Metathesis or metal exchange reaction)
	FeS_2	
7.37	FeS or $[Fe(Me)S] + S(0) \rightarrow Fe(Me)S_2$	(Pyrite formation and metal inclusion)
7.38	FeS or $[Fe(Me)S] + H_2S \rightarrow Fe(Me)S_2 + H_2$	
7.39	$FeS_2 + Me^{2+} \rightarrow Fe\text{-}S\text{-}S\text{–}Me$	(Metal adsorption onto pyrite)
7.40	$Fe\text{-}S\text{-}S\text{–}Me \rightarrow Fe(Me)S_2$	(Metal inclusion into pyrite)

^{14}C is mostly used for two purposes. The first is for age dating and obtaining sediment accumulation rates. The second is incubation of ^{14}C-labelled organic compounds in sediments in order make rate measurements on their decomposition. The primary use of ^{35}S is to make measurements of the rates of sulphate reduction using radiolabelled sulphate (e.g. Jorgensen 1978). It is additionally widely used to investigate the biogeochemical cycling of sulphur (see review of Fossing 1995).

7.3 Sedimentary Geochemistry of Carbonate and Sulphide Systems

7.3.1 "Normal" Marine Sediments

7.3.1.1 *Relationship Between Organic-C and Pyrite-S Burial*

The term "normal marine" is often used to describe fairly common coastal and even to some degree estuarine sediments as a depositional environment for sedimentary sulphides and carbon. Normal marine alone refers to sediments overlain by oxic sea water of typical oceanic salinity. The type of sediment of interest is typically a fine-grained siliciclastic sediment. Excluded from the larger definition are coarse, sandy sediments, $CaCO_3$-rich sediments (hence the adjective "siliciclastic"), and sediments overlain by freshwater and anoxic marine waters (e.g. Berner 1982; Berner and Raiswell 1984). Because of the virtual absence of sulphur, also excluded are sediments that do not become sufficiently anoxic for sulphate reduction to become a major process near the sediment-water interface (e.g. many deep-sea sediments and sediments containing very low concentrations of organic-C).

Fine-grained, normal marine siliciclastic sediments have a relatively narrow range of C/S ratios. Consequently, organic-C and pyrite-S must covary in a close to linear

Fig. 7.7. Sulphur isotope ratio changes as a function of the extent of sulphate reduced under closed system conditions (after Thode 1991)

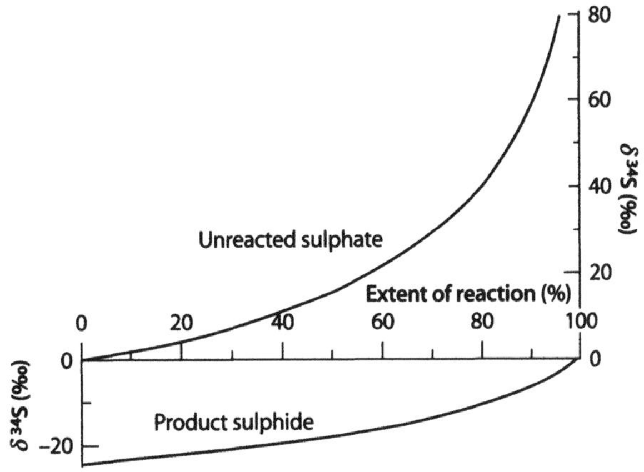

manner. Building on the earlier work of Goldhaber and Kaplan (1974) and Volkov and Rosanov (1983), as well as his own observations and the observations of several other studies, Berner (1982) obtained a value of 2.8 ±0.8 for the C/S weight ratio in normal marine sediments. Since then, other major studies (e.g. Lin and Morse 1991; Raiswell and Berner 1986) have lent substantial support to the concept that the variation in the C/S ratios of fine-grained normal marine siliciclastic sediments is usually within fairly narrow limits.

In order to attempt to explain this occurrence and the processes responsible for producing the observed relationships, Morse and Berner (1995) produced a mathematical model. The basic elements of this model are shown in Fig. 7.8. A simple expression (Eq. 7.42) was derived for sediments that relates the fraction of the organic carbon that is destroyed by sediment metabolism (f_M) to the fraction of sulphate reduction that is fixed (f_{CS}) buried as pyrite (f_{SP}). The *molar* organic-C to pyrite-S ratio (R) for buried sediment is then:

$$R = \frac{2\left(\dfrac{1}{f_M} - 1\right)}{f_{CS} f_{SP}} \tag{7.42}$$

Because the metabolized organic fraction is a function of sedimentation rate, the fraction of the sulphur that is fixed as pyrite can also be shown to be correlated directly with burial rates (Fig. 7.9).

A major question in understanding sulphide geochemistry in sediments is what ultimately limits authigenic iron sulphide mineral formation. In normal marine sediments, Berner (e.g. 1984) has found that metabolizable organic matter usually limits iron sulphide production. However, the fraction of sedimentary organic matter that is metabolizable is hard to determine (e.g. Boudreau 1991; Canfield 1994) and not simply related to total organic matter (e.g. Morse and Emeis 1990). The question of what constitutes "reactive" iron during early diagenesis has also evolved into a complex

Fig. 7.8. Schematic flow diagram illustrating the relationships among the major components of the model. Note O_2 respir. includes suboxic processes such as nitrate and metal oxide reduction as well as oxic respiration (redrawn from Morse and Berner 1995)

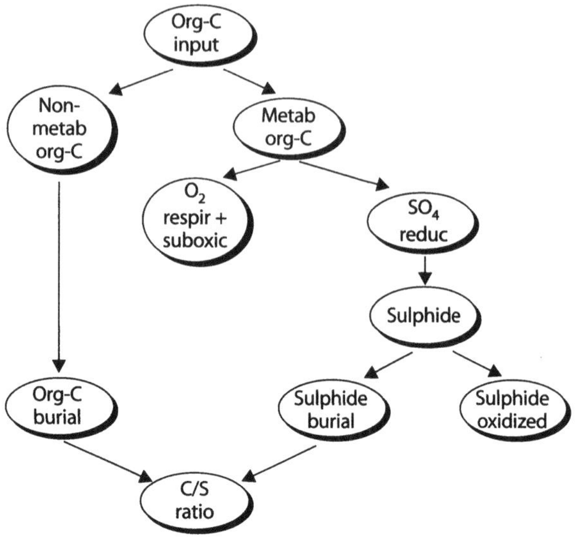

Fig. 7.9. The log-log dependence of f_{SP} on sediment burial rate in the Gulf of Mexico. The *circled data point* was not used in curve fitting or computations, as it is believed to represent slumped sediment from the upslope. The *line* is a linear least squares fit to data (after Morse and Berner 1995)

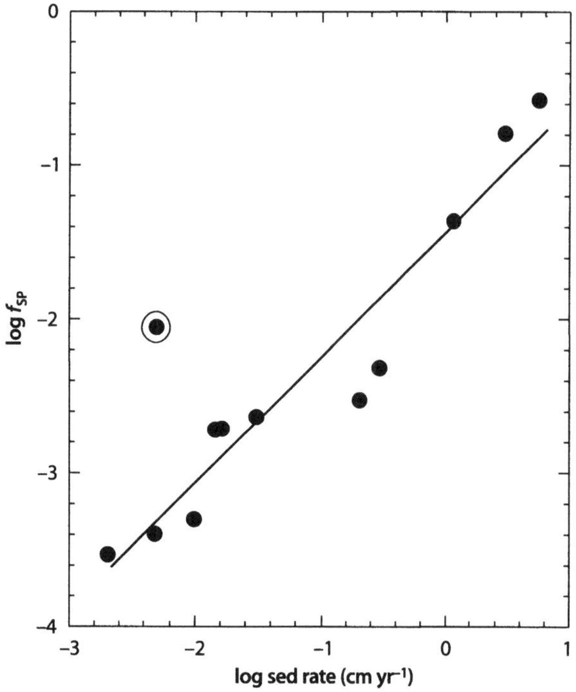

question involving the reaction kinetics between dissolved sulphide and various iron oxides (e.g. Canfield 1988, 1989). It now can be argued that in sediments where dissolved sulphide is present, "reactive" iron is limiting the rate of iron sulphide mineral formation. A further complication that is little discussed in the literature is that in sediments where dissolved sulphate is totally reduced in the top few cm of sediment, its rate of transport into the sediment is the limiting factor for sulphide production in the sediment (Arvidson and Morse, to be published). In the following section, the influence of biologically driven transport of oxidants on sedimentary sulphides and carbonates will be discussed.

7.3.1.2
Influences of Biologically-Driven Transport

Oxidation of sedimentary sulphides via reaction with iron and manganese oxides can be an important or even dominant reaction pathway (Aller and Rude 1988). However, the active transport of oxygen to several cm below the sediment-water interface by infaunal organisms (Fig. 7.10) can also be a very important process (e.g. Aller 1988). It is primarily through this process that most sulphides are often oxidized. Indeed, in measuring benthic oxygen demand, biologists often assume this is the major indirect pathway for oxygen consumption and that any sulphide not oxidized is of insignificant importance (e.g. Pamatmat 1971).

Just as the oxidation of pyrite in soils or sulphide minerals in mine tailings can lead to highly acidic waters, the oxidation of sedimentary sulphides produces acid as well (Eq. 7.14). This can lower the saturation state of pore waters with respect to carbonate minerals and lead to their extensive dissolution. Dissolution is highest in highly bioturbated and bioirrigated sediments, where these processes supply oxygen and remove accumulated alkalinity from dissolution (Aller 1982). In vegetated sediments, plants such a seagrasses can cause active transport of oxygen to sediments in their root zones. Based on extensive field data that included carbon isotope studies, Eldridge and Morse (2000) constructed a diagenetic model for this process. They found that the oxygen flux from roots was important for keeping sulphide concentrations below toxic levels for the seagrasses and that generally over half of the dissolved inorganic carbon (*DIC*) in the pore waters came from carbonate mineral dissolution rather than oxidation of organic matter.

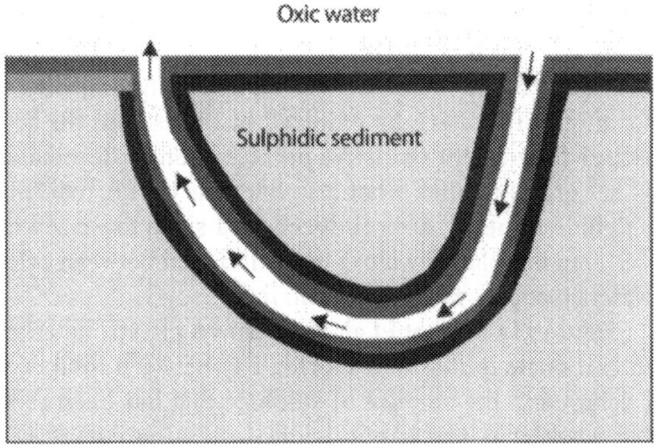

Fig. 7.10. Schematic representation of burrow and its influence on sediment redox chemistry. *Dark grey* is suboxic zone, *black* is zone of nitrate and manganese reduction (based on concepts in Aller 1988)

7.3.1.3
Carbonate Precipitation in Sulphidic Sediments

In Fig. 7.3, it was shown that sulphate reduction that initially results in a major decrease in pore water saturation with respect to calcium carbonate can lead again to supersaturation with increasing extent of sulphate reduction. In sediments and sedimentary rocks, this is a well-known mechanism for producing authigenic calcium carbonate. For example, the massive carbonate cap rocks on salt domes are a result of this process.

An example from modern sediments comes from Baffin Bay, Texas, which is a hypersaline negative estuary that has been studied because of its similarities to environments that may have existed in ancient epicontinental seas (Morse et al. 1992). Figure 7.11 shows the relationship of the loss of dissolved sulphate and calcium to the increase in total carbon dioxide. The decrease in dissolved calcium occurs because of the precipitation of calcium carbonate. If the loss of carbon dioxide from precipitation is corrected for by the loss of calcium, then a close to ideal two to one increase in dissolved carbon dioxide with sulphate loss is observed (Fig. 7.12).

For the precipitation of calcium carbonate to be quantitatively significant, it is necessary for there to be exceptionally high quantities of metabolizable organic matter available to produce the needed carbon dioxide. An example of such a situation is at hydrocarbon seeps in the Gulf of Mexico (Arvidson and Morse, to be published). There, the organic content of the sediments and concentration of calcium carbonate show a close association. (Fig. 7.13). Because large amounts of sulphide are also produced by this process, total dissolved hydrogen sulphide can exceed 10 mM.

7.3.2
Carbonate-Rich Sediments

During the last decade, Lynn Walter and her associates (Walter and Burton 1990; Walter et al. 1993; Ku et al. 1999) have done much to demonstrate the major quantitative importance of dissolution of calcium carbonate from shallow carbonate sediments (on the order of up to 50%) driven by the previously discussed sulphate reduction-sulphide oxidation process. Such sediments account for about a third of all ocean carbonate production (Milliman 1993) and are therefore of considerable significance in understanding the global carbon cycle and impact of fossil fuel CO_2.

The basic "problem" for the sulphide system in calcium carbonate-rich sediments is the general lack of reactive iron to produce iron sulphide minerals. The sulphide that is produced by sulphate reduction then basically can only be buried in dissolved form in pore waters, be oxidized or diffuse out the sediments. For most carbonate-rich sediments, the oxidative process strongly dominates the fate of the sulphide. This is not grossly unlike what also happens in "normal" marine sediments. However, no sulphide minerals are produced, and since these sediments are carbonate-rich they will remain relatively close to equilibrium between calcium carbonate and their pore waters (see Fig. 7.4; Morse et al. 1985).

Figure 7.14 (Walter et al. 1993) shows the strong relationship that generally occurs in carbonate muds from Florida Bay between total carbon dioxide, excess dissolved calcium and the amount of sulphate that has been reduced. This figure is similar to Fig. 7.11, except instead of calcium decreasing, it increases indicating dissolution rather

Fig. 7.11. The relationships of the change in sulphate and calcium concentrations to the change in total dissolved carbon dioxide in Baffin Bay, Texas sediments. (From Morse et al. 1992)

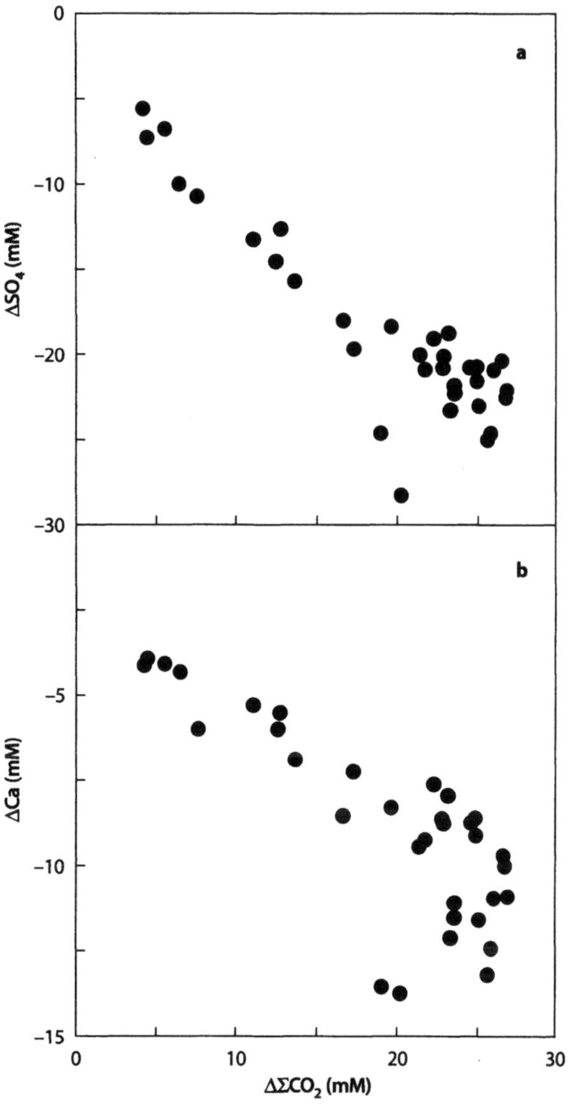

than precipitation. A probable explanation is that in the carbonate-rich sediments only typical for about 10% of the sulphate was reduced, whereas in the Baffin Bay, sediments close to total sulphate reduction often occurred. Referring to Fig. 7.3, it can be observed that the Florida Bay sediments are in the undersaturated region, whereas the Baffin sediments are in the supersaturated region. It is also noteworthy that the burrowed banks show much more extensive increases in calcium than the other mudbanks. This is in good agreement with the Aller and Rude (1988) observations for Long Island Sound sediments, that increased bioturbation leads to increased sulphide oxidation and carbonate dissolution.

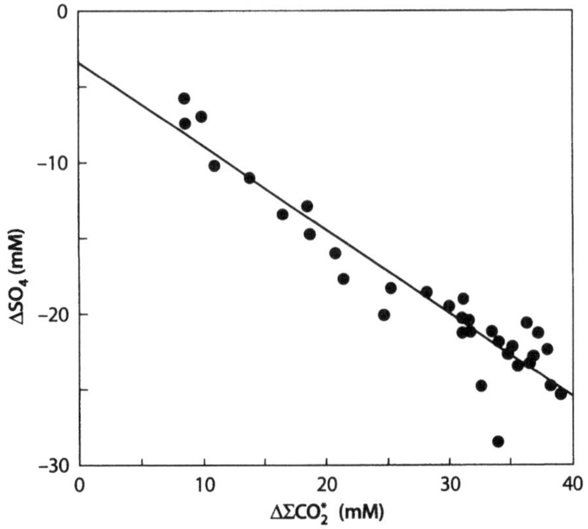

Fig. 7.12. The relationship of the change in sulphate concentration to the change in total dissolved carbon dioxide in Baffin Bay, Texas sediments when calcium carbonate precipitation is corrected for. The *line* is the ideal 1:2 ratio (redrawn from Morse et al. 1992)

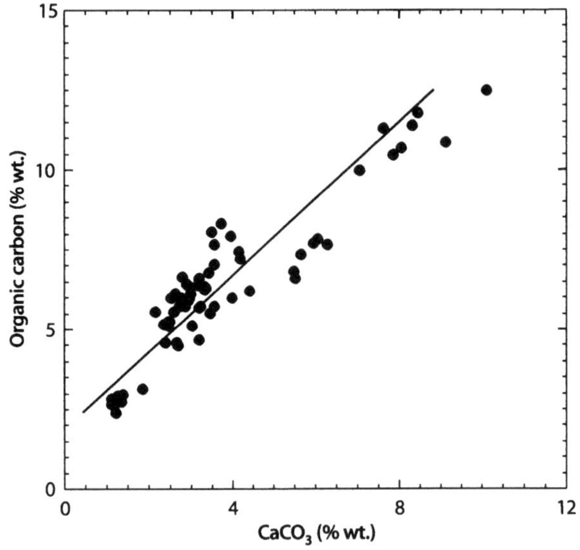

Fig. 7.13. The relationship between organic-C and calcium carbonate in Gulf of Mexico cold seep communities (redrawn from Arvidson and Morse, to be published)

7.4 Interactions of Toxic Metals with Sulphides in Anoxic Sediments

7.4.1 General Considerations

The study of the relationship of toxic metals to sedimentary sulphides has primarily focused on three major areas. Early observations indicated that the concentrations of

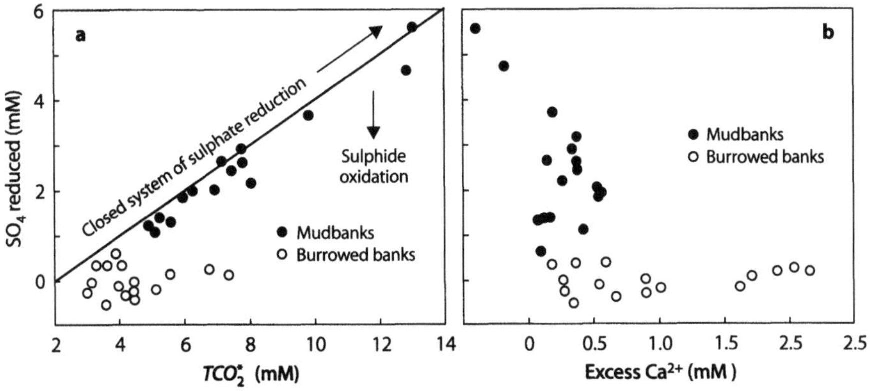

Fig. 7.14. The relationships of the change in sulphate to the change in; **a** total dissolved carbon dioxide; **b** calcium in Florida Bay sediments (redrawn from Walter et al. 1993)

toxic metals were often at elevated levels in sulphidic sediments (see Morse et al. 1987 for review). The observations of Di Toro et al. (1990) indicating that the toxicity of some metals was greatly reduced in sulphidic sediments stimulated interest in the topic Huerta-Diaz and Morse (1992) found that for many metals the "reactive" fraction was substantially "pyritized" during very early diagenesis. In the ensuing years, a rather vast literature has sprung up around this topic. This is in large part because of its practical application to assessing potential influences of toxic metals on ecosystems; a topic that remains highly controversial (e.g. Lee et al. 2000). Earlier (Sect. 7.2.3.3), it was pointed out that in sediments where there is both dissolved sulphide and carbonate, almost always the metal sulphides will limit metal solubility. Therefore, the discussion will be confined to metal sulphides here.

7.4.2 "Pyritization" of Trace Metals

Huerta-Diaz and Morse (1990) were able to determine trace metals that were co-leached with pyrite-Fe using a modification of the pyrite-Fe extraction method of Lord (1982). Since the development of this technique, thousands of analyses have been performed on sediments from numerous locations (e.g. Huerta-Diaz and Morse 1992; Morse et al. 1993), and clear patterns of behaviour for different trace metals have emerged. For several trace metals, it has been observed that there is often a fairly regular increase in the concentration of the metal occurring in the pyrite-extracted fraction with depth below the sediment-water interface.

In interpreting this behaviour, it has been found useful to apply a measure of the extent to which the operationally defined "reactive" fraction has become transformed into the fraction that extracts with pyrite-Fe. This is done in the same manner as the degree of pyritization (*DOP*) is calculated for Fe [*DOP* = pyrite-Fe / (pyrite-Fe + "reactive"-Fe)] and is referred to as the degree of trace metal pyritization (*DTMP*) (Huerta-Diaz and Morse 1990). Plots of *DTMP* against *DOP* have proven to be useful in establishing relationships for the relative degrees to which different metals are pyritized compared to for-

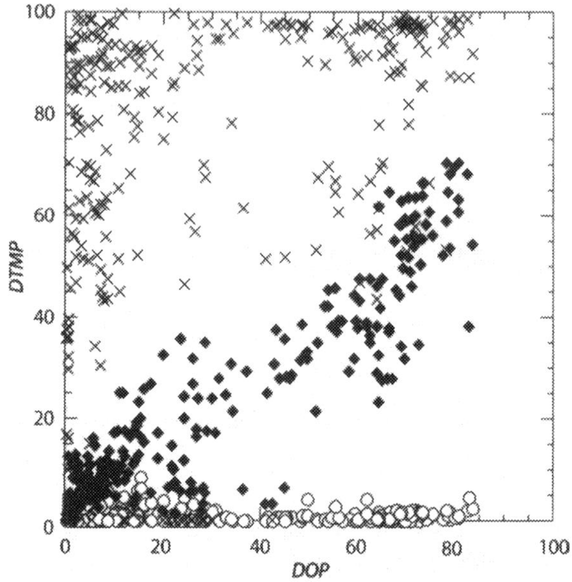

Fig. 7.15. Examples of different apparent degrees of trace metal pyritization (*DTMP*) vs. the degree of pyritization (*DOP*) of Fe (redrawn from Morse and Luther 1999) (x = Hg; ♦ = Co; O = Cd)

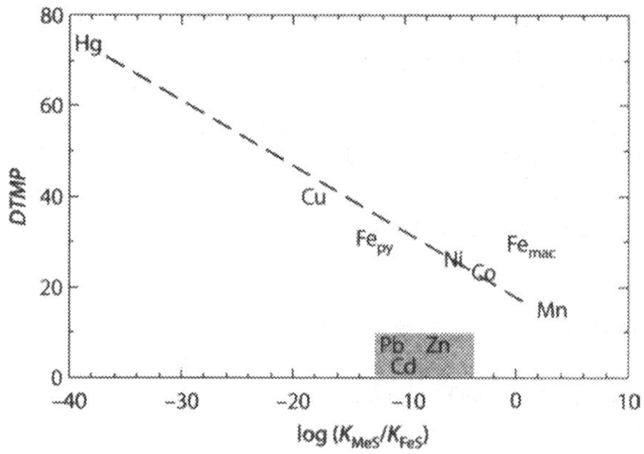

Fig. 7.16. Degree of trace metal pyritization (*DTMP*) vs. $\log(K_{MeS}/K_{mackinawite,\,FeS})$. Note that pyrite falls more closely to the line for all metals except CdS, PbS and ZnS that have *DTMPs* which fall well below those for other metals (redrawn from Morse and Luther 1999)

mation of the primary sedimentary sulphide mineral pyrite (Fig. 7.15). The observed pattern *DTMP* for different metals in decreasing order is Hg > As = Mo > Cu = Fe > Co > Ni >> Mn > Zn > Cr = Pb > Cd.

A plot (Fig. 7.16) of $-\log(K_{MeS}/K_{mackinawite,\,FeS})$ vs. *DTMP* produces a close to linear increase of MnS, CoS, NiS, Cu_2S, and HgS, which occurs with increasing *DTMP*, indicating a good relationship between metal sulphide solubility and *DTMP*. However, ZnS, CdS and PbS fall well below the line obtained for the other metals (Morse and Luther 1999). Cooper and Morse (1998) measured the extraction efficiency of pure metal sulphides in different solutions. Results provide at least a partial explanation for this difference in behaviour. The extent to which pure metal sulphides dissolve in cold HCl is: NiS_2 1%, HgS 1%, CuS 12%, Cu_2S 18%, NiS 23%, Ni_3S_2 28%; whereas CdS, PbS and

ZnS extract completely. For MeS compounds, this leads to Hg < Cu < Ni << Zn = Pb = Cd. This is close to the relationship observed for *DTMP*. Therefore, if Zn, Cd and Pb largely make their own sulphide, they would appear to have a low *DTMP* because they extract in the HCl fraction. Unfortunately, this means that it is not possible to determine how extensively Zn, Cd and Pb are sulphidized. The other metals may be coprecipitates with pyrite or occur as discrete phases.

Acknowledgements

The US National Science Foundation, Office of Naval Research, and the Texas Sea Grant Program have supported much of the work presented in this chapter. Many colleagues, post-doctoral fellows, graduate and undergraduate students participated in these studies. Dr. Morse was supported in his efforts to prepare and produce this chapter by funds from the Louis and Elizabeth Scherck Chair in Oceanography.

References

Aller RC (1982) Carbonate dissolution in near-shore terrigenous muds: The role of physical and biological reworking. J Geol 90:79–95

Aller RC (1988) Benthic fauna and biogeochemical processes in marine sediments: The role of burrow structures. In: Blackburn TH, Sorensen J (eds) Nitrogen cycling in coastal marine environments. John Wiley and Sons, New York, pp 301–338

Aller RC, Rude PD (1988) Complete oxidation of solid phase sulfides by manganese and bacteria in anoxic marine sediments. Geochim Cosmochim Acta 52:751–765

Arvidson RS, Morse JW (to be published) Controls on rates of sulfate reduction in chemosynthetic cold seep communities, Gulf of Mexico, USA. Geochim Cosmochim Acta

Ben-Yaakov S (1973) pH buffering of pore water of recent anoxic sediments. Limnol Oceanogr 18:86–94

Berner RA (1971) Principles of chemical sedimentology. McGraw-Hill, New York

Berner RA (1982) Burial of organic carbon and pyrite sulfur in the modern ocean: Its geochemical and environmental significance. Am J Sci 282:451–473

Berner RA (1984) Sedimentary pyrite formation: An update. Geochim Cosmochim Acta 48:605–615

Berner RA, Raiswell R (1984) C/S method for distinguishing freshwater from marine sedimentary rocks. Geology 12:365–368

Bernstein LD, Morse JW (1985) The steady-state calcium carbonate ion activity product of recent shallow water carbonate sediments in seawater. Mar Chem 15:311–326

Boudreau BP (1991) On a reactive continuum representation of organic matter diagenesis. Am J Sci 291:507–538

Canfield DE (1988) Sulfate reduction and the diagenesis of iron in marine sediments. PhD dissertation, Yale University

Canfield DE (1989) Reactive iron in marine sediments. Geochim Cosmochim Acta 53:619–632

Canfield DE (1994) Factors influencing organic carbon preservation in marine sediments. Chem Geol 114:315–329

Cooper DC, Morse JW (1998) Biogeochemical controls on trace metal cycling in anoxic marine sediments. Environ Sci Technol 32:327–330

Davison W (1991) The solubility of iron sulfides in synthetic and natural waters at ambient temperatures. Aquat Sci 53/54:309–329

Di Toro DM, Mahony JD, Hansen DJ, Scott KJ, Hicks MB, Mayr SM, Redmond MS (1990) Toxicity of cadmium in sediments: The role of acid volatile sulfide. Environ Sci Technol 9:1487–1502

Di Toro DM, Mahony JD, Hansen DJ, Scott KJ, Carlson AR, Ankley GT (1992) Acid volatile sulfide predicts the acute toxicity of cadmium and nickel in sediments. Environ Sci Technol 26:96–101

Eldridge PM, Morse JW (2000) A diagenetic model for sediment-seagrass interactions. Mar Chem 70:89–104

Fossing H (1995) ^{35}S-radiolabeling to probe biogeochemical cycling of sulfur. In: Vairavamurthy MA, Schoonen MAA (eds) Geochemical transformations of sedimentary sulfur. ACS Press, Washington, D.C. (ACS Symp. Ser 162, pp 348–364)

Goldhaber MM, Kaplan IR (1974) The sulfur cycle. In: Goldberg ED (ed) The sea, vol V. John Wiley and Sons, New York, pp 596–657

Hoehler TM, Alperin PI, Albert DB, Martens CS (1994) Field and laboratory studies of methane oxidation in an anoxic marine sediment: Evidence for a methanogen-sulfate reducer consortium. Global Biogeochem Cycles 8:451–463
Huerta-Diaz MA, Morse JW (1990) A quantitative method for determination of trace metal concentration in sedimentary pyrite. Mar Chem 29:119–144
Huerta-Diaz MA, Morse JW (1992) The pyritization of trace metals in anoxic marine sediments. Geochim Cosmochim Acta 56:2681–2702
Jorgensen BB (1978) A comparison of methods for the quantification of bacterial sulfate reduction in coastal marine sediments. I. Measurement with radiotracer techniques. J Geomicrobiol 1:11–27
Jorgensen BB (1990) The thiosulfate shunt in the sulfur cycle of marine sediments. Science 249:152–154
Krauskopf K B (1956) Factors controlling the concentrations of thirteen rare metals in sea-water. Geochim Cosmochim Acta 9:1–32
Ku TCW, Walter LM, Coleman ML, Blake RE, Martini AM (1999) Coupling between sulfur recycling and syndepositional carbonate dissolution: Evidence from oxygen and sulfur isotope composition of pore water sulfate, South Florida Platform, USA. Geochim Cosmochim Acta 63:2529–2546
Lee BG, Groscom SB, Lee JS, Choi HJ, Koh CH, Luoma SN, Fisher NS (2000) Influences of dietary uptake and reactive sulfides on metal bioavailability from aquatic sediments. Science 287:282–284
Lin S, Morse JW (1991) Sulfate reduction and iron sulfide mineral formation in Gulf of Mexico anoxic sediments. Am J Sci 291:55–89
Lord CJ III (1982) A selective and precise method for pyrite determination in sedimentary materials. J Sed Petrol 52:664–666
Luther GW III (1991) Pyrite synthesis via polysulfide compounds. Geochim Cosmochim Acta 55:2839–2849
Manheim FT (1961) A geochemical profile in the Baltic Sea. Geochim Cosmochim Acta 25:52–70
Milliman JD (1993) Production and accumulation of calcium carbonate in the ocean: Budget of a nonsteady state. Global Biogeochem Cycles 7:927–957
Morse JW (1994) Interactions of trace metals with authigenic sulfide minerals: Implications for their bioavailability. Mar Chem 46:1–6
Morse JW, Berner, RA (1995) What controls sedimentary C/S ratios? Geochim Cosmochim Acta 59:1073–1077
Morse JW, Cornwell JC (1987) Analysis and distribution of iron sulfide minerals in recent anoxic marine sediments. Mar Chem 22:55–69
Morse JW, Emeis KC (1990) Controls on C/S ratios in hemipelagic sediments. Am J Sci 290:1117–1135
Morse JW, Luther GW III (1999) Chemical influences on trace metal-sulfide interactions in anoxic sediments. Geochim Cosmochim Acta 63:3373–3379
Morse JW, Mackenzie FT (1990) Geochemistry of sedimentary carbonates. Elsevier, Amsterdam
Morse JW, Zullig JJ, Bernstein LD, Millero FJ, Milne P, Mucci A, Choppin GR (1985) Chemistry of calcium carbonate-rich shallow water sediments in the Bahamas. Am J Sci 285:147–185
Morse JW, Millero FJ, Cornwell J, Rickard D (1987) The chemistry of the hydrogen sulfide and iron sulfide systems in natural waters. Earth-Sci Rev 24:1–42
Morse JW, Cornwell, JC, Arakaki T, Lin S, Huerta-Diaz MA (1992) Iron sulfide and carbonate mineral diagenesis in Baffin Bay, Texas. J Sed Petrol 62:671–680
Morse JW, Presley BJ, Taylor RJ, Benoit G, Santschi P (1993) Trace metal chemistry of Galveston Bay: Water, sediments and biota. Mar Environ Res 36:1–37
Pamatmat M (1971) Oxygen consumption by the seabed. IV. Shipboard and laboratory measurements. Limnol Oceanogr 16:536–550
Pyzik AJ, Summer JE (1981) Sedimentary iron monosulfides: Kinetics and mechanism of formation. Geochim Cosmochim Acta 45:687–698
Raiswell R, Berner RA (1986) Pyrite and organic matter in Phanerozoic normal marine shales. Geochim Cosmochim Acta 50:1967–1976
Redfield AC, Ketchum BH, Rickard FA (1963) The influence of organisms on the composition of seawater. In: Hill MN (ed) The Sea, vol II. John Wiley and Sons, New York, pp 27–77
Rickard DT (1975) Kinetics and mechanism of pyrite formation at low temperatures. Am J Sci 275:636–652
Rickard DT (1997) Kinetics of pyrite formation by the H_2S oxidation of iron(II) monosulfide in aqueous solutions between 25 °C and 125 °C: The rate equation. Geochim Cosmochim Acta 61:115–134
Rickard DT, Luther GW III (1997) Kinetics of pyrite formation by the H_2S oxidation of iron(II) monosulfide in aqueous solutions between 25 °C and 125 °C: The mechanism. Geochim Cosmochim Acta 61:135–147
Rickard DT, Schoonen MAA, Luther GW III (1995) Chemistry of iron sulfides in sedimentary environments. In: Vairavamurthy MA, Shoonen MAA (eds) Geochemical transformations of sedimentary sulfur. ACS Press, Washington, D.C. (ACS Symp. Ser 162, pp 168–193)
Stumm W, Morgan JJ (1996) Aquatic geochemistry. John Wiley and Sons, New York

Thode HG (1991) Sulfur isotopes in nature and the environment: an overview. In: Krouse HR, Grinenko VA (eds) Stable isotopes: Natural and anthropogenic sulphur in the environment, Scope 43. John Wiley and Sons, New York, pp 1–26

Volkov II, Rosanov AG (1983) The sulfur cycle in the oceans. In: Volkov II, Rosanov AG (eds) The global biogeochemical sulfur cycle, Scope 19. John Wiley and Sons, New York pp 357–447

Walter LM, Burton EA (1990) Dissolution of recent platform carbonate sediments in marine pore fluids. Am J Sci 290:601–643

Walter LM, Bischof SA, Patterson WP, Lyons TW (1993) Dissolution and recrystallization in modern shelf carbonates: Evidence from pore water and solid phase chemistry. Phil Trans R Soc Lond A 344:27–36

Part II

Chemical Equilibria and Speciation in Sea Water

Chapter 8

Speciation of Metals in Natural Waters

F. Millero · D. Pierrot

8.1 Introduction

The reactions of many trace metals in natural waters are affected by their speciation or form. This will affect the biological uptake (Anderson and Morel 1982) and toxicity (Sunda and Ferguson 1983) as well as the solubility (Liu and Millero 1999). For example, Fe(II) and Mn(II) are biologically available for marine organisms, while Fe(III) and Mn(IV) are not normally available. Although the form of an element in natural waters can be four phases (solid, gas, colloid and dissolved), we will only consider the form of a metal in the dissolved state. The definition of a dissolved metal is defined by the filter size used to separate solid and colloidal phases from the soluble form. In the past, this separation was made with a 0.45 µm filter, and in more recent work smaller size filters are used (~0.2 µm). The speciation of metals is controlled by ionic interactions of the metals with inorganic (Cl^-, OH^-, CO_3^{2-}, etc.) and organic (fulvic and humic acids) ligands. The dissolved forms of a metal like Fe in sea water can include:

- Free ions: Fe^{2+}, Fe^{3+}
- Inorganically Complexed: $Fe(OH)^+$, $Fe(OH)_2$, $FeCO_3$ and $Fe(CO_3)_2^{2-}$, $FeCl^{2+}$, $FeCl_2^+$, $FeSO_4^+$, $Fe(SO_4)_2^{2-}$, $Fe(OH)^{2+}$, $Fe(OH)_2^+$, $Fe(OH)_3$, $FeCO_3^+$, and $Fe(CO_3)_2^{2-}$
- Organically Complexed: FeL, where L can be a wide range of unknown natural ligands (fulvic and humic acids, siderifores, etc.)

The speciation can also affect the rates of redox reactions (Millero 1990a,b) in natural water. Cu^+ is rapidly oxidized with O_2 or H_2O_2, while CuCl and $CuCl_2^-$ are nearly inert to the oxidation (Sharma and Millero 1988, 1989). $CuHS^-$ is rapidly oxidized with O_2, while $Zn(HS)_2$ is not oxidized (Vasquez et al. 1989). Cu^{2+} is rapidly reduced by H_2O_2, while $CuCO_3$ or Cu-EDTA is not reduced (Millero et al. 1991, 1992).

Some of the factors that can control the equilibrium speciation in natural waters are:

1. The oxidation potential (Eh)
2. The pH
3. The inorganic ligands (OH^-, CO_3^{2-})
4. The organic ligands (humics)

The ionic strength, temperature and pressure are also important. The importance of Eh on the speciation of Fe is shown in Fig. 8.1. At low pH and Eh, iron is predomi-

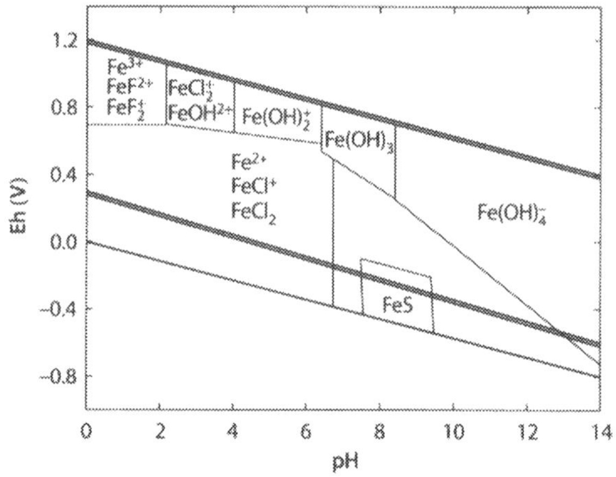

Fig. 8.1. The effect of pH and Eh on the speciation of Fe in natural waters (Millero 1996)

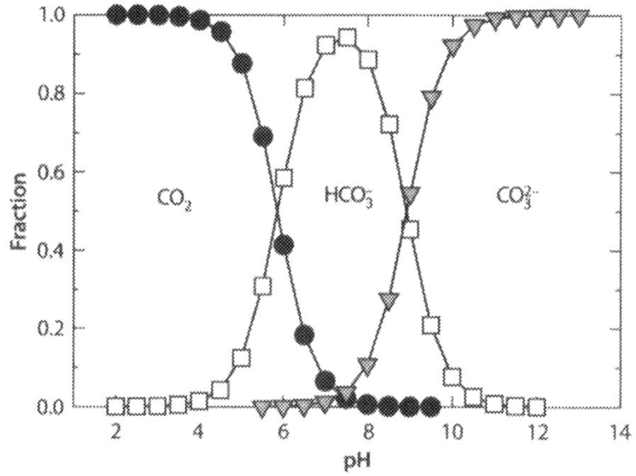

Fig. 8.2. The effect of pH on the speciation of carbonate ions in sea water (Millero 1996)

nantly in the ferrous form, while at high pH and Eh it is in the ferric form. Since the ligands are also a function of pH (Fig. 8.2), it can also be important in controlling the amount of ligands available. At high pH, the more negatively charged ligands (CO_3^{2-}, PO_4^{3-}) will form stronger complexes. On a molecular level, the hydration of a metal can also change its equilibrium structure. This can include the formation of covalently bonded, contact, solvent shared, and solvent separated (Fig. 8.3) ion-pairs. The type of inorganic ligand can be important for different types of metals. Heavy metals (Cu^+, Ag^+, Hg^{2+}) form strong complexes with halides Cl^-, Br^-, etc., while most divalent and trivalent metals form strong complexes with OH– and CO_3^{2-}. Most of the transition metals form strong complexes with organic ligands. The importance of the formation of organic complexes on the behaviour of metals in natural waters can be demonstrated by the effect of Cu(II) on the growth of natural bacterial populations. The uptake of tritiated (^3H) amino acid as a function the total Cu with various levels of NTA ligand is shown in Fig. 8.4 (Sunda and Ferguson 1983). The addition of Cu to the solution suppresses the uptake of the amino acid. The addition of NTA allows one to add more

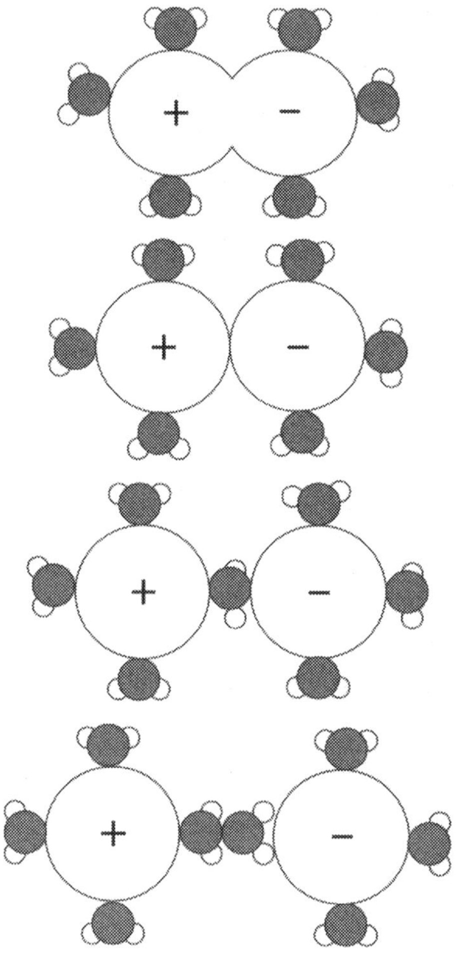

Fig. 8.3. Types of ion-pairs for cation and anion interactions (Millero 1996)

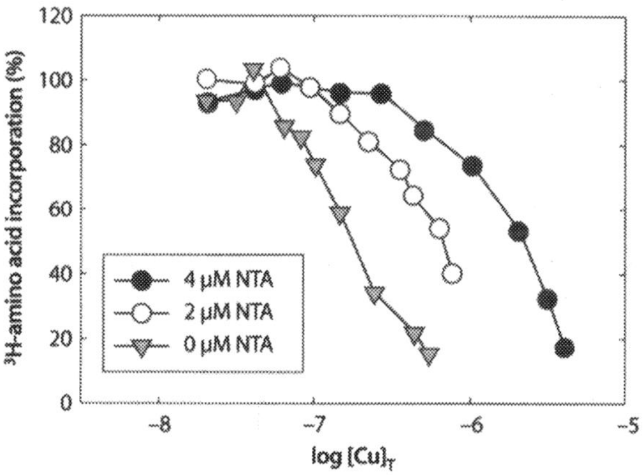

Fig. 8.4. The uptake of tritiated amino acids in sea water by natural populations of bacteria as a function of added total Cu^{2+} (Sunda and Ferguson 1983)

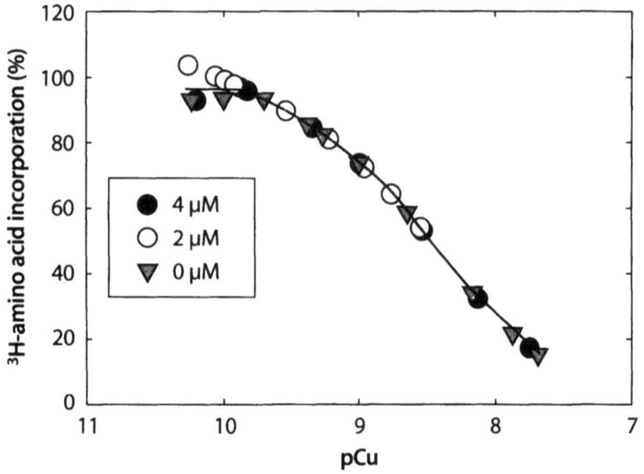

Fig. 8.5. The uptake of tritiated amino acids in sea water by natural populations of bacteria as a function of free Cu^{2+} (Sunda and Ferguson 1983)

Cu to the solution before the effect is seen. If one plots the effect as a function of the free Cu (Fig. 8.5) in the solution, all of the results fall on the same curve (Sunda and Ferguson 1983). Thus, free copper is toxic to the natural bacterial population, while organically complexed Cu is non-toxic.

8.2 Effect of Inorganic Speciation on the Solubility of Metals

The effect of the speciation of a metal on its solubility can be considered for Fe(III) in natural waters. The solubility of Fe(III) in sea water is controlled by (Byrne and Kester 1976; Millero et al. 1995):

$$Fe(OH)_3\,(s) + 3H^+ \leftrightarrow Fe^{3+} + 3H_2O \tag{8.1}$$

At equilibrium, the thermodynamic equilibrium constant is given by:

$$K_{Fe(OH)_3} = a_{Fe} a_{H_2O}{}^3 \,/\, a_H{}^3 \tag{8.2}$$

where $a_i = [i]\gamma_i$ are the activities of species i ($[i]$ is the concentration and γ_i is the activity coefficient of species i) and a_{H2O} is the activity of water. At a given ionic strength, the value of a_{Fe} in sea water is given by

$$a_{Fe} = \gamma_{Fe}[Fe^{3+}] = \gamma_{Fe}[Fe(III)] \,/\, (1 + \Sigma\beta_i^*[H^+]^{-n} + K^*_{FeX_i}[X_i]) \tag{8.3}$$

where the values of $[Fe^{3+}]$ and [Fe(III)] are the concentrations of free iron and total dissolved iron, respectively. The cumulative thermodynamic hydrolysis reactions are given by:

$$\beta_1\!: Fe^{3+} + H_2O = Fe(OH)^{2+} + H^+ \tag{8.4}$$

$$\beta_2\text{: } Fe^{3+} + 2H_2O = Fe(OH)_2^+ + 2H^+ \quad (8.5)$$

$$\beta_3\text{: } Fe^{3+} + 3H_2O = Fe(OH)_3^0 + 3H^+ \quad (8.6)$$

$$\beta_4\text{: } Fe^{3+} + 4H_2O = Fe(OH)_4^- + 4H^+ \quad (8.7)$$

At a given ionic strength, hydrolysis constants are given by:

$$\beta_j^* = [Fe(OH)_j^{(3-j)}][H^+]^j / a_{H_2O}^j[Fe^{3+}] = \beta_j\{a_{H2O}^j \gamma_{Fe} / \gamma_{Fe(OH)j}\gamma_H^j\} \quad (8.8)$$

where β_j is the thermodynamic hydrolysis constant, $[i]$, a_i, and γ_i are the concentration, activity and activity coefficient of species i in the ionic medium. The $[H^+]$ is defined on the free hydrogen ion molality scale (Byrne and Kester 1976; Millero 1986).

One also needs to know the stability constants for the formation of Fe(III) complexes with Cl^-, SO_4^{2-}, CO_3^{2-}, etc.

$$Fe^{3+} + nX_i^k = Fe(X_i)_n^{3+nk} \quad (8.9)$$

where $X_i = Cl^-$, etc. The stability constant at a given ionic strength can be represented by:

$$K_{FeXi}^* = [FeX_i] / [Fe^{3+}][X_i]^n = K_{FeX_i}\gamma_{Fe}\gamma_{X_i}^n / \gamma_{FeX_i} \quad (8.10)$$

where K_{FeX_i} is the thermodynamic stability constant.

The equilibrium solubility of [Fe(III)] can be determined using:

$$[Fe(III)] = K_{Fe(OH)_3}\gamma^3{}_H[H^+]^3 / (\alpha_{Fe}a^3{}_{H_2O}\gamma_{Fe}) \quad (8.11)$$

where the fraction of free Fe^{3+} is given by:

$$\alpha_{Fe} = [Fe^{3+}] / [Fe(III)] = 1 / (1 + \Sigma\beta_i^*[H^+]^{-i} + \Sigma K_{FeX_i}^*[X_i]) \quad (8.12)$$

The activity coefficient for free iron γ_{Fe} can be estimated from measurements made in a solution where no complex formation occurs ($NaClO_4$) at the same ionic strength of the solution or calculated from mean activity coefficient measurements in a solution that does not form strong interactions with Fe(III) ($NaClO_4$ or NaCl). The equilibrium constant ($K_{Fe(OH)_3}$) for the formation of amorphous iron hydroxide, $Fe(OH)_3$, in various ionic media has been estimated by a number of researchers (Baes and Mesmer 1976; Millero et al. 1995; Stumm and Morgan 1996; Liu and Millero 1999; Byrne and Luo 2000). The importance of the hydrolysis constants in controlling the solubility of Fe(III) in NaCl (0.7 m) and sea water ($S = 35$) is demonstrated in Figs. 8.6 and 8.7. The hydrolysis constants in NaCl give a reasonable representation of the solubility measurements (Liu and Millero 1999) over a wide range of pH values. The measurements for sea water (Byrne and Kester 1976; Kuma et al. 1996) are also well represented by the model, but are restricted to pH values below 8.5 due to the precipitation of $Mg(OH)_2(s)$.

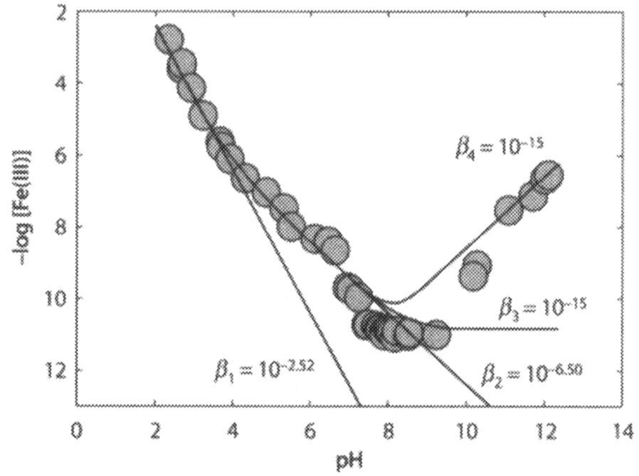

Fig. 8.6. The effect of pH on the solubility of Fe(III) in 0.7 M NaCl at 25 °C

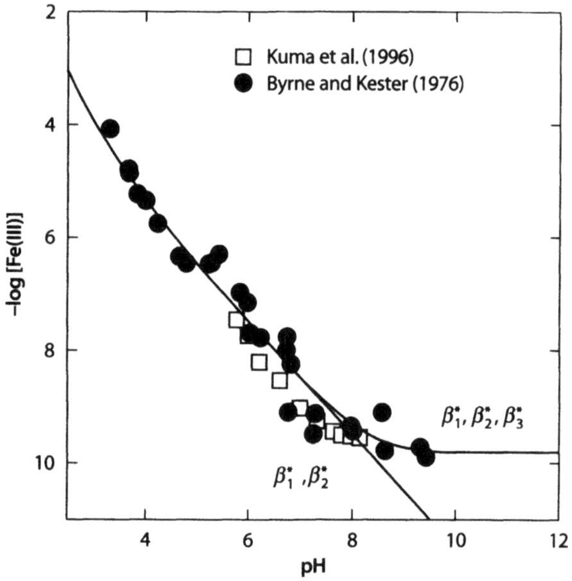

Fig. 8.7. The effect of pH on the solubility of Fe(III) in sea water at 25 °C and $S = 35$

This brief examination of the method that can be used to determine the solubility of Fe(III) in natural waters points out the need to have methods that can be used to estimate the activity coefficients of ions as a function of composition and ionic strength. Once the stability constants are known at a given ionic strength, the fraction of the free metal, $[M]_F$, and ligand, $[X]_F$, can be calculated from:

$$\alpha_M = [M]_F / [M]_T = 1 / (1 + \sum K^*_{FeX_i}[X_i]_F) \tag{8.13}$$

$$\alpha_X = [X]_F / [X]_T = 1 / (1 + \sum K^*_{FeX_i}[M_i]_F) \tag{8.14}$$

The fraction of the metal and ligand complexed respectively to various ligands (X_i) and metals (M_i) are given by:

$$\alpha_{MX_i} = \alpha_M K^*_{MX_i}[X_i] \tag{8.15}$$

$$\alpha_{M_iX} = \alpha_X K^*_{M_iX}[M_i] \tag{8.16}$$

These equations can be solved by a series of iterations. Although trace metals do not affect the concentration of the major ligands in the solution (Cl, OH, SO_4, CO_3), the major cations (Mg, Ca) must be considered. This is normally done first so that the concentration of the free major ligands can be estimated before the iteration of the trace metal speciation. In the next section, we will outline the model that we have developed to calculate the trace activity coefficients in natural waters over a wide range of composition (Na, Mg, Ca, K, Sr, Cl, SO_4, HCO_3, Br, CO_3, $B(OH)_4$, F) with major (Cl, SO_4, CO_3) and trace (OH, PO_4) ligands.

8.3
Estimation of the Activity Coefficients of Ions in Natural Waters

To estimate the activity coefficients of ions in mixed electrolyte solutions, a self-consistent model valid over a wide range of ionic strengths and composition is needed (Millero 1982, 2001). One would also like to know the form or speciation of metals in waters of interest. The estimation of the activity coefficients of ions in natural waters can be determined by using the ion pairing model (Garrels and Thompson 1962; Truesdale and Jones 1969; van Breeman 1973; Dickson and Whitfield 1981; Turner et al. 1981; Millero and Schreiber 1982) and the specific interaction model (Pitzer 1979, 1991 Harvie and Weare 1980; Harvie et al. 1984; Millero 1982; Millero and Roy 1997; Millero and Pierrot 1998). The recent progress in using these models to estimate the activity of ionic solutes and speciation of metals is described elsewhere (Millero 2001). The specific interaction model as formulated by Pitzer (1979, 1991) is used to estimate the activity coefficients in our model. The general equation is given by:

$$\ln \gamma_i = \text{D.H.} + \Sigma_{ij} m_i m_j B^{\gamma}_{ij} + \Sigma_{ijk} m_i m_j m_k C^{\gamma}_{ijk} \tag{8.17}$$

where D.H. is a form of the Debye-Hückel limiting law. The B^{γ}_{ij} and C^{γ}_{ijk} parameters are related to the binary (ions i and j) and ternary (ions i, j and k) interactions and can be a function of ionic strength. The activity coefficient of a cation (M) in a mixed electrolyte is given by:

$$\ln \gamma_M = Z^2_M f^{\gamma} + 2\Sigma_a m_a (B_{Ma} + EC_{Ma}) + Z^2_M R + Z_M S + \Sigma_c m_c (2\Theta_{Mc} + \Sigma_a m_a \Psi_{Mca})$$
$$+ \Sigma_a \Sigma_{a'} m_a m_{a'} \Psi_{aa'M} + \Sigma_c m_c 2^E\Theta_{Mc} + Z^2_M R_1 + Z^2_M R_2 \tag{8.18}$$

It can be attributed to six contributions:

1. The Debye-Hückel term ($Z^2_M f^{\gamma}$) is the limiting law, which is only a function of ionic strength.

2. The interaction parameters of M with the major anions (a) in the solution ($2\Sigma_a m_a(B_{Ma} + EC_{Ma})$ are determined from binary solutions of Ma (e.g. NaCl).
3. The interaction parameters of M with the major cations (c) ($\Sigma_c m_c 2\,\Theta_{Mc}$) are determined from ternary solutions (Ma + ca) (e.g. NaCl-$MgCl_2$).
4. The triplet interaction parameters of M with the major cations and anions ($\Sigma_c m_c \Sigma_a m_a \Psi_{Mca}$ and $\Sigma_a \Sigma_{a'} m_a m_{a'} \Psi_{aa'M}$) are determined from ternary solutions (Ma + ca).
5. The media terms for the major components ($Z_M^2 R + Z_M S$) are determined from binary solutions of the major components of the solution.
6. The higher order electrical terms ($\Sigma_c m_c 2^E\Theta_{Mc} + Z_M^2 R_1 + Z_M^2 R_2$) for the interactions of ions of different charge (Mg-Na) are a function of ionic strength.

As shown elsewhere (Millero 1982) for the trace components (H^+, Cu^{2+}, etc.) of the solution, only the major cation interaction parameters (2, 3, 4) are needed to make reasonable estimates of the "trace" activity coefficients (Millero 1982). The trace components of the solution do not contribute significantly to the media terms. Although this simplifies the estimation of the activity coefficients of trace constituents, many trace constituents form strong interactions with the major (SO_4^{2-}) and minor components (OH^-, CO_3^{2-}, etc.) and cannot be accounted for using reasonable values for the interaction parameters. To correct for these strong interactions, one must consider the formation of an ion pair between the cation and anion. This leads to parameters for the ion pairs that are model dependent.

At a given temperature, the model requires the interaction terms β^0_{MX}, β^1_{MX}, β^2_{MX}, and C^ϕ_{MX} for all the major electrolytes (MX) that make up the solution. These parameters account for the binary interactions of the individual components of each electrolyte (M-M, X-X, and MX) in the mixture (Pitzer and Mayorga 1973, 1974). The values of β^0_{MX}, β^1_{MX}, β^2_{MX}, C^ϕ_{MX} for the major components of natural waters are available at 25 °C and are tabulated elsewhere (Pitzer 1991). The Pitzer activity coefficient parameters for a number of electrolytes important in natural systems (HCl, NaCl, KCl, NaOH, $MgCl_2$, $CaCl_2$, Na_2SO_4, K_2SO_4, $MgSO_4$, $CaSO_4$) are known (Møller 1988; Greenberg and Møller 1989; Spencer et al. 1990; Pabalan and Pitzer 1987) over a wide range of temperatures (0 to 250 °C) and have been fitted to equations of a form (Pabalan and Pitzer 1987; Møller 1988; Greenberg and Møller 1989) such as:

$$P(T) = a_1 + a_2(1/T - 1/T_R) + a_3 \ln(T/T_R) + a_4(T - T_R) + a_5(T^2 - T_R^2) \tag{8.19}$$

P is β^0_{MX}, β^1_{MX}, or C^ϕ_{MX} and T_R is a reference temperature (298.15 K) and a_i are adjustable parameters given elsewhere (Millero and Pierrot 1998). Data over a more limited temperature range (0 to 50 °C) (Simonson et al. 1987a,b, 1988) are fitted to equations of the form:

$$\beta^0_{MX}(T) = a + b(T - T_R) + c(T - T_R)^2 \tag{8.20}$$

One can make reasonable estimates of activity coefficients over smaller ranges of temperature (0 to 50 °C) by using heat capacity (Criss and Millero 1996, 1999) and enthalpy (Silvester and Pitzer 1978) data at 25 °C (Millero 1979a). The

Pitzer parameter relative to a reference temperature T_R (298.15 K) can be determined by:

$$\beta^0 = \beta^0_R + q_1(1/T - 1/T_R) + q_2(T^2 - T^2_R) \tag{8.21}$$

where

$$q_1 = (\beta^{0J}/3)T^3_R - T^2_R\beta^{0L}_R \tag{8.22}$$

$$q_2 = \beta^{0J}/6 \tag{8.23}$$

Similar equations can be derived for β^1 and C^ϕ (Millero and Pierrot 1998).

The interactions of like charged ions (M-N and X-Y) are related to mixing parameters (Θ_{MN} and Θ_{XY}), which are determined (Pitzer and Kim 1974; Pitzer 1975; Harvie and Weare 1980) from solubility or activity measurements on ternary solutions (MX + NX and MX + MY). The values of Θ_{ij} and Ψ_{ijk} terms are usually determined from regression analysis of experimental data from simple ternary systems for which the pure electrolyte parameters are known (Pitzer 1991). For a ternary mixture MX-NX, where M and N are cations and X is a common anion, the values of Θ_{ij} and Ψ_{ijk} are determined from:

$$(\ln\gamma_{MX} - \ln\gamma^0_{MX}) / m_N = \Theta_{MN} + ½(m_M + m_X)\Psi_{MNX} \tag{8.24}$$

where γ^0_{MX} is the activity coefficient of MX calculated using only the pure electrolyte parameters and γ_{MX} is the experimental activity coefficient. The values of Θ_{MN} and Θ_{XY} are assumed to be independent of the common ion (X or Y and M or N) (called Young's Second Rule). Triplet interactions (M-N-X and M-X-Y) that may occur in these mixtures are accounted for by the addition of the parameters Ψ_{MNX} and Ψ_{MXY}. It should be pointed out that the Θ_{MN} and Ψ_{MNX} terms are normally not large and do not contribute much in dilute solutions (Millero 1982). The values of Θ_{MN} and Ψ_{MNX} are model dependent, since they require known values of β^0_{MX}, β^1_{MX}, β^2_{MX}, C^ϕ_{MX}, β^0_{NX}, β^1_{NX}, β^2_{NX}, and C^ϕ_{NX} and depend on the experimental data used for their evaluation (activity or solubility). It is thus important to take care in mixing the parameters determined by various researchers.

The activity coefficient of non-electrolytes (N) in mixed electrolyte solutions is determined from:

$$\ln\gamma_N = \Sigma_c m_c(2\lambda_{Nc}) + \Sigma_a m_a(2\lambda_{Na}) + \Sigma_c\Sigma_a m_c m_a \zeta_{Nca} \tag{8.25}$$

The values of λ_{Nc}, λ_{Na} and ζ_{Nca} are for the interactions of non-electrolytes (N) with various cations (c = Na^+) and anions (a = Cl^-) (Pitzer 1991; Harvie and Weare 1980; Millero 2001).

For a solution containing the major components of sea water (Na^+, Mg^{2+}, Cl^-, and SO_4^{2-}), one needs to know 13 binary parameters (β^0_{NaCl}, β^1_{NaCl}, C^ϕ_{NaCl}; $\beta^0_{Na_2SO_4}$, $\beta^1_{Na_2SO_4}$, $C^\phi_{Na_2SO_4}$; $\beta^0_{MgCl_2}$, $\beta^1_{MgCl_2}$, $C^\phi_{MgCl_2}$; $\beta^0_{MgSO_4}$, $\beta^1_{MgSO_4}$, $\beta^2_{MgSO_4}$, C^ϕ_{MgSO4}) and 6 ternary parameters (Θ_{NaMg}, Θ_{ClSO_4}, Ψ_{NaMgCl}, Ψ_{NaMgSO_4}, Ψ_{ClSO_4Na}, Ψ_{ClSO_4Mg}). The addition of the other major sea water components (K^+, Ca^{2+}, Sr^{2+}, HCO_3^-, Br^-, CO_3^{2-} and F^-) requires more parameters, but they do not contribute much to the media term of the Pitzer equation.

8.4 The MIAMI Ionic Interaction Model

A sketch of the model we use is given in Fig. 8.8. The input can be any of the major components of sea water or the salinity. The detailed comments on the structure of our Pitzer model (Pitzer 1979, 1991) are given elsewhere (Millero and Pierrot 1998). The model is an extension of those developed by others (Harvie et al. 1984; Felmy and Weare 1986; Millero 1982; Pabalan and Pitzer 1987; Møller 1988; Greenberg and Møller 1989; Spencer et al. 1990; Campbell et al. 1993; Clegg and Whitfield 1991, 1995; Millero and Roy 1997; Millero and Pierrot 1998). The backbone of the model accounts for the interactions of the major cations (H^+, Na^+, K^+, Mg^{2+}, Ca^{2+}, Sr^{2+}) and anions (F^-, Cl^-, Br^-, OH^-, HCO_3^-, $B(OH)_4^-$, SO_4^{2-}, CO_3^{2-}, CO_2, $B(OH)_3$, HF) of sea water from 0 to 50 °C and $I = 0$ to 6 m). Since parameters for the major components (Pabalan and Pitzer 1987;

Fig. 8.8. Sketch of ionic interaction model

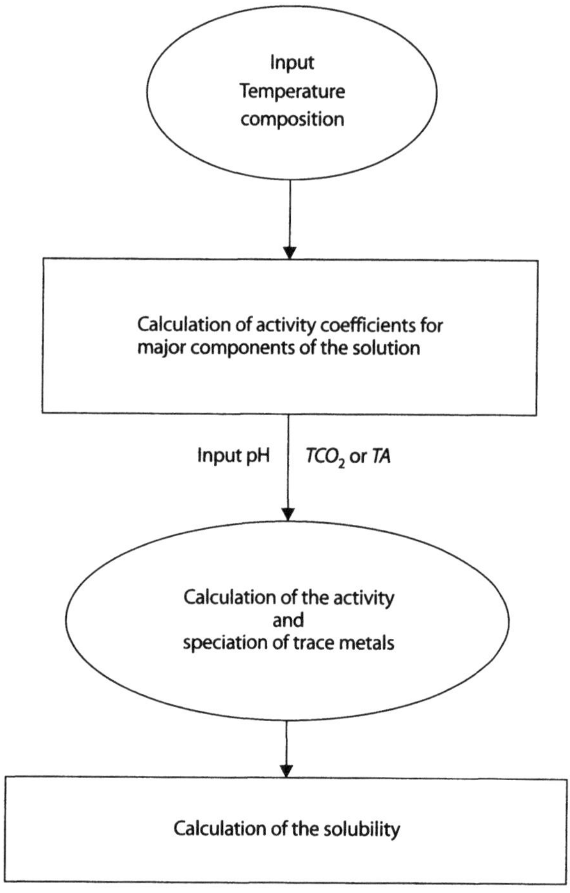

Møller 1988; Greenberg and Møller 1989) of sea water to 250 °C are in the model, it can be used to make estimates for these ions at higher temperatures (Millero and Pierrot 1998). The model includes the formation of a number of ion pairs (HF, MgF^+, CaF^+, HSO_4^-, $MgOH^+$, $MgB(OH)_4^+$, $CaB(OH)_4^+$, $MgCO_3$, $CaCO_3$, $SrCO_3$). Most of the binary terms (β^0_{MX}, β^1_{MX}, β^2_{MX}, C^{ϕ}_{MX}) come from activity coefficient measurements (Pitzer 1991). The higher order interaction terms (Θ_{ij}, Ψ_{ijk}) have been determined from emf and solubility measurements. A number of other cations and anions (see Table 8.1) have been added to the model (Millero 1982; Millero and Pierrot 1998). In this paper, we report on the addition of a number of divalent (Millero and Hawke 1992) and trivalent (Millero 1992) metals to our model. The temperature coefficients come mainly from enthalpy and heat capacity data as discussed earlier. The ionic strength range is limited to 2 m due to the lack of data that can be used to determine the activity coefficients of the metal ion pairs. It should be pointed out that our model directly calculates activity coefficients as well as dissociation and association constants and speciation. It does not directly calculate the solubility of minerals.

The thermodynamic constants for the dissociation of acids and the formation of ion pairs have been fitted to equations of the form (Millero 1979b, 1995):

$$\ln K = A + B / T + C \ln T + DT \tag{8.26}$$

where the parameters A, etc. are given elsewhere (Millero and Pierrot 1998; Millero 2001).

8.5 Reliability of the Model

The model has been used to estimate the activity coefficients of a number of cations and anions as a function of temperature for $S = 35$ sea water. The results for some cations and anions are given in Table 8.2. The values for most of the ions have large changes between 0 to 20 °C, but do not change much above 25 °C. The reliability of these activity coefficients can be examined by comparing them with the measured values (Millero and Pierrot 1998). This can best be done for sea water solutions at 25 °C, where a large number of studies have been made. The comparisons shown in Table 8.3 demonstrate that the model predicts reliable activity coefficients for a wide range of ions. Com-

Table 8.1. Ions considered in the Miami model

Monovalent cations	H, Li, Na, K, Rb, Cs, NH_4, Cu, Ag
Neutral solutes	NH_3, $B(OH)_3$, H_3PO_4, H_2S, SO_3, CO_2, HF
Divalent cations	Be, Mg, Ca, Sr, Ba, Mn, Fe, Co, Ni, Cu, Zn, UO_2, Cd, Pb, Hg
Trivalent cations	Fe, Al, Ga, In, La, Ce, Pr, Nd, Pm, Sm, Eu, Gd, Tb, Dy, Ho, Er, Tm, Yb, Lu, Y
Monovalent anions	F, Cl, Br, I, OH, HCO_3, $B(OH)_4$, ClO_3, ClO_4, BrO_3, CSN, NO_2, NO_3, HSO_4, HS, H_2PO_4, H_2AsO_4, Acetate
Divalent anions	SO_4, CO_3, SO_3, HPO_4, $HAsO_4$
Trivalent anions	PO_4, AsO_4

Table 8.2. Activity coefficients of some ions at different temperatures (Millero and Pierrot 1998)

Ion	0 °C	5 °C	10 °C	15 °C	20 °C	25 °C	30 °C	35 °C	40 °C
H^+	0.687	0.664	0.641	0.615	0.588	0.559	0.528	0.495	0.462
Li^+	0.737	0.733	0.729	0.725	0.721	0.717	0.713	0.709	0.704
Na^+	0.625	0.630	0.634	0.637	0.638	0.639	0.639	0.639	0.638
K^+	0.590	0.592	0.594	0.595	0.596	0.596	0.595	0.595	0.594
Rb^+	0.583	0.586	0.588	0.590	0.591	0.592	0.592	0.593	0.592
Cs^+	0.531	0.535	0.538	0.541	0.543	0.544	0.546	0.546	0.547
NH_4^+	0.588	0.592	0.595	0.597	0.599	0.601	0.602	0.604	0.604
$MgOH^+$	0.918	0.913	0.908	0.903	0.898	0.892	0.887	0.881	0.876
Mg^{2+}	0.222	0.218	0.215	0.211	0.207	0.203	0.199	0.195	0.191
Ca^{2+}	0.203	0.206	0.205	0.203	0.200	0.198	0.195	0.191	0.188
Sr^{2+}	0.203	0.201	0.199	0.197	0.195	0.193	0.190	0.187	0.184
F^-	0.313	0.314	0.316	0.318	0.319	0.321	0.322	0.324	0.325
Cl^-	0.684	0.688	0.690	0.691	0.692	0.692	0.692	0.691	0.690
Br^-	0.709	0.712	0.715	0.716	0.717	0.718	0.718	0.717	0.716
I^-	0.753	0.755	0.757	0.758	0.759	0.759	0.758	0.757	0.755
OH^-	0.312	0.304	0.296	0.288	0.281	0.273	0.266	0.259	0.253
HS^-	0.691	0.691	0.690	0.689	0.688	0.687	0.686	0.684	0.682
NO_2^-	0.660	0.656	0.653	0.650	0.647	0.643	0.640	0.636	0.632
NO_3^-	0.604	0.611	0.617	0.622	0.627	0.630	0.633	0.635	0.636
HCO_3^-	0.592	0.593	0.594	0.594	0.595	0.596	0.596	0.596	0.597
HSO_4^-	1.062	0.999	0.941	0.889	0.842	0.799	0.761	0.726	0.694
HSO_3^-	0.818	0.801	0.785	0.769	0.754	0.739	0.725	0.711	0.698
$B(OH)_4^-$	0.386	0.391	0.394	0.397	0.398	0.399	0.398	0.396	0.392
$H_2PO_4^-$	0.537	0.537	0.536	0.536	0.535	0.534	0.533	0.532	0.531
$H_2AsO_4^-$	0.819	0.816	0.812	0.807	0.803	0.799	0.795	0.790	0.786
$Acetate^-$	0.672	0.668	0.665	0.661	0.657	0.653	0.649	0.645	0.641
SO_4^{2-}	0.120	0.118	0.116	0.114	0.112	0.110	0.107	0.105	0.102
SO_3^{2-}	0.161	0.153	0.145	0.138	0.131	0.125	0.118	0.112	0.107
CO_3^{2-}	0.052	0.050	0.049	0.047	0.045	0.043	0.042	0.040	0.039
HPO_4^{2-}	0.060	0.060	0.060	0.059	0.059	0.059	0.058	0.058	0.057
$HAsO_4^{2-}$	0.113	0.111	0.109	0.106	0.104	0.102	0.099	0.097	0.094
PO_4^{3-}	3.3×10^{-5}	3.0×10^{-5}	2.9×10^{-5}	2.8×10^{-5}	2.6×10^{-5}	2.5×10^{-5}	2.5×10^{-5}	2.4×10^{-5}	2.3×10^{-5}
AsO_4^{3-}	9.9×10^{-4}	9.5×10^{-4}	9.0×10^{-4}	8.6×10^{-4}	8.2×10^{-4}	7.8×10^{-4}	7.3×10^{-4}	6.9×10^{-4}	6.5×10^{-4}
CO_2	1.208	1.197	1.188	1.179	1.171	1.162	1.153	1.143	1.134

parisons over a wider range of temperature can be made for only a few ions (H^+, Cl^-, NH_4^+, OH^-, HCO_3^-, CO_3^{2-}, $B(OH)_4^-$). The most extensive and reliable measurements are

Table 8.3. Comparison of measured and calculated activity coefficients of ions in sea water at 25 °C and $S = 35$ (based on $\gamma_{Cl} = 0.666$) (Millero and Pierrot 1998)

Ion	Measured	Calculated	Δ	Reference
H^+	0.590	0.581	0.009	Khoo et al. (1977a)
	0.589		0.008	Dickson (1990b); Campbell et al. (1993)
Na^+	0.668	0.664	0.004	Johnson and Pytkowicz (1981)
	0.670		0.006	Gieskes (1966)
	0.678		0.014	Platford (1965)
K^+	0.625	0.619	0.006	Whitfield (1975)
NH_4^+	0.616	0.624	−0.005	Khoo et al. (1977b)
	0.592		0.032	Johansson and Wedborg (1980)
Mg^{2+}	0.240	0.219	0.021	Thompson (1966)
Ca^{2+}	0.203	0.214	−0.011	Mucci (1983)
	0.180		−0.034	Culberson et al. (1978)
Sr^{2+}	0.190	0.208	−0.018	Culberson et al. (1978)
F^-	0.296	0.309	−0.013	Culberson et al. (1970)
Cl^-	0.666	0.666	0	Assigned value
OH^-	0.255	0.263	−0.008	Millero (1995)
	0.242		−0.021	Hansson (1972)
	0.254		−0.009	Culberson and Pytkowicz (1973)
	0.254		−0.009	Dickson and Riley (1979a)
HS^-	0.617	0.661	−0.044	Yao and Millero (1995)
HCO_3^-	0.570	0.574	−0.004	Mehrbach et al. (1973)
	0.587		0.013	Hansson (1973)
	0.586		0.012	Goyet and Poisson (1989)
	0.583		0.009	Roy et al. (1993)
HSO_4^-	0.782	0.770	0.012	Dickson (1990b)
$B(OH)_4^-$	0.390	0.384	0.006	Dickson (1990a)
	0.419		0.035	Hansson (1973)
	0.398		0.014	Byrne and Kester (1974)
	0.351		−0.033	Millero et al. (1993)
$H_2PO_4^-$	0.453	0.514	−0.061	Yao and Millero (1995)
	0.492		−0.022	Dickson and Riley (1979b)
	0.352		−0.162	Kester and Pytkowicz (1967)
SO_4^{2-}	0.104	0.102	0.002	Khoo et al. (1977)
	0.101		−0.001	Culberson et al. (1970)
	0.112		0.010	Whitfield (1975)
	0.121		0.019	Platford and Dafoe (1965)
	0.121		0.019	Culberson et al. (1978)

Table 8.3. *Continued*

Ion	Measured	Calculated	Δ	Reference
CO_3^{2-}	0.039	0.040	−0.001	Mehrbach et al. (1973)
	0.039		−0.001	Hansson (1973)
	0.037		−0.003	Roy et al. (1993)
	0.037		−0.003	Goyet and Poisson (1989)
HPO_4^{2-}	0.043	0.054	−0.011	Yao and Millero (1995)
PO_4^{3-}	0.00002	0.00002	0	Yao and Millero (1995)

Table 8.4. Comparison of measured and calculated values of pK* (total scale) in sea water at 25 °C and $S = 35$

Acid		Measured[a]	Calculated	Δ	Reference
H_2CO_3	pK_0	1.547	1.546	0.001	Weiss (1974)
	pK_1	5.837	5.832	0.005	Mehrbach et al. (1973)
		5.850		0.018	Hansson(1973)
		5.849		0.017	Dickson and Millero (1987)
		5.846		0.014	Goyet and Poisson (1989)
		5.847		0.015	Roy et al. (1993)
	pK_2	8.955	8.954	0.001	Mehrbach et al. (1973)
		8.942		−0.012	Hansson(1973)
		8.945		−0.009	Dickson and Millero (1987)
		8.919		−0.035	Goyet and Poisson (1989)
		8.915		−0.039	Roy et al. (1993)
NH_4^+		9.256	9.235	0.021	Yao and Millero (1995)
H_2O		13.211	13.218	−0.007	Millero (1995)
H_2S	pK_1	6.510	6.533	−0.023	Millero et al. (1988)
HF		2.611	2.552	0.059	Dickson and Riley (1979a)
H_2SO_3	pK_1	1.58	1.477	0.103	Millero et al.(1989)
	pK_2	6.13	6.159	−0.029	Millero et al.(1989)
$B(OH)_3$		8.582	8.596	−0.014	Dickson (1990a)
H_3PO_4	pK_1	1.591	1.605	−0.014	Yao and Millero (1995)
	pK_2	5.955	6.003	−0.048	Yao and Millero (1995)
	pK_3	8.783	8.746	0.037	Yao and Millero (1995)
HSO_4^-		0.983	0.993	−0.010	Dickson (1990b)
Calcite		6.367	6.397	−0.030	Mucci (1983)
Aragonite		6.186	6.221	−0.035	Mucci (1983)

[a] Millero and Pierrot (1998).

for the mean activity coefficient for HCl in sea water. Emf measurements have been made in artificial sea water with (Khoo et al. 1977a; Dickson 1990b; Campbell et al. 1993) and without SO_4^{2-} (Khoo et al. 1977a). The mean activity coefficient of HCl, $\gamma_\pm(HCl) = (\gamma_H\gamma_{Cl})^{0.5}$, determined from the model has been shown to be in good agreement with the measured values (Campbell et al. 1993). The deviations are generally within 0.005, with the exception of the 40 °C data that is reproduced within ±0.01 units (Millero and Pierrot 1998).

The model yields reliable dissociation constants for the ionization of acids in sea water from 0 to 50 °C (Millero and Roy 1997). Comparisons of the measured and calculated p*K*s for a number of acids at 25 °C are given in Table 8.4. The differences are close to the experimental error of the measurements and demonstrate the reliability of the model. Comparisons of the measured (Mehrbach et al. 1973) and calculated values of the pK_1 and pK_2 for carbonic acid in sea water as a function of salinity and temperature are shown in Figure 8.9. Again the comparisons are quite good over the salinity and temperature range of the measurements (0 to 35 °C and $S = 10$ to 40).

8.6 Speciation of Metals

As stated earlier in this paper, we have added divalent and trivalent metals to the model. The metals added are listed in Table 8.1. The effect of temperature on the activity coefficients of metal cations is given in Table 8.5. As with the major cations, the changes in the activity coefficients with temperature are much larger between 0 to 25 than 25 to 50 °C. The calculations of the speciation of some metals in sea water are shown in Figs. 8.10 to 8.12. Pb^{2+}, Cd^{2+}, and Hg^{2+} form strong complexes with Cl^- (Fig. 8.10). Be^{2+} is strongly complexed with OH^-, while UO_2^{2+} is highly complexed with CO_3^{2-} (Fig. 8.11). The transition metals (Fig. 8.12) Mn, Fe, Co, Ni and Zn are not strongly complexed,

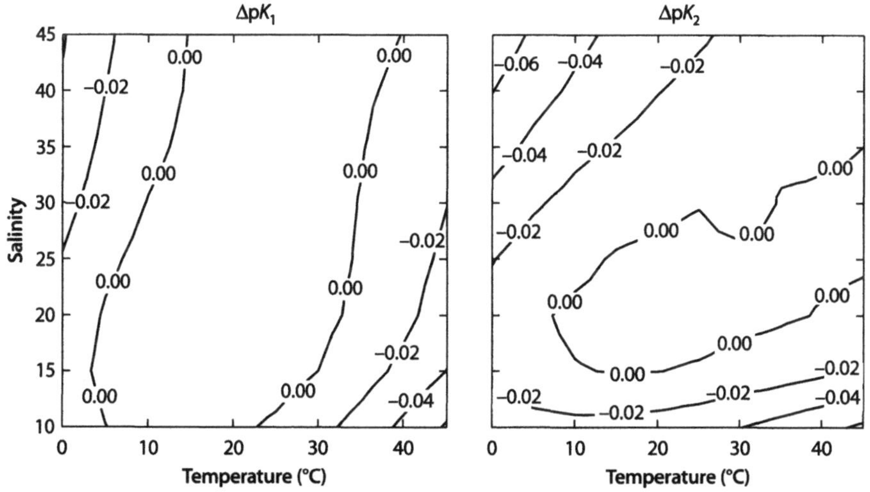

Fig. 8.9. Comparison of the measure (Millero and Roy 1997) and calculated p*K*'s for carbonic acid in sea water obtained using the ionic interaction model

Table 8.5. Activity coefficients of some trace monovalent, divalent and trivalent metals in sea water ($S=35$) at different temperatures

Ion	0 °C	5 °C	10 °C	15 °C	20 °C	25 °C	30 °C	35 °C	40 °C
Li^{+}	0.7404	0.7367	0.7329	0.7126	0.7250	0.7210	0.7168	0.7126	0.7082
Rb^{+}	0.5862	0.5888	0.5910	0.5958	0.5941	0.5950	0.5956	0.5958	0.5956
Cs^{+}	0.5338	0.5375	0.5407	0.5494	0.5457	0.5474	0.5486	0.5494	0.5496
Cu^{+}	0.5345	0.5318	0.5290	0.5136	0.5231	0.5200	0.5169	0.5136	0.5102
Mn^{2+}	0.2095	0.2053	0.2010	0.1776	0.1919	0.1873	0.1825	0.1776	0.1727
Fe^{2+}	0.2302	0.2255	0.2209	0.1965	0.2113	0.2064	0.2015	0.1965	0.1914
Co^{2+}	0.2317	0.2271	0.2224	0.1979	0.2128	0.2079	0.2029	0.1979	0.1928
Ni^{2+}	0.2373	0.2323	0.2272	0.2020	0.2171	0.2121	0.2071	0.2020	0.1969
Cu^{2+}	0.2021	0.1983	0.1943	0.1738	0.1863	0.1822	0.1780	0.1738	0.1695
Zn^{2+}	0.2162	0.2120	0.2077	0.1852	0.1989	0.1944	0.1898	0.1852	0.1805
UO_2^{2+}	0.2575	0.2524	0.2471	0.2199	0.2364	0.2310	0.2254	0.2199	0.2142
Be^{2+}	0.0860	0.0843	0.0825	0.0734	0.0790	0.0771	0.0753	0.0734	0.0715
Cd^{2+}	0.0988	0.0968	0.0948	0.0844	0.0907	0.0886	0.0865	0.0844	0.0822
Pb^{2+}	0.0322	0.0315	0.0309	0.0275	0.0295	0.0288	0.0282	0.0275	0.0268
Hg^{2+}	0.0815	0.0798	0.0782	0.0696	0.0748	0.0731	0.0713	0.0696	0.0678
Ba^{2+}	0.1812	0.1806	0.1795	0.1690	0.1763	0.1742	0.1717	0.1690	0.1659
La^{3+}	0.0567	0.0548	0.0530	0.0442	0.0494	0.0476	0.0459	0.0442	0.0425
Ce^{3+}	0.0588	0.0571	0.0554	0.0467	0.0520	0.0503	0.0485	0.0467	0.0450
Pr^{3+}	0.0579	0.0563	0.0546	0.0460	0.0512	0.0495	0.0478	0.0460	0.0443
Nd^{3+}	0.0577	0.0560	0.0544	0.0459	0.0510	0.0493	0.0476	0.0459	0.0442
Pm^{3+}	0.0579	0.0563	0.0546	0.0460	0.0512	0.0495	0.0478	0.0460	0.0443
Sm^{3+}	0.0585	0.0569	0.0553	0.0464	0.0519	0.0501	0.0483	0.0464	0.0445
Eu^{3+}	0.0599	0.0580	0.0561	0.0469	0.0524	0.0506	0.0487	0.0469	0.0450
Gd^{3+}	0.0605	0.0585	0.0565	0.0469	0.0526	0.0507	0.0488	0.0469	0.0451
Tb^{3+}	0.0603	0.0585	0.0567	0.0472	0.0529	0.0510	0.0491	0.0472	0.0453
Dy^{3+}	0.0609	0.0590	0.0571	0.0474	0.0532	0.0513	0.0493	0.0474	0.0454
Ho^{3+}	0.0605	0.0589	0.0572	0.0477	0.0535	0.0516	0.0497	0.0477	0.0457
Er^{3+}	0.0609	0.0590	0.0570	0.0475	0.0532	0.0513	0.0494	0.0475	0.0456
Tm^{3+}	0.0604	0.0586	0.0568	0.0474	0.0531	0.0512	0.0494	0.0474	0.0455
Yb^{3+}	0.0607	0.0588	0.0568	0.0474	0.0530	0.0512	0.0493	0.0474	0.0456
Lu^{3+}	0.0610	0.0589	0.0570	0.0473	0.0531	0.0511	0.0492	0.0473	0.0455
Y^{3+}	0.0605	0.0589	0.0571	0.0477	0.0535	0.0516	0.0497	0.0477	0.0457
Al^{3+}	0.0035	0.0034	0.0032	0.0025	0.0029	0.0028	0.0026	0.0025	0.0023
Ga^{3+}	0.0035	0.0034	0.0032	0.0025	0.0029	0.0028	0.0026	0.0025	0.0023
In^{3+}	0.0035	0.0034	0.0032	0.0025	0.0029	0.0028	0.0026	0.0025	0.0023
Fe^{3+}	0.1066	0.1027	0.0987	0.0782	0.0906	0.0864	0.0823	0.0782	0.0741

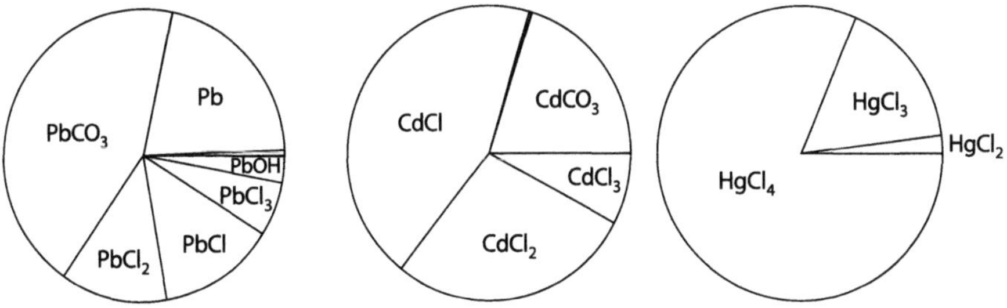

Fig. 8.10. The speciation of Pb, Cd and Hg in sea water at 25 °C and $S = 35$

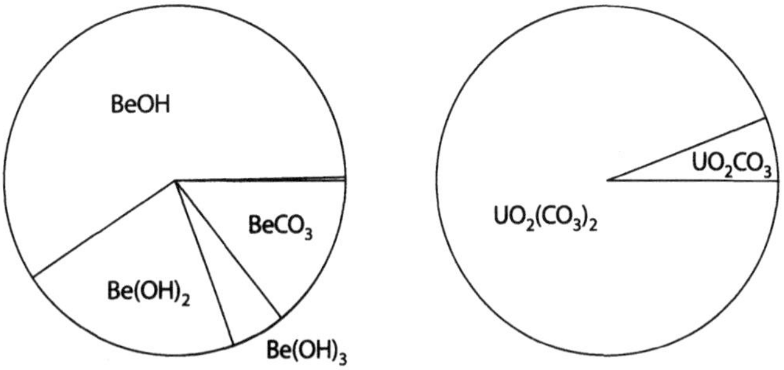

Fig. 8.11. The speciation of Be and UO_2 in sea water at 25 °C and $S = 35$

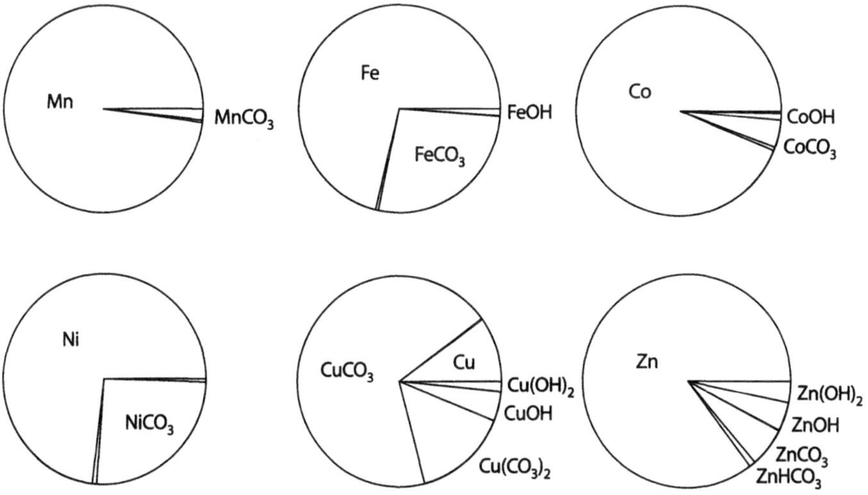

Fig. 8.12. The speciation of Mn, Fe, Co, Ni, Cu and Zn in sea water at 25 °C and $S = 35$

while Cu^{2+} forms strong carbonate complexes. The effect of pH on the speciation of Fe(III) in NaCl and sea water as a function of pH is shown in Figs. 8.13 and 8.14. In NaCl above a pH = 2, hydrolysis dominates the speciation. At a pH ~ 8, the dominant forms are $Fe(OH)_2^+$ and $Fe(OH)_4^-$. In sea water at low pH, the speciation is affected by the formation of F^- and SO_4^{2-} complexes. Near a pH of 8, the $Fe(OH)_3$ species appears to dominate. Since the hydrolysis constants for the formation of $Fe(OH)_4^-$ are not known, this may be an artefact. If the β_4 for NaCl is used for the sea water calculation, the $Fe(OH)_4^-$ becomes important.

Fig. 8.13. The effect of pH on the speciation of Fe(III) in 0.7 mNaCl at 25 °C

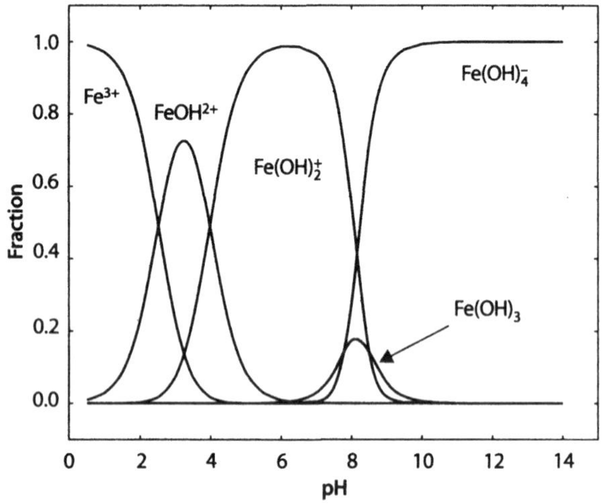

Fig. 8.14. The effect of pH on the speciation of Fe(III) in sea water at 25 °C and $S = 35$

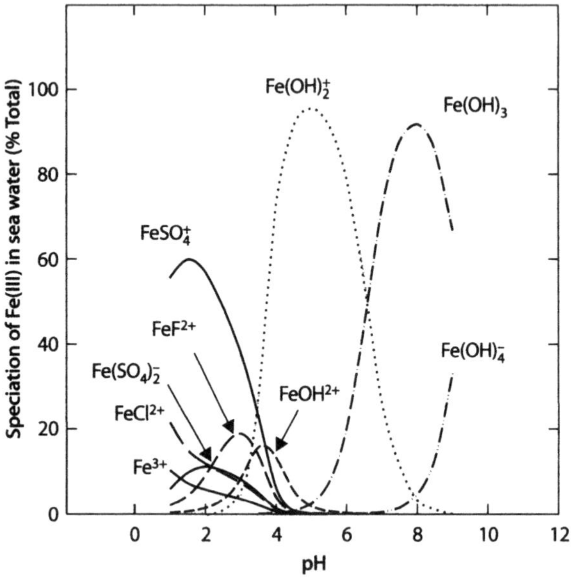

8.7 Formation of Metal Organic Complexes

When one examines the solubility of Fe(III) in natural waters (Liu and Millero 1999, 2002) one finds that the solubility in $S = 35$ sea water (200 pM) is much greater than in 0.7 m NaCl (10 pM) at a pH near 8 (Fig. 8.15). Since measurements made in the major sea salts and artificial sea water ($I = 0.7$ m) are similar to the values in NaCl (Fig. 8.16), one can attribute this to the effect of organic ligands (L) increasing the solubility due to the formation of Fe^{3+} complexes (FeL) (Liu and Millero 2002). The dilution of sea water with NaCl (Fig. 8.17) and the UV irradiation of sea water yield solubilities equal

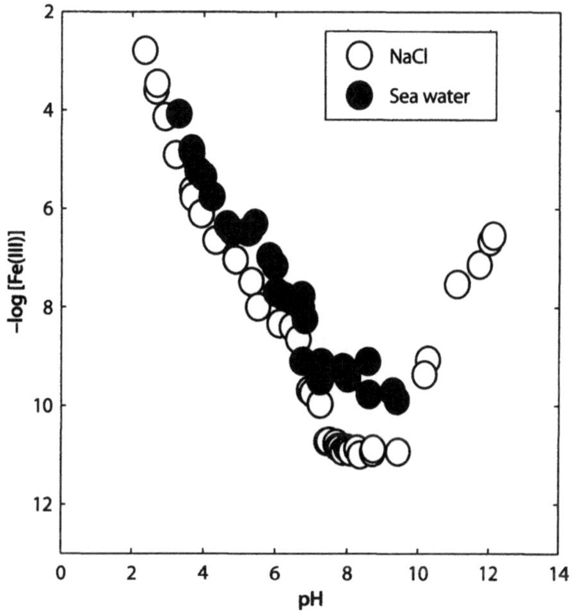

Fig. 8.15. Comparison of the solubility of Fe(III) in 0.7 M NaCl and sea water ($S = 35$) at 25 °C

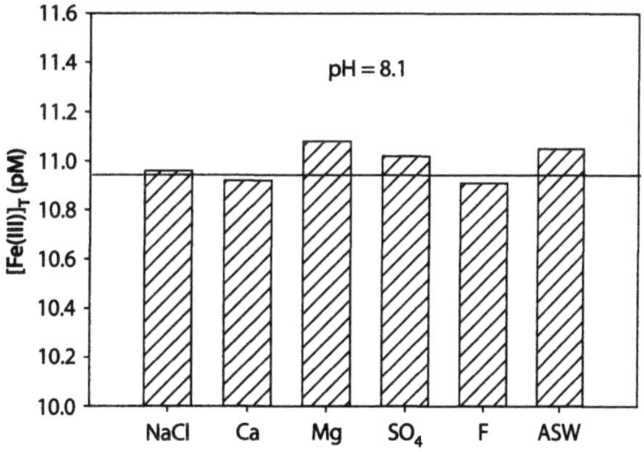

Fig. 8.16. The solubility of Fe(III) in the major sea salts and artificial sea water at I = 0.7 m, 25 °C and pH = 8

to the values in NaCl, supporting this notion (Liu and Millero 2002). This example makes it clear that the formation of metal complexes with natural organics can affect the solubility as well as the toxicity shown earlier. With this in mind, we have made it possible to use our model to examine the competition of inorganic and organic ligands with various metals. Although natural organic ligands have an unknown structure, researchers (Mantoura et al. 1978; Sohn and Hughes 1981) have been able to determine stability constants for the formation of metals with humic type ligands and those of unknown structure. Much of our knowledge of the concentration and strength of the metal organic ligands in sea water has come from using voltametric methods (Gledhill and van den Berg 1994; Wu and Luther 1995; Rue and Bruland 1995; van den Berg 1995).

Before we discuss these results, it is useful to briefly examine the methods used to study the formation of the complexes in natural waters between a metal (M) and organic ligand (L):

$$M + L = ML \tag{8.27}$$

Because the measurements are made directly in sea water, the formation constant used is defined in terms of easily measurable quantities:

$$K'_{ML} = [ML] / [M'][L'] \tag{8.28}$$

where [ML] is the concentration of the complex, [M'] is the concentration of the metal not complexed by L, and [L'] is the concentration of the free ligand not complexed by M. The values of [M'] and [L'] are related to the total concentrations by:

$$[M]_T = [M'] + [ML] \tag{8.29}$$

$$[L]_T = [L'] + [ML] \tag{8.30}$$

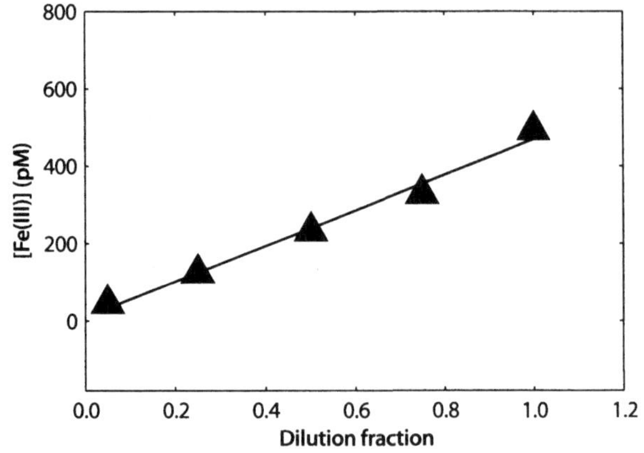

Fig. 8.17. The solubility of Fe(III) in sea water ($S = 35$) diluted with 0.7 m NaCl at 25 °C and pH = 8

where the subscript T is used to denote the total analytical concentrations. The values of [M'] and [L'] are related to the free metal and ligand by:

$$[M'] = [M^{n+}] + \Sigma MX_i \tag{8.31}$$

$$[L'] = [L^{n-}] + \Sigma N_iL \tag{8.32}$$

where X_i are the inorganic ligands (OH^-, CO_3^{2-}, etc.) that complex the metal and N_i are the cations that can complex the ligand (Mg^{2+}, Ca^{2+}, and other trace metals). Because natural organic ligands cannot normally be studied in simple solutions, it is not possible to determine the free ligand concentrations [L^{n-}], and the values of [L'] are normally reported. The concentration of the free metal not complexed to inorganic ligands can be determined by the methods discussed above:

$$[M^{n+}] = [M']\,\alpha_M \tag{8.33}$$

where α_M is the fraction of free metal in the solution without the organic ligand determined from:

$$\alpha_M = 1 / (1 + \Sigma K_{MX_i}[X]_i) \tag{8.34}$$

To examine the competition between organic and inorganic ligands, it is more appropriate to use the stability constant defined in terms of the free metal:

$$K_{ML} = [ML] / [M^{n+}][L'] \tag{8.35}$$

The free metal concentration in the solution with the organic ligand can be determined from:

$$[M^{n+}] = [M]_T\alpha^*_M \tag{8.36}$$

where

$$\alpha^*_M = 1 / (1 + \Sigma K_{MX_i}[X]_i + K_{ML}[L']) \tag{8.37}$$

Because the concentrations of the inorganic ligands are normally much higher than that of the organic ligands, one can use the fraction of free metals determined in sea water without organics to make a reasonable estimate of the value from:

$$K_{ML} \approx K'_{ML}\alpha^*_M. \tag{8.38}$$

The values of α^*_M, the fraction of free metal in sea water, with various concentrations of inorganic and organic ligands can be estimated from Eq. 8.37. For the organic complexes to dominate the speciation of a metal $K_{ML}[L'] > \Sigma K_{MX_i}[X]_i$. This can occur when K_{ML} or [L'] is large. In more simple terms, if the value of $[L'] > 1 / K_{ML}$ or $K_{ML} > 1 / [L']$, organic complexation can be important. For example, if $K_{ML} = 10^9\ M^{-1}$, the concentration of [L'] must be greater than 1 nM to start to affect the speciation. The competition between inorganic and organic ligands can be examined quickly by using the speciation programmes described above in Visual Basic.

Because fulvic and humic acids are the most abundant organic material in natural waters, a number of researchers have determined the stability constants for the formation of metal complexes with extracted fulvics and humics. The constants are frequently determined at low concentration (0.01 M) and pH (6) in an ionic medium ($NaClO_4$). Although these organics may not be the most important ligands in natural waters, they give one an idea of the magnitude of the effect organics can have on the speciation of metals. A summary of the average stability constants of metals with extracted organic matter is given in Fig. 8.18 (Mantoura et al. 1978). The stability constants for individual metals with different source material are in reasonable agreement. The results for a given extraction show a behaviour that is similar to that predicted by the Irving Williams order. Mercury forms quite strong complexes with humic material. Because the exact compositions of fulvic and humic acids are not known, it is not possible to discuss in detail the significance of these constants. One would expect that the known functional groups OH^-, $COOH^-$, and NH^- are components of the portion of the humic complexes with the metals.

Recent voltametric studies have shown that a number of metals are strongly complexed with natural organic ligands present in sea water. The metals Cu^{2+}, Fe^{3+}, Zn^{2+}, and Pb^{2+} are thought to be highly complexed (60–99%) with organic ligands in sea water. The most widely studied metals are Cu and Fe. The concentration and stability constants for the formation of the Cu complexes are given in Table 8.6. Many of the studies indicate that at least two Cu binding ligands ($L_1 = 2$–60 nM and $L_2 = 10$–300 nM) are present. The stronger ligand L_1 ($\log K_1 = 10$–12) is present in surface waters, and the weaker ligand L_2 ($\log K_2 = 8.5$–10.2) is present throughout the water column. The L_1 ligand is thought to be related to the production of phytoplankton in surface waters. The effect of these organic ligands on the speciation of Cu can be demonstrated (Millero 1990b) by assuming the concentrations $L_1 = 5$ nM and $L_2 = 150$ nM with stability constants of $\log K_1 = 12$ and $\log K_2 = 9$. In sea water without organics at a pH = 8.1,

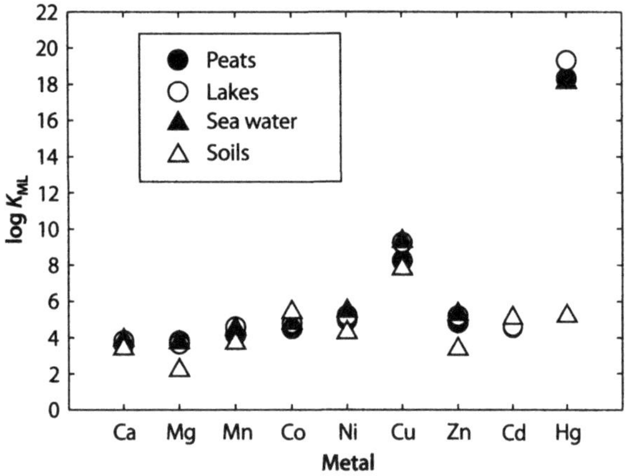

Fig. 8.18. Metal complexes with humic material collected in various locations

the fractions of Cu^{2+} are $\alpha_{Cu} = 0.039$, $\alpha_{CuOH} = 0.049$, $\alpha_{Cu(OH)_2} = 0.022$, $\alpha_{CuSO_4} = 0.010$, $\alpha_{CuCO_3} = 0.738$, $\alpha_{Cu(CO_3)_2} = 0.142$, and $\alpha_{CuHCO_3} = 0.001$. Combining the estimates of the ligand concentrations and stability constants with the inorganic speciation, we get $\alpha_{Cu} = 0.0002$, $\alpha_{CuCO_3} = 0.0034$, $\alpha_{CuL_1} = 0.9929$, $\alpha_{CuL_2} = 0.0030$, and $\alpha_{Cu(CO_3)_2} = 0.0005$. So in sea water without organics, 3.9% of Cu^{2+} is free, while natural organic ligands can lower this value to as low as 0.02%. As shown in Fig. 8.19, the difference between the speciation of Cu(II) with and without the inclusion of organic ligands is quite striking. We have added a provision to our computer code that allows one to input the concentration of a trace inorganic (HS^-, PO_4^{3-}) or organic ligand (humic), the concentration of the trace metal, and concentration of the ligand and the stability constant. This allows one to examine the competition between inorganic and organic ligands with some of the major anions (OH^-, CO_3^{2-}).

A number of researchers (Gledhill and van den Berg 1994; Wu and Luther 1995; Rue and Bruland 1995; van den Berg 1995) have used voltammetric techniques to examine the concentration and strength of natural organic ligands capable of complexing Fe(III) in sea water. These studies (Table 8.7) yield ligand concentrations $[L]_T$ of 0.4–13 nM, total iron concentrations $[Fe]_T$ of 0.2–8.5 nM, and apparent stability constants of $K_{FeL} = 10^{19}$–10^{23}. At high ligand concentrations and a pH near 8, Fe(III) is almost completely complexed with this ligand. As discussed above, this ligand is thought to be responsible for increasing the solubility of Fe(III) in sea water. The fraction of free Fe(III) in natural waters is a function of the pH, $[L]_T$, and $\log K_1$. Near a pH of 8, the fractions of complexes can range from 70 to 99% depending on the levels of $[L]_T$, [Fe(III)], and the $\log K$ selected (Millero 1998).

Bruland (1989) also determined the complexation of natural organic ligands with Zn in the North Pacific. He determined $\log K_{ZnL} = 11$ and $[L'] = 1.2$ nM and estimated that more than 98% of dissolved zinc in surface waters is complexed by organic ligands. The concentration of the ligand was quite uniform from the surface to deep. Because the total dissolved Zn varies from 0.1 nM in the surface to 3 nM in the deep waters, the inorganic Zn^{2+} varies from 2 pM to 2 nM from the surface to deep. Future work is needed to determine the structure of the various natural organic ligands that complex trace metals in natural waters.

Bruland (1992) determined the complexation of natural organic ligands with Cd in the North Pacific. He found values of $\log K_{CdL} = 12$ and $[L'] = 0.1$ nM. He estimated that 70% of dissolved Cd in surface waters is complexed to organic ligands. The organic

Table 8.6. Concentration and stability constants for metal organic complexes in sea water

Metal	M (nM)	L_1 (nM)	L_2 (nM)	$\log K_1$	$\log K_2$
Cu(II)[a]	1–10	2–60	10–300	11–12	8.5–10.2
Cd(II)[b]	0.002–0.8	0.1		12	
Zn(II)[c]	0.1–2	1.2		11	

[a] Coale and Bruland (1988); Moffett and Zika (1987); Hering et al. (1987); Sunda and Hanson (1987); Sunda and Ferguson (1983); van den Berg (1984); Anderson et al. (1984); Sunda et al. (1984); Kramer and Duinker (1984); van den Berg (1982).
[b] Bruland (1992).
[c] Bruland (1989).

Fig. 8.19. The inorganic and organic speciation of Cu(II) in sea water at $S = 35$ and 25 °C

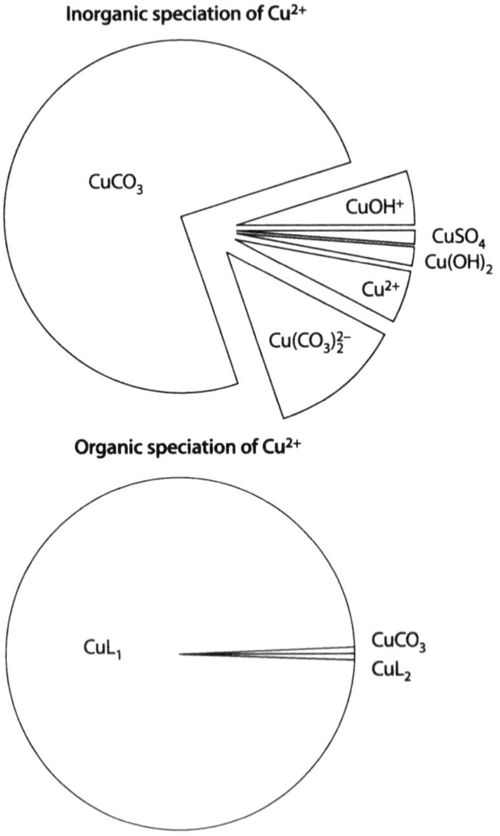

$$Cu^{2+} + L_1 = CuL_1 \quad K_1 = 10^{12}$$
$$Cu^{2+} + L_2 = CuL_2 \quad K_2 = 10^{9}$$

Table 8.7. Concentration and stability constants for Fe(III) organic complexes in sea water

Region	$[L]_T$ (nM)	$[Fe]_T$ (nM)	$\log K_{FeL}$	Reference
Menai Straits	4–10	3.5–8.5	21.3	Gledhill and van der Berg (1994)
Atlantic	3–5	0.8–1.8	19.3	Gledhill and van der Berg (1994)
Atlantic	0.4–0.6	0.17–0.27	23.2	Wu and Luther (1995)
Pacific	0.4	0.2–0.8	23.1	Rue and Bruland (1995)
Mediterranean	4–13	3.1	21.8	van den Berg (1995)

ligand was only present in surface waters. Because the concentration of inorganic forms of Cd varies from 0.7 to 800 pM in surface to deep waters, the concentration of free Cd ranges from 20 fM in surface waters to 22 pM in the deep.

8.8
Future of the Model

The present model provides reliable activity coefficients for a number of cations and anions from 0 to 50 °C. It can also be used to determine the speciation of divalent and trivalent metals in natural waters. The effect of the addition of organic ligands with a known or estimated stability constant can also be examined. The model has been shown to be reliable for the major components of natural waters, but little data is available to examine the speciation of metals in natural waters. Solubility measurements of trace metal oxides and carbonates in sea water as a function of pH would be useful in testing the reliability of the model. New measurements of the stability constants for the complexation of OH^-, CO_3^{2-}, etc. with divalent and trivalent metals in NaCl solutions over a wide range of temperature (0 to 50 °C) and ionic strength (0 to 6 m) are needed to derive more reliable Pitzer coefficients for trace metal ion pairs. Extensions of the model for the minor anions of natural waters at the present time are difficult due to the lack of reliable activity coefficient data for Na^+, K^+, Mg^{2+} and Ca^{2+} salts over a wide range of temperatures. The carbonic acid p*K* measurements of He and Morse (1993) can be used to extend the carbonate model to 100 °C. More measurements of this kind are needed to extend the model to higher temperatures for other acids and trace metal ion pairs.

Acknowledgements

The authors would like to acknowledge the support of the Oceanographic Section of the National Science Foundation for supporting our marine physical chemistry work.

References

Anderson MA, Morel FM (1982) The influence of aqueous iron chemistry on the uptake of iron by the coastal diatom *Thalassiosira weissflogii*. Limnol Oceanogr 27:789–813

Baes CF, Mesmer RE (1976) The hydrolysis of cations. John Wiley and Sons, New York

Berg CMG van den (1982) Determination of copper complexation with natural organic ligands in seawater by equilibration with MnO_2. II. Experimental procedures and application to surface seawater. Mar Chem 11:323–342

Berg CMG van den (1984) Determination of the complexing capacity and conditional stability constants of complexes of copper(II) with natural organic ligands in seawater by cathodic stripping voltammetry of copper-catechol complexions. Mar Chem 15:1268–1274

Berg CMG van den (1995) Evidence for organic complexation of iron in seawater. Mar Chem 50:139–157

Breeman N van (1973) Calculation of activity coefficients in natural waters. Geochim Cosmochim Acta 37:101–107

Bruland KW (1989) Complexation of zinc by natural organic ligands in the central North Pacific. Limnol Oceanogr 34:269–285

Bruland KW (1992) Complexation of cadmium by natural organic ligands in the central North Pacific. Limnol Oceanogr 37:1008–1017

Byrne RH Jr, Kester DR (1974) Inorganic speciation of boron in seawater. J Mar Res 32:119–127

Byrne RH, Kester DR (1976) Solubility of hydrous ferric oxide and iron speciation in seawater. Mar Chem 4:255–274

Byrne RH, Luo Y-R (2000) Direct observations of nonintegral hydrous ferric oxide solubility products: $K^*_{SO} = [Fe^{3+}][H^+]^{-2.86}$. Geochim Cosmochim Acta 64:1873–1877

Campbell DM, Millero FJ, Roy R, Roy L, Lawson M, Vogel KM, Moore CP (1993) The standard potential for the hydrogen – silver, silver chloride electrode in synthetic seawater. Mar Chem 44:221–233

Clegg SL, Whitfield M (1991) Activity coefficients in natural waters. In: Pitzer KS (ed) Activity coefficients in electrolyte solutions. CRS, Boca Raton, FL, pp 279–434

Clegg SL, Whitfield M (1995) A chemical model of seawater including dissolved ammonia and the stoichiometric dissociation constant of ammonia in estuarine water and seawater from -2 to 40 °C. Geochim Cosmochim Acta 59:2403–2421
Coale KH, Bruland KW (1988) Copper complexation in the Northeast Pacific. Limnol Oceanogr 33:1084–1101
Criss C, Millero FJ (1996) Modeling the heat capacities of aqueous 1-1 electrolyte solutions with Pitzer's equations. J Phys Chem 91:1288–1294
Criss C, Millero FJ (1999) Modeling the heat capacities of high valence-type electrolyte solutions with Pitzer's equations. J Solution Chem 28:849–864
Culberson C, Pytkowicz RM (1973) Ionization of water in seawater. Mar Chem 1:309–316
Culberson C, Pytkowicz RM, Hawley JE (1970) Seawater alkalinity determination by the pH method. J Mar Res 28:15–21
Culberson C, Latham G, Bates RG (1978) Solubilities and activity coefficients of calcium and strontium sulfates in synthetic seawater at 0.5 and 25 °C. J Phys Chem 82:2693–2699
Dickson AG (1990a) Thermodynamics of the dissociation of boric acid in synthetic seawater from 273.15 to 318.15 K. Deep-Sea Res 37:755–766
Dickson AG (1990b) Standard potential of the reaction: AgCl(s) + 1/2H_2(g) = Ag(s) + HCl(aq), and the standard acidity constant of the HSO_4^- in synthetic sea water from 273.15 to 318.15 K. J Chem Thermodyn 22:113–127
Dickson AG, Riley JP (1979a) The estimation of acid dissociation constants in seawater from potentiometric titrations with strong base. I. The ion product of water – K_W. Mar Chem 7:89–99
Dickson AG, Riley JP (1979b) The estimation of acid dissociation constants in seawater from potentiometric titrations with strong base. II. The dissociation of phosphoric acid. Mar Chem 7:101–109
Dickson AG, Whitfield M (1981) An ion-association model for estimating acidity constants (at 25 °C and 1 atm total pressure) in electrolyte mixtures related to seawater (ionic strength < 1 mol kg^{-1} H_2O). Mar Chem 10:315–333
Felmy AR, Weare JH (1986) The prediction of borate mineral equilibria in natural waters: Application to Searles Lake, California. Geochim Cosmochim Acta 50:2771–2783
Garrels RM, Thompson ME (1962) A chemical model for seawater at 25 °C and one atmosphere total pressure. Am J Sci 260:57–66
Gieskes JMT (1966) The activity coefficients of sodium chloride in mixed electrolyte solutions at 25 °C. Physik Chemie Neue Folge 50:78–90
Gledhill M, Berg CMG van den (1994) Determination of complexation of iron(III) with natural organic complexing ligands in seawater using cathodic stripping voltammetry. Mar Chem 47:41–54
Goyet C, Poisson A (1989) New determination of carbonic acid dissociation constants in seawater as a function of temperature and salinity. Deep-Sea Res 36:1635–1654
Greenberg JP, Møller N (1989) The prediction of mineral solubilities in natural waters: A chemical equilibrium model for the Na-K-Ca-Cl-SO_4-H_2O system to high concentration from 0 to 250 °C. Geochim Cosmochim Acta 53:2503–2518
Hansson I (1972) An analytical approach to the carbonate system in seawater. PhD dissertation, University of Göteborg, Sweden
Hansson I (1973) A new set of acidity constants for carbonic acid and boric acid in seawater. Deep-Sea Res 20:461–478
Harvie CE, Weare JH (1980) The prediction of mineral solubilities in natural waters: The Na-K-Mg-Ca-SO_4-Cl-H_2O system from zero to high concentration at 25 °C. Geochim Cosmochim Acta 44:981–997
Harvie CE, Møller N, Weare JH (1984) The prediction of mineral solubilities in natural waters: The Na-K-Mg-Ca-H-Cl-SO_4-OH-HCO_3-CO_3-CO_3-H_2O system to high ionic strengths at 25 °C. Geochim Cosmochim Acta 48:723–752
He S, Morse JW (1993) The carbonic acid system and calcite solubility in aqueous Na-K-Ca-Mg-Cl-SO_4 solutions from 0 to 90 °C. Geochim Cosmochim Acta 57:3533–3554
Hering JG, Sunda WG, Ferguson RL, Morel FMM (1987) A field comparison of two methods for the determination of copper complexation: Bacterial bioassay and fixed-potential amperometry. Mar Chem 20:299–312
Johansson O, Wedborg M (1980) The ammonia-ammonium equilibrium in seawater at temperatures between 5 and 25 °C. J Solution Chem 9:37–44
Johnson KE, Pytkowicz RM (1981) The activity of NaCl in seawater of 10–40‰ salinity and 5–25 °C at 1 atmosphere. Mar Chem 10:85–91
Kester DR, Pytkowicz RM (1967) Determination of the apparent dissociation constants of phosphoric acid in seawater. Limnol Oceanogr 12:243–252
Khoo KH, Ramette RW, Culberson CH, Bates RG (1977a) Determination of hydrogen ion concentrations in seawater from 5 to 40 °C: standard potentials at salinities from 20 to 45‰. Anal Chem 49:29–34.
Khoo KH, Culberson CH, Bates RG (1977b) Thermodynamics of ammonium ion in seawater from 5 to 40 °C. J Solution Chem 6:281–290

Kramer CJM, Duinker JC (1984) Complexation capacity and conditional stability constants for copper of sea and estuarine waters, sediment extracts and colloids. In: Kramer CJM, Duinker JC (eds) Complexation of trace metals in natural waters. Nijhoff/Junk, The Hague, The Netherlands, pp 217–228
Kuma K, Nishioka J, Matsunaga K (1996) Controls on iron(III) hydroxide solubility in seawater: The influence of pH and natural organic chelators. Limnol Oceanogr 41:396
Liu SX, Millero FJ (1999) The solubility of iron in sodium chloride solutions. Geochim Cosmochim Acta 63:3487–3497
Liu SX, Millero FJ (2002) The solubility of iron in seawater. Mar Chem 17:43–54
Mantoura RFC, Dickson A, Riley JP (1978) The complexation of metals with humic materials in natural waters. Est Coastal Mar Sci 6:387–408
Mehrbach C, Culberson CH, Hawley JE, Pytkowicz RM (1973) Measurement of the apparent dissociation constants of carbonic acid in seawater at atmospheric pressure. Limnol Oceanogr 18:897–907
Millero FJ (1979a) Effects of pressure and temperature on activity coefficients. In: Pytkowicz RM (ed) Activity coefficients in electrolyte solutions, vol II. CRC Press, Boca Raton, FL, pp 63–151
Millero FJ (1979b) The thermodynamics of the carbonate system in seawater. Geochim Cosmochim Acta 43:1651–1661
Millero FJ (1982) Use of models to determine ionic interactions in the natural waters. Thalassia Jugoslavica 1–4:253–291
Millero FJ (1986) The pH of estuarine waters. Limnol Oceanogr 31(4):839–847
Millero FJ (1990a) Effect of speciation on the rates of oxidation of metals. In: Melchior D, Bassett R (eds) Chemical modeling in aqueous systems II. ACS Books, Washington D.C., pp 447–460
Millero FJ (1990b) Marine solution chemistry and ionic interactions. Mar Chem 30:205–229
Millero FJ (1992) Stability constants for the formation of rare earth inorganic complexes as a function of ionic strength. Geochim Cosmochim Acta 56:3123–3132
Millero FJ (1995) Thermodynamics of carbon dioxide system in the oceans. Geochim Cosmochim Acta 59:661–677
Millero FJ (1998) Solubility of Fe(III) in seawater. Earth Planet Sci Lett 154:323–330
Millero FJ (1996) Chemical oceanography. CRC Press, Boca Raton, FL
Millero FJ (2001) The physical chemistry of natural waters. Wiley Scientific, N.Y.
Millero FJ, Hawke DJ (1992) Ionic interactions of divalent metals in natural waters. Mar Chem 40:19–48
Millero FJ, Pierrot D (1998) A chemical model for natural waters. Aquatic Geochem 4:153–199
Millero FJ, Roy R (1997) A chemical model for the carbonate system in natural waters. Croatia Chemica Acta 70:1–38
Millero FJ, Schreiber DR (1982) Use of the ion pairing model to estimate activity coefficients of the ionic components of natural waters. Am J Sci 282:1508–1540
Millero FJ, Plese T, Fernandez M (1988) The dissociation of hydrogen sulfide in seawater. Limnol Oceanogr 33:269–274
Millero FJ, Sharma VK, Karn B (1991) The rate of reduction of Cu(II) with hydrogen peroxide in seawater. Mar Chem 36:71–83
Millero FJ, Johnson R, Vega C, Sharma VK, Sotolongo S (1992) The effect of ionic interactions on the rates of reduction of Cu(II) with H_2O_2 in aqueous solutions. J Solution Chem 21:1271–1287
Millero FJ, Yao W, Aicher J (1995) The speciation of Fe(II) and Fe(III) in natural waters. Mar Chem 50:21–39
Moffett JW, Zika RG (1987) Solvent extraction of copper acetylacetonate in studies of copper (II) speciation in seawater. Mar Chem 21:301–313
Møller N (1988) The prediction of mineral solubilities in natural waters: A chemical equilibrium model for the Na-Ca-Cl-SO_4-H_2O system, to high temperature and concentration. Geochim Cosmochim Acta 52:821–837
Mucci A (1983) The solubility of calcite and aragonite in seawater at various salinities, temperatures and one atmosphere total pressure. Am J Sci 283:780–799
Pabalan RT, Pitzer KS (1987) Thermodynamics of concentrated electrolyte mixtures and the prediction of mineral solubilities to high temperature for mixtures in the system Na-K-Mg-Cl-SO_4-OH-H_2O. Geochim Cosmochim Acta 51:2429–2443
Pitzer KS (1975) Thermodynamics of electrolytes. V. Effects of higher order electrostatic terms. J Solution Chem 3:249–265
Pitzer KS (1979) Theory: Ion interaction approach. In: Pytkowicz RM (ed) Activity coefficients in electrolyte solutions, vol I. CRC Press, Boca Raton, FL, pp 157–208
Pitzer KS (1991) Theory: Ion interaction approach: Theory and data collection. In: Pitzer KS (ed) Activity coefficients in electrolyte solutions, 2nd edn, vol I. CRC Press, Boca Raton, FL, pp 75–153
Pitzer KS, Kim JJ (1974) Thermodynamics of electrolytes. IV. Activity and osmotic coefficients for mixed electrolytes. J Am Chem Soc 96:5701–5707
Pitzer KS, Mayorga G (1973) Thermodynamics of electrolytes. II. Activity and osmotic coefficients for strong electrolytes with one or both ions univalent. J Phys Chem 77:2300–2308

Pitzer KS, Mayorga G (1974) Thermodynamics of electrolytes. III. Activity and osmotic coefficients for 2-2 electrolytes. J Solution Chem 3:539–546
Platford RF (1965) The activity coefficient of sodium chloride in seawater. J Mar Res 23:55–62
Platford RF, Dafoe T (1965) The activity coefficient of sodium sulfate in seawater. J Mar Res 23:63–68
Roy RN, Roy LN, Lawson M, Vogel KM, Porter-Moore C, Davis W, Millero FJ, Campbell DM (1993) The dissociation constants of carbonic acid in seawater at salinities 5 to 45 and temperatures 0 to 45 °C. Mar Chem 44:249–259
Rue EL, Bruland KW (1995) Complexation of iron(III) by natural organic ligands in the central North Pacific as determined by competitive equilibration/adsorptive cathodic stripping voltammetric method. Mar Chem 50:117–138
Sharma VK, Millero FJ (1988) Oxidation of Copper(I) in seawater. Environ Sci Technol 22:768–771
Sharma VK, Millero FJ (1989) The oxidation of Cu(I) with H_2O_2 in natural waters. Geochim Cosmochim Acta 53:2269–2276
Silvester LF, Pitzer KS (1978) Thermodynamic of electrolytes. X. Enthalpy and the effect of temperature on the activity coefficients. J Solution Chem 7:327–337
Simonson JM, Roy RN, Gibbons JJ (1987a) Thermodynamics of aqueous mixed potassium carbonate, bicarbonate, and chloride solutions to 368 K. J Chem Eng Data 32:41–45
Simonson JM, Roy RN, Connole J, Roy LN, Johnson DA (1987b) The thermodynamics of aqueous borate solutions. II. Mixtures of boric acid with calcium or magnesium borate and chloride. J Solution Chem 16:791–803
Simonson JM, Roy RN, Mrad D, Lord P, Roy LN, Johnson DA, (1988) The thermodynamics of aqueous borate solutions, I. Mixtures of boric acid with sodium or potassium borate and chloride. J Solution Chem 17:435–446
Sohn ML, Hughes MC (1981) Metal complex formation constants of some sedimentary humic acids with Zn(II), Cu(II) and Cd(II). Geochim Cosmochim Acta 45:2393–2399
Spencer RJ, Møller N, Weare JH (1990) The prediction of mineral solubilities in natural waters: A chemical equilibrium model for the Na-K-Ca-Mg-Cl-SO_4-H_2O system at temperatures below 25 °C. Geochim Cosmochim Acta 54:575–590
Stumm W, Morgan JJ (1996) Aquatic chemistry: Chemical equilibria and rates in natural waters, 3rd edn. Wiley-Interscience, New York
Sunda WG, Ferguson RL (1983) Sensitivity of natural bacterial communities to additions of copper and to cupric ion activity: A bioassay of copper complexation in seawater. In: Wong CS, Boyle E, Bruland KW, Burton JD, Goldberg ED (eds) Trace metals in seawater. Plenum Press, New York, pp 871–891
Sunda WG, Hanson AK (1987) Measurement of free cupric ion concentration in seawater by a ligand competition technique involving copper sorption onto C18 SEP-PAK cartridges. Limnol Oceanogr 32:537–551
Sunda WG, Klaveness D, Palumbo AV (1984) Bioassays of cupric ion activity and copper complexation. In: Kramer CJM, Duinker JC (eds) Complexation of trace metals in natural waters. Nijhoff/Junk, The Hague, The Netherlands, pp 399–409
Thompson ME (1966) Magnesium in sea water: An electrode measurement. Science 153:866–867
Truesdale AH, Jones BF (1969) Ion association of natural brines. Chem Geol 4:1–62
Turner DR, Whitfield M, Dickson AG (1981) The equilibrium speciation of dissolved components in freshwater and seawater at 25 °C and 1 atm pressure. Geochim Cosmochim Acta 45:855–881
Vazquez F, Zhang JZ, Millero FJ (1989) Effect of trace metals on the oxidation rates of H_2S in seawater. Geophys Res Lett 16:1363–1366
Whitfield M (1975) The extension of chemical models for seawater to include trace components. Geochim Cosmochim Acta 39:1545–1557
Wu J, Luther GW (1995) Complexation of Fe(III) by natural organic ligands in the Northwest Atlantic Ocean by a competitive ligand equilibration method and kinetic approach. Mar Chem 50:159–177
Yao W, Millero FJ (1995) The chemistry of the anoxic waters in the Framvaren Fjord, Norway. Aquatic Chem 1:53–88

Binding Ability of Inorganic Major Components of Sea Water towards some Classes of Ligands, Metal and Organometallic Cations

C. De Stefano · C. Foti · A. Gianguzza · D. Piazzese · S. Sammartano

9.1 Introduction

"*Sea water chemical system*" can be thought of as a heterogeneous multi-component solution whose aqueous phase, which also contains particulate matter and colloids, is in contact with the sediments and the atmosphere (see Millero, this volume). In such a system, all chemical elements are present as different chemical species in a very wide range of concentrations, from nmol l^{-1} to mmol l^{-1}. A possible classification of sea water components, based on the types of substances present (inorganic and organic ions and molecules, organometallic compounds, colloids, etc.) and on their concentrations, is shown in Table 9.1.

Most biogeochemical processes occur in surface sea water (up to a few hundred metres in depth) and involve dissolved and particulate organic matter (productivity and degradation) and nutrients, as well as essential trace metals. These processes are generally called "open processes," because they depend on the environmental and geographical conditions of the seas. This means that the strength of these processes, as well as the amounts of substances involved ("*non conservative components*"), is extremely variable. Despite their importance to marine life and their high number, the total concentration of these components is generally low in aqueous solution (μg l^{-1}), contributing to sea water salinity by less than 0.5%.

The most abundant solutes in sea water are major and minor ions (Table 9.1). These are present in constant proportions ("*conservative components*"), because their concentrations are largely controlled by physical processes, such as evaporation/precipitation, by the transportation of water masses and by geological exchange processes between water and sediments. Taken together, these ions constitute over 99.5% of the total content of solutes dissolved in sea water, of which the major ions Na^+, K^+, Ca^{2+}, Mg^{2+}, Cl^-, SO_4^{2-} make up 99.8%. The sodium and chloride ions alone account for 86%. Due to their abundance, the major ions cause sea water to have very high concentrations of positive and negative charges, which can determine electrostatic interactions both with each other (internal interactions) and with other acid-base systems present in the solution (external interactions).

9.2 Artificial Sea Water

Considering the above, it can be stated that an artificial solution containing major and minor ions (Table 9.1) can usefully represent total sea water salinity and can be used as a salt ionic medium for thermodynamic studies in sea water. Artificial sea water has been widely used over the past few decades for two main purposes: (*a*) as a refer-

Table 9.1. Substances and their concentration range in sea water

Components	Substances and Ions	Concentration Range
Conservative components		
Major ions	Na^+, K^+, Ca^{2+}, Mg^{2+}, Cl^-, SO_4^{2-}	11–550 mmol l^{-1} [a]
Minor ions	Br^-, F^-, $B(OH)_3$, Sr^{2+}, HCO_3^-	0.1–2 mmol l^{-1} [a]
Non conservative components		
Nutrients	N-NH_3, N-NO_2, N-NO_3, P-PO_4, SiO_2	μg l^{-1} [b]
Trace metals	Fe, Mn, Zn, Cd, Cu, Co, Pb, etc.	<0.05 μmol l^{-1} [b]
Organometallic compounds	R_xSn(IV), R_xHg(II), R_xAs(III or V), R_xSb(V), etc.	ng l^{-1} to μg l^{-1} [b]
Dissolved organic matter (*DOM*)	Amino-acids saccharydes, lipids, humic and fulvic acids	ng l^{-1} to μg l^{-1} [b]
Particulate organic matter (*POM*)	Phyto- and zooplankton, animal and vegetable dead tissues, clay, etc	μg l^{-1} to mg l^{-1} [b]
Colloids	Organic and/or inorganic	

[a] Concentration values for sea water at $S = 35$.
[b] Order of magnitude.

ence solution in which to take salinity measurements (Lyman and Fleming 1940; Kester et al. 1967) and (*b*) as an ionic medium for equilibrium analysis (Culberson et al. 1970; Dickson and Riley 1979; Fiol et al. 1995a,b; Garrels and Thompson 1962; Hansson 1972; Johnson and Pytkowicz 1979; Johansson and Wedborg 1985; Kester et al. 1967; Khoo et al. 1977; Millero 1996; Pytkowicz and Hawley 1974). Table 9.2 shows the composition (mol l^{-1}) of the most widely used artificial sea waters.

9.2.1 The Major Components of Sea Water as a Single Sea Salt: *The "Single Salt Approximation"*

For equilibrium analysis studies of the acid-base systems about to be described, an artificial sea water containing the six major ions (Na^+, K^+, Ca^{2+}, Mg^{2+}, Cl^-, SO_4^{2-}) was used as an ionic medium. Table 9.3 shows the composition of artificial sea water (*SSWE, Synthetic Sea Water for Equilibrium studies*) for different salinities at 25 °C.

As pointed out in the introduction, the high concentration of positive and negative charges of the ion components in this ionic medium (SSWE) produces a network of interactions (*internal interactions*), which must be examined before studying any acid-base equilibrium in the same ionic medium. Some investigations performed to evaluate the strength of these interactions (De Robertis et al. 1994; De Stefano et al. 1994), showed that: (*a*) very weak ($NaCl^0$, KCl^0, $Na(OH)^0$ and $K(OH)^0$), (*b*) weak ($MgCl^+$, $CaCl^+$, $Ca(OH)^+$ and $Na(SO_4)^-$), and (*c*) fairly stable ($Mg(OH)^+$, $Mg(SO_4)^0$, $Ca(SO_4)^0$ and HSO_4^-) species are formed. As can be seen, without considering the hydroxo species of sodium, potassium and calcium and the protonated species of sulphate (these species are formed at pH values out of the range of natural fluids), at least 8 species have to be considered as the basic complexation model for every acid-base system in sea water.

Table 9.2. Recipes for artificial sea water by different authors ($S = 35$; $t = 25\,°C$)

Component	Composition (mol kg^{-1})								
	(a)	(b)	(c)	(d)	(e)	(f)	(g)	(h)	(i)
Na^+	0.48527	0.4752	0.48532	0.49152	0.49528	0.4822	0.48617	0.4822	0.46893
Mg^{2+}	0.05529	0.0540	0.05522	0.05504	0.05595	0.05489	0.05475	0.05459	0.05282
Ca^{2+}	0.01049	0.0104	0.01071	0.01074	0.01036	0.01063	0.01065	0.01063	0.01028
K^+	0.01013	0.0100	0.01026	–	–	0.01062	0.01058	0.01062	0.01021
Sr^{2+}	0.00016	–	0.00009	–	–	0.00009	–	–	0.00009
Cl^-	0.56572	0.5543	0.56579	0.56478	0.56988	0.5622	0.56825	0.5657	0.56725
Br^-	0.00082	–	0.00086	–	–	0.00087	–	–	0.00084
SO_4^{2-}	0.02914	0.0284	0.02925	0.02915	0.02901	0.02906	0.02927	0.02906	0.02824
F^-	0.00007	–	0.00005	–	–	0.00007	–	–	0.00007
HCO_3^-	0.00241	0.00238	0.00241	–	–	0.00213	–	–	0.00205
CO_3^{2-}	–	0.00024	–	–	–	0.00017	–	–	–
$B(OH)_3$	0.00046	–	0.00044	–	–	–	–	–	0.00041

(a) Lyman and Fleming (1940); *(b)* Garrels and Thompson (1962); *(c)* Kester and Pytkowicz (1967); *(d)* Culberson et al. (1970); *(e)* Hansson (1972); *(f)* Pytkowicz and Hawley (1974); *(g)* Dickson and Riley (1979); *(h)* Johnson and Pytkowicz (1979); *(i)* Millero (1996).

Table 9.3. Composition of the six components artificial sea water at $S = 35$ and $t = 25\,°C$

Component	C (mol l^{-1})	
NaCl	0.4221	(0.42740)[a]
Na_2SO_4	0.0288	(0.02919)
KCl	0.011	(0.01112)
$CaCl_2$	0.0111	(0.01121)
$MgCl_2$	0.0548	(0.05552)
I	0.717	(0.726)
BA	0.5751	(0.58240)

[a] In parentheses the concentrations for $S = 35$ are expressed in mol kg^{-1}; molalities at other salinities can be calculated by the equation: $m_s = m_{35}\,27.56572S\,/\,(1\,000 - 1.005714S)$

In order to simplify the study of anion and cation interactions, it is useful to consider the major inorganic components of sea water as a single salt BA, whose cation B and anion A are representative of all major cations (Na^+, K^+, Ca^{2+} and Mg^{2+}) and anions (Cl^- and SO_4^{2-}), respectively. The concentration of the single salt BA (see Table 9.3) was calculated as a mean ionic concentration, and the resulting ionic charge was ±1.117 (De Stefano et al. 1998). The use of the *single salt approximation* allows us to considerably reduce the number of internal interactions between the components of the ionic medium and simplifies the calculations used to define the chemical model for SSWE.

Indeed, only three species [BA, HA and B(OH)] have to be considered. Moreover, only the self association of BA plays a significant role in marine chemistry studies (at $S = 35$, ~15% of the sea water salt is associated), since the protonation of anion A and the hydrolysis of cation B are outside the pH range under investigation. Table 9.4 shows equilibrium constants for the self association of the BA salt, $\log K_{BA}$, as well as protonation constants for the anion A^{z-}, $\log K_{HA}$, at different temperatures and salinities. The hydrolysis constant [species B(OH)] can be obtained by the simple equation:

$$\log K_{BOH} = -12.75 + 0.234 I^{1/2}(2 + 3 I^{1/2})^{-1} - 0.205 I \tag{9.1}$$

The thermodynamic parameters in Table 9.4 allow us to complete the definition of the salt BA, and therefore, to calculate formation parameters for complexes formed by the minor and trace components of sea water (ligands and /or metal ions) with B^{z+} and /or A^{z-}, giving a general picture of the cumulative binding ability of the inorganic components of sea water.

Table 9.4. Thermodynamic parameters for the self association of BA, for the protonation of A^{z-} and the hydrolysis of B^{z+}, at different temperatures and ionic strengths

Reaction	$\log {}^{T}K$ [a]	ΔH (kJ mol^{-1}) [a]	ΔC_p (J K^{-1} mol^{-1}) [a]
$B^{z+} + A^{z-} = AB^0$	−0.03	0	0
$H^+ + A^{z-} = HA^{(1-z)}$	0.24	14.6	100
$B^{z+} = B(OH)^{(z-1)} + H^+$	−12.75	65.5	−102

T	S	$\log K_{BA}$	$\log K_{HA}$	T	S	$\log K_{BA}$	$\log K_{HA}$
5	5	−0.27	−0.11	5	35	−0.34	−0.24
15	5	−0.27	−0.03	15	35	−0.36	−0.15
25	5	−0.27	0.06	25	35	−0.39	−0.07
35	5	−0.28	0.14	35	35	−0.41	0.02
45	5	−0.28	0.23	45	35	−0.43	0.10
5	15	−0.34	−0.18	5	45	−0.32	−0.25
15	15	−0.35	−0.10	15	45	−0.34	−0.16
25	15	−0.36	−0.01	25	45	−0.37	−0.08
35	15	−0.37	0.07	35	45	−0.40	0.01
45	15	−0.38	0.16	45	45	−0.43	0.09
5	25	−0.35	−0.22				
15	25	−0.37	−0.13				
25	25	−0.39	−0.05				
35	25	−0.40	0.04				
45	25	−0.42	0.12				

[a] At $t = 25$ °C and $I = 0$ mol l^{-1}.

Table 9.5. Effective free concentrations rounded to the fourth decimal place, in mol (kg H_2O)$^{-1}$

S	*I*	[Na]	[K]	[Mg]	[Ca]	[Cl]	[SO_4]	[BA]
5	0.092	0.065	0.0015	0.0066	0.0013	0.078	0.0029	0.077
10	0.181	0.129	0.0029	0.0121	0.0025	0.153	0.0045	0.151
15	0.265	0.191	0.0043	0.0167	0.0035	0.225	0.0056	0.223
20	0.346	0.251	0.0056	0.0206	0.0043	0.296	0.0064	0.292
25	0.423	0.309	0.0069	0.0239	0.0051	0.365	0.0069	0.359
30	0.497	0.366	0.0081	0.0268	0.0057	0.431	0.0074	0.423
35	0.567	0.421	0.0093	0.0294	0.0063	0.496	0.0077	0.486
40	0.635	0.475	0.0105	0.0316	0.0068	0.559	0.0080	0.546
45	0.701	0.527	0.0116	0.0337	0.0073	0.620	0.0083	0.604

In this picture, once the "internal interactions" have been defined, the effective free concentrations of the major components must be considered (see Table 9.5), in order to investigate the interactions of the ligands and/or trace metal ions in sea water. As can be seen, in comparison with the analytical concentration values at $S = 35$ (Table 9.3), free ion concentrations are much lower, particularly in regards to magnesium and sulphate ions, which are the most interactive of all the major sea water components.

9.3 Interactions of Acid-Base Systems with the Components of Artificial Sea Water

Micro and trace components of sea water interact to different extents with the major inorganic constituents (Na^+, K^+, Mg^{2+}, Ca^{2+}, Cl^- and SO_4^{2-}). Metal ions are complexed by Cl^-, SO_4^{2-} CO_3^{2-} both as free ions (Martell and Smith 1997), i.e. M^{z+}, and as hydroxo species $M(OH)_n^{(z-n)}$ (these laboratories, unpublished results). Analogous behaviour is shown by some organometallic cations, such as organomercury(II) (De Robertis et al. 1998a) and organotin(IV) (De Stefano et al. 1999b,c, 2000a; Foti et al. 1999, 2000). O-ligands [(poly)carboxylate, phenols, hydroxycarboxylates] form weak species with Na^+ and K^+, and fairly stable complexes with Mg^{2+} and Ca^{2+} (Daniele et al. 1994 and references reported therein). Amines form weak complexes with Cl^- and SO_4^{2-} in their protonated form (Casale et al. 1998; Daniele et al. 1995; De Robertis et al. 1993) and with Mg^{2+} and Ca^{2+} (De Stefano et al. 1999a). Amino acids show an intermediate behaviour (De Stefano et al. 1995, 2000b).

These interactions have been quantified in two ways: (*a*) by studying the effect of the major components of sea water on the activity coefficients of metal ions and ligands (Pitzer 1991); (*b*) by building up appropriate complex formation models based on all significant binary interactions (Garrels and Thompson 1962; Millero 1974, 1990). Hybrid models have also been proposed (Millero and Schreiber 1982). For these studies, the above artificial sea waters and others (Demianov et al. 1995; De Robertis et al. 1997; De Stefano et al. 1994 ; Fiol et al. 1995a,b) were used.

Using results previously obtained from equilibrium analysis studies performed on various acid base-systems in artificial sea water (SSWE) both as six components and as single salt, we can now show a comprehensive picture of the binding capacity of SSWE towards O- and N-ligands and organometallic cations, with the aim (*a*) of drawing up a rigorous speciation picture of the acid-base systems under investigation, and (*b*) of defining general relationships useful in estimating the behaviour of a whole class of ligands.

The interactions of carboxylate, amine, amino acid and phosphate classes of ligands and of organotin(IV) cations with the components of SSWE as a single salt BA will be discussed. The binding of some divalent cations by the anion of BA was also considered using predictive relationships.

9.3.1 Organic Ligands

9.3.1.1 *Carboxylic Ligands*

In speciation studies of natural waters, different classes of ligands have to be considered. Among these, carboxylates are the most common and ubiquitous naturally occurring organic complexants, present in all the fractions of natural organic matter, particularly in fulvic compounds (Buffle 1988). It can be estimated that the main binding sites of aquagenic refractory organic matter and fulvic acids are -COO^- (2–10 mmol g^{-1}), often of aliphatic nature, and phenolic -OH (1–5 mmol g^{-1}). A number (about 25 µg of C atoms l^{-1}) of carboxylic groups are also present in sea waters as both free and combined hydrolysable amino acids (Daumas 1976; Lee and Wakeham 1989; Stumm and Brauner 1975; Williams 1971) and as fatty acids (about 40 µg C l^{-1}, free and combined) (Stumm and Brauner 1975). Some important di- and tri-carboxylic acids such as pyruvic, succinic and citric acids are always present in natural waters and derive from the biochemical processes of living organisms. Therefore, the importance of carboxylic and polycarboxylic ligands in the general picture of organic complexation in natural waters is evident. A number of data concerning the binding capacity of carboxylic ligands (Martell and Smith 1997; Sillén and Martell 1964, 1971; Pettit and Powell 1997) are reported in literature, but no data are reported for the complex formation of these ligands in sea water. The binding capacity of carboxylic ligands towards alkali and alkaline earth metals has been extensively studied at different ionic strengths (Daniele et al. 1985, 1994 and references reported therein). The stability of different complexes of various polycarboxylic anions with Na^+, Ca^{2+} and Mg^{2+} follows a regular trend with respect to the charges involved in the formation reaction, as shown in Fig. 9.1, where the formation constants of ML species are plotted vs. $[(z_a z_c)^{2/3} - 1]$.

Linearity is quite good and ensures good predictive power. Moreover, this homogeneous trend indicates that sodium, magnesium and calcium cations have very similar binding capacities towards carboxylic anions.

With the aim of giving a comprehensive picture of the binding capacity of low molecular weight carboxylic ligands in sea water, apparent protonation constants and complex formation constants with the cationic macrocomponents of sea water were determined for the following carboxylic ligands (see formulas and abbreviations in

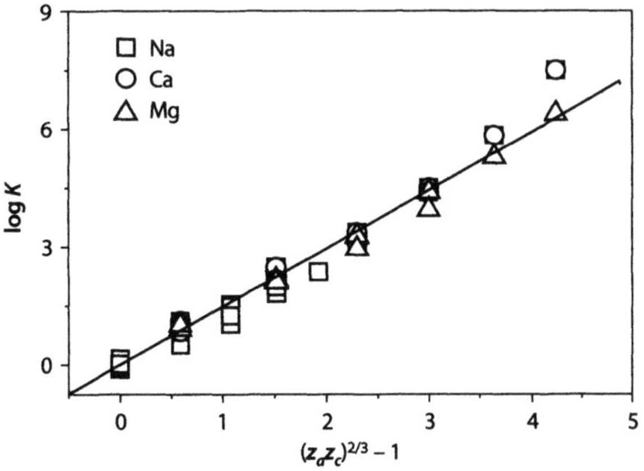

Fig. 9.1. Stability of ML species vs. $[(z_a z_c)^{2/3} - 1]$ at $I = 0$ mol l^{-1} and $t = 25$ °C; z_a = charge of carboxylic anion; z_c = charge of metal; L = mono-, di-, tri-, … hexa-carboxylic anions

Sect. A9.1 of appendix): acetate, malonate, succinate, malate, tartrate, azelate, oxydiacetate, trioxydiacetate, tricarballylate, citrate, methyltricarballylate, butanetetracarboxylate, and mellitate (Daniele et al. 1985, 1994; De Robertis et al. 2000; De Stefano et al. 1999d, 2000c, and references reported therein)

Potentiometric data can be firstly interpreted in terms of apparent protonation constants, $\log\beta_j^{H*}$, obtained by the extrapolation to zero ligand concentration of the conditional protonation constants $\log\beta_j^{H'}$. Conditional protonation constants depend on ligand concentration, and $\log\beta_j^{H*}$ are given by:

$$\log\beta_j^{H*} = \lim_{C\to 0} (\log\beta_j^{H'}) \tag{9.2}$$

where C = carboxylic ligand concentration and $\beta_j^{H'}$ = conditional protonation constant. The extrapolation function is:

$$\log\beta_j^{H*} = \log\beta_j^{H'} - (a/S)C^j \tag{9.3}$$

with S = salinity; a and j = empirical parameters. As an example, the effect of citrate concentration on $\log\beta_j^{H'}$ is reported in Fig. 9.2.

Apparent protonation constants for all the above carboxylate ligands in synthetic sea water are reported in Tables A9.1 and A9.2, Section A9.2 of the appendix. Their values are lower than those determined in NaCl (De Robertis et al. 2000).

The dependence of $\Delta\log\beta_j^{H}$ values $[\Delta\log\beta_j^{H} = \log\beta_j^{H*}(\text{NaCl}) - \log\beta_j^{H*}(\text{SSWE})]$ on salinity is a steadily decreasing function, as shown in Fig. 9.3, where $\Delta\log\beta_j^{H} = \log\beta_j - \log^{T}\beta_j^{H}$ is plotted vs. S, for the protonation of tca (tricarballylate). The sharply decreasing trend can be explained by considering the high complexing ability of carboxylic ligands towards alkaline earth metal cations and to a lesser extent towards alkali metal cations (De Robertis et al. 2000 and references reported therein; De Stefano et al. 1999d, 2000c, and references reported therein).

This is particularly evident for ligands containing more than two carboxylic groups. For mellitic acid, only the first five protonation constants can be calculated; in the pres-

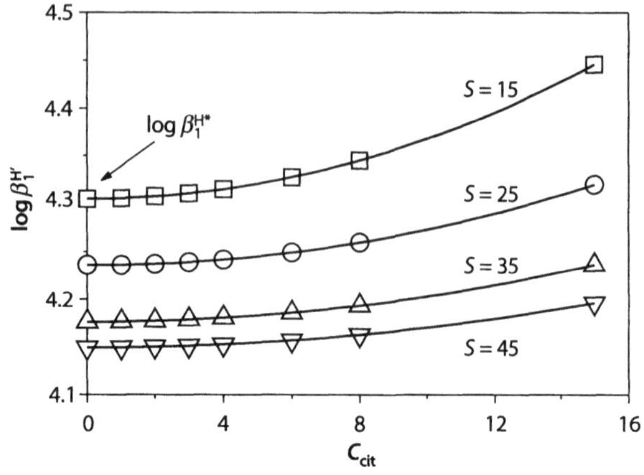

Fig. 9.2. $\log\beta_1^{H'}$ of citric acid at different salinities vs. concentration

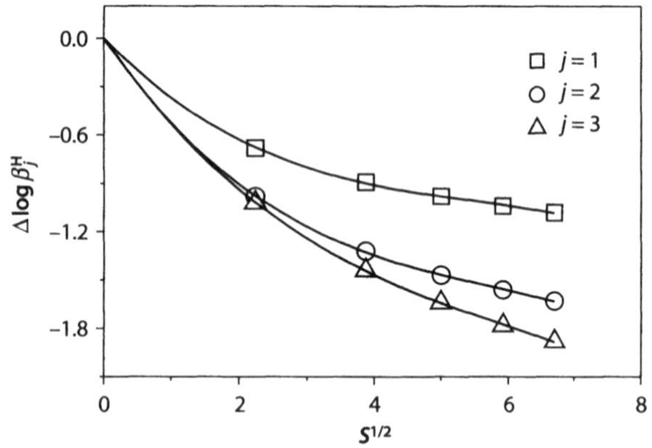

Fig. 9.3. $\Delta\log\beta_j^H = \log\beta_j - \log^T\beta_j^H$ vs. $S^{1/2}$, for 1,2,3-propanetricarboxylate (t = 25 °C)

ence of the artificial sea water salt, the last carboxylic group behaves like a strong acid. The apparent protonation constants of mono, di and tri-carboxylic ligands can be expressed as a function of salinity by the equation:

$$\log\beta_j^{H*} = \log^T\beta_j^H + a_1 S^{1/2} + a_2 S + a_3 S^{3/2} \tag{9.4}$$

For tetra- and hexacarboxylic ligands Eq. 9.4 is inadequate, and it has been found that it is possible to use the relationship:

$$\log\beta_j^{H*} = \log^T\beta_j^H + b_1 I + b_0 z^* \log(1 + b_2 I) \tag{9.5}$$

with

$$z^* = \Sigma(\text{charge})^2_{\text{reactants}} - \Sigma(\text{charge})^2_{\text{products}}$$

Table 9.6. Empirical parameters for the dependence on salinity of the apparent protonation constants of carboxylic ligands

Ligand	j	$\log {}^T\beta_j^H$	a_1 [a]	a_2 [a]	a_3 [a]
ac	1	4.75	−0.134	0.0213	−0.0015
mal	1	5.70	−0.423	0.0561	−0.0027
	2	8.57	−0.550	0.0732	−0.0038
succ	1	5.64	−0.297	0.0419	−0.0021
	2	9.85	−0.434	0.0631	−0.0036
mala	1	5.10	−0.301	0.0411	−0.0021
	2	8.57	−0.439	0.0652	−0.0038
tar	1	4.43	−0.308	0.0411	−0.0020
	2	7.46	−0.434	0.0603	−0.0033
aza	1	5.49	−0.268	0.0508	−0.0031
	2	10.04	−0.398	0.0657	−0.0040
toda	1	4.25	−0.337	0.0459	−0.0023
	2	7.56	−0.462	0.0629	−0.0034
tca	1	6.49	−0.434	0.0680	−0.0041
	2	11.40	−0.608	0.0880	−0.0050
	3	15.09	−0.602	0.0749	−0.0040
mtca	1	7.58	−0.439	0.0607	−0.0035
	2	12.71	−0.630	0.0773	−0.0039
	3	16.13	−0.634	0.0682	−0.0034

Ligand	j	$\log {}^T\beta_j^H$	b_1 [b]	b_2 [b]	b_0 [b]
btc	1	7.18	0.0096	74	0.0720
	2	13.01	0.0135	43	0.0720
	3	17.54	0.0154	29	0.0720
	4	20.92	0.0160	23	0.0720
mlt	1	7.85	0.0097	2.2×10^5	0.0396
	2	14.30	0.0103	7.5×10^4	0.0396
	3	19.39	0.0192	4.8×10^4	0.0396
	4	22.86	0.0262	2.3×10^4	0.0396

[a] Eq. 9.4.
[b] Eq. 9.5.

Values of empirical parameters a_1, a_2, a_3, b_0, b_1 and b_2, are reported in Table 9.6.

As concerns the empirical parameters of Eq. 9.4 for mono-, di- and tri-carboxylic ligands, we observed a regular trend as a function of z^*, according to the simplest relationship, valid for all the ligands:

$$a_i = c_i z^*$$

with

$$a_1 = -0.00648\ (\pm 0.0029)\, z^* \qquad (9.4a)$$

$$a_2 = -0.00872\ (\pm 0.00054)\, z^* \qquad (9.4b)$$

$$a_3 = -0.00047\ (\pm 0.00003)\, z^* \qquad (9.4c)$$

The empirical parameters of Eq. 9.5 can very probably also be related to some specific parameters for carboxylic ligands; in particular, b_2 can be related to the mean stability of the alkali and alkaline metal complexes of fully deprotonated or partially protonated ligands.

The sharp decrease in apparent protonation constants with increasing salinity can also be interpreted in terms of complex formation between the carboxylic anion and the cation $B^{1.117+}$ of artificial sea water. For all the systems, complex species are formed with both unprotonated and partially protonated ligands. Formation constants, reported in Table 9.7, are strictly related to the number of carboxylic groups involved in the co-ordination, n.

Table 9.7. Formation constants[a] for carboxylic ligand complexes with $B^{1.117+}$ ($I = 0$ mol l^{-1}; $t = 25$ °C)

Ligand	$\log\beta_{110}$	$\log\beta_{111}$	$\log\beta_{112}$	$\log\beta_{113}$	$\log\beta_{114}$	$\log\beta_{210}$	$\log\beta_{121}$
ac	0.30	–	–	–	–	–	–
mal	1.68	6.05 (0.35)[b]	–	–	–	–	–
succ	1.26	6.04 (0.39)	–	–	–	–	–
mala	1.25	5.37 (0.27)	–	–	–	–	–
tar	1.29	4.69 (0.26)	–	–	–	–	–
oda	1.80	4.69 (0.33)	–	–	–	–	–
aza	0.95	6.00 (0.51)	–	–	–	–	–
toda	1.41	4.63 (0.38)	–	–	–	–	–
tca	1.83	7.52 (1.03)	11.60 (0.19)[b]	–	–	2.10 (0.3)[c]	–
mtca	1.90	8.75 (1.17)	12.94 (0.23)	–	–	2.6 (0.7)	–
cit	3.24	7.82 (1.40)	11.30 (0.13)	–	–	3.57 (0.3)	–
btc	3.05	9.12 (1.94)	14.01 (1.00)	17.65 (0.11)[b]	–	3.18 (0.13)	–
mlt	7.05	13.00 (5.15)	18.22 (3.92)	22.28 (2.89)	25.26 (2.4)[b]	8.81 (1.76)	14.25 (1.25)[d]

[a] β_{pqr}: $pB^{1.117+} + qL^{z-} + rH^+ \leftrightarrow B_pL_qH_r^{(1.117p+r-qz)}$.

[b] In parentheses partial formation constants (K_{pqr}) refer to the reaction: $pB^{1.117+} + H_rL^{(r-z)} \leftrightarrow B_pLH_r^{(1.117p+r-z)}$.

[c] Partial formation constants refer to the reaction: $B^{1.117+} + BL^{(1.117-z)} \leftrightarrow B_2L^{(2.234-z)}$.

[d] Partial formation constants refer to the reaction: $B^{1.117+} + BHL^{(2.117-z)} \leftrightarrow B_2HL^{(3.234-z)}$.

For example:

$$btc^{4-} + B^{1.117+} = B(btc)^{2.883-} \qquad n = 4$$

$$H_3(mlt)^{3-} + B^{1.117+} = BH_3(mlt)^{1.883-} \qquad n = 3$$

The simplest relationship between $\log K$ and n is:

$$\log K = 0.37\,(\pm 0.01)\,n^{3/2} \qquad (1 \le n \le 5) \tag{9.6}$$

Inspection of stability data revealed some other interesting trends: (*i*) stability is inversely proportional to the length of the alkyl chain. If we consider malonate, succinate and azelate we have $\beta_{110} = 48$, 18 and 9 M^{-1}, respectively: the decreasing function is not linear, owing to the higher flexibility of longer chains; (*ii*) the higher stability of oxydiacetate and trioxydiacetate is due to the involvement of the ethereal group in the co-ordination. This is consistent with previous findings on the stability of calcium complexes of oxydiacetate and trioxydiacetate (De Stefano et al. 1999d, 2000c).

Figure 9.4 shows the speciation diagram vs. pH for the tca system in artificial sea water (BA) at $S = 35$. In this system, it can be observed that all the species show high yields (>20%) and that at pH > 7, the sum of $B(tca)^{1.883-}$ plus $B_2(tca)^{0.766-}$ is >90%.

The same holds for the btc-BA system with a higher yield (except for the tri-protonated complex), and at pH > 7 we have ~100% of the ligand as $B(btc)^{2.883-}$ and $B_2(btc)^{1.766-}$. Formation percentages are strongly dependent on the number of carboxylic groups in the ligand; to illustrate this trend quantitatively, in Fig. 9.5 we plotted Σ(species)% vs. salinity, at pH 8.2 for mono-, di-, tri- and tetra-carboxylic ligands. Percentages increase with n: for mellitate, at pH > 5, a 100% yield is always observed.

9.3.1.2
Phenols

A study carried out on the interaction of phenol and some of its derivatives in artificial sea water (Demianov et al. 1995) shows that phenols in artificial sea water form the weak species $BL^{0.117+}$ (Table 9.8), with a mean stability of 1.3 M^{-1}. This stability is somewhat lower than that of mono-carboxylic ligands, but is still comparable (same order of magnitude).

9.3.1.3
Amines

Amines are quite an important class of ligands, present as trace components in all biological fluids and natural waters. In particular, open chain polyamines have been widely studied owing to their strong ability to bind several metal cations (Perrin 1979; Pettit and Powell 1997; Martell and Smith 1997) and to form fairly stable species, in their protonated form, with organic and inorganic polyanions (Daniele et al. 1997). Amino compounds play an important role in reactions with sugars (or their derivatives) in natural waters leading to polymerized products (Bremmer 1967; Stevenson and Butler 1969) containing several amino groups along a linear chain, whose composition is

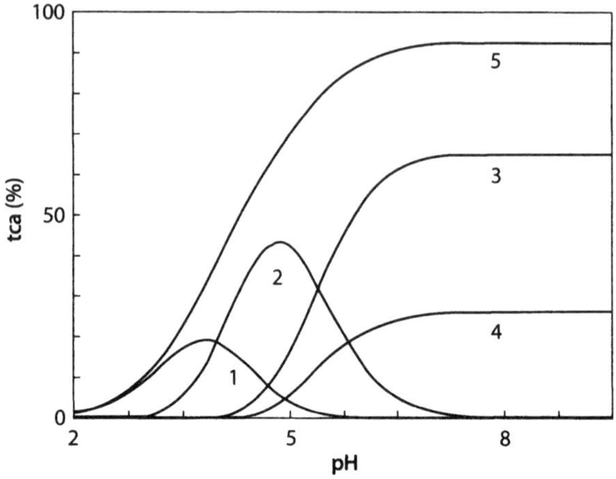

Fig. 9.4. Speciation diagram of tca in artificial sea water single-salt ($S = 35$, $t = 25\,°C$); $C_{BA} = 0.5752\ mol\ l^{-1}$; $C_{tca} = 10^{-4}\ mol\ l^{-1}$. Curves: *1.* $B(tca)H_2$; *2.* $B(tca)H$; *3.* $B(tca)$; *4.* $B_2(tca)$; *5.* $\Sigma_{species}\%$

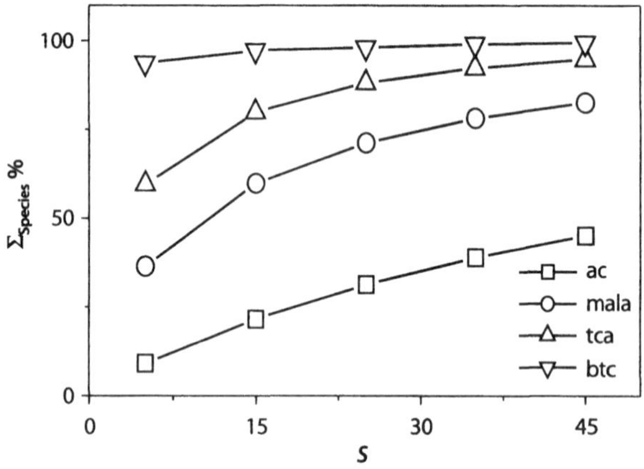

Fig. 9.5. Sum of species percentages vs. salinity (S) of carboxylic ligands in SSWE as single salt, at pH = 8.2

Table 9.8. Formation constant for phenol complexes with $B^{1.117+}$ ($I = 0\ mol\ l^{-1}$; $t = 25\,°C$)

Ligand	log *K* [a]
phenol	0.3
o-cresol	0.1
p-cresol	0.3
o-nitrophenol	0.2
p-nitrophenol	−0.1
Mean value	0.15

[a] *K* refers to the reaction : B + L ↔ BL.

very similar to that of the humic water-soluble fraction. Therefore, polyamines can be considered as a representative model of aminic sites in the humic acids, whose NMR spectra analysis shows intense dialkylamine bands (Mikita et al. 1981). Chloride and sulphate anions form weak complexes with partially or fully protonated polyamines, whose stability is strictly dependent on the charge of the polyammonium cation (Daniele et al. 1997). Mg^{2+} and Ca^{2+} also form complexes with polyamines (De Stefano et al. 1999a). The stability of these complexes is not very high and shows a regular increasing trend as a function of the number of amino groups. Moreover, for the reaction

$$M^{2+} + H_iA^{i+} = M(H_iA)^{(i+2)+}$$

it is possible to express stability using the general equation:

$$\log K_i^{M} = b_0 + b_1 n + b_2 n_H \qquad (9.7)$$

(with $n_H = i$ = number of protons in the complex species). The values of empirical parameters are for Mg^{2+} complexes: $b_0 = 0.39$, $b_1 = 0.44$ and $b_2 = -0.16$; and for Ca^{2+} complexes: $b_0 = -0.95$, $b_1 = 0.55$ and $b_2 = -0.05$.

In order to check the binding capacity of low molecular weight aminic ligands towards the major components of sea water, we examined the available literature data (Bates and Calais 1981; Bates and Erickson 1986; Czerminski et al. 1982) and the results of an extensive potentiometric investigation carried out at $t = 25$ °C on the following amines in artificial sea water at different salinities (De Stefano et al. 2000d): ethylenediamine, diethylenetriamine, triethylenetetramine, tetraethylenepentamine and spermine (formulas and abbreviations are reported in Sect. A9.1 of appendix). Apparent protonation constant, or $\log \beta_i^{H*}$ values obtained from potentiometric data by extrapolation to zero ligand concentration are shown in Table A9.3 of the appendix (Sect. A9.2). As an example, apparent protonation constants of triethylenetetramine ($n_H = 1$ to 4) are plotted as a function of ionic strength (Fig. 9.6).

For these ligands too, $\log K_i^{H*}$ values can be expressed as a function of salinity by Eq. 9.4. Empirical parameters a_1, a_2 and a_3 are shown in Table 9.9, together with thermodynamic protonation constants $\log {}^{T}K^{H}$.

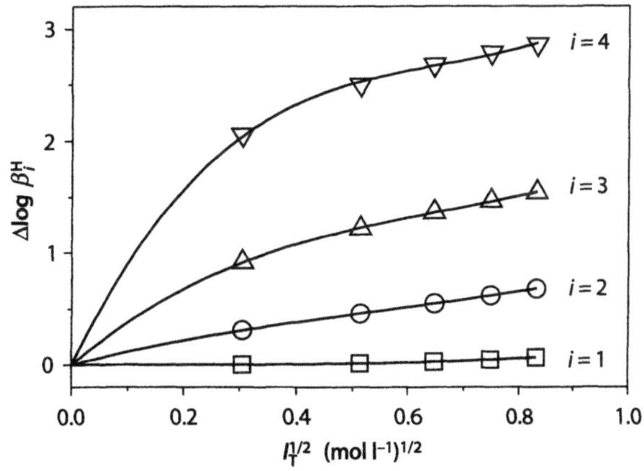

Fig. 9.6. $\Delta \log \beta_i^H$ vs. total ionic strength for triethylenetetramine, at 25 °C

Table 9.9. Parameters of Eq. 9.4 for the dependence on salinity of apparent protonation constants[a] of amines, at $t = 25$ °C

Amine	i	$\log^T K_i^H$	a_i	a_2	a_3
en	1	9.91	–	0.003714	–
dien	1	9.80	–	0.003178	–
trien	1	9.67	–	0.001211	–
tetren	1	9.83	–	−0.000768	–
sper	1	10.70	–	0.000056	–
en	2	6.86	0.1594	−0.01958	0.00114[b]
dien	2	8.74	0.1481	−0.01865	0.00114
trien	2	8.87	0.1756	−0.02001	0.00114
tetren	2	9.01	0.1324	−0.01895	0.00114
sper	2	9.70	0.1260	−0.01932	0.00114
dien	3	3.66	0.4452	−0.06523	0.00344[b]
trien	3	6.12	0.3824	−0.06102	0.00344
tetren	3	7.73	0.3251	−0.05445	0.00344
sper	3	8.32	0.3368	−0.05403	0.00344
trien	4	2.38	0.8332	−0.1675	0.01095
tetren	4	3.90	0.7031	−0.1301	0.00812
sper	4	7.22	0.6003	−0.1114	0.00688
tetren	5	1.88	1.0958	−0.2321	0.01562

[a] Refer to the reaction: $iH^+ + A^0 \leftrightarrow H_iA^{i+}$.
[b] The parameter a_3 was constrained, for $i = 2, 3$, to be constant for all the amines.

The parameters of Eq. 9.4 depend on: (*i*) the protonation step (mainly), i.e. the charge of the protonated species; and (*ii*) the number of aminogroups in the amine, n_N. For $i = 1,\ldots 3$, we have the following relationships, which are valid for all the amines studied:

$$\log K_1^{H*} = \log^T K_1^{H*} + S(0.00722 - 0.00159\, n_N) \tag{9.8}$$

$$\log K_2^{H*} = \log^T K_2^{H*} + S^{1/2}(0.179 - 0.0077\, n_N) - 0.0217\, S + 0.00143\, S^{3/2} \tag{9.8a}$$

$$\log K_3^{H*} = \log^T K_3^{H*} + S^{1/2}(0.500 - 0.0346\, n_N) - 0.0514\, S + 0.00246\, S^{3/2} \tag{9.8b}$$

The decrease in apparent protonation constants with increasing salinity was accounted for by considering the formation of complex species between protonated amines with $A^{1.117-}$. In all the amine-BA systems, the species $B(am)^{1.117+}$ and $(am)AH_i^{(1.117-i)}$ (with $i = 1, \ldots, n$; n = number of aminogroups) are formed. Formation constants for amine-BA complex species are shown in Table 9.10.

Table 9.10. Formation constants for amine complexes with the cation and the anion of marine salt BA, at $I = 0$ mol l^{-1} and $t = 25$ °C

Amine	i	$\log \beta_i$ [a]	$\log K_i$ [b]	$\log K_B$ [c]	Amine	i	$\log \beta_i$ [a]	$\log K_i$ [b]	$\log K_B$ [c]
en	1	10.4	0.5	0.14	tetren	1	10.3	0.5	0.42
	2	17.88	1.1			2	19.73	0.9	
						3	28.10	1.5	
dien	1	10.3	0.5	0.18		4	32.98	2.5	
	2	19.60	1.1			5	35.92	3.6	
	3	24.23	2.0						
					sper	1	11.2	0.5	0.40
trien	1	10.2	0.5	0.35		2	21.23	0.8	
	2	19.77	1.2			3	30.31	1.6	
	3	26.59	1.9			4	38.28	2.3	
	4	29.96	2.9						

[a] β_i: $am^0 + iH^+ + A^{1.117-} \leftrightarrow (am)AH_i^{(i-1.117)}$.
[b] K_i: $(am)H_i^+ + A^{1.117-} \leftrightarrow (am)AH_i^{(i-1.117)}$.
[c] K_B: $am^0 + B^{1.117+} \leftrightarrow (am)B^{1.117+}$.

The relative formation constants of cation complexes are fairly low, ranging from 1.4 to 2.6 M^{-1}. For anion complexes, real stability is given by the partial equilibrium between the protonated amine and the anion; the stability of these species is considerably higher than that of cationic ones, ranging from 3 to 4 000 M^{-1}. Moreover, $\log K_i$ is strictly dependent on i, i.e. on the charge of the protonated amine.

The $\log K_B$ constant for the formation of cationic species can be related to the relative formation constants of amine complexes with Mg^{2+} and Ca^{2+}. To this end, we may consider a mean formation constant for alkaline earth metal species, K_{MgCa}, which takes into account the concentration of Mg^{2+} and Ca^{2+} in the artificial sea water:

$$K_{MgCa} = (K_{Mg}C_{Mg} + K_{Ca}C_{Ca}) / (C_{Mg} + C_{Ca}) \tag{9.9}$$

having the following values: $\log K_{MgCa} = 0.35, 0.87, 1.39, 1.71, 1.63$ for en, dien, trien, tetren and sper, respectively. The simple relationship between the two constants is:

$$\log K_B = 0.27 \log K_{MgCa} \tag{9.10}$$

As regards to anionic species $(am)AH_i^{(i-1.117)}$, for $i > 1$, their formation constants, $\log K_i$ (Table 9.10) are strictly dependent on protonated amine charges, according to the relationship:

$$\log K_i = -0.65 + 0.817 I \tag{9.11}$$

Further inspection of formation data for the different amines reveals the en ~ dien > trien > tetren > sper trend, which can be explained by considering the ratio

$R_{N/C}$ = (number of aminogroups) / (number of -CH_2-groups) = 1, 0.75, 0.67, 0.625, 0.4, for en, dien, trien, tetren and sper, respectively. By considering the $R_{N/C}$ term we obtain:

$$\log K_i = -1.335 + 0.857\,i + 0.914\,R_{N/C} \qquad (9.12)$$

By using the parameter $R_{N/C}$, it is also possible to obtain a general equation for apparent protonation constants. Calculations performed using all the protonation data for $i \geq 2$ made it possible to formulate the following general relationships for the empirical parameters of Eq. 9.4:

$$a_1 = 0.0317 + 0.0177\,n_N \qquad (9.4d)$$

$$a_2 = -0.00507\,n_N + 0.00379\,R_{N/C} \qquad (9.4e)$$

$$a_3 = 0.000285\,n_N \qquad (9.4f)$$

Both apparent protonation constants and BA complex formation constants can be calculated from protonation constants at different ionic strengths and the formation constants of $(am)ClH_i^{i-1}$, $(am)(SO_4)H_i^{i-2}$, $Mg(am)^{2+}$ and $Ca(am)^{2+}$ complexes. This is certainly possible for $S \leq 15$ ($I_T \leq 0.3$ mol l^{-1}) and $i \leq 3$, but for higher salinities and for amines containing several aminogroups, the probability of the formation of ternary species such as $(am)Cl(SO_4)H_i^{i-3}$ or $MgCa(am)^{4+}$ becomes significant. For purposes of comparison, Table 9.11 shows experimental $\log \beta_i^{H*}$ values and the relative calculated ones at $S = 35$. Deviation, $\delta = \log \beta_i^{H*}{}_{(calc)} - \log \beta_i^{H*}{}_{(exp)}$, increases consistently with i, and mean deviation, ε, can be expressed as a linear function of z^*, $\varepsilon = 0.03 + 0.026\,z^*$.

On the basis of the above observations, by way of example, Fig. 9.7 shows a speciation diagram for the spermine system in artificial sea water as a single salt (BA). High yields of complex species of different forms of protonated spermine with the sea water anion component (A) are observed in a wide range of pH values. The most important species in terms of formation percentages is the one with the higher amine charge,

Table 9.11. Differences between experimental and calculated values of apparent protonation constants of amines

Amine	δ_1 [a]	δ_2 [a]	δ_3 [a]	δ_4 [a]	δ_5 [a]
en	0.05	0.11	–	–	–
dien	0.03	0.11	0.19	–	–
trien	0.02	0.07	0.14	0.20	–
tetren	–0.01	0.01	0.08	0.15	0.21
sper	–0.03	0.01	0.09	0.20	–
ε [b]	0.03	0.06	0.13	0.18	(0.21)

[a] $\delta_i = \log \beta_i^{H*}(\text{calc}) - \log \beta_i^{H*}(\text{exp})$.
[b] ε = mean values of $|\delta_i|$.

i.e. $(sper)AH_4^{2.883}$. In our conditions for standard sea water ($S = 35$, $pH = 8.2$), 88% of total spermine is present as $(sper)AH_4^{2.883}$ (~60%) and $(sper)AH_3^{1.883}$ (~30%). Fully protonated spermine complexes are also present at $pH > 8.5$.

All the polyamines with amino groups separated by longer alkyl chains display the same behaviour as spermine (biogenic amines belong to this class of amines). As regards the unprotonated cationic species, significant yields only occur at very high pH values ($pH > 10.5$), and in the conditions necessary for the speciation of sea water their presence is negligible. In general, the total binding capacity of sea water salt towards amines is quite significant, as shown in Fig. 9.8, where Σspecies % (total percentage of complex species) is reported vs. S salinity, at $pH = 8.2$: for all the amines Σspecies % > 40 at $S = 35$.

The en < dien < tetren < trien < sper trend is mainly due to the presence, at $pH = 8.2$, of different protonated species of amines. The stability of $(am)AH_i^{(i-1.117)}$ com-

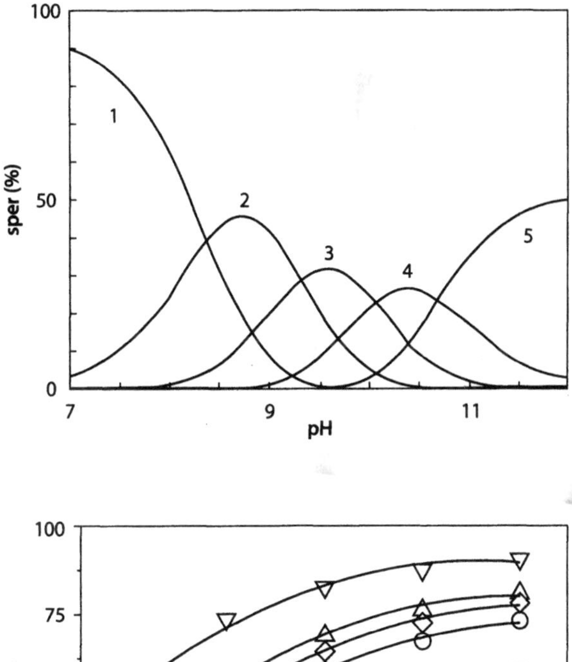

Fig. 9.7. Speciation diagram of spermine in artificial sea water, 35 salinity, vs. pH, at $t = 25\,°C$. Curves:
1. $(sper)AH_4$; 2. $(sper)AH_3$; 3. $(sper)AH_2$; 4. $(sper)AH$; 5. $B(sper)$

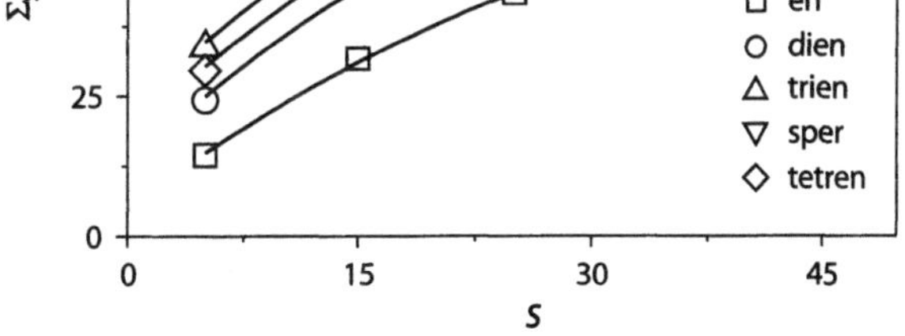

Fig. 9.8. Σspecies % (total percentages of complex species) vs. salinity, at $pH = 8.2$

Table 9.12. Formation constants of polyammonium cation complexes $[(am)XH_n]^{(n-z)}$ with $X = A^{1.117-}$, Cl^- and SO_4^{2-}, at $I = 0$ mol l^{-1} and $t = 25$ °C

Amine	$\log K_n$ [a]		
	$A^{1.117-}$	Cl^-	SO_4^{2-}
en	1.1	0.7	2.4
dien	2.0	1.5	3.8
trien	2.9	1.7	5.2
tetren	3.6	2.2	5.6
sper	2.3	1.5	3.9

[a] Refer to the reaction: $H_n(am)^{n+} + X^{z-} \leftrightarrow (am)XH_n{}^{n-z}$; $X^{z-} = A^{1.117-}$, Cl^-, SO_4^{2-} (n = maximum protonation degree).

plexes can be compared with that of analogous Cl^- and SO_4^{2-} complexes (De Robertis et al. 1998b). As an example, Table 9.12 shows the formation constants of $(am)AH_n^{(n-1.117)}$, $(am)ClH_n^{(n-1)}$, and $(am)(SO_4)H_n^{(n-2)}$ species.

The stability of BA sea salt complexes is intermediate between that of Cl^- and SO_4^{2-} species, as expected on the basis of the electrostatic model for the binding of anions by polyammonium cations.

9.3.1.4
Amino Acids

Natural waters contain a wide range of individual (free amino acids, *FAA*) and hydrolysable combined amino acids (*HAA*) in solution, which constitute the most important fraction of the dissolved organic nitrogen matter. They can be found in natural waters as the excretion products of living organisms and/or as the hydrolysis products of pedogenic polypeptides. Moreover, *FAA* can be formed in natural waters by the reaction of ammonia, obtained during the biological fixation process of elemental nitrogen, with some naturally occurring carboxylic acids, in the presence of NADPH. Glutamic acid is formed from the reaction of ammonia with α-ketoglutaric acid, and most other amino acids can be formed from it by transamination: α-alanine, for example, is formed by the transamination of glutamic acid reacting with pyruvic acid. Therefore, amino acids are particularly abundant in waters with high productivity. In some oligotrophic lakes, dissolved hydrolysable amino acids are reported to make up 30–40% of total dissolved organic nitrogen (Tuschall and Brezonik 1980). Together with sugars, amino acids represent an important food and energy source for heterotrophic microorganisms (Campbell and Goldstein 1972). Accumulated amino acids participate in the synthetic and respiratory metabolism of the organisms. The assimilation and release of amino acids by marine organisms has been studied using ^{14}C-labelled amino acids and by colourimetrically measuring their disappearance from the medium and their appearance in various chemical fractions of the different organisms, and vice versa (Stephens 1972). After hydrolysis, the following percentages of combined amino acids (*HAA*): 33, 55–75, 55.4 and 13–26 were measured with respect to dry organic matter in marine zoo- and phytoplankton, in bacteria, in molluscs, and in macrophytes, respectively (Buffle 1988). Salinity seems to be an important factor in regulating the assimilation of amino acids. Investigations carried out on some organism tests showed that uptake of glycine and α-alanine continues until salinity decreases to about 1/3 of normal sea water, i.e. a salinity of about 12. At lower salinities, the animals survive but can no longer acquire free amino acids from solution, probably because the osmotic and chloride regulation processes of the body fluid are reduced (Stephens 1972). Moreover, increasing salinity determines a lowering of amino acid assimilation, probably owing to their reduced availability due to complexation with the cation macro components of sea water. This last point further demonstrates the importance of chemical speciation studies in natural waters.

Besides their biological role, among the low molecular weight ligands present in natural waters, amino acids are interesting because of their acid-base behaviour, which can be considered to be intermediate between that of carboxylic acids and amines (De Stefano et al. 1995). Of the twenty amino acids that commonly occur in proteins, about half contain side chain donor atoms that are potentially capable of forming stable complex species with metal ions, and several studies have been described in literature on this subject (Sillén and Martell 1964, 1971; Martell and Smith 1997). Most of these studies concern the co-ordination chemistry of transition metal ions, and relatively few data regarding their complexation with alkali and alkaline earth metal ions have been published (Evans and Guevremont 1979; Casale et al. 1989; De Robertis et al. 1991; De Stefano and Gianguzza 1991; De Stefano et al. 1995). While taking into account the above considerations on the dependence of amino acid assimilation on sea water salinity, because of their importance in defining the chemical speciation of amino acids in natural waters, we also thought it would be worthwhile examining the interactions of this class of compounds with sodium, calcium and magnesium (i.e. the cation macro constituents of natural waters).

The protonation constants of glycine, alanine, histidine, aspartic, glutamic and iminodiacetic acids were studied in different aqueous media and at different ionic strengths (De Stefano et al. 1995, 2000b). These studies showed that: (*a*) carboxylic groups interact weakly with Na^+ and rather strongly with Ca^{2+} and Mg^{2+}, (*b*) unprotonated aminogroups interact weakly with Ca^{2+} and Mg^{2+} and (*c*) protonated aminogroups interact weakly with Cl^- and SO_4^{2-}. Moreover, by studying the dependence of protonation constants on ionic strength and complexing ability, amino acids were grouped into three different types: the first includes glycine, alanine and serine; the second lysine and histidine; and the third aspartic, glutamic and iminodiacetic acids.

Apparent protonation constants ($\log\beta^{H*}$) for glycine, alanine, histidine, and glutamic and aspartic acids (De Stefano et al. 1995, 2000b) and for serine, valine, leucine, threonine and methionine (Fiol et al. 1995a,b, 1998) determined in artificial sea water are shown in Tables A9.4 and A9.5, respectively, in Sect. A9.2 of the appendix (in Sect. A9.1 of the appendix formulas and abbreviations of all amino acids considered here are reported).

The dependence on salinity of $\Delta\log\beta_i^H$ values [$\Delta\log\beta_i^H = \log\beta_i^{H*}(NaCl) - \log\beta_i^{H*}(SSWE)$] is shown in Fig. 9.9, for the first (Fig. 9.9a) and the second (Fig. 9.9b) protonation step of glycine, histidine and glutamic acid. As can be seen, gly and his behave in a similar manner (in particular as regards $\log\beta_1^{H*}$), whilst protonation constants for glutamic acid are significantly lower. This is due to the presence of a second carboxylic group, which binds cations (see formation constants for alkali and alkaline earth complexes below) more strongly than aminogroups.

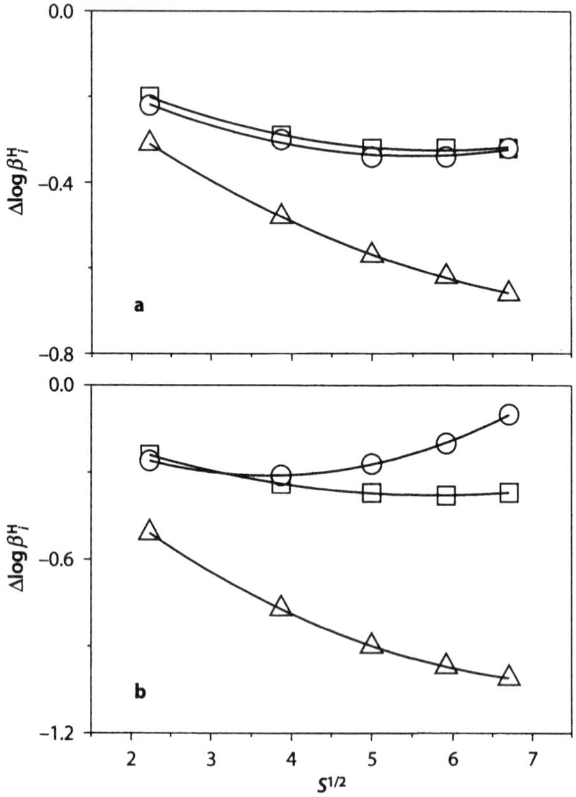

Fig. 9.9. $\Delta\log\beta_i^H$ values vs. $S^{1/2}$; **a** for the first; **b** for the second protonation step of glycine (□), histidine (○) and glutamic acid (Δ). [$\Delta\log\beta_i^H = \log\beta_i^{H*}(\text{NaCl}) - \log\beta_i^{H*}(\text{SSWE})$]

Apparent protonation constants obtained in artificial sea water can be expressed as a function of salinity by the equation:

$$\log\beta_i^H = \log{}^T\beta_i^H + aS^{1/2} + bS \tag{9.13}$$

The values of the empirical parameters of Eq. 9.13 were quite similar for the same amino acid class and the same protonation step. In particular, parameter b can be considered a "common parameter" for glycine and alanine and for aspartic and glutamic acids without any significant increment in the σ of fit. For parameters a and b it was also possible to give general parameters for each amino acid group:

type 1	$a_1 = -0.098 \pm 0.015$	$b_1 = 0.009 \pm 0.003$
	$a_2 = -0.105 \pm 0.019$	$b_2 = 0.009 \pm 0.003$
type 3	$a_1 = -0.186 \pm 0.030$	$b_1 = 0.010 \pm 0.005$
	$a_2 = -0.292 \pm 0.027$	$b_2 = 0.019 \pm 0.005$
	$a_3 = -0.316 \pm 0.023$	$b_3 = 0.022 \pm 0.004$

Table 9.13. Complex species formation of amino acid in SSWE as a single salt (BA), at $I = 0$ mol l^{-1} and $t = 25$ °C

Reaction	log *K*				
	gly	ala	his	glu	asp
B + L = BL	0.8 (0.7)[a]	0.65 (0.6)[a]	0.8 (0.6)[a]	1.34 (1.3)[a]	1.52 (1.6)[a]
B + HL = BHL	0.3 (0.2)	0.25 (0.2)	0.2 (0.1)	0.71 (0.6)	0.5 (0.4)
$B + H_2L = BH_2L$	–	–	–	0.1	0.0
$H_2L + A = H_2LA$	0.4	0.4	0.6	–	–
$H_3L + A = H_3LA$	–	–	0.7	–0.1	0.0

[a] In parentheses values calculated by using formation constants of weak complexes (see De Stefano et al. 1995).

The dependence on medium of protonation constants can be interpreted in terms of complex formation. We found the following species are formed: BL, BHL and H_2LA (for L = gly and ala); BL, BHL, H_2LA and H_3LA (for L = his); BL, BHL and BH_2L (for L = glu and asp). Formation constants are shown in Table 9.13.

The stability of BL and BHL species for gly, ala and his, and of BL, BHL and BH_2L species for glu and asp is fairly constant, and the following mean values of formation constants can be used: $\log K^{BL} = 0.75$, $\log K^{BHL} = 0.25$ (for L = gly, ala and his); $\log K^{BL} = 1.43$, $\log K^{BHL} = 0.605$, and $\log K^{BH_2L} = 0.05$ (for L = glu and asp). Table 9.13 also shows formation constant values for the simplest species BL and BHL, calculated from the formation constants of binary systems.

The calculated values are fairly similar to experimental ones but are consistently lower: this may indicate that ternary interaction cannot be neglected. As examples, the distribution diagrams of glycine, histidine and glutamic acid vs. pH are shown in Fig. 9.10.

At pH = 8, the complexation of gly and glu is higher than 50% and for his > 40%. For glycine, the most important species at the pH of natural waters is B(gly)H; for histidine, mainly B(his)H (with small percentages of $A(his)H_2$ and B(his)) must be considered; for glutamic acid, B(glu)H (with small percentages of B(glu)H) predominates.

9.3.2 Inorganic ligands

9.3.2.1 *Phosphates*

The average concentration of phosphorus in sea water is about 2.5 μmol l^{-1} in the lower layers, while this falls to zero in the upper 100 m because of biological activity. In spite of its very low concentration, phosphorus can be considered a master element in the marine ecosystem, since it is involved in most important biological processes, such as

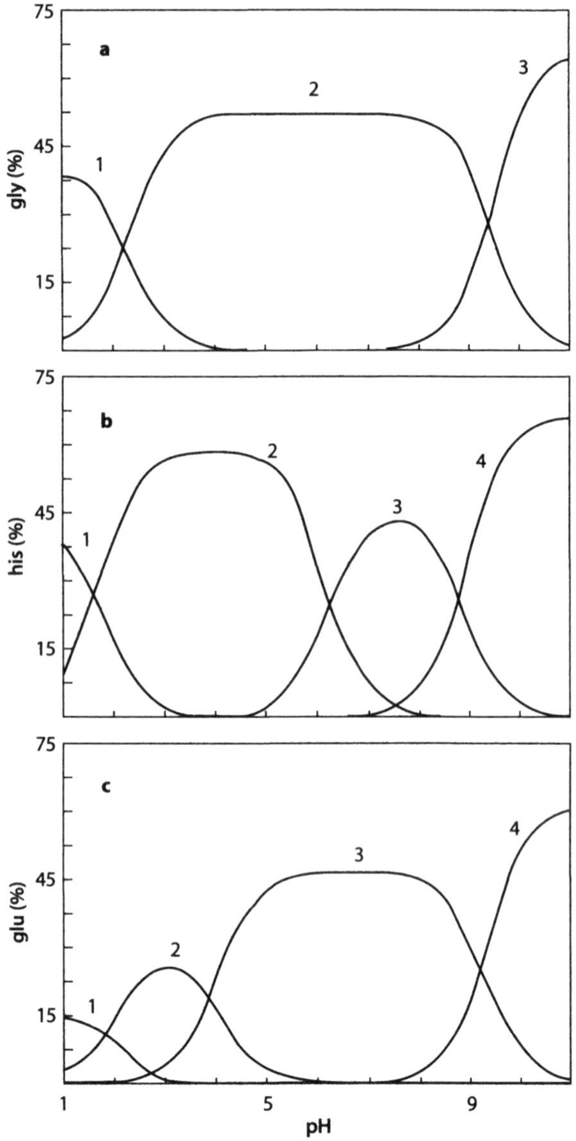

Fig. 9.10. Distribution diagram of; **a** glycine; **b** histidine; **c** glutamic acid vs. pH, in artificial sea water as a single salt BA and $t = 25\,°C$. Species: *a*: 1. $A(gly)H_2$, 2. $B(gly)H$, 3. $B(gly)$; *b*: 1. $A(his)H_3$, 2. $A(his)H_2$, 3. $B(his)H$, 4. $B(his)$; *c*: 1. $A(glu)H_3$, 2. $B(glu)H_2$, 3. $B(glu)H$, 4. $B(glu)$. Curves relative to simple protonated species and free components are not reported

marine primary productivity (Fogg 1975; Spencer 1975). In that process, as well as in other biochemical processes, the reactive form of phosphorus is phosphate, which derives from both the particulate and the dissolved fraction. The latter exists as soluble inorganic phosphorus and as soluble organic phosphorus compounds. Soluble inorganic phosphate has been regarded as being present exclusively as orthophosphate, i.e. as $H_2PO_4^-$ and HPO_4^{2-}, with negligible amounts of PO_4^{3-} and H_3PO_4. Dissolved orthophosphate groups can also derive from the organic fraction (sugar phosphates, phospholipids, phosphonucleotides) as a consequence of enzymatic hydrolytic processes. Thus, the study of phosphorus compounds in sea water is mainly concerned with the chemical behaviour of the phosphate group. A comprehensive picture of the distribution of the different forms of phosphorus in sea water was drawn up by Armstrong (1965). Although there is no evidence of the presence of dissolved polyphosphates in sea water slowly leading to the formation of phosphate groups as a result of hydrolytic processes, it is well-known that phosphoric acid can undergo condensation to form poly-acids such as pyrophosphoric acid or poly-meta-phosphoric anions, $(HPO_3)_n$, with $n = 3$ up to 70 per chain. Polyphosphates can also be present in coastal sea waters containing sewage and industrial wastes. Therefore, these compounds too must be taken into account in chemical speciation studies of phosphorus compounds in sea water. Studies of the solution chemistry of the phosphate group in natural waters are complicated by interactions with macroconstituent cations such as Ca^{2+} and Mg^{2+}, which, under certain pH and free concentration conditions, can lead to the formation of insoluble species.

The speciation of phosphate in sea water has been investigated by few authors (Kester and Pytkowicz 1967; Atlas et al. 1976; Dickson and Riley 1979; Johansson and Wedborg 1979). The shift in the apparent protonation constants of phosphoric acid with changes in the composition of the ionic medium (containing macrocomponents of sea water, Na^+, Mg^{2+} and Ca^{2+}) was interpreted by the authors in terms of the association of orthophosphate with medium cations using ion association models. Hershey et al. (1989) studied the ion pairing of phosphate with Mg^{2+} and evaluated Pitzer interaction parameters for the major components of sea water. Table A9.6 of the appendix (Sect. A9.2) shows apparent protonation constants, $\log \beta_j^{H*}$, for $P_iO_{(3i+1)}^{(i+2)-}$ in SSWE. For purposes of comparison, speciation diagrams for different forms of phosphate ligand in SSWE ($S = 35$) and in NaCl (0.75 mol l^{-1}) are shown in Fig. 9.11.

As can be seen, curves for the different protonated species in SSWE have shifted significantly to lower pH values in comparison with those in the NaCl medium. The lowering effect on $\log \beta_j^{H}$ can be simply explained by taking into account the formation of simple and mixed protonated $B^{1.117+} - P_iO_{(3i+1)}^{(i+2)-}$ species (Table 9.14).

The speciation diagrams obtained by also including species formation as a result of the interaction of phosphate ligands with the sea water cation (B) are shown in Fig. 9.12.

9.3.3 Metals and Organometallic Compounds

9.3.3.1 *Divalent Metal Ions*

Trace metals interact significantly with both SO_4^{2-} and Cl^-. Many divalent cations, such as Mn^{2+}, Fe^{2+}, Co^{2+}, etc., form ion pairs with SO_4^{2-} ($K = 400$ m^{-1}) and with Cl^- ($K = 1$ M^{-1}). The ion pairs $CdCl^+$ and $CdCl_2^0$, whose stability is much higher, constitute an exception. Moreover, most divalent and trivalent cations undergo strong hydrolysis and often hydrolytic species are very important at the pH value of sea water. To give an ex-

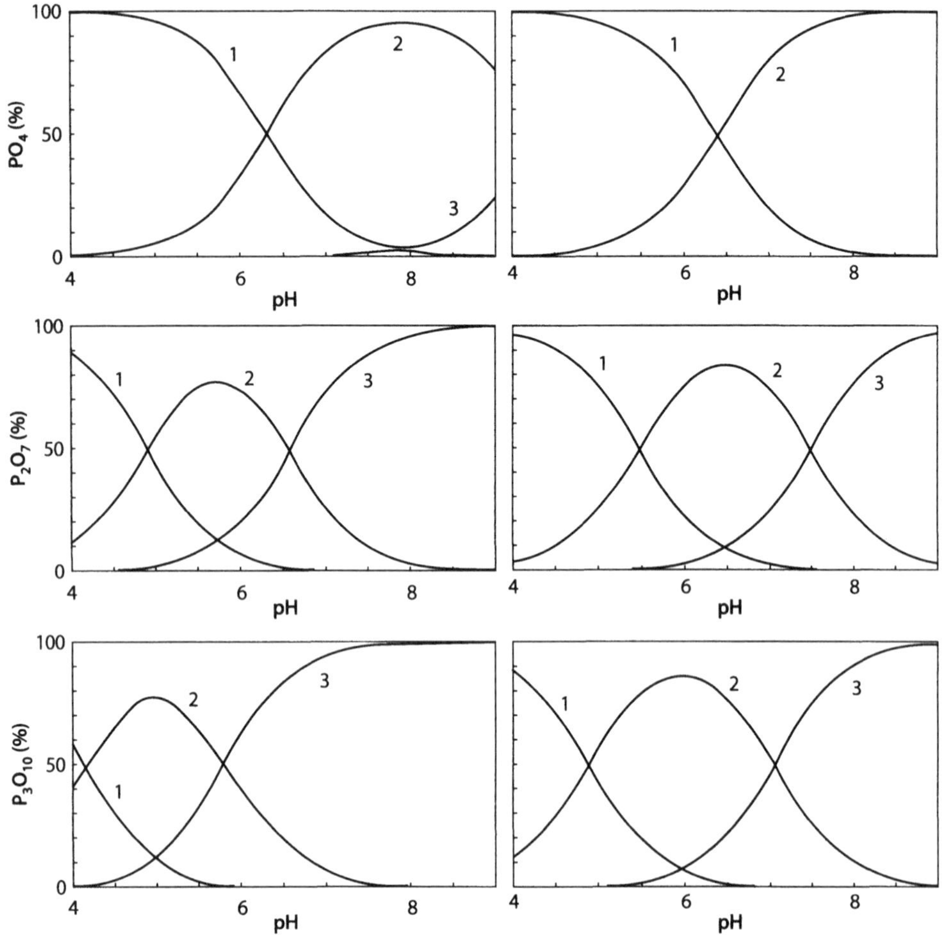

Fig. 9.11. Speciation diagrams for PO_4^{3-}, $P_2O_7^{4-}$ and $P_3O_{10}^{5-}$, by considering the apparent protonation constants only, in SSWE (*right*) ($S = 35$), and in NaCl 0.75 mol l^{-1} (*left*), at $t = 25$ °C. Species: *1*: H_2L; *2*: HL; *3*: L

Table 9.14. Formation constants of $B^{1.117+}$-PO_4^{3-}, -$P_2O_7^{4-}$ and -$P_3O_{10}^{5-}$ complexes, at $t = 25$ °C and $I = 0$ mol l^{-1}

L	$\log\beta_0$ [a]	$\log\beta_1$ [a]	$\log\beta_2$ [a]
PO_4^{3-}	4.75	13.90 (1.57)[b]	20.3 (0.8)[b]
$P_2O_7^{4-}$	5.81	12.82 (3.22)	17.96 (1.6)
$P_3O_{10}^{5-}$	6.36	12.80 (3.36)	17.3 (1.4)

[a] Overall formation constants relative to the reaction (charges omitted): $B + L + rH \leftrightarrow BLH$.

[b] In parentheses values of equilibrium constants for the reaction: $B + H_rL \leftrightarrow BLH_r$.

Fig. 9.12. Speciation diagrams for PO_4^{3-}, $P_2O_7^{4-}$ and $P_3O_{10}^{5-}$, by considering the formation of complexes with $B^{1.117+}$, in SSWE, $S = 35$ and $t = 25$ °C. Species: *1.* H_2L; *2.* HL; *3.* BH_2L; *4.* BHL; *5.* BL

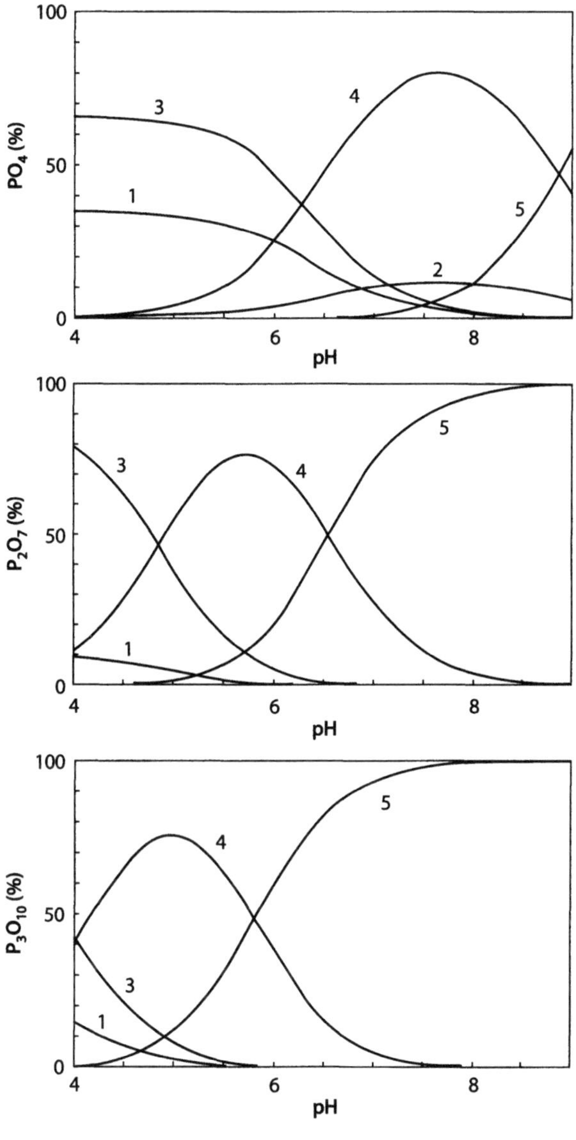

ample of speciation for this class of trace components, we considered the divalent cations Mn^{2+}, Fe^{2+}, Co^{2+}, Ni^{2+}, Cu^{2+}, Zn^{2+}, Cd^{2+}. In some cases, this group of cations is very homogeneous, but in general its speciation profiles display significant differences. Using the formation constants obtained for the Cl^- and SO_4^{2-} complexes of divalent cation under examination, it is quite simple to obtain formation constants for $A^{1.117-}$ complexes, as reported in Table 9.15. For Mn^{2+}, Fe^{2+}, Co^{2+}, Ni^{2+}, Cu^{2+} and Zn^{2+}, stabil-

Table 9.15. Equilibrium constants for the interactions of some divalent cations with SSWE as single salt, at $I = 0$ mol l^{-1} and $t = 25$ °C

Metal cation	$\log K_1$ [a]	$\log \beta_2$ [a]	$\log \beta_{M(A)(OH)}$ [b]
Mn^{2+}	0.63	0.3	3.8
Fe^{2+}	0.69	0.4	5.0
Co^{2+}	0.64	0.3	5.3
Ni^{2+}	0.64	0.3	5.1
Cu^{2+}	0.74	0.5	7.65
Zn^{2+}	0.80	0.6	6.2
	0.7 ±0.1	0.4 ±0.2	
Cd^{2+}	1.94	2.9	5.7

[a] β_i values refer to the reaction $M^{2+} + iA^{1.117-} \leftrightarrow MA_i^{(2-i1.117)}$.
[b] β values refer to the reaction $M^{2+} + A^{1.117-} + OH^- \leftrightarrow MA(OH)^{0.117-}$.

ity is fairly constant, $K_1 = 5 \pm 1$ M^{-1} and $\beta_2 = 2.5 \pm 1.2$ M^{-2}, whilst Cd^{2+} species are stronger ($K_1 = 90$ M^{-1} and $\beta_2 = 800$ M^{-2}).

This means that, at low pH values (pH < 5), ~50% of M^{2+} is complexed by $A^{1.117-}$, but for Cd^{2+} (100%). Nevertheless, at pH > 6, hydrolysis takes place in two ways: (*a*) with the formation of M(OH) and $M(OH)_2$ simple hydrolytic species and (*b*) with the formation of mixed MA(OH) species, though this may not be important. The formation constants for the species $M(OH)_j$ are available (Martell and Smith 1997; Pettit and Powell 1997), and those of mixed species can be obtained from the statistical value K_{stat} for the reaction:

$$MA_2 + M(OH)_2 = 2MA(OH)$$

Different approaches were used to estimate K_{stat}, and we used the equation:

$$K_{stat} = 2 + \frac{K_1^A}{K_1^{OH}}\left(\frac{K_2^{OH}}{K_2^A}\right)^{0.5} + \frac{K_1^{OH}}{K_1^A}\left(\frac{K_2^A}{K_2^{OH}}\right)^{0.5} \tag{9.14}$$

Therefore, $\beta_{MA(OH)}$ can be obtained by:

$$\beta_{MA(OH)} = ½\,(K_{stat}\beta_{M(OH)_2}\beta_{MA_2}) \tag{9.15}$$

These estimated constants are shown in Table 9.15. In these cases, the different cations display very different stabilities. Table 9.16 shows calculated species percentages at different pH values. As can be seen, quite different speciation profiles are obtained: at the pH value of sea water (~8), the main species for Mn^{2+}, Fe^{2+}, Co^{2+} are MA and MA_2 (~50%), as they are for Cd^{2+} (~100%); for Zn^{2+} there is 10% mixed species, and 0.5% of Cu^{2+} is present as CuA(OH). Small but significant differences can also be observed for the other species.

Table 9.16. Metal species percentages in SSWE as single salt, at different pH ($S = 35$)

Metal cation	pH	MA	MA_2	M(OH)	$M(OH)_2$	MA(OH)
Mn^{2+}	7	43.1	6.8	–	–	0.0
	8	43.1	6.8	0.1	–	0.1
	9	42.6	6.7	0.7	–	0.6
Fe^{2+}	7	45.7	7.9	0.1	–	0.1
	8	45.0	7.8	0.8	–	0.8
	9	39.0	6.8	7.3	0.1	7.4
Co^{2+}	7	43.6	6.7	0.1	–	0.2
	8	42.6	6.6	0.6	–	1.8
	9	34.1	5.3	4.5	3.1	14.4
Ni^{2+}	7	43.6	6.7	–	–	0.1
	8	43.0	6.6	0.4	–	1.1
	9	37.1	5.7	3.1	2.2	9.9
Zn^{2+}	7	49.4	10.5	0.2	–	1.1
	8	43.7	9.3	2.0	0.6	10.1
	9	16.5	3.5	7.6	21.0	38.4
Cd^{2+}	7	24.2	74.4	–	–	0.0
	8	24.1	74.3	–	–	0.1
	9	23.9	73.5	0.1	–	1.3
Cu^{2+}	6	45.7	8.9	0.8	–	3.4
	7	33.2	6.4	5.5	0.1	24.9
	8	8.7	1.7	14.4	2.5	65.0
	9	0.8	0.2	13.5	23.7	61.1

9.3.3.2
Organometallic Compounds

Among the organometallic compounds, organotin(IV) and organomercury(II) are the most widespread in the aquatic environment as a result of their use in industry (organotin in fungicides and acaricides in agriculture, in wood and stone preservatives, in antifouling agents in paints for ships, in stabilizers and catalysts in PVC and foam production; organomercury in fungicides for paper and wood and in antibacterial agents in medicine, etc.) and of the bioalkylation processes of inorganic metals by means of a variety of bacterial substrates (Craig and Miller 1997, and references therein).

Organotin(IV) compounds include a variety of organometallic moieties characterized by a central tin atom covalently bonded to various organic groups (methyl, ethyl, propyl, butyl, octyl, phenyl, etc.) through one or more carbon atoms. Their general formula can be written: $R_nSnX_{(4-n)}$ (R = organic group; X = halide, nitrate, acetate, hydroxide, etc.; $n = 1$ to 4). Organotin cations in aqueous solution are considered as acids of different hardness on the Lewis scale depending on the groups bonded to the tin(IV) (Tobias et al. 1966). They therefore show a strong tendency to hydrolysis in aqueous solution, resulting in some cases in condensation reactions of the monomeric conjugate bases with the formation of polynuclear hydroxo-complexes in solution according to the following general reaction (Tobias et al. 1966):

$$q\mathrm{R}_n\mathrm{Sn}(\mathrm{OH}_2)_x^{z+} + p\mathrm{H}_2\mathrm{O} = p\mathrm{H}_3\mathrm{O}^+ + (\mathrm{R}_n\mathrm{Sn})_q(\mathrm{OH})_p(\mathrm{OH}_2)_y^{(qz-p)+}$$

The hydrolysis processes of mono-, di- and trimethyltin(IV) cations have recently been reviewed, extensively investigated and defined in different ionic media, including SSWE, in a wide range of ionic strengths and salinities (De Stefano et al. 1999b 1999c 2000a; Foti et al. 1999, 2000). Hydrolysis constants for mono-, di- and trimethyltin(IV) cations are shown in Table A9.7 of appendix (Sect. A9.2)

Once hydrolytic equilibria have been defined for all the organotin(IV) cations investigated, the interactions of simple and hydroxo-organotin(IV) species with the anion A of SSWE as a single salt can be considered. On the basis of the results obtained by potentiometric measurements, we formulated the complex formation model shown in Table 9.17.

Mixed hydroxo-species are formed in all the systems. This indicates the strength of the hydrolytic species to be stronger than the simple association of organotin cations with the anionic component A of SSWE. By way of example, Fig. 9.13 shows a speciation diagram for the species CH_3Sn^{3+} in artificial sea water as a single salt (BA). As can be seen, the main species formed are mixed hydroxo species where the sea water anion (A) is representative of chloride and sulphate anions. In particular, at pH = 8 the predominant species is $[(CH_3Sn)_2A(OH)_5]^{0.117-}$ (more than 60% formation, Curve 6), while the simple hydrolytic species $[CH_3Sn(OH)_3]^0$ achieves about 24% formation (Curve 2) at the same pH value, confirming the strength of the hydrolysis processes.

9.4
Discussion and Conclusions

In the preceding sections, we illustrated two very important features of the interactions of low molecular weight ligands and metal cations, namely: (*a*) that these inter-

Table 9.17. Interactions of mono-, di- and triorganotin compounds in SSWE as single salt BA, at $I = 0$ mol l^{-1} and $t = 25$ °C

System	Species	log β [a]	log K [b]
$(CH_3)_3Sn$-BA	$[(CH_3)_3Sn(A)]^{0.117-}$	0.15	0.15
$(CH_3)_2Sn$-BA	$[(CH_3)_2SnA]^{0.883+}$	0.90	0.90
	$[(CH_3)_2SnA(OH)]^{0.117-}$	−3.05	−0.2
CH_3Sn-BA	$[CH_3SnA(OH)]^{0.883+}$	0.7	2.2
	$[CH_3SnA(OH)_2]^{0.117-}$	−1.45	2.0
	$[(CH_3Sn)_2A(OH)_5]^{0.117-}$	−5.80	1.89

[a] β refers to the reaction: $M + H_2O + A \leftrightarrow M_x(OH)_yA$, with $M = (CH_3)_xSn^{(4-x)+}$.
[b] K refers to the reaction: $M_x(OH)_y + A \leftrightarrow M_x(OH)_yA$, with $M = (CH_3)_xSn^{(4-x)+}$.

actions are quite significant, both in terms of stability constant values and in terms of the formation percentages of the various species; (*b*) that the behaviour of some major classes of ligands is generally homogeneous. To permit comparison of different ligands and metal cations, the percentages of complexes formed with the anion and the cation of the marine salts are reported in Fig. 9.14.

Only mono-carboxylic anions gave yields <50% whilst in the other cases, yields were always ≥50% and in some cases 100%. These percentages include protonated species for the different ligands and mixed hydroxo complexes for metal cations. Note that these speciation data are strictly valid only for the artificial sea water used (SSWE) and can differ significantly for other sea waters, in particular for those containing carbonate and fluoride. These differences are quite small for the ligands (amines in their protonated form interact with F^-, HCO_3^- and CO_3^{2-}) (De Stefano et al., work in progress)

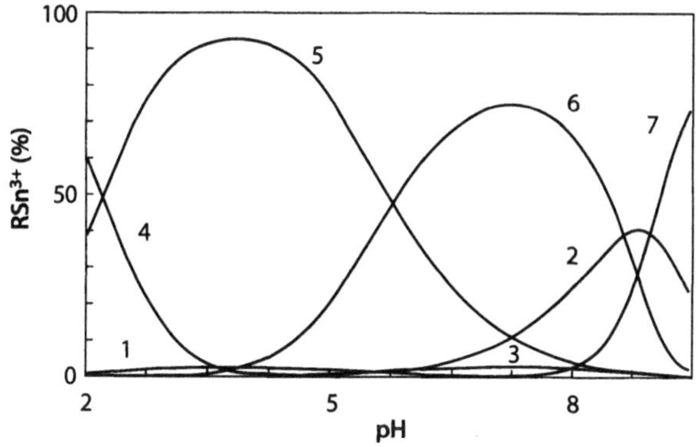

Fig. 9.13. Speciation diagram for the system CH_3Sn^{3+}-BA at 25 °C. $C_{RSn^{3+}}$ = 0.5 mM; C_{BA} = 0.575 M. Species: *1.* $RSn(OH)_2$; *2.* $RSn(OH)_3$; *3.* $(RSn)_2(OH)_5$; *4.* $RSnA(OH)$; *5.* $RSnA(OH)_2$; *6.* $RSn)_2A(OH)_5$

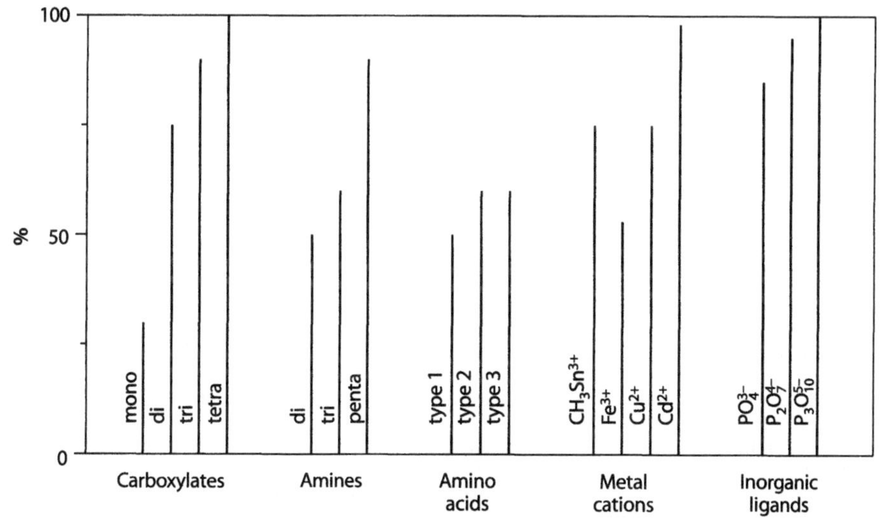

Fig. 9.14. Percentages of the component X (carboxylates, amines, amino acids, metal cations, inorganic ligands) complexed by the cation or the anion of marine salt BA, $S = 35$, at pH 8

but can be considerable for the metallic and organometallic cations. A recent investigation (Foti et al. 2000) showed that the addition of carbonate ($[CO_3^{2-}] + [HCO_3^-]$ $= 2.7$ mmol l^{-1}, for sea water 35 *S*) to the basic inorganic speciation model of organotins in synthetic sea water leads to the formation (at pH = 8) of a very stable mixed species $(CH_3)Sn(CO_3)(OH)_2$ in the monomethyltin system, whilst it is absolutely negligible for the other dimethyl and trimethyltin systems at the same pH value (see Table A9.8, Sect. A9.2 of appendix). Interactions of organotin cations and hydroxo species with floride ions ($[F^-] = 0.07$ mmol l^{-1}, for sea water 35 *S*) are very weak, and no complex species are formed.

As concerns the possibility of having predictive relationships, we showed that polycarboxylic anions form complexes whose stability follows very regular trends. Moreover, apparent protonation constants can be expressed as a function of salinity by simple polynomial equations, whose empirical parameters mainly depend on the protonation step.

References

Armstrong FAJ (1965) Phosphorus. In: Riley JP, Skirrow G (eds) Chemical oceanography, 1st edn, vol I. Academic Press, New York, pp 323–364

Atlas E, Culberson C, Pytkowicz RM (1976) Phosphate association with Na^+, Ca^{2+} and Mg^{2+} in seawater. Mar Chem 4:243–254

Bates RG, Calais JG (1981) Thermodynamics of the dissociation of $BisH^+$ in seawater from 5 to 40 °C. J Sol Chem 10(4):269–279

Bates RG, Erickson WP (1986) Thermodynamics of the dissociation of 2-aminopyridinium ion in synthetic seawater and a standard for pH in marine systems. J Sol Chem 15(11):891–901

Bremmer JM (1967) In: Mc Laren AD, Peterson GH (eds) Soil Biochemistry, vol I. Marcel Dekker, New York

Buffle J (1988) Complexation reactions in aquatic systems – an Analytical approach. Ellis Horwood Series in Analytical Chemistry, Chichester, England

Campbell JW, Goldstein L (eds) (1972) Nitrogen metabolism and the environment. Academic Press, London

Casale A, De Robertis A, De Stefano C, Gianguzza A (1989) Thermodynamic parameters for the formation of calcium α-alalinate complexes in aqueous solution. Thermochim Acta 140:59–66

Casale A, Foti C, Sammartano S, Signorino G (1998) Thermodynamic parameters for the protonation of some polyamines $C_{(2n-2)}N_nH_{(5n-2)}$ in NaCl aqueous solution at different ionic strengths. Ann Chim (Rome) 88:55–70

Craig PJ, Miller D (1997) Analysis and speciation of organometallic compounds in the marine environment. In: Gianguzza A, Pelizzetti E, Sammartano S (eds) Marine chemistry – an environmental analytical chemistry approach. Kluwer Academic, Dordrecht (Water Science and Technology Library, vol XXV, pp 161–172)

Culberson C, Pytkowicz RM, Hawley JE (1970) Seawater alkalinity determination by the pH method. J Mar Res 28:15–21

Czerminski JB, Dickson AG, Bates RG (1982) Thermodynamics of the dissociation of morpholinium ion in seawater: J Sol Chem 11(2):79–89

Daniele PG, De Robertis A, De Stefano C, Sammartano S, Rigano C (1985) On the possibility of determining the thermodynamic parameters for the formation of weak complexes using a simple model for the dependence on ionic strength of activity coefficients. Na^+, K^+ and Ca^{2+} complexes of low molecular weight ligands in aqueous solution. J Chem Soc Dalton Trans 2353–2361.

Daniele PG, De Stefano C, Prenesti E, Sammartano S (1994) Weak complex formation in aqueous solution. Curr Top Sol Chem 1:95–106

Daniele PG, Prenesti E, De Stefano C, Sammartano S (1995) Formation and stability of proton-amine-inorganic anion complexes in aqueous solution. J Solution Chem 24:325–341

Daniele PG, Prenesti E, De Robertis A, De Stefano C, Foti C, Giuffrè O, Sammartano S (1997) Binding of inorganic and organic polyanions by protonated open chain polyamines in aqueous solution. Ann Chim (Rome) 87:415–447 (Errata corrige (1998) 88:447–448)

Daumas RA (1976) Variations of particulate proteins and dissolved amino acids in coastal seawater. Mar Chem 4:225–242
Demianov P, De Stefano C, Gianguzza A, Sammartano S (1995) Equilibrium studies in natural waters: Speciation of phenolic compounds in synthetic seawater at different salinities. Env Tox Chem 14(5):767–773
De Robertis A, De Stefano C, Gianguzza A (1991) Salt effects on the protonation of L-histidine and L-aspartic acid: A complex formation model. Thermochim Acta 177:39–57
De Robertis A, De Stefano C, Patanè G, Sammartano S (1993) Effects of salt on the protonation in aqueous solution of triethylenetetramine and tetraethylenepentamine. J Solution Chem 22:927–940
De Robertis A, De Stefano C, Sammartano S, Gianguzza A (1994) Equilibrium studies in natural fluids. A chemical speciation model for the major constituents of seawater. Chem Spec Bioav 6:65–84
De Robertis A, Foti C, Sammartano S, Gianguzza A (1997) Chemical speciation of some classes of low molecular weight ligands in seawater. In: Gianguzza A, Pellizzetti E, Sammartano S (eds) Marine chemistry – an environmental analytical chemistry approach. Kluwer Academic Publishers, Dordrecht, pp 59–69
De Robertis A, Foti C, Patanè G, Sammartano S (1998a) Hydrolysis of $(CH_3)Hg^+$ in different ionic media: Salt effects and complex formation. J Chem Eng Data 43:957–960
De Robertis A, De Stefano C, Gianguzza A, Sammartano S (1998b) Binding of polyanions by biogenic amines. I. Formation and stability of protonated putrescine and cadaverine complexes with inorganic anions. Talanta 46:1085–1093
De Robertis A, De Stefano C, Foti C, Gianguzza A, Piazzese D, Sammartano S (2000) Protonation constants and association of polycarboxylic ligands with the major components of seawater J Chem Eng Data 45(6):996–1000
De Stefano C, Gianguzza A (1991) A complex formation model for the salt effects on the protonation of lysine in aqueous sodium and calcium chlorides and tetraethylammonium iodide solutions. Ann Chim (Rome) 81:119–130
De Stefano C, Foti C, Gianguzza A, Rigano C, Sammartano S (1994) Equilibrium studies in natural fluids. Use of synthetic seawater and other media as background salts. Ann Chim (Rome) 84:159–75
De Stefano C, Foti C, Gianguzza A, Sammartano S (1995) Chemical speciation of amino acids in electrolyte solutions containing major components of natural fluids. Chem Spec Bioavail 7(1):1–8
De Stefano C, Foti C, Gianguzza A, Sammartano S (1998) The single salt approximation for the major components of seawater: Association and acid-base properties. Chem Spec Bioavail 10(1):27–29
De Stefano C, Foti C, Gianguzza A, Sammartano S (1999a) Interaction of polyamines with Mg^{2+} and Ca^{2+}. J Chem Eng Data 44(4):744–749
De Stefano C, Foti C, Gianguzza A, Marrone F, Sammartano S (1999b) Hydrolysis of ,ethyltin(IV) trichloride in aqueous NaCl and $NaNO_3$ solutions at different ionic strengths and temperatures. Appl Organomet Chem 13:805–811
De Stefano C, Foti C, Gianguzza A, Millero FJ, Sammartano S (1999c) Hydrolysis of $(CH_3)_3Sn^+$ in various salt media. J Solution Chem 28(7):959–972
De Stefano C, Gianguzza A, Piazzese D, Sammartano S (1999d) Speciation of low molecular weight carboxylic ligands in natural fluids: Protonation constants and association with major components of seawater of oxydiacetic and citric acids. Anal Chim Acta 398:103–110
De Stefano C, Foti C, Gianguzza A, Sammartano S.(2000a) Hydrolysis processes of organotin(IV) compounds in sea water. In: Gianguzza A, Pellizzetti E, Sammartano S (eds) Chemical processes in marine environments. Springer-Verlag, Berlin, pp 231–228
De Stefano C, Foti C, Gianguzza A, Sammartano S (2000b) The interaction of amino acids with the major constituents of natural waters at different ionic strengths. Mar Chem 72:61–76
De Stefano C, Gianguzza A, Piazzese D (2000c) Complexes of azelaic and diethylenetrioxydiacetic acids with Na^+, Mg^{2+} and Ca^{2+} in NaCl aqueous solutions, at 25 °C. J Chem Eng Data 45(1):15–19
De Stefano C, Foti C, Gianguzza A, Sammartano S (2000d) Speciation of low molecular weight ligands in natural fluids: Protonation constants and association of open chain polyamines with the major components of seawater. Anal Chim Acta 418:43–51
Dickson AG, Riley JP (1979) The estimation of acid dissociation constants in seawater media from potentiometric titrations with strong base. II The dissociation of phosphoric acid. Mar Chem 7:101–109
Evans CA, Guevremont R (1979) Metal complexes of aspartic acid and glutamic acid. In: Siegel H (ed) Metal ions in biological systems, vol IX. Marcel Dekker, New York, pp 41–69
Fiol S, Brandariz I, Sastre de Vicente ME (1995a) The protonation constants of glycine in artificial seawater at 25 °C. Mar Chem 49:215–219
Fiol S, Brandariz I, Herrero RF, Vilariño T, Sastre de Vicente ME (1995b) Protonation constants of amino acids in artificial sea water at 25 °C. J Chem Eng Data 40:117–119

Fiol S, Vilariño T, Herrero RF, Sastre de Vicente ME, Arce F (1998) Protonation constants of valine, serine, and β-alanine in artificial seawater at 25 °C. J Chem Eng Data 43:393–395
Fogg GF (1975) Primary productivity. In: Riley JP, Skirrow G (eds) Chemical oceanography, 2nd edn. Academic Press, New York, pp 346–444
Foti C, Gianguzza A, Millero FJ, Sammartano S (1999) The speciation of $(CH_3)_2Sn^{2+}$ in electrolyte solution containing the major components of natural waters. Aquatic Geochem 5:381–398
Foti C, Gianguzza A, Piazzese D, Trifiletti G (2000) Inorganic speciation of organotin(IV) cations in natural waters with particular references to seawater. Chem Spec Bioavail 12(2):1–12
Garrels RM, Thompson ME (1962) A chemical model for sea water at $T = 25$ °C and one atmosphere total pressure. Am J Sci 260:57–66
Hansson I (1972) An analytical approach to the carbonate system seawater. Thesis, Gøteborg
Hershey JP, Millero FJ, Fernandez M (1989) The ionization of phosphoric acid in NaCl and NaMgCl solutions at 25 °C. J Solution Chem 18:875–892
Johansson O, Wedborg M (1979) Stability constants of phosphoric acid in seawater of 5–40‰ salinity and temperatures of 5–25 °C. Mar Chem 8:57–69
Johansson O, Wedborg M (1985) Determination of the stability constant for acetic acid in synthetic seawater media at various temperatures and salinities. J Solution Chem 14:431–439
Johnson KS, Pytckowicz RM (1979) Activity coefficients in electrolyte solutions, vol I. CRC Press, Boca Raton, FL, pp 1–61
Kester DR, Pytkowicz RM (1967) Determination of the apparent dissociation constants of phosphoric acid in seawater. Limnol Oceanogr 12:243–252
Kester DR, Duedall IW, Connors DN, Pytkowicz RM (1967) Preparation of artificial seawater. Limnol Oceanogr 12:176–178
Khoo KH, Ramette RW, Culberson CH, Bates RG (1977) Determination of hydrogen ion concentrations in seawater from 5 to 40 °C: Standard potentials at salinities from 20 to 45. Analytical Chem 49(1):29–34
Lee C, Wakeham SG (1989) Organic matter in sea water: Biogeochemical processes. In: Riley JP (ed) Chemical oceanography, vol IX. Academic Press, New York
Lyman J, Fleming RH (1940) Composition of seawater. J Mar Res 3:134–146
Martell AE, Smith RM (1997) Stability constants of metal complexes, NIST PC-based database. National Institute of Standards and Technology, Gaithersburg, MD
Mikita MA, Steenlink C, Wershaw RL (1981) Carbon-13 enriched Nuclear Magnetic Resonance method for the determination of hydroxyl functionality in humic substances. Anal Chem 53:1715
Millero FJ (1974) Seawater as a multi-component electrolyte solution. In: Goldberg ED (ed) The Sea, vol V. Wiley, New York
Millero FJ (1990) Marine solution chemistry and ionic interactions. Mar Chem 30:205–229
Millero FJ (1996) Chemical oceanography, 2nd edn. CRC Press, Boca Raton, FL
Millero FJ, Schreiber DR (1982) Use of the ion pairing model to estimate activity coefficients of the ionic components of natural waters. Am J Sci 282:1508–1540
Perrin DD (1979) Stability constants of metal ions complexes. Part B: Organic ligands. IUPAC (Chemical Data Series No. 22)
Pettit LD, Powell KJ (1997) IUPAC stability constants database. Academic Software, Outlay, UK
Pitzer KS (1991) Activity coefficients in electrolyte solutions, 2nd edn. CRC Press, Boca Raton, FL
Pytkowicz RM, Hawley JE (1974) Bicarbonate and carbonate ion-pairs and a model of seawater at 25 °C. Limnol Oceanogr 19:223–234
Sillén LG, Martell AE (1964) Stability constants. The Chemical Society, London (Special Publication No. 17)
Sillén LG, Martell AE (1971) Stability constants. The Chemical Society, London (Special Publication No. 25)
Spencer CP (1975) The micronutrient elements. In: Riley JP, Skirrow G (eds) Chemical oceanography, 2nd edn. Academic Press, New York, pp 245–295
Stephens GC (1972) Amino acid accumulation and assimilation in marine organism. In: Campbell JW, Goldstein L (eds) Nitrogen metabolism and the environment. Academic Press, London, pp 155–184
Stevenson FJ, Butler JHA (1969) In: Eglinton G, Murphy MTJ (eds) Organic geochemistry. Springer-Verlag, Berlin
Stumm W, Brauner PA (1975) Chemical speciation. In: Riley JP, Skirrow G (eds) Chemical oceanography, 2nd edn, vol I. Academic Press, New York, pp 173–234
Tobias RS, Farrer H, Hughes M, Nevett BA (1966) Hydrolysis of the aquo ions R_3Sn+ and R_2Sn^{2+}: Steric effects on the dissociation of aquo acids. Inorg Chem 5:2052–2055
Tuschall JR, Brezonik PL (1980) Characterization of organic nitrogen in natural water: Its molecular size, protein content, and interaction with heavy metals. Limnol Oceanogr 25:495–504
Williams PM (1971) Organic compounds in aquatic environment. Marcel Dekker, New York

Appendix

A9.1
Abbreviations and Formulas

A9.1.1
Carboxylate Ligands

Ligand name and abbreviation		Formula and charge of the fully deprotonated ligand
Acetate	ac	$[CH_3COO]^-$
Malonate	mal	$[OOC\text{-}CH_2\text{-}COO]^{2-}$
Succinate	succ	$[OOC\text{-}(CH_2)_2\text{-}COO]^{2-}$
Malate	mala	$[OOC\text{-}CH_2\text{-}CH(OH)\text{-}COO]^{2-}$
Tartrate	tar	$[OOC\text{-}CH(OH)\text{-}CH(OH)\text{-}COO]^{2-}$
Nonanedioate (azelate)	aza	$[OOC\text{-}(CH_2)_7\text{-}COO]^{2-}$
Diglycolate (oxydiacetate)	oda	$[O(CH_2COO)]^{2-}$
Diethylenetrioxydiacetate	toda	$[OOC\text{-}CH_2\text{-}O(CH_2)_2\text{-}O\text{-}(CH_2)_2\text{-}O\text{-}CH_2COO]^{2-}$
Citrate	cit	$[HOC(COO)(CH_2COO)_2]^{3-}$
1,2,3-Propanetricarboxylate (tricarballylate)	tca	$[OOC\text{-}CH_2\text{-}CH(COO)\text{-}CH_2COO]^{3-}$
Methyltricarballylate	mtca	$[OOC\text{-}CH(CH_3)\text{-}CH(COO)\text{-}CH_2COO]^{3-}$
1,2,3,4-Butane-tetracarboxylate	btc	$[OOCCH_2\text{-}(CHCOO)_2\text{-}CH_2COO]^{4-}$
Benzene-hexacarboxylate	mlt	$[C_6\text{-}(COO)_6]^{6-}$

A9.1.2
Amine Ligands

Ligand name and abbreviation		Formula
Ethylenediamine	en	$H_2NCH_2CH_2NH_2$
Diethylenetriamine	dien	$H_2NCH_2CH_2NHCH_2CH_2NH_2$
Triethylenetetramine	trien	$H_2NCH_2CH_2NHCH_2CH_2NHCH_2CH_2NH_2$
Tetraethylenepentamine	tetren	$H_2NCH_2CH_2NHCH_2CH_2NHCH_2CH_2NHCH_2CH_2NH_2$
Spermine	sper	$H_2N(CH_2)_3NH(CH_2)_4NH(CH_2)_3NH_2$

A9.1.3
Amino Acid Ligands

Ligand name and abbreviation		Formula
Glycine	gly	$H_2NCH_2CO_2H$
Alanine	ala	$H_2NCH_2CH_2\ CO_2H$
Histidine	his	$C_6H_9N_3O_2$
Aspartic acid	asp	$HO_2CCH_2CH(NH_2)CO_2H$
Glutamic acid	glu	$HO_2CCH_2CH_2CH(NH_2)CO_2H$
Iminodiacetic acid	ida	$HN(CH_2CO_2H)_2$
Serine	ser	$OHCH_2CH(NH_2)CO_2H$
Valine	val	$(CH_3)_2CHCH(NH_2)CO_2H$
Leucine	leu	$(CH_3)_2CHCH_2(NH_2)CO_2H$
Threonine	thr	$CH_3CH(OH)CH(NH_2)CO_2H$
Methionine	met	$CH_3SCH_2CH_2CH(NH_2)CO_2H$

A9.2
Tables

Table A9.1. Apparent protonation constants[a] of mono and dicarboxylic ligands in artificial sea water, at $t = 25$ °C

S	$\log K_1^{H*}$	$\log \beta_2^{H*}$	S	$\log K_1^{H*}$	$\log \beta_2^{H*}$
ac			tar		
5	4.54	–	5	3.92	6.75
15	4.47	–	15	3.74	6.50
25	4.43	–	25	3.66	6.38
35	4.40	–	35	3.62	6.32
45	4.37	–	45	3.60	6.27
mal			aza		
5	5.00	7.66	5	5.11	9.43
15	4.75	7.32	15	5.04	9.26
25	4.64	7.17	25	5.04	9.19
35	4.59	7.08	35	5.05	9.15
45	4.56	7.02	45	5.06	9.12
succ			toda		
5	5.16	9.15	5	3.70	6.80
15	5.00	8.91	15	3.50	6.52
25	4.94	8.80	25	3.43	6.39
35	4.91	8.73	35	3.39	6.32
45	4.90	8.68	45	3.37	6.26
mala			oda		
5	4.61	7.87	5	3.57	6.34
15	4.43	7.63	15	3.28	5.96
25	4.36	7.53	25	3.17	5.80
35	4.33	7.46	35	3.10	5.68
45	4.30	7.41	40	3.08	5.65

[a] β_j^H refer to the reaction: $L^{z-} + jH^+ \leftrightarrow H_jL^{j-z}$.

Table A9.2. Apparent protonation constants[a] of tri, tetra and hexacarboxylic ligands in artificial sea water, at $t = 25$ °C

S	$\log K_1^{H*}$	$\log \beta_2^{H*}$	$\log \beta_3^{H*}$	$\log \beta_4^{H*}$	$\log \beta_5^{H*}$
tca					
5	5.81	10.42	14.07		
15	5.60	10.08	13.65		
25	5.51	9.93	13.45		
35	5.45	9.84	13.31		
45	5.41	9.77	13.21		
mtca					
5	6.86	11.64	15.01		
15	6.59	11.21	14.51		
25	6.46	11.00	14.24		
35	6.37	10.87	14.06		
45	6.30	10.78	13.93		
cit					
15	4.31	8.28	11.10		
25	4.24	8.12	10.94		
35	4.18	8.00	10.79		
45	4.15	7.91	10.63		
btc					
5	5.74	10.74	14.81	18.03	
15	5.56	10.36	14.34	17.47	
25	5.55	10.31	14.23	17.35	
35	5.57	10.28	14.20	17.30	
45	5.56	10.31	14.18	17.28	
mlt					
5	5.03	9.50	13.10	15.79	16.6
15	4.89	9.17	12.70	15.35	15.8
25	4.89	9.10	12.66	15.33	15.6
35	4.92	9.08	12.68	15.39	15.5
45	4.96	9.07	12.72	15.46	15.5

[a] β_j^{H} refer to the reaction: $L^{z-} + jH^+ \leftrightarrow H_jL^{j-z}$.

Table A9.3. Apparent protonation constants[a] of amines in artificial sea water, at $t = 25\,°C$

Amine	S	$\log\beta_1^{H*}$	$\log\beta_2^{H*}$	$\log\beta_3^{H*}$	$\log\beta_4^{H*}$	$\log\beta_5^{H*}$
en	5	9.94	17.07	–	–	–
	15	9.97	17.22	–	–	–
	25	10.01	17.32	–	–	–
	35	10.04	17.40	–	–	–
	45	10.07	17.46	–	–	–
dien	5	9.82	18.82	23.20	–	–
	15	9.86	18.95	23.55	–	–
	25	9.88	19.04	23.72	–	–
	35	9.91	19.11	23.84	–	–
	45	9.94	19.18	23.93	–	–
trien	5	9.68	18.85	25.58	29.12	–
	15	9.68	19.00	25.88	29.57	–
	25	9.70	19.09	26.02	29.75	–
	35	9.71	19.16	26.12	29.85	–
	45	9.73	19.22	26.20	29.93	–
tetren	5	9.82	19.05	27.27	32.19	35.55
	15	9.80	19.10	27.47	32.59	36.09
	25	9.80	19.14	27.57	32.75	36.26
	35	9.80	19.17	27.64	32.84	36.34
	45	9.82	19.21	27.70	32.91	36.40
sper	5	10.70	20.60	29.44	37.53	–
	15	10.69	20.65	29.66	37.91	–
	25	10.69	20.68	29.77	38.07	–
	35	10.70	20.71	29.85	38.16	–
	45	10.72	20.74	29.92	38.23	–

[a] β_i^H refer to the reaction: $iH^+ + A^0 \leftrightarrow H_iA^{i+}$.

Table A9.4. Apparent protonation constants[a] of amino acids in artificial sea water at different salinities, at $t = 25$ °C

S	*i*	$\text{Log}\,\beta_i^{H*}$ Glycine	Alanine	Histidine	Glutammic acid	Aspartic acid
5	1	9.58	9.71	9.08	9.67	9.58
	2	11.90	12.06	15.03	13.86	13.28
	3	–	–	16.62	16.00	15.18
15	1	9.49	9.65	9.00	9.50	9.35
	2	11.80	12.00	14.98	13.60	12.96
	3	–	–	16.61	15.72	14.86
25	1	9.46	9.64	8.96	9.41	9.21
	2	11.77	11.99	15.02	13.47	12.80
	3	–	–	16.69	15.59	14.71
35	1	9.46	9.64	8.96	9.36	9.12
	2	11.76	11.99	15.09	13.40	12.71
	3	–	–	16.81	15.53	14.63
45	1	9.46	9.65	8.98	9.32	9.06
	2	11.77	11.99	15.19	13.36	12.66
	3	–	–	16.95	15.51	14.60

[a] Refer to the reaction: $A^{z-} + iH^+ \leftrightarrow H_iA^{z-i}$.

Table A9.5. Apparent protonation constants of amino acids in SSWE at different salinities

S	Glycine[a]		Serine[b]		Valine[b]		β-alanine[b]	
	$\log K_1$	$\log K_2$	$\log K_1$	$\log K_2$	$\log K_1$	$\log K_2$	$\log K_1$	$\log K_2$
5	2.426	9.496	2.292	9.072	2.341	9.480	3.495	9.923
5.5	2.393	9.539						
9			2.252	9.011	2.276	9.465		
10	2.363	9.447						
10.6							3.448	9.907
12.5	2.371	9.447						
14.5			2.247	8.951	2.297	9.431	3.515	9.853
17.5			2.186	8.773	2.286	9.393	3.472	9.866
20			2.210	8.791	2.315	9.287	3.451	9.905
21				8.883	2.298	9.340	3.465	9.852
22.6			2.186	8.858	2.276	9.361		
24.7				8.845	2.304	9.341	3.498	9.884
25	2.365							
26.5	2.370	9.339						
29			2.172	8.727	2.281	9.338	3.488	9.837
32			2.124	8.821	2.332	9.378		
35	2.358	9.392	2.177	8.793	2.316	9.319	3.476	9.836

S	α-alanine[c]		Leucine[c]		Threonine[c]		Methionine[c]	
	$\log K_1$	$\log K_2$	$\log K_1$	$\log K_2$	$\log K_1$	$\log K_2$	$\log K_1$	$\log K_2$
4.5	2.478	9.715	2.424	9.583	2.293	8.985	2.251	9.087
13.5	2.415	9.626	2.414	9.514	2.261	8.842	2.271	9.010
22.5	2.414	9.614	2.388	9.474	2.223	8.767	2.261	8.973
27.7	2.422	9.581	2.396	9.452	2.290	8.791	2.272	8.963
32.4	2.465	9.589	2.473	9.482	2.342	8.790	2.384	9.001

[a] Fiol et al. (1995a).
[b] Fiol et al. (1998).
[c] Fiol et al. (1995b).

Table A9.6. Apparent protonation constants[a] of phosphate, di-phosphate and tri-phosphate in synthetic sea water, at $t = 25$ °C and at different salinities

S	$\log \beta_1$ [a]	$\log \beta_2$ [a]	$\log \beta_3$ [a]
PO_4^{3-}			
5	9.10	15.73	17.61
15	8.93	15.40	17.12
25	8.91	15.34	16.96
35	8.92	15.34	16.89
45	8.94	15.36	16.85
$P_2O_7^{4-}$			
5	6.54	11.58	13.15
15	6.48	11.37	12.62
25	6.51	11.39	12.47
35	6.55	11.45	12.42
45	6.60	11.53	12.41
$P_3O_{10}^{5-}$			
5	5.79	10.29	12.06
15	5.71	9.94	11.48
25	5.75	9.91	11.34
35	5.80	9.96	11.31
45	5.86	10.04	11.32

[a] Refer to the reaction: $P_iO_{(3i+1)}^{(i+2)-} + jH^+ \leftrightarrow H_jP_iO_{(3i+1)}^{(i+2-j)-}$.

Table A9.7. Hydrolysis constants of $(CH_3)_xSn^{(4-x)+}$ cations at $I = 0$ mol l^{-1} and $t = 25$ °C

Species	$\log \beta_{pq}$ [a,b]		
(*pq*)	$x = 1$	$x = 2$	$x = 3$
1-1	−1.5 (12.5)[b]	−2.86 (11.14)	−6.14 (7.86)
1-2	−3.46 (24.54)	−8.16 (19.84)	−18.88 (9.12)
1-3	−9.09 (32.91)	−19.35 (22.65)	–
1-4	−20.47 (35.53)	–	–
2-2	–	−4.99 (23.01)	–
2-3	–	−9.06 (32.94)	–
2-5	−7.69 (62.31)	–	–

[a] β_{pq} refer to the reaction: $pM + qH_2O \leftrightarrow M_p(OH)_q + rH$ [M = methyltin(IV) cation; charges omitted].
[b] In parenthesis, formation constants refer to the reaction: $pM + qOH \leftrightarrow M_p(OH)_q$.

Table A9.8. Formation percentages of organotin(IV) species in SSWE (as single salt, BA) including carbonate ligand at different pH values

Species	pH	$x = 1$	$x = 2$	$x = 3$
$(CH_3)_xSn\text{-}OH$[a]	4	7	54	–
	6	5	78	18
	8	–	97	97
$(CH_3)_xSn\text{-}A$[b]	4	64	20	32
	6	5	3	28
	8	<1	<1	<1
$(CH_3)_xSn\text{-}CO_3$[c]	4	<1	19	–
	6	15	19	–
	8	93	3	<1

[a] Sum of percentages of simple hydrolytic species.
[b] Sum of percentages for the species containing $A^{1.117-}$ (including mixed -OH species).
[c] Sum of percentages for carbonate species (including mixed -A and -OH species).

Chapter 10

Equilibrium Analysis, the Ionic Medium Method and Activity Factors

I. Grenthe

10.1 Introduction

Chemical models are used extensively to describe environmental systems, transport of pollutants, sorption phenomena, etc. The following chapter describes two different models for ionic medium/ionic strength corrections of equilibrium constants used in chemical speciation calculations. Published equilibrium constant data often refer to ionic media very different from those encountered in nature and in engineering systems, and they have therefore to be recalculated to the conditions relevant for the system under study. In addition, one may be interested to compare experimental data obtained in different ionic media in order to ascertain their quality. It is important to notice that models used in science and technology are provisional; they are used as tools, and it is their "practicality" that often decides which model one selects. The prime requirement of a model is that it has a scientific basis and that it is internally consistent. Before describing the scientific basis and structure of ion interaction theories, we will give a short discussion of equilibrium analysis, the method used to obtain precise information of the composition and equilibrium constants, i.e. the data used in speciation calculations.

10.2 Equilibrium Analysis

Equilibrium analysis is an analytical method used to study the species present in multi-component solution systems, most commonly aqueous solutions where the different species are in rapid equilibrium with one another, as exemplified by the following two-component system:

$$pM + nL \leftrightarrow M_pL_n\,; \quad 1 \leq p \leq P \text{ and } 1 \leq n \leq N \tag{10.1}$$

The method has some similarity with the classical elemental analysis; the analytical data only provide information on constitution and the amounts of the different species, but not on their structure, the presence of isomers, etc. Equilibrium analysis is the first step in the scientific exploration of problems in solution co-ordination chemistry such as the structure of complexes, chemical bonding and reactivity. The results of equilibrium analysis is given as equilibrium constants, the numerical values of which are valid at a certain ionic strength/ionic medium composition and temperature. Data of this type are useful in many applications, e.g. to determine the *chemical speciation* of complex chemical systems, such as surface and ground waters, where the total concentrations of the components are known. In addition, one must have access to a chemical database that contains equilibrium constants for the predominant chemical reactions in the system. The chemical speciation describes the composition and the concentrations of the species as a function of one or more *master variables* such as the free hydrogen ion concentration, the free ligand concentration and the redox potential of the system.

When using an analytical tool, it is necessary to ascertain its range of validity, as well as its precision and accuracy; this is true also for equilibrium analysis. The following list indicates some of the factors necessary to consider when making equilibrium analytical experiments and when assessing the quality of equilibrium data reported in the scientific literature.

- Is the chemical system studied at equilibrium, or not? It is fairly easy to decide this in most experimental studies. It is much more difficult to do so in complex systems in engineering or in nature. Some equilibria may be rapidly attained, e.g. many complex formation reactions, while others may take a very long time, e.g. redox reactions involving the transfer of more than one electron and for which there are major changes in composition between the reduced and oxidized forms of a certain component. A typical example is the redox chemistry of sulphur, where most equilibrium reactions are extremely sluggish at room temperature.
- The equilibrium analysis provides information on the *concentration* of the species formed from experimental values of the *total* concentrations of the different components and the *equilibrium* concentration of one or more species, often the free metal ion or ligand concentration. The free concentration is often measured with an electrode that responds to the activity, and we must therefore be able to transform this quantity to concentrations. This is done using the following relationship:

$$a_i = [A_i]\gamma_i \tag{10.2}$$

  where a_i and γ_i are the activity and the activity coefficient, respectively of species i. How can we design our experiments so that the activity and concentrations are equal, or related by a *constant* factor? This is where the ionic medium method enters the stage.
- The activity of the species present in solution varies with the total concentration of the components. It is well-known that:

$$\lim a_i = [A_i] \;\; ; [A_i] \rightarrow 0 \tag{10.3}$$

  From Eq. 10.3, one must not draw the conclusion that an equilibrium analysis should be performed in very dilute solutions. In order to analyse even very simple equilibrium systems it is necessary to vary the concentrations of the components over a large concentration range, hence we cannot approach the limiting value of an extremely dilute solution. Instead, we add a strong electrolyte to the chemical system at a concentration that is much higher than that of the reactants. The strong electrolyte must not be a reactant, i.e. it must not interfere chemically in any significant way with the reaction studied. The activity coefficients of the reactants in these systems will then mainly be determined by the ionic medium electrolyte, as explained by the ion-ion interaction models described in the following text and in

Sect. 10.4. As the concentration of the ionic medium varies fairly little in a properly designed equilibrium analytical experiment, we can assume that the activity coefficients of all reactants and products are constant. We *define* this constant as unity in a given ionic medium. This means that comparison of equilibrium constants determined in different ionic media is not trivial; it requires a method to calculate activity coefficients of all reactants and products. This will be discussed later.

- Equilibrium analytical methods are described by Rossotti and Rossotti[1], and only a brief summary is given below, using a two-component system as an example:

The experimental data consist of sets of total concentrations of the two components M, and L, and the concentration of the one or both of the concentrations of free M, and L, denoted *m*, and *l*. There are two mass-balance conditions:

$$M_i = m + [ML] + [ML_2] + \ldots + [ML_N] = m(1 + \beta_1 l + \beta_2 l^2 + \ldots + \beta_N l^N) \quad (10.4)$$

$$L_i = l + [ML] + 2[ML_2] + \ldots + N[ML_N] = l + \beta_1 l + 2\beta_2 l^2 + \ldots + N\beta_N l^N \quad (10.5)$$

where β_n in this example denotes the different stability constants for the different mononuclear complexes formed.

The stoichiometry and the equilibrium constants can be deduced using methods described in Rossotti and Rossotti (1961). Recently, the graphical methods have often been replaced by least-square techniques, where a chemical model consisting of stoichiometry and equilibrium constants for different assumed complexes is fitted to the experimental data. The least-square analysis provides a measure of the agreement between the model and the experimental data and an estimate of the precision of the deduced equilibrium constants. There are several reasons to be careful when using these methods: false minima may appear in the least-squares refinement, resulting in erroneous chemical conclusions; systematic errors in the experimental data may be described by introducing unrealistic chemical species. It is difficult to avoid such errors in equilibrium analytical experiments; an example is small variations in the activity coefficients resulting from the change in composition of the test solutions during an experiment. In order to ascertain the quality of equilibrium analytical data, it is essential to make an analysis of the amounts of the different complexes formed throughout the experiment. Complexes that are present in small amounts should be looked upon with suspicion, especially if they are formed in concentration ranges where large changes in the composition of the test solutions have been made. The least squares assigned uncertainty of the reported equilibrium constants are often underestimated. They depend on the weights of the individual measurements and the number of data points that are used to determine a particular equilibrium constant, and these factors are not always analysed. In addition, the uncertainty estimate represents an estimate of the *precision* of the experiment, not its *accuracy*. A measure of the latter can only be obtained by studying the same equilibrium system with different experimental methods. Recent advances in NMR-spectroscopy and laser-based spectroscopic methods and structure information based on synchrotron light sources (EXAFS) makes it possible to attain much more precise insight into equilibrium systems than offered by the "classical" methods; some examples are given by Szabo et al. 1997, Farkas et al. 2000, and Moll et al. 1999.

Conclusion: Users of equilibrium analytical data should be aware of the sources of error present in experiments of this type. The stoichiometric coefficients determined in an equilibrium analysis are not "fitting parameters," they are related to the known co-ordination chemistry of the system, and one should therefore always investigate if the chemical model suggested by an equilibrium analysis is consistent with other types of chemical information or not. The users of tabulated equilibrium data are often faced with a number of different values of equilibrium constants, often determined in ionic media of different composition. In order to compare these data with one another, it is necessary to have a method for calculating activity coefficients. The same is true if one needs to use existing equilibrium constants for a certain metal complex determined, e.g. in 3 M $NaClO_4$, to describe the chemistry in another system, e.g. marine waters.

10.3 Activity Factors in Multi-Component Electrolyte Systems

The ionic medium method dates back to the first decades of the last century (Bodländer and Storbek 1902a,b; Bodländer and Fittig 1902; Bodländer and Eberlin 1904; von Euler 1903a,b,c; Grossman 1905). The early use was based on empirics, which was later given a theoretical foundation through the work of Debye and Hückel (1923).

Theoretical background. All electrolyte models are based on macroscopic physicochemical descriptions of the interactions between dissolved ions, and sometimes between these and the solvent. However, all models used so far are provisional in the sense that they contain parameters that have to be determined experimentally; an exception is the Debye-Hückel limiting law.

The classical Debye-Hückel model takes only electrostatic interactions between ions of opposite charge into account. It provides a fairly accurate description of the variation of the mean-activity coefficients of single charged 1:1 electrolytes up to concentrations of 0.01 M; for higher charged electrolytes the range of validity is much smaller. The Debye-Hückel *limiting law* is:

$$\log \gamma_{\pm} = -|Z_+ Z_-| \mathrm{A} \sqrt{I_m} \tag{10.6}$$

where A is a known constant equal to 0.5100 $mol^{-1/2}\,kg^{1/2}$ at 25° C. The range of validity of the Debye-Hückel model can only be extended to higher ionic strengths by the introduction of an electrolyte-dependent "effective" size of the ions. This parameter must be determined from experimental mean-activity coefficients. The extended Debye-Hückel equation is:

$$-\log \gamma_{\pm} = \frac{|Z_+ Z_-| \mathrm{A} \sqrt{I_m}}{1 + \mathrm{B} a_d \sqrt{I_m}} \tag{10.7}$$

where B is a constant determined by properties of the solvent (temperature, pressure, density and dielectric constant), and a_d is the "effective" distance between anion and cation. The extended equation is valid up to I_m approx. 0.03 M for single charged 1:1 electrolytes.

In order to extend the range of validity of the Debye-Hückel model further and to take experimentally observed individual electrolyte characteristics into account, the following methods have been used:

- Introduction of non-electrostatic short-range interactions by adding terms proportional to the concentrations of the various ions, or the ionic strength. An example is the Davies equation, written below for the single activity coefficient of an ion, γ_i:

$$\log \gamma_i = -0.5100 Z^2 \left(\frac{\sqrt{I_m}}{1+\sqrt{I_m}} - 0.3 I_m \right) \qquad (10.8)$$

  The Davies equation works fairly well up to an ionic strength of 0.1 M, and has a formal resemblance to the expression for single ion activity coefficients obtained from the specific ion interaction model to be described later. However, the activity factor expression does not take electrolyte specific effects into account. To describe these, one is forced to introduce the concept of *ion pairing*, described by equilibrium constants between the single electrolyte anion and cations. In ionic media, it is sufficient to take only the complex formation between the reactants/products and the anion and cation of the ionic medium into account.

  Helgeson et al. (1981) have developed more elaborate ion pairing models that have found extensive use for the modelling of geochemical systems.
- The individual characteristics of electrolytes may also be described by using *specific ion-interaction* models. We will discuss two of these: the *Brønsted-Guggenheim-Scatchard* specific ion-interaction model, in the following denoted the SIT-model, and the *Pitzer* ion interaction model. Both are semi-empirical; they are extensions of the Debye-Hückel model, but contain parameters that take non-electrostatic interactions into account. The parameters have a theoretical basis; however, they cannot be calculated ab initio, but have to be determined from experimental, mean-activity or osmotic coefficient data. A more detailed description of the structure of these models is given in Sect. 10.4.

10.4 The Pitzer and the Brønsted-Guggenheim-Scatchard Ion Interaction Models

Pitzer (1973, 1991) considered his model as an extension of the simple but general approach, presented by Guggenheim (1935), who proposed the following equation describing the concentration dependence of the activity coefficient of a cation M in a mixture:

$$\log \gamma_M = -\frac{A Z_M^2 I^{1/2}}{1+I^{1/2}} + \sum_a B_{Ma} m_a \qquad (10.9)$$

where A is the Debye-Hückel parameter (note that $A = 3A_\Phi = 0.5100 \text{ mol}^{-1/2}\text{ kg}^{1/2}$ at 25 °C, where A_Φ is the corresponding parameter in the Pitzer model). The summation involves all anions, a, present in solution. B_{Ma} is an interaction parameter specific for each cation-anion pair, Ma. In accordance with the Brønsted (1922) postulate on specific interaction between ions, the terms for ions of the same charge sign are equal to zero. The analogous expression for the anion X is obtained by changing the subscripts M and a for X and c, respectively, where c denotes a cation in general. A detailed discussion of the use of the Guggenheim model for describing the concentration dependence of the osmotic coefficient and the mean activity coefficients in both single and mixed electrolyte solutions are given in Pitzer and Brewer (1961).

Scatchard (1959, 1976) suggested that the denominator $(1 + I^{1/2})$ should be replaced by $(1 + 1.5I^{1/2})$ to decrease the concentration dependence of the interaction coefficients at low ionic strengths, giving the following expression for the activity coefficient of the reactants in an ionic medium:

$$\log \gamma_i = -\frac{A Z_i^2 I^{1/2}}{1 + 1.5 I^{1/2}} + \sum_j \varepsilon_\gamma(i,j)\, m_j \tag{10.10}$$

where $\varepsilon(i,j)$ is the specific ion interaction coefficient for ion i and the different counter-ions j. Equation 10.10 is known as the Brønsted-Guggenheim-Scatchard (SIT) model.

The SIT-model ignores both binary interactions between species of the same charge, and the contribution of ternary interactions to the activity coefficients.

The constancy of the ion interaction coefficients in the SIT-model at high molality was recognized long ago. However, the parameter is concentration-dependent at low molality (Pitzer and Brewer 1961; Pitzer 1973). These variations give only a small contribution to the accuracy of the calculated activity coefficients because of the product where $\varepsilon(i,j)\, m_j$ makes only a small contribution at low molality, c.f. Eq. 10.10. The concentration-dependence of the interaction parameters reflects the concentration-dependence of the sum of the radial distribution functions for like-charged and unlike-charged ions, see Pitzer (1973, 1991) and Scatchard (1959).

In the Pitzer formalism, the concentration dependence of the activity coefficient of a cation M (the corresponding equation for an anion L is obtained by interchanging L for M, a for c, and c for a throughout, where a and c stand for anion and cation in general) in a mixed solution containing a number of different ions and neutral species (for notation c.f. Pitzer 1991, Eq. 63) is:

$$\begin{aligned}\ln \gamma_M = {} & Z_M^2 F + \sum_a m_a (2B_{Ma} + ZC_{Ma}) + \sum_c m_c \left(2\phi_{Mc} + \sum_a m_a \psi_{Mca}\right) \\ & + \sum_a \sum_{a'} m_a m_{a'} \psi_{Maa'} + |Z_M| \sum_c \sum_a m_c m_a C_{Ca} + 2\sum_n m_n \lambda_{nM}\end{aligned} \tag{10.11}$$

The subscript n denotes neutral species; B_{Ma} is the virial coefficient describing the interactions between a cation, M, and an anion, a; C_{Ma} and C_{Ca} are defined by Eq. 10.17, ϕ_{Mc} is the virial coefficient arising from binary interaction between a specific cation and the other cations; λ_{nM} is the virial coefficient representing the interactions between a specific cation and neutral species; ψ_{ijk} is the virial coefficient representing interactions between ions i, j, k (where i and j are different anions and k is a cation or when i and j are different cations and k is an anion). The parameters ψ and λ are assumed to be independent of ionic strength. The quantity F includes the Debye-Hückel, f^{γ}, and other terms as follows:

$$F = f^{\gamma} + \sum_{c}\sum_{a} m_c m_a B'_{ac} + \sum_{c}\sum_{c'} m_c m_{c'} \phi'_{cc'} + \sum_{a}\sum_{a'} m_a m_{a'} \phi'_{aa'} \tag{10.12}$$

where

$$f^{\gamma} = -A_{\phi}\left\{\frac{\sqrt{I_m}}{1 + b\sqrt{I_m}} + \frac{2}{b}\ln\left(1 + b\sqrt{I_m}\right)\right\} \tag{10.13}$$

and ϕ' and B' are the ionic strength derivatives of ϕ and B, respectively (see below). B and B' are equal to:

$$Z = \sum_{i} m_i |Z_i| \tag{10.14}$$

$$B_{Ma} = \beta^{(0)}_{Ma} + \frac{2\beta^{(1)}_{Ma}}{\alpha^2 I}\left\{1 - (1 + \alpha I^{1/2})\exp(-\alpha I^{1/2})\right\} = \beta^{(0)}_{Ma} + \beta^{(1)}_{Ma} g(\alpha I^{1/2}) \tag{10.15}$$

and

$$B'_{Ma} = -\frac{2\beta^{(1)}_{Ma}}{\alpha^2 I^2}\left\{1 - (1 + \alpha I^{1/2} + \frac{\alpha^2 I}{2})\exp(-\alpha I^{1/2})\right\} = \beta^{(1)}_{Ma}\frac{g'(\alpha I^{1/2})}{I} \tag{10.16}$$

C_{ca} and the virial coefficient ϕ_{ij} are defined as follows:

$$C_{Ca} = \frac{C^{\Phi}_{Ca}}{2|Z_c Z_a|^{1/2}} \qquad C_{Ma} = \frac{C^{\Phi}_{Ma}}{2|Z_c Z_a|^{1/2}} \tag{10.17}$$

$$\phi_{ij} = \theta_{ij} + {}^{E}\theta_{ij}(I) \tag{10.18}$$

C^{Φ}_{ca}, are tabulated quantities for each cation-anion pair (Pitzer 1991), ${}^{E}\theta_{ij}(I)$ is a function of the ionic strength only; it is zero except for unsymmetrical mixing of ions of the same sign, i.e. when the charges of i and j are of the same sign but different magnitude (this term is given by theory and its values can calculated numerically as described in (Pitzer 1991, Appendix B). $g(\alpha I^{1/2})$ and $g'(\alpha I^{1/2})$ are known functions of I; α is an empirical parameter equal to 2.0 $kg^{1/2}\,mol^{-1/2}$.

The features of the Pitzer and the SIT-models given above provide the rationale for the simplification of the Pitzer equations presented below.

10.5 Comparison of the SIT and Pitzer Models

How many empirical parameters is it advisable to use when calculating activity factors? It is wise to keep the old scholastic principle called Occam's razor in mind. This maxim says, "Entities are not to be multiplied without necessity," or in Bertrand Russel's (1959) words: "If everything in some science can be interpreted without assuming this or that hypothetical entity, there is no ground assuming it." In the following, we will discuss some possible simplifications of the Pitzer model. The most common simplification is to neglect the contribution of the third virial coefficient. Indeed, a check of the typical values of C^{Φ} from the available compilation (Pitzer 1991) shows that this contribution is significant only at high ionic strength, typically above 5–6 mol kg^{-1}. Hence, $C^{\Phi} = 0$ is a reasonable assumption when the concentrations of reactants/products are small in comparison with the ionic medium concentration.

Another simplification of the Pitzer model was proposed by Millero (1983), who suggested the estimate $\beta^{(1)} = 0$, for complexes. This simplification may lead to errors as discussed by Plyasunov et al. (1998) and illustrated by Fig. 10.1.

From Fig. 10.1 it is apparent that the one-parameter Pitzer model is inferior to the SIT-model. Another important observation is that the values of $\beta^{(0)}$ obtained using

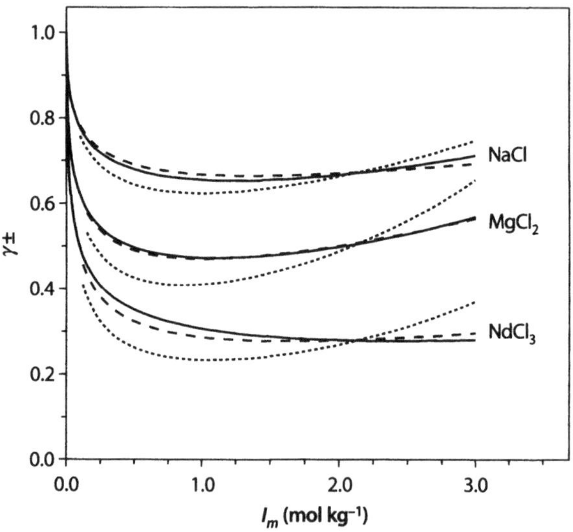

Fig. 10.1. The experimental and fitted mean activity coefficients of NaCl, $MgCl_2$ and $NdCl_3$ using $\beta^{(1)} = 0$ and neglecting triple ion interactions, plotted as a function of the ionic strength up to $I = 3$ mol kg^{-1}. The *thick full-drawn curves* represent the experimental data. The calculated activity coefficients using the complete Pitzer model coincide with the experimental values within 0.2%. The *short-dashed curve* shows the mean activity coefficients calculated using the Pitzer model with only the $\beta^{(0)}$ parameter. The *long-dotted curves* are calculated using the SIT-model with a fitted value of ε_γ

$\beta^{(1)} = 0$ always differ from the tabulated values from Pitzer (1991) given in parenthesis for the electrolytes in Fig. 10.1; NaCl: $\beta^{(0)} = 0.110$ (0.0765); $MgCl_2$: $\beta^{(0)} = 0.557$ (0.352) and $NdCl_3$: $\beta^{(0)} = 1.247$ (0.568). The error increases with the charge type of the electrolyte. To conclude: The Pitzer equations *must* be used with both the $\beta^{(0)}$ and $\beta^{(1)}$ terms.

An examination of the values of $\beta^{(1)}$ at 298.15 K from Pitzer (1973) shows that they are correlated with the charge type of the electrolyte: for most 1-1 electrolytes the values of $\beta^{(1)}$ fall in a range 0.20 ±0.20, for most 2-1 electrolytes in the range 1.4 ±0.6, and for most 3-1 electrolytes in the range 5.2 ±1.2. These averages may be used as "fixed" values of $\beta^{(1)}$ in the Pitzer equations, thus reducing the number of unknown parameters.

It may be important to be able to transfer parameters from one ion interaction model to the other. We will now demonstrate how this can be done in the ionic strength range, $0 < I < 3\text{–}4\ m$, where the SIT and Pitzer models are approximately equivalent. The analytical statements for the mean activity coefficient in the Pitzer model without the C^{Φ} term is:

$$\ln \gamma_{\pm} = -|Z_M Z_L| A_{\Phi} \left\{ \frac{I^{1/2}}{1 + bI^{1/2}} + \frac{2}{b} \ln(1 + bI^{1/2}) \right\} + m \frac{2\nu_M \nu_L}{\nu} (2\beta_{ML}^{(0)} + 2\beta_{ML}^{(1)} X) \tag{10.19}$$

where

$$X = \frac{1}{\alpha^2 I} \left\{ 1 + (1 + \alpha I^{1/2} - \frac{\alpha^2 I}{2}) \exp(-\alpha I^{1/2}) \right\} \tag{10.20}$$

For the SIT-model we obtain:

$$\ln \gamma_{\pm} = -\frac{A|Z_M Z_L| I^{1/2}}{1 + 1.5 I^{1/2}} + m \frac{2\nu_M \nu_L}{\nu} \varepsilon_{\gamma}(M, L) \tag{10.21}$$

Taking into account that $A = 3A_{\Phi}$ and making elementary transformations, we obtain:

$$Y = -\frac{A_{\phi}|Z_M Z_L|\nu}{4\nu_M \nu_L m} \left\{ \frac{3I^{1/2}}{1 + 1.5 I^{1/2}} - \frac{I^{1/2}}{1 + bI^{1/2}} - \frac{2}{b} \ln(1 + bI^{1/2}) \right\} = (\beta_{ML}^{(0)} - \frac{\varepsilon_{\gamma}(M, L)}{2}) + \beta_{ML}^{(1)} X \tag{10.22}$$

X and Y are known quantities that depend only on the charge type of the electrolyte, the ionic strength/molality and A_Φ. The linear function Y of X has the intercept $(\beta^{(0)}_{ML} - \varepsilon_\gamma(M,L) / 2)$ and the slope $\beta^{(1)}_{ML}$. The functions $Y(X)$ are plotted in Fig. 10.2 for 1-1, 2-1 and 3-1 ion combinations.

The linearity is very good, indicating that the SIT-model is approximately equivalent to the Pitzer model without the C^Φ term and with a constant value of $\beta^{(1)}$ for each charge type. The relationships between the two sets of parameters given in Table 10.1 for different ion combinations may be used to convert the large set of ε_γ values already available (Pitzer 1991, Guggenheim 1935) for complexes, into $\beta^{(0)}$ and $\beta^{(1)}$ values. Note that ε values, not ε_γ, are tabulated in Pitzer (1991) and Guggenheim (1935); the relationship between them is $\varepsilon_\gamma = \varepsilon \ln(10)$. The difference is due to the use of decadic and e logarithms, respectively.

The estimated values of $\beta^{(1)}$ for 3-1 and 4-1 interactions ($\beta^{(1)} = 4.3$ for 3-1 and 1-3, and $\beta^{(1)} = 8.9$ for 4-1 and 1-4 interactions) seem to be slightly lower than the "averaged" values from Pitzer (1991), $\beta^{(1)} = 5.2 \pm 1.2$ and $\beta^{(1)} = 11 \pm 2$, respectively.

It is not straightforward to determine Pitzer parameters from experimental mean activity factors, and osmotic coefficient data. The interaction parameters are often strongly correlated, and different experiments may not always give concordant results. It is therefore advisable to investigate how sensitive the calculated activity coefficients are for variations in the Pitzer parameters. Example 1 demonstrates the use of the two models for a simple protolytic equilibrium and also how sensitive the Pitzer model is for the numerical values of the parameters.

Use of the Pitzer model in equilibrium analysis. Here we are faced with slightly different problems from those described above. Pitzer parameters for complexes can rarely be determined from mean activity coefficients; they have to be deduced from the ionic strength/ionic medium dependence of equilibrium constants. This requires a very large experimental effort and very precise values of the equilibrium constants. Very few data of this type are available. A simplifying factor for the Pitzer analysis is when the reactants/products are present in such a low concentration that the main ion-ion interactions are those between the medium ions themselves and between the medium ions and the reactants/products.

Complex formation equilibria are in general described using concentration equilibrium constants; these are "true" thermodynamic quantities as long as concentrations and activities for reactants/products are proportional to one another in the ionic medium used. Each ionic medium may thus be considered as a particular solvent. A practical way to compare equilibrium constants obtained in different ionic media is to use *one* reference medium, usually pure water. A simplified Pitzer model for the

Table 10.1. Quantitative relationship between the Pitzer parameters $\beta^{(0)}$ and $\beta^{(1)}$ and the SIT parameter ε_γ for different ion combinations

Ion combination	$(\beta^{(0)} - \varepsilon_\gamma/2)$	$\beta^{(1)}$
M^+, X^-	0.035	$0.34 \approx 0.3$
M^{2+}, X^- and M^+, X^{2-}	0.150	$1.56 \approx 1.6$
M^{3+}, X^- and M^+, X^{3-}	0.366	$4.29 \approx 4.3$
M^{4+}, X^- and M^+, X^{4-}	0.754	$8.89 \approx 8.9$

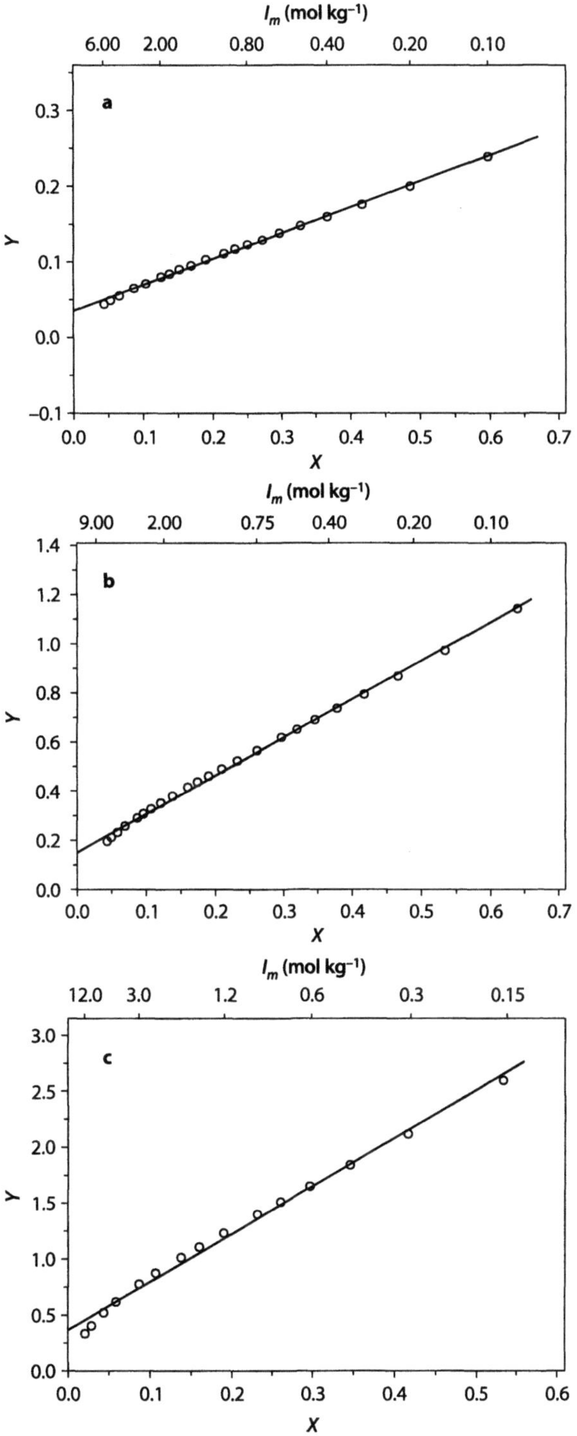

Fig. 10.2a-c. The relationship between the SIT and Pitzer parameters for 1:1, 2:1 and 3:1 electrolytes

concentration/ionic medium dependence of equilibrium constants in ionic media is described in the following section.

For a chemical reaction with reactants/products Q_i, we have

$$\sum_i p_i Q_i + r\mathrm{H_2O(l)} = 0 \tag{10.23}$$

and

$$\ln K^0 = \sum_i p_i \ln m_i + \sum_i p_i \ln \gamma_i + r \ln a_{\mathrm{H_2O}} = \ln K + \sum_i p_i \ln \gamma_i + r \ln a_{\mathrm{H_2O}} \tag{10.24}$$

Most experimental studies of complex formation reactions have been performed in ionic media (Rossotti and Rossotti 1961), with a concentration that is much higher than that of the reactants/products. In this way, the "trace" activity coefficients of reactants/products remain very near constant, even when the total concentrations of reactants/products are varied. At "trace" concentrations of reactants and products, all summations in Eqs. 10.11 and 10.12 reduce to terms that include only the molality of the ionic medium electrolyte; the others are negligible. Hence, the Pitzer model in a 1:1 electrolyte ionic medium, NX, results in the following analytical statement:

$$\begin{aligned}\ln K^0 &= \ln K + r \ln a_{\mathrm{H_2O}} + \sum_i p_i Z_i^2 (f^\gamma + m^2 B'_{NX}) + 2m \sum_i p_i B_{ij} + 2m^2 \sum_i p_i C_{ij} \\ &\quad + 2m \sum_i p_i \phi_{ii'} + m^2 \sum_i p_i \psi_{ii'j} + m^2 \sum_i p_i |Z_i| C_{NX} \\ &= \ln K + r \ln a_{H_2O} + \Delta Z^2 (f^\gamma + m B'_{NX}) + m^2 \Delta |Z| C_{NX} \\ &\quad + 2m(\Delta B + \Delta\phi) + 2m^2 (\Delta C + \frac{\Delta\psi}{2})\end{aligned} \tag{10.25}$$

where the index i refers to species i, and i' and j stand for ionic medium ions, having the same and opposite charge signs, respectively, as i. The definitions of ΔZ^2, ΔB, etc. are clear from Eq. 10.25; m is the molality of ionic medium 1-1 electrolyte. The ionic strength dependence of parameter ΔB is

$$\Delta B = \Delta\beta^{(0)} + \Delta\beta^{(1)} g(\alpha I^{1/2}) \tag{10.26}$$

where

$$\Delta\beta^{(0)} = \sum_i p_i \beta_i^{(0)} \text{ and } \Delta\beta^{(1)} = \sum_i p_i \beta_i^{(1)} .$$

Equation 10.25 shows that the coefficients for the terms in m and m^2 for the reactants/products contain $\Delta\beta^{(0)}$ and $\Delta\phi$, and ΔC and $\Delta\psi$, respectively. Hence, it is not possible to obtain the individual Pitzer parameters $\Delta\beta^{(0)}$ and ΔC from this equation alone. When using Eq. 10.25 in a regression analysis, it is convenient to rewrite it as:

$$\begin{aligned}\ln K^0 &= \ln K + r\ln a_{H_2O} + \Delta Z^2(f^{\gamma} + m^2 B'_{NX}) + m^2\Delta|Z|C_{NX} + 2mX_1 \\ &\quad + 2mg(\alpha I^{1/2})X_2 + 2m^2X_3\end{aligned} \tag{10.27}$$

where $X_1 = \Delta\beta^{(0)} + \Delta\phi$, $X_2 = \Delta\beta^{(1)}$, $X_3 = \Delta C + 1/2\Delta\psi$.

The use of this equation is demonstrated in Example 2.

Using the SIT formalism, the concentration dependence of the equilibrium constant for Eq. 10.21, studied at trace concentration of the reaction participants in a constant ionic medium, is equal to:

$$\begin{aligned}\ln K^0 &= \ln K + r\ln a_{H_2O} - \frac{AI^{1/2}}{1+1.5I^{1/2}}\sum_i p_i Z_i^2 + m\sum_i p_i\varepsilon_\gamma(i,j) \\ &= \ln K + r\ln a_{H_2O} - \frac{A\Delta Z^2 I^{1/2}}{1+1.5I^{1/2}} + m\Delta\varepsilon_\gamma\end{aligned} \tag{10.28}$$

where m is the molality of the ionic medium electrolyte. The definitions of ΔZ^2 and $\Delta\varepsilon_\gamma$ are clear from Eq. 10.28.

It is not trivial to use this equation in a regression analysis as discussed in Plyasunov et al. (1998).

For chemical equilibria studied in the presence of an ionic medium ($I < 4$ mol kg^{-1}), one may neglect all parameters accounting for triple ionic interactions (C_{NX}) and binary higher-order mixing terms (X_3). These approximations will be discussed later. For 1-1 ionic media, Eq. 10.27 can be simplified to the following statement for the ionic medium dependence of $\ln K$:

$$\begin{aligned}\ln K^0 &= \ln K + r\ln a_{H_2O} - \Delta Z^2 A_\Phi\left[\frac{I^{1/2}}{1+bI^{1/2}} + \frac{2}{b}\ln(1+bI^{1/2})\right] \\ &\quad + \Delta Z^2 m^2 B'_{NX} + 2mX_1 + 2mg(\alpha I^{1/2})X_2\end{aligned} \tag{10.29}$$

The corresponding relation for the SIT-model is given by Eq. 10.28. After elementary transformations, we obtain:

$$\begin{aligned}Y &= \left\{A_\Phi\left[\frac{I^{1/2}}{1+bI^{1/2}} + \frac{2}{b}\ln(1+bI^{1/2}) - \frac{3I^{1/2}}{1+1.5I^{1/2}}\right] - m^2B'_{NX}\right\}/2m \\ &= \frac{1}{\Delta Z^2}(X_1 - \frac{\Delta\varepsilon_\gamma}{2}) + \frac{X_2}{\Delta Z^2}g(\alpha I^{1/2})\end{aligned} \tag{10.30}$$

i.e. Y is a linear function of $g(\alpha I^{1/2})$, with the slope $X_2 / \Delta Z^2$ and the intercept $(X_1 - \Delta\varepsilon_\gamma / 2) / \Delta Z^2$. α is a fixed empirical Pitzer parameter equal to 2.0 $kg^{1/2}\,mol^{-1/2}$ for all electrolytes. Y can be calculated from the known ionic strength, the Debye-Hückel parameter A_Φ and B'_{NX}, i.e. the known value of $\beta^{(1)}$ for the 1-1 ionic medium electrolyte. The values of $g(\alpha I^{1/2})$ are obtained from the ionic strength of the solution.

In Fig. 10.3 we have plotted the values of Y for some common 1-1 ionic media.

The linearity is good for all electrolytes considered, and the values of the parameters $(X_1 - \Delta\varepsilon_\gamma / 2) / \Delta Z^2$ and $X_2 / \Delta Z^2$ determined for each electrolyte are given in Table 10.2. As these parameters, especially the slope, do not vary much with the nature of the 1-1 electrolyte, one can use all the data to determine one common set of parameters (see the last line of Table 10.2) where the uncertainties are given as 3σ. This find-

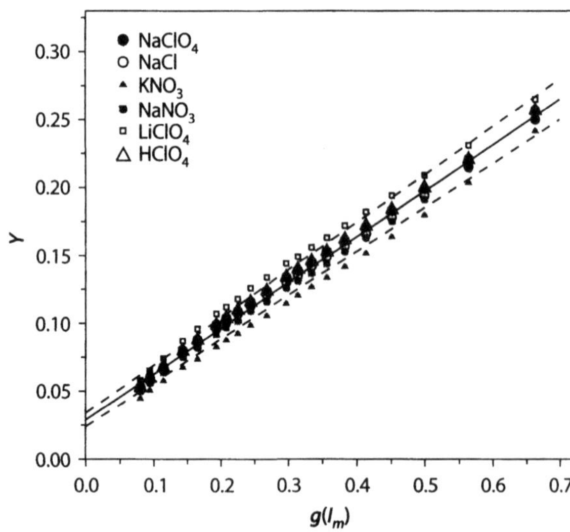

Fig. 10.3. The relationship between the SIT parameters $\Delta\varepsilon$ and the Pitzer parameters $\Delta\beta^{(0)}$ and $\Delta\beta^{(1)}$ for reactions studied in different 1:1 ionic media. The symbols denote calculated values of Y as a function of $g(I_m)$ using Eq. 10.30

Table 10.2. Quantitative relationship between the Pitzer and SIT model parameters for chemical reactions in different ionic media at 298.15 K

1:1 ionic medium electrolyte	$(X_1 - \Delta\varepsilon_\gamma / 2) / \Delta Z^2$	$X_2 / \Delta Z^2$
$NaClO_4$	0.031	0.340
$LiClO_4$	0.038	0.348
$HclO_4$	0.032	0.342
$NaNO_3$	0.026	0.334
KNO_3	0.019	0.327
NaCl	0.030	0.332
All data	0.029 ±0.005	0.337 ±0.014

ing is very convenient for the estimation of the Pitzer parameter $\Delta\beta^{(1)}$ for reactions, because it only requires the value of the sum of squared charges of the ions participating in the reaction, ΔZ^2. One can show that for isocoulombic reactions, where $\Delta Z^2 = 0$, the proposed result is consistent with $\Delta\beta^{(1)} = 0$.

Using the data in Table 10.2 we can now estimate the values of X_1 and X_2, c.f. Example 2, for the protonation of sulphate, for which we obtain $X_1 = -0.11 \pm 0.07$ and $X_2 = -1.35 \pm 0.05$, in good agreement with the results in Table 10.4.

Equation 10.25 was obtained by neglecting the contribution of the terms for higher-order electrostatic unsymmetrical mixing. By including these terms, the slope of the function Y is changed somewhat, particularly for ions of charge 3 or higher.

Because the determination of the Pitzer parameters for a reaction (or for complexes) from $\log K$ data is an ill-conditioned problem, it is rarely possible to determine more than one interaction parameter. We therefore suggest the following strategy when using $\log K$ data determined in 1-1 electrolyte ionic media to determine the Pitzer parameters for complexes:

1. Use the SIT equation to obtain $\log K^0$.
2. Estimate X_2 from the ΔZ^2 value for the reaction if the charge of the reactants/products do not exceed 2 and calculate X_1 using X_2 as a fixed parameter. The terms $m^2\Delta|Z|C_{NX}$ and higher-order electrostatic unsymmetrical mixing terms for the ionic medium ions should be included for data at high ionic strength.
3. Calculate the Pitzer parameters for the complexes from the values of with $\Delta\beta^{(0)}$ and with $\Delta\beta^{(1)}$ and the corresponding quantities $\beta^{(0)}$ and $\beta^{(1)}$ for the reactants.
4. In order to describe equilibrium data at higher ionic strengths or mixed electrolyte systems, it is necessary to determine additional interaction parameters. This can only be achieved by additional equilibrium constant measurements under these conditions.

10.6 Determination of Interaction Parameters

The interaction parameters in the two models must be determined from experimental mean activity coefficients, osmotic coefficients and/or concentration equilibrium constants. The accuracy of these data are typically ± 0.005 in $\log \gamma_{\pm}$ or Φ, and ten to fifty times lower for concentration equilibrium constants. These constants have usually been determined at only a few ionic strengths. For the users of thermodynamic data, it is essential to be aware of the limitations of the methods used to make activity corrections and the consequences of approximations in the models. It is useful to find relationships, c.f. Eq. 10.22, between the interaction parameters in the two models and between the corresponding quantities for reactions. This is practical when one wishes to use the extensive compilations of Pitzer parameters for strong electrolytes together with the compilation of SIT parameters for complexes.

Examples of the use of the specific ion interaction models are in equilibrium analysis.

In the following, I will present three examples of the use of specific ion interaction methods.

Example 1. This is a comparison of the SIT and Pitzer models for the modelling of the ionic strength dependence for the reaction $CO_2(aq) + H_2O(l) \Leftrightarrow HCO_3^- + H^+$ in a NaCl medium.
We have:

$$\log K^0 = \log K + \log \gamma_{H^+} + \log \gamma_{HCO_3^-} - \log \gamma_{CO_2(aq)} - \log a_{H_2O}$$

From the Pitzer equations, we obtain the following values for the single-ion activity coefficients:

$$\log \gamma_{H^+} = F + m_{Cl}(2B_{H,Cl} + 2m_{Cl}C_{H,Cl}) + m_{Na}(2\phi_{H,Na} + m_{Cl}\Psi_{H,Na,Cl}) + m_{Na}m_{Cl}C_{Na,Cl}$$

$$\log \gamma_{HCO_3^-} = F + m_{Na}(2B_{Na,HCO_3} + 2m_{Na}C_{Na,HCO_3}) + m_{Cl}(2\phi_{Cl,HCO_3} + m_{Na}\Psi_{Na,Cl,HCO_3})$$
$$+ m_{Na}m_{Cl}C_{Na,Cl}$$

where

$$F = -A_\phi\left\{\frac{\sqrt{I_m}}{1 + b\sqrt{I_m}} + \frac{2}{b}\ln\left(1 + b\sqrt{I_m}\right)\right\} + m_{na}m_{Cl}B'_{NaCl}$$

$$\log a_{H_2O} = \Phi v m M_W / 1000$$

where Φ is the osmotic coefficient of the electrolyte, M_W the mol mass of water, and $v = 2$ for NaCl.

The activity coefficient for $CO_2(aq)$ is equal to:

$$\log \gamma_{CO_2(aq)} = 2(\lambda_{CO_2,Na} + \lambda_{CO_2,Cl})$$

where λ represents the interaction coefficient between the ionic medium electrolyte ions and $CO_2(aq)$. All the necessary Pitzer parameters for the equations above are known and listed in Table 10.3.

Note that the various interaction parameters in the example can be determined from experimental mean activity data of the pure electrolytes. It is not necessary to deter-

Table 10.3. Pitzer parameters for the equations in Example 1

$\beta^{(0)}_{H,Cl}$	=	0.1775	$C^{\Phi}_{H,Cl}$	=	0.00080	$C^{\Phi}_{H,Cl}$	= 0.00060
$\beta^{(0)}_{Na,Cl}$	=	0.0765	$\beta^{(1)}_{Na,Cl}$	=	0.2664	$C^{\Phi}_{Na,Cl}$	= 0.00127
$\beta^{(0)}_{Na,HCO_3}$	=	0.0277	$\beta^{(1)}_{Na,HCO_3}$	=	0.0411	C^{Φ}_{Na,HCO_3}	= 0.0
$\theta_{H,Na}$	=	0.036	θ_{Cl,HCO_3}	=	0.03		
$\psi_{H,Na,Cl}$	=	−0.004	ψ_{Na,Cl,HCO_3}	=	−0.015		
$\lambda_{CO2,Na} + \lambda_{CO2,Cl}$	=	0.096					

mine them from experimental concentration equilibrium constants in various ionic media. However, the fitting is very sensitive even for small parameter variations. Figure 10.5 illustrates this.

The fit is not satisfactory without the mixing terms. The *dotted line*, obtained by including the mixing terms θ_{ClO_4,HCO_3}, θ_{ClO_4,CO_3}, ψ_{Na,ClO_4,HXO_3} and ψ_{Na,ClO_4,CO_3}, which have been fitted to the experimental equilibrium constants, results in a good agreement between model and experimental data. One can discuss whether this is a satisfactory

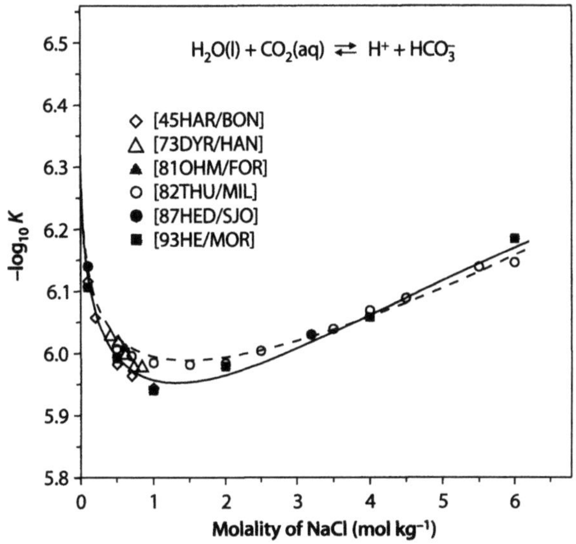

Fig. 10.4. Comparison of experimental (*symbols*) and calculated equilibrium constants $\log K$ for the reaction $CO_2(aq) + H_2O(l) \leftrightarrow HCO_3^- + H^+$ in a NaCl medium using the SIT (*dashed line*) and Pitzer (*full-drawn line*) models

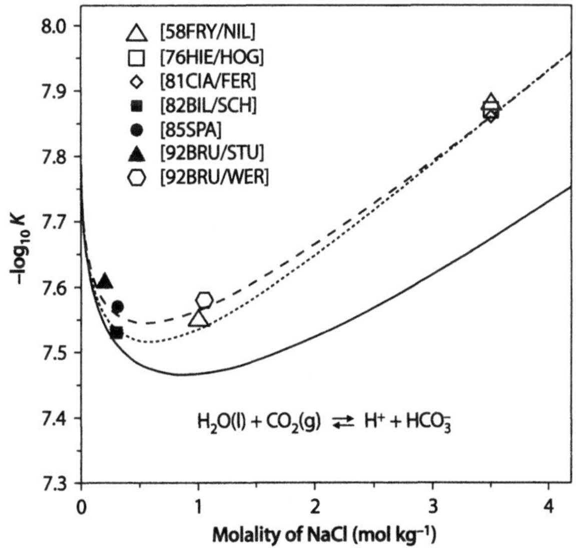

Fig. 10.5. Comparison of experimental and calculated ionic strength dependence of the first dissociation constant of carbonic acid. The *full-drawn curve* has been calculated using the Pitzer model with available literature values for the interaction parameters, but without including mixing terms for which no data are available. The *dotted curve* is obtained using mixing terms estimated from the experimental equilibrium constants. The *dashed curve* has been calculated using the SIT-model

method; in order to be confident about the mixing terms, they should be confirmed by an independent experiment.

The first example concerns a problem where we have a simple equilibrium system and where the equilibrium constant at zero ionic strength can be determined by a simple extrapolation method. The situation is very different if both equilibrium constants and Pitzer parameters have to be determined from experimental concentration equilibrium constants. The second example illustrates this:

Example 2. This example refers to the determination of $\log_{10}K^0$ and the Pitzer and SIT parameters from the experimental values of $\log_{10}K$ for the first protonation constant of the sulphate ion, $H^+ + SO_4^{2-} \Leftrightarrow HSO_4^-$, studied in $NaClO_4$ media. References to the original data are given in Grenthe et al. (1997), where more details about the procedure are given.

The experimental data are fitted to an equation of type Eq. 10.23, which in this case is equal to:

$$\ln K^0 = \ln K + r \ln a_{H_2O} + \Delta Z^2 (f^{\gamma} + mB'_{NX}) + m^2 \Delta|Z| C_{NX} + 2mX_1$$
$$+ 2mg(I_m)X_2 + 2m^2 X_3$$

where

$$\Delta Z^2 = \Sigma v_i Z_i^2;\ \Delta|Z| = \Sigma v_i |Z_i|$$

$$X_1 = \Delta\beta^{(0)} + \Delta\phi = \Sigma v_i \beta^{(0)}{}_{ij} + \Sigma v_i \phi_{ij'}$$

$$X_2 = \Delta\beta^{(1)} = \Sigma v_i \beta_{ij}^{(1)}$$

$$X_3 = \Delta C + 1/2\Delta\psi = \Sigma v_i C_{ij} + 1/2 \Sigma v_i \psi_{ii'j}$$

It is obvious that one cannot determine individual Pitzer parameters from these data, even if they are precise enough to permit the evaluation of all fitting terms. The results of different fitting models are given in Table 10.4 and Fig. 10.6.

- Model I. Determination of the whole set of parameters $\log_{10}K^0$, X_1, X_2 and X_3.
- Model II. Determination of the parameters $\log_{10}K^0$, X_1, and X_2, neglecting ternary interactions.
- Model III. Determination of the parameters $\log_{10}K^0$, X_1 and X_3, assuming $\beta^{(1)} = 0$ for all species.
- Model IV. Determination of the whole set of parameters $\log_{10}K^0$ and X_1, the smallest number of parameters in the Pitzer model. The quantities X_1, X_2 and X_3 are in this case known by independent activity data. This is a situation that is rare.

The results of the various fitting models are shown in Fig. 10.6.

The parametrization of the SIT and different choices of the Pitzer model are illustrated below for the reaction $H^+ + SO_4^{2-} \Leftrightarrow HSO_4^-$ in $NaClO_4$ ionic media at 25 °C.

Table 10.4. Regression data for the determination of Pitzer and SIT parameters

The SIT model	The Pitzer model parameters for models I–IV					
	Parameter	I	II	III	IV	Accepted values
$-\log_{10}K^0 = 1.989 \pm 0.084$	$\log_{10}K^0$	2.10 ± 0.27	2.04 ± 0.20	2.13 ± 0.13	2.23 ± 0.11	1.987 ± 0.009
$\Delta\varepsilon = 0.003 \pm 0.084$	X_1	-0.45 ± 0.88	-0.19 ± 0.16	-0.56 ± 0.21	-0.34 ± 0.08	-0.15 ± 0.05
	X_2	-0.46 ± 3.40	-1.39 ± 1.34	(0)	(0)	-0.995
	X_3	0.05 ± 1.02	(0)	0.07 ± 0.07	(0)	-0.006 ± 0.010

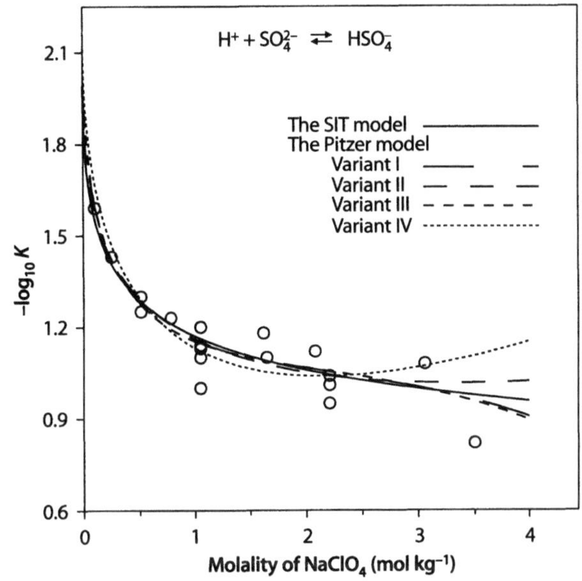

Fig. 10.6. The parametrization of the SIT model and different choices of the Pitzer model for the reaction $H^+ + SO_4^{2-} \leftrightarrow HSO_4^-$ in $NaClO_4$ ionic media at 25 °C

An important problem in equilibrium analysis is to decide if a certain complex may be an artefact due to ionic medium variations, or not. An estimate of the variation of the activity coefficients of reactants and products provides important clues. It is evident that this problem is particularly important when weak complexes are formed. We can use the SIT and Pitzer models both to design the experiments and to evaluate the magnitude of the activity factor variations of reactants and products as a result of varying the composition of the equilibrium solutions.

Example 3. Assume that we are investigating the formation of fairly weak complexes, e.g. the Cu(II)-Cl^- system, using a Cu-reversible electrode. Should the experiment be made in a medium with constant ionic strength, constant concentration of Na^+ or ClO_4^-? What is the magnitude of the variation of the activity factors if we exchange 25% of the $NaClO_4$ ionic medium for NaCl? I will use the SIT model, because it demonstrates the principles without too much computational effort.

In the experiments, we measure the concentration of free or non-complex bound Cu^{2+} with the electrode. It is highly desirable that the activity coefficient for this species is constant throughout the experiment. Using the SIT single ion activity coefficient for Cu^{2+}, we have:

$$\log \gamma_{Cu^{2+}} = -0.5100 Z^2_{Cu^{2+}} \frac{\sqrt{I}}{1+1.5\sqrt{I}} + \varepsilon(Cu^{2+}, ClO_4^-) m_{ClO_4} + \varepsilon(Cu^{2+}, Cl^-) m_{Cl}$$

From this equation, we can see that the activity coefficient of Cu^{2+} will not be constant throughout the experiment. The interaction coefficients are $\varepsilon(Cu^{2+}, ClO_4^-) = 0.32$ and $\varepsilon(Cu^{2+}, Cl^-) \approx 0.15$. Hence, a replacement of 25% of a 3.5 molal $NaClO_4$ ionic medium with NaCl will result in a change of $\log_{10} \gamma_{Cu^{2+}}$ equal to 0.14, which is not negligible. We could have studied the same system with a chloride electrode. In this case, it would have been advantageous to keep the sodium ion concentration constant, because the activity coefficients of a particular ion are mainly influenced by the ions of opposite charge. The example illustrates some of the problems encountered when interpreting equilibrium analytical data when large changes have been made in the ionic medium. An attempt to use the SIT method to interpret data in a mixed ionic medium is described in Grenthe and Lagerman (1993). A more detailed discussion of ionic medium effects with several different examples is given in Grenthe et al. (1997).

References

Bodländer G, Storbeck O (1902a) Z Anorg Chem 31:1
Bodländer G, Storbeck O (1902b) Z Anorg Chem 31:458
Bodländer G, Fittig R (1902) Z Physik Chem Leipzig 39:597
Bodländer G, Eberlin W (1904) Z Anorg Chem 39:197
Brønsted JN (1922) J Am Chem Soc 44:877
Debye P, Hückel E (1923) Physik Z 24:185
Euler H von (1903a) Ber 36:1854
Euler H von (1903b) Ber 36:2878
Euler H von (1903c) Ber 36:3400
Farkas I, Bányai I, Szabo Z, Wahlgren U, Grenthe I (2000) Inorg Chem 39:799
Grenthe I, Lagerman B (1993) Radiochim Acta 61:169
Grenthe I, Spahiu K, Plyasunov A (1997) Estimation of medium effects on thermodynamic data. In: Grenthe I, Puigdomenech I (eds) Modelling in aquatic chemistry. Nuclear Energy Agency, OECD, Paris
Grossman H (1905) Z Anorg Chem 43:356
Guggenheim EA (1935) Philos Mag 19:588
Helgeson HC, Kirkham DH, Flowers GC (1981) Am J Sci 281:1249
Millero FJ (1983) Geochim Cosmochim Acta 47:2121
Moll H, Denneke MA, Jalilehvand F, Sandström M, Grenthe I (1999) Inorg Chem 38:1795
Pitzer KS (1973) J Phys Chem 77:268
Pitzer KS (1991) Ion interaction approach: Theory and data correlation. In: Pitzer KS (ed) Activity coefficients in electrolyte solutions, 2nd edn. CRC Press, Boca Raton, FL, pp 75–153
Pitzer KS, Brewer L (1961) Revision of Lewis and Randall's thermodynamics. McGraw-Hill, New York
Plyasunov A, Fanghänel T, Grenthe I (1998) Acta Chem Scand 52:250
Rossotti FJC, Rossotti H (1961) The determination of stability constants. McGraw-Hill, New York
Russel B (1959) A history of western philosophy, 12th edn. Simon and Schuster
Scatchard G (1959) The interpretation of activity and osmotic coefficients. In: Hamer WJ (ed) The structure of electrolytic solutions.
Scatchard G (1976) Equilibrium in solution: Surface and colloid chemistry. Harvard University Press, Cambridge, MA
Szabo Z, Aas W, Grenthe I (1997) Inorg Chem 36:5369

Acid-Base Equilibria in Saline Media: Application of the Mean Spherical Approximation

M. E. Sastre de Vicente · T. Vilariño

11.1 Introduction

The simplest mathematical formulation of the hypothesis that chemical equilibrium exists in the bulk of a saline solution is based on the equilibrium constant, $K^T = K^* Q(\gamma_i)$, where K^T is the thermodynamic equilibrium constant, K^* the stoichiometric constant and $Q(\gamma_i)$ the ratio of activity coefficients (γ_i) associated with the equilibrium. Studies on the effect of ionic strength on stoichiometric constants are based on modelling of the $Q(\gamma_i)$ term (Sastre de Vicente 1997; Daniele et al. 1997).

With regard to this equation, the following points can be mentioned:

- Most treatments rely on the Debye-Hückel limiting law, although their final purpose is their possible application to moderate and high ionic strengths, as shown through many studies devoted to the analysis of changes in activity and osmotic coefficients for strong electrolytes.
- The vast majority of the treatments lead to explicit expressions for pK^* as a function of I, which is clearly an advantage; however, interpreting the parameters associated to the fitting of a pK^* vs. I curve is often far from easy.
- This kind of study is carried out in a constant ionic medium, which has been traditionally employed as a synonym for "background electrolyte;" the principal weakness of the method arises from the assumption that the added salt is inert to the equilibria involved. If the background electrolyte cannot be considered as inert, ion association must be included and the equilibria redefined in the system.

11.2 Acid-Base Equilibria in Saline Media

For the case of acid-base equilibrium reactions in saline media, we should consider the most significant acid-base reactions that take place in solution, which can be classified as follows (Edsall 1943; Izatt et al. 1995):

- Type I:

$$AH = A^- + H^+ \tag{11.1}$$

where the charge separation involved usually has a strong effect on H_2O structure.

- Type II:

$$AH_2^+ = AH + H^+ \tag{11.2}$$

This reaction is usually termed “isocoulombic.” Because the net charge of reactants and products is the same, the change in water structure is usually small.
- Type III:

$$^-AH_2^+ = AH + H^+ \tag{11.3}$$

This reaction can be considered to be a particular case of Type I and involves the zwitterionic species $^-AH^+$, which usually behaves as a neutral molecule or as two separate ions depending on the separation between positive and negative charge.
- Type IV:

Type IV refers to complex molecules. It is worth mentioning that the study of the different aspects of the above simple equilibria I and II is the first step to encompass the analysis of acid-base equilibria of polyelectrolytic molecules.

The equilibrium constant for types I and II can be formulated as:

$$K_I^T = \frac{[A^-][H^+]}{[AH]} \frac{\gamma_{A^-}\gamma_{H^+}}{\gamma_{AH}} \tag{11.4}$$

$$K_{II}^T = \frac{[AH]\ [H^+]}{[AH_2^+]} \frac{\gamma_{AH}\,\gamma_{H^+}}{\gamma_{AH_2^+}} \tag{11.5}$$

Taking logarithms in Eqs. 11.4 and 11.5 yields

$$pK_I^T = pK_I^* - \log \frac{\gamma_{A^-}\gamma_{H^+}}{\gamma_{AH}} \tag{11.6}$$

$$pK_{II}^T = pK_{II}^* - \log \frac{\gamma_{AH}\,\gamma_{H^+}}{\gamma_{AH_2^+}} = pK_{II}^* - \log \gamma_{AH} \frac{\gamma_{H^+}\,\gamma_{X^-}}{\gamma_{AH_2^+}\,\gamma_{X^-}} \tag{11.7}$$

As shown above, the dependence of pK^* on the ionic strength is clearly a function of the way the $Q_i(\gamma)$ term (i.e. the activity coefficient term), or rather its logarithm is modelled. Equations 11.4–11.7 include two types of activity coefficient, viz., that for the ions and that for neutral species. Obviously, the form of such coefficients used in the different theoretical approaches associated to the theory of electrolytes explicitly dictates the form of the final equation.

11.3 pK^* vs. Ionic Strength Equations

In Table 11.1, it is shown that virtually all the expressions proposed over the years are of the type:

$$\mathrm{p}K^{\mathrm{T}} = \mathrm{p}K^{*} + f(I, \text{parameters}) \tag{11.8}$$

The different equations refer to the equilibria types I and II but based on different models for the activity coefficients $\log \gamma_i$ and $\log \gamma_{\text{neutral}}$: Pitzer, Guggenheim, or other semi-empirical modifications of the Debye-Hückel equation. Also, a non-Debye-Hückel based equation is included at the bottom of the Table 11.1. This simple cubic-root equation is based on a classical quasi-lattice model, which exhibits this characteristic dependence.

Different authors(for example Grenthe and Plyasunov 1997; Millero and Pierrot 1998; Salvatore et al. 1986; Partanen and Juusola 2000; Daniele et al. 1997; Sastre de Vicente

Table 11.1. pK vs. ionic strength dependence for equilibria I and II according to different models for activity coefficients

Model I (Pitzer)

$$f^{(4)} = -0.392\left[\frac{\sqrt{I}}{1+1.2\sqrt{I}} + \frac{2}{1.2}\ln\left(1+1.2\sqrt{I}\right)\right],\quad f^{(5)} = -1 + \left(1 + 2\sqrt{I} + 2I\right)e^{-2\sqrt{I}}$$

Model II (Scatchard)

$$\mathrm{p}K_i^* = \mathrm{p}K_i^T - z\frac{1.018\sqrt{I}}{1+1.5\sqrt{I}} + P_i\, I + Q_i\, I^2 + R_i\, I^3$$

Model III (Guggenheim)

$$\mathrm{p}K_i^* = \mathrm{p}K_i^T + \varepsilon_i\, I - z\frac{1.018\sqrt{I}}{1+1.5\sqrt{I}}$$

Model IV (De Stefano et al.)

$$\mathrm{p}K_i^* = \mathrm{p}K_i^T - z\frac{2\sqrt{I}}{2+3\sqrt{I}} + b_i\, I + a_i\, I^{3/2}$$

Model V (quasi-lattice)

$$\mathrm{p}K_i^* = \mathrm{p}K_i^T - z\, a_i\sqrt[3]{I} + b_i\, I$$

$$z = i - 1$$

Taken from (Sastre de Vicente 1997).

1997; De Robertis et al. 1997) some of which are contributors in this volume, traditionally have been using these equations to model pK^* vs. ionic strength data.

Some authors have derived equations of the following type from studies over wide temperature and ionic strength ranges:

$$\log Q = f(I, T, \text{param.}) \tag{11.9}$$

which includes the dependence on both the ionic strength and temperature (Kettler et al.1995)

With regard to these explicit equations in ionic strength, in particular Pitzer's, which are the more general and widely used, it must be stated that these equations:

- Are able to reproduce the measured properties within the experimental error;
- Can be applied to single electrolytes and mixtures, i.e. natural waters (Millero and Pierrot 1998), up to high concentrations;
- However, coefficients obtained from linear regression analysis of pK^* vs. I data have no clear physical meaning. Other problems are for example ill-conditioning (Fiol et al. 1994), which leads to high errors in the regression parameters.

11.4 The Mean Spherical Approximation: Estimation of $Q(\gamma)$ Term by Use of the MSA Model

An alternative to modelling pK-I data is the mean spherical approximation (MSA) (Blum 1975; Friedman 1985), the application of acid-base equilibria on the basis of the "constant ionic medium" approach, is described below.

From a general point of view, in an electrolyte solution there must be a balance among different types of interactions (Israelichvili 1991):

a Forces of a purely electrostatic nature that arise from Coulomb's law (e.g. interactions between charges, permanent dipoles or quadrupoles, etc.);

b Polarization forces between dipoles induced by the action of permanent charges and dipoles on atoms and molecules. This type of force is present in any solvent medium (particularly in electrolyte solutions).

c Quantum mechanical forces consisting of an attractive part responsible for chemical bonding and a repulsive part of the steric type or in the form of exchange interactions that equilibrate attractive forces over very short distances.

A frequent way of describing such interactions in qualitative terms is based on the distinction between long-range (coulomb) interactions and short-range interactions. So, schematically, the interaction between two ions in solution can be expressed as:

$$\text{Interaction between particles } i, j = \text{Coulomb Law} + \text{other kind of interactions} \tag{11.10}$$

This dichotomy in the conceptualization of solution interactions has prevailed in modelling electrolyte solutions since the beginning of the development of theories early in this century.

Also, with MSA, the activity coefficient for an individual ionic species of charge z_i, diameter σ_i and numerical density ρ_i in an arbitrary mixture can be resolved into two contributions, Fig. 11.1 (Blum and Hoye 1977; Corti 1987; Sanchez Castro and Blum 1989; Triolo et al. 1976; Turq et al. 1992; Waisman and Lebowitz 1972), namely:

$$\ln \gamma_i = \ln \gamma_i^{\text{el}} + \ln \gamma_i^{\text{hs}} \tag{11.11}$$

where the first term on the right-hand side describes the electrostatic contribution and the second the hard sphere contribution.

The use of MSA to estimate the log $Q(\gamma)$ term entails estimating $\log \gamma_i$ for each ionic species (AH_2^+, A^-, H^+) involved in the particular acid–base equilibrium from electrostatic and hard-sphere contributions, the expressions for which in the general case of a mixture of charged hard spheres of arbitrary ionic radii a_i and a_j, and charges z_i and z_j, are (Cartallier et al. 1992) as shown in Fig. 11.1.

$$\ln \gamma_i = \ln \gamma_i^{\text{el}} + \ln \gamma_i^{\text{hs}}$$

Electrostatic contribution

$$\ln \gamma_i^{\text{el}} = -\frac{\alpha^2 z_i^2}{4\pi}\,\frac{\Gamma}{1+\Gamma\sigma_i} - \frac{\alpha^2 z_i \sigma_i}{4(1+\Gamma\sigma_i)}\,\frac{P_n}{\Delta} + \frac{\pi\alpha^2\sigma_i^3}{16}\left(\frac{1}{1+\Gamma\sigma_i} - \frac{1}{3}\right)\frac{P_n^2}{\Delta^2}$$

Hard sphere contribution

$$\ln \gamma_i^{\text{hs}} = -\ln\Delta + \frac{\sigma_i^3 X_0 + 3\sigma_i^2 X_1 + 3\sigma_i X_2}{\Delta} + \frac{3\sigma_i^3 X_1 X_2 + 9/2\,\sigma_i^2 X_2^2}{\Delta^2} + \frac{3\sigma_i^3 X_2^3}{\Delta^3}$$

$$2\Gamma = \left[\alpha^2 \sum_{i=1}^{n} \rho_i \frac{\left(z_i - \frac{\pi}{2\Delta}\sigma_i^2 P_n\right)^2}{(1+\Gamma\sigma_i)^2}\right]^{1/2} \qquad \alpha^2 = \frac{4\pi\beta e^2}{\varepsilon_0\varepsilon_r} \qquad X_k = \frac{\pi}{6}\sum_i \rho_i\sigma_i^k \qquad \Delta = 1 - \frac{\pi}{6}\zeta_3$$

$$P_n = \frac{1}{\Omega}\sum_k \frac{\rho_k\sigma_k z_k}{(1+\Gamma\sigma_k)} \qquad \Omega = 1 + \frac{\pi}{2\Delta}\sum_k \frac{\rho_k\sigma_k^3}{(1+\Gamma\sigma_k)} \qquad \zeta_n = \sum \rho_k\sigma_k^n \;(n = 0, 1, 2, 3)$$

Fig. 11.1. Scheme where β is the reciprocal temperature, $1/kT$ (where k is the Boltzmann constant and T the absolute temperature, in K), e the electron charge, ε_0 the permittivity of free space, and ε_r the relative permittivity of the solvent

The most relevant aspect in the expression of log γ by using the MSA is that all parameters appearing in the equations shown in Fig. 11.1 are an explicitly simple function of charges, diameters, concentrations, dielectric constant and temperature.

11.5 Comparison With the Pitzer Model

At this point, let us recall the expression for the activity coefficient of an ion, using, for example, a Pitzer equation.

The corresponding expression for the more general case of an electrolyte mixture is (Pitzer 1973, 1991) given as:

$$\begin{aligned}\ln\gamma_{M^+} &= z_M^2 f^\gamma + 2\sum_a \left[B_{Ma} + \left(\sum mz\right)C_{Ma}\right] + 2\sum_c m_c\theta_{Mc} \\ &+ \sum_c\sum_a m_c m_a\left(z_M^2 B_{ca} + z_M C_{ca} + \psi_{Mca}\right) \\ &+ \frac{1}{2}\sum_a\sum_{a'} m_a m_{a'}\left(z_M^2\theta'_{aa'} + \psi_{Maa'}\right) + \frac{z_M^2}{2}\sum_c\sum_{c'} m_c m_{c'}\Theta'_{cc'}\end{aligned} \tag{11.12}$$

For a negative ion, it suffices to replace anions with cations.

Function f^γ includes long-range electrostatic interactions and is dependent on the ionic strength, of which B is also a function – in contrast with the previous theories, where all parameters were independent of I. Parameter B takes account of dual interactions (e.g. B_{aM} represents the interactions a-a, M-M and a-M), and C accounts for triple interactions in an electrolyte (e.g. C_{Ma} represents the interactions M-M-a and a-a-M). ψ, θ and θ' are mixing parameters; thus, θ is associated with the interactions between two ions of the same sign (e.g. θ_{Mc} accounts for the interactions M-c, M-M and c-c, M and c being two cations), while θ' is its derivative with respect to the ionic strength ($\theta' = d\theta / dI$).When chemical equilibria are involved, these parameters appear combined in the $Q(\gamma)$ term, and the meaning of the coefficients thus obtained become more difficult.

Conversely, the MSA approach entails the use, in all equations, of easily understandable parameters as charge, dielectric constant, diameters, concentration and temperature, and no direct mention to dual or triple interactions is done.

11.6 Neutral Molecules

Until now, we have described the activity coefficient of present ions. It should be noted that the activity coefficient for the neutral species (AH, A) is assumed to obey the Setschenow equation (Long and McDevitt 1952), i.e. to be a linear function of ionic strength and hence of the molar concentration:

$$\log\gamma_s = k_s c \tag{11.13}$$

11.7 Data We Need for Working With the Mean Spherical Approximation

1. *Electrical charges*, z_i, are defined by the corresponding equilibrium model.
2. *Numerical densities*, ρ_i. Taking into account that a constant ionic medium requires that the supporting electrolyte be present at a much higher concentration than the substance whose ionization equilibria is being studied, the sole numerical densities that will appreciably contribute to the different parameters to be determined can be assumed to be those for the electrolyte ions.
3. *Ionic diameters* (σ_i)
   Using the MSA model to calculate activity coefficients requires the prior knowledge of the diameters of all species present in the system:
   a Background electrolyte. The diameters of the ions that make up the most usual electrolytes can be replaced with Pauling diameters as shown by Marcus (1988), who found ionic radii of dissolved species to be largely consistent with Pauling's crystal radii. Also, taking into account the well-known fact that cation sizes increase to a greater extent with increasing concentration than do anion sizes, the former are often assumed to vary with the concentration, while the latter remain constant and equal to Pauling values (Triolo et al. 1976, 1977, 1978).

$$\sigma_+(c) = \sigma_0(1 + \Sigma a_n c^n) \tag{11.14}$$

   where a_n are coefficients to be determined from experimental data.
   b Species involved in the equilibria. Owing to the few reported diameters (Kielland 1937) for the different ionic species that result from the ionization of organic compounds, we often estimate such diameters from Van der Waals volumes calculated from group contributions (Bondi 1964).
4. *Dielectric constant*
   Although the MSA model can use the dielectric constant for the medium at zero electrolyte concentration, the use of a dielectric constant dependent on the concentration of the particular electrolyte improves the results as it accounts for the fact that dielectric shielding must be smaller in the more concentrated solutions, where ions, on average, are closer to one another and hence less strongly shielded by the solvent.
   The expression for the dielectric constant of an electrolyte solution is usually a simple polynomial, whether linear or slightly more complex (Barthel et al. 1998; Fawcett and Tikanen 1996; Whitfield and Turner 1981), which relates ε to the molar concentration, hence considering the decrease in the constant with the increase in concentration.

$$\varepsilon(c) = \varepsilon_0(1 + \Sigma \beta_n c^n) \tag{11.15}$$

where β_n are coefficients to be determined from experimental data.

11.8
An Example: Fitting pK^* vs. I Plot by Use of MSA for an Isocoulombic Equilibrium

If we take as an example the isocoulombic equilibrium: $AH^+ = A + H^+$, the first step is always the estimation of the scaling parameter, Γ. The way of obtaining parameter 2Γ from the general expression (see Fig. 11.1) depends on the number of ions present in solution; the procedure usually involves solving an algebraic equation of a high order. Thus, the equation for a binary salt is of sixth order and must be resolved numerically. For this reason, various explicit approximations to Γ have been developed, one of the most frequently used is the truncated expression:

$$2\Gamma = \left[\alpha^2 \sum_{i=1}^{n} \frac{\rho_i z_i^2}{(1+\Gamma\sigma_i)^2}\right]^{1/2} \tag{11.16}$$

which assumes that $P_n = 0$, i.e. that ionic diameters are all identical. This expression holds in most cases – provided ionic sizes are fairly similar and/or the concentrations of the species involved are about or smaller than 1 M (Lee 1988; Sun et al. 1994).

Equation 11.16 can readily be solved either by using the Newton-Raphson method, by starting at very low concentrations, or by using an iterative method. In both cases, the inverse Debye length is used as the starting value:

$$2\Gamma_0 = \kappa = \left[\alpha^2 \sum_{i} \rho_i z_i^2\right]^{1/2} \tag{11.17}$$

As can be seen from the previous expressions, estimating the scaling parameter requires the prior knowledge of both the dielectric constant, which is included in parameter α^2, and ion sizes.

As stated above, both the dielectric constant of the medium and the diameters of the different ions of the electrolytes present in it can be assumed to be fixed or concentration-dependent.

Moreover, it should be noted that the experimental data are given by the Lewis-Randall (LR) theory and that MSA expresses thermodynamic quantities in terms of molar concentrations in the framework of the McMillan–Mayer (MM) theory of solutions. Thus, data must be converted from LR to MM scales, that is to say from molality to molarity (Friedman 1972; Simonin 1996; Simonin and Blum 1996).

Once the electrolyte diameter and dielectric constant are known, the activity coefficients for the ionic species involved in the equilibria studied can readily be calculated from the MSA electrostatic and hard-sphere contributions, then we can rearrange the expressions for the equilibrium constants in order that the left-hand side of the resulting equation can be calculated numerically at each concentration. This yields to:

$$AH_2^+ = AH + H^+$$

$$K_1^{\mathrm{T}} = \frac{[\mathrm{AH}][\mathrm{H}^+]}{[\mathrm{AH}_2^+]}\frac{\gamma_{\mathrm{AH}}\gamma_{\mathrm{H}^+}}{\gamma_{\mathrm{AH}_2^+}} = K_1^* Q_1(\gamma_i)$$

$$\ln K_1^{\mathrm{T}} = \ln K_1^* + \ln\gamma_{\mathrm{AH}^\pm} + \ln\gamma_{\mathrm{H}^+} - \ln\gamma_{\mathrm{AH}_2^+}$$

$$\ln\gamma_i = \ln\gamma_i^{\mathrm{el}} + \ln\gamma_i^{\mathrm{hs}}$$

$$\ln\gamma_{\mathrm{AH}^\pm} = 2\lambda m = \lambda' c_{\mathrm{MX}}$$

$$\ln K_1^{\mathrm{T}} = \ln K_1^* + \lambda' c_{\mathrm{MX}} + \ln\gamma_{\mathrm{H}^+}^{\mathrm{el}} + \ln\gamma_{\mathrm{H}^+}^{\mathrm{hs}} - \ln\gamma_{\mathrm{AH}_2^+}^{\mathrm{el}} - \ln\gamma_{\mathrm{AH}_2^+}^{\mathrm{hs}}$$

$$\mathrm{p}K_1^* + \frac{1}{\ln 10}\left[\ln\gamma_{\mathrm{H}^+}^{\mathrm{el}} - \ln\gamma_{\mathrm{AH}_2^+}^{\mathrm{el}} + \ln\gamma_{\mathrm{H}^+}^{\mathrm{hs}} - \ln\gamma_{\mathrm{AH}_2^+}^{\mathrm{hs}}\right] = \mathrm{p}K_1^{\mathrm{T}} + \lambda' c_{\mathrm{MX}}$$

$$Y(\sigma_{\mathrm{VdW}}(\mathrm{optim.})) = \mathrm{p}K_1^{\mathrm{T}} + \lambda' c_{\mathrm{MX}}$$

$\mathrm{p}K^*$ values are obtained experimentally and the electrostatic and hard-sphere terms are worked out analytically by using MSA. Then, based on the tabulated salting coefficient for the neutral species, the diameters of the charged ions involved in the equilibrium are optimized in order to obtain the best fit of a plot of the left-hand side expression against the electrolyte concentration, which will provide the thermodynamic $\mathrm{p}K^{\mathrm{T}}$ as the intercept of the fitted line. As noted earlier, the initial diameters of organic species are calculated from Van der Waals volumes that are in turn derived from Bondi (Bondi 1964) tabulated data.

Once $\mathrm{p}K^{\mathrm{T}}$ has been calculated, experimental $\mathrm{p}K^*$ vs. electrolyte concentration plots can readily be fitted.

One of the analyzed systems was the acid-base equilibria of glycine in artificial sea water (Vilariño and Sastre de Vicente 1999). Table 11.2 shows the recipe of sea water and other parameters used for calculations.

In Fig. 11.2 different aspects associated to MSA and other approaches are compared. Perhaps the most interesting difference between them is connected to the possibility of studying size effects.

Because the MSA expression for the activity coefficient consists of an electrostatic term and a hard-sphere term, the weight of each contribution to the overall value at variable ion sizes was studied. The electrostatic contribution increases very little with increasing diameter, consistent with the expectations, as it depends mostly on ion charges. On the other hand, the hard-sphere contribution increases markedly with increasing diameter and is roughly proportional to the ionic strength of the solution, similarly to salting coefficients.

In this context, it is worth mentioning that the most recent work of our group on this topic (Taboada-Pan et al. 2001) refers to the study of the isocoulombic equilibrium of three alkylamines in saline media. On the one hand in that study, the Pitzer

Table 11.2. Recipe of sea water and other parameters used for calculations

Sea water ions	Ionic Concentration c_i (mol l^{-1})	$\sigma = \sigma_0 + ac_i + bc_i^2$		
		σ_0 (Å)	a (Å mol^{-1} l)	b (Å mol^{-2} l^2)
Na^+	0.92146 c_{sw}	3.77	−0.190	0.0246
K^+	0.02034 c_{sw}	3.42	−0.327	0.0503
Mg^{2+}	0.10573 c_{sw}	6.41	−0.214	0.00448
Ca^{2+}	0.002068 c_{sw}	5.78	−0.0107	0.0201
Cl^-	1.08246 c_{sw}	3.056		
SO^-	0.05608 c_{sw}	2.981		

Dielectric constant: $\varepsilon_r = \varepsilon_0 - \delta_s c_s + b_s{}^{3/2}$ with $\delta = 19.57 \pm 0.08$ l mol^{-1} and $b = 4.6 \pm 0.1$ l$^{3/2}$ mol$^{-3/2}$ (from Vilariño and Sastre de Vicente 1999).

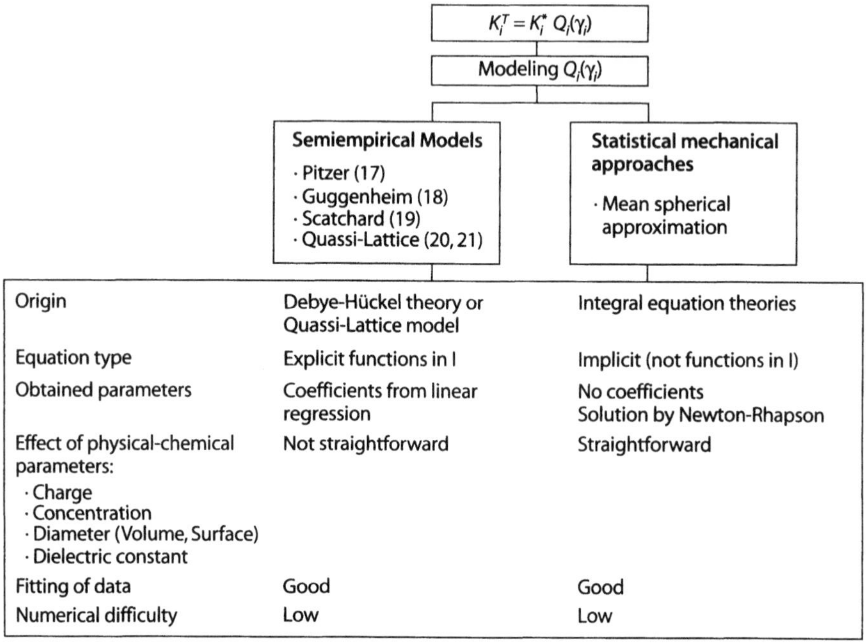

Fig. 11.2. Comparison of different aspects associated to MSA and other approaches

theory is used for the calculation of the salting coefficient of amines from pK^* data. On the other hand, the MSA approximation is applied to predict the size of alkylamines in that media, assuming a hard sphere contribution for the activity coefficient of the neutral molecule of alkylamine. These data are compared to those calculated from molar volumes proposed by several authors and taking neutral molecules as spheres.

As stated above, data show that the Bondi diameters are the closest values to those obtained from MSA calculations.

Finally, it should be mentioned that equations for the equilibria AH= AH + H^+ and AH = A^- + H^+, typical of many processes (viz. those involving simple acid–base model systems such as carboxylic acids, amines and amino acids) were derived by using the mean spherical approximation. Detailed descriptions of those studies are provided in the references (Sastre de Vicente et al. 1998; Vilariño and Sastre de Vicente 1996, 1997, 1999; Vilariño et al. 1997, 1998)

Acknowledgements

M. S. V. wishes to acknowledge the University of A Coruña and the Xunta de Galicia for their financial support of Projects XUGA 10303B90, XUGA 10303B92, XUGA 10303B94, XUGA 10310B97, PGIDT00PXI10308PR, and Dr. J. L. Barriada for his help with the word-processing.

References

Barthel JMG, Krienke H, Kunz W (1998) Physical chemistry of electrolyte Solutions. Steinkopff-Springer, Darmstadt

Blum L (1975) Mean spherical model for asymmetric electrolytes. I. Method of solution. Mol Phys 30:1529–1535

Blum L, and Høye JS (1977) Mean spherical model for asymmetric electrolytes. 2. Thermodynamic properties and the pair correlation function. J Phys Chem 81:1311–1316

Bondi A (1964) Van der Waals volumes and radii. J Phys Chem 68(3):441–451

Cartallier T, Turq P, Blum L, Condamine N (1992) Thermodynamics of ion association in the mean spherical approximation. J Phys Chem 96:6766–6772

Corti HR (1987) Prediction of activity coefficients in aqueous electrolyte mixtures using the mean spherical approximation. J Phys Chem 91:686–689

Daniele PG, De Stefano C, Foti C, Sammartano S (1997) The effect of ionic strength and ionic medium on the thermodynamic parameters of protonation and complex formation. Current Topics in Solution Chemistry 2:253–274

De Robertis A, Foti C, Sammartano S, Gianguzza A (1997) Chemical speciation of some classes of low molecular weight ligands in seawater. In: Gianguzza A, Pelizzetti E, Sammartano S (eds.) Marine chemistry: An environmental analytical chemistry approach. Kluwer Academic Pub., Dordrecht, The Netherlands

Edsall JT (1943) Chap. 4. In: Cohn, Edsall (eds) Proteins, amino acids and peptides. Wiley, Reinhold, New York

Fawcett WR, Tikanen AC (1996) Role of solvent permittivity in estimation of electrolyte activity coefficients on the basis of the mean spherical approximation. J Phys Chem 100:4251–4255

Fiol S, Brandariz I, Herrero R, Vilariño T, Sastre de Vicente M (1994) The protonation constants of glycine in NaCl at 25 °C based on the Pitzer and Scatchard models: Data analysis by ridge regression. Ber Bunsen Phys Chem 98(2):164–171

Friedman HL (1972) Lewis-Randall to McMillan-Mayer conversion for the thermodynamic solutions. J Solution Chem 1:387–431

Friedman HL (1985) A course in statistical mechanics. Prentice Hall, Englewood Cliffs, NJ, pp 137–156

Grenthe I, Plyasunov A (1997) On the use of semiempirical electrolyte theories for the modeling of solution chemical data. Pure Appl Chem 69(5):951–958

Israelichvili J (1991) Intermolecular and surface forces. Academic Press, London

Izatt RM, Oscarson JL, Gillespie SE, Cheng X, Wang P, Watt GD (1995) A calorimetric study of ligand interactions with protons and metal ions in the 100 to 400 °C range. Pure Appl Chem 67:543–549

Kettler RM, Palmer DA, Wesolowsky DJ (1995) Dissociation quotient of benzoic acid in aqueous solution chloride media to 250 °C. J Solution Chem 24(4):385–407

Kielland J (1937) Individual activity coefficients of ions in aqueous solutions. J Am Chem Soc 59:1675–1678

Lee LL (1988) Molecular thermodynamics of nonideal fluids. Butterworths
Long FA, McDevit WF (1952) Activity coefficients of nonelectrolyte solutes in aqueous salt solutions. Chem Rev 51:119–169
Marcus Y (1988) Ionic radii in aqueous solutions. Chem Rev 88:1475–1498
Millero FJ, Pierrot D (1998) A chemical equilibrium model for natural waters. Aquatic Geochem 4:153–199
Partanen JI, Juusola PM (2000) Comparison of different methods for calculation of the stoichiometric dissociation constant of acetic acid from results of potentiometric titrations at 298.15 K in aqueous sodium or potassium chloride solutions. Fluid Phase Equilibria 169:149–166
Pitzer KS (1973) Thermodynamics of electrolytes. I. Theoretical basis and general equations. J Phys Chem 77:268–277
Pitzer KS (1991) Ion interaction approach: Theory and data correlation. In: Pitzer KS (ed) Activity coefficients in electrolyte solutions, 2nd edn. CRC Press, Boca Raton, FL, pp 75–153
Salvatore F, Ferri D, Palombari R (1986) Salt effect on the dissociation constant of acid-base indicators. J Solution Chem 15(5):423–431
Sanchez-Castro C, Blum L (1989) Explicit approximation for the unrestricted mean spherical approximation for ionic solutions. J Phys Chem 93:7478–7482
Sastre de Vicente ME (1997) Ionic strength effects on acid-base equilibria. A review. Current Topics in Solution Chemistry 2:157–181
Sastre de Vicente ME, Vilariño T, Brandariz I (1998) Application of the mean spherical approximation to the study of ionic strength effects on acid-base equilibria. Recent Res Devel Phys Chem 2:489–500
Simonin JP (1996) Study of experimental-to-McMillan-Mayer conversion of thermodynamic excess functions. J Chem Soc Faraday Trans 92:3519–3523
Simonin JP, Blum L (1996) Departures from ideality in pure ionic solutions using the mean spherical approximation. J Chem Soc Faraday Trans 92:1533–1536
Simonin JP, Blum L, Turq P (1996) Real ionic solutions in the mean spherical approximation. 1. Simple salts in the primitive model. J Phys Chem B 100:7704–7709
Sun T, Lenard JL, Teja AS (1994) A simplified mean spherical approximation for the prediction of the osmotic coefficients of aqueous electrolyte solutions. J Phys Chem 98:6870–6875
Taboada-Pan C, Brandariz I, Barriada JL, Vilariño T, Sastre de Vicente ME (2001) The salting coefficient and size of alquilamines in saline media at different temperatures: estimation from Pitzer equations and the mean spherical approximation. Fluid Phase Equilibria 180:313–325
Triolo R, Grigera JR, Blum L (1976) Simple electrolytes in the mean spherical approximation. J Phys Chem 80:1858–1861
Triolo R, Blum L, Floriano MA (1977) Simple electrolytes in the mean spherical approximation. 3. A workable model for aqueous solutions. J Chem Phys 67:5956–5959
Triolo R, Blum L, Floriano MA (1978) Simple electrolytes in the mean spherical approximation. 2. Study of a refined model. J Phys Chem 82:1368–1370
Turq P, Barthel J, Chemla M (1992) Transport, relaxation and kinetic processes in electrolyte solutions. Springer-Verlag, Berlin (Lectures Notes in Chemistry 57, pp 74–109)
Vilariño T, Sastre de Vicente ME (1996) Protonation of glycine in saline media: Evaluation of the effect of the ionic strength by use of the mean spherical approximation. J Phys Chem 100:16378–16384
Vilariño T, Sastre de Vicente ME (1997) Theoretical calculation of the ionic strength dependence of the ionic product of water based on a mean spherical approximation. J Solution Chem 26 (9):833–846
Vilariño T, Sastre de Vicente ME (1999) The mean spherical approximation methodology applied to the acid-base equilibria of glycine in artificial seawater. Phys Chem Chem Phys 1:2453–2456
Vilariño T, Brandariz I, Fiol S, Armesto XL, Sastre de Vicente ME (1997) Study of the effect of ionic strength on the protonation equilibrium of various amino acids by using the mean spherical approximation. J Chem Soc Faraday Trans II 93:413–417
Vilariño T, Alonso P, Armesto XL, Rodriguez P, Sastre de Vicente ME (1998) Effect of ionic strength on the kinetics of the oxidation of ascorbic acid by hexacyanoferrate(III): Comparison between specific interaction theories and the mean spherical approximation. J Chem Res (S) 9:558–559
Waisman E, Lebowitz JL (1972) Mean spherical model integral equation for charged hard spheres. J Chem Phys 56:3086–3099
Whitfield M, Turner DR (1981) Sea water as an electrochemical medium. In: Whitfield M, Turner DR (eds) Marine chemistry. A practical introduction. John Wiley and Sons, Rochester, pp 3–66

Chapter 12

Modelling of Natural Fluids: Are the Available Databases Adequate for this Purpose?

D. Ferri · C. Manfredi · E. Vasca · C. Fontanella · V. Caruso

12.1 Introduction

The research activity in our laboratory does not directly focus on marine chemistry. However it is mainly devoted to the study of equilibrium reactions that take place in natural systems, of which the sea is certainly one of the most important, and to determination of their behaviour.

Let us start by stating that the guidelines adopted in our laboratory are those developed by L. G. Sillén and his school in Stockholm. The primary rule we abide to is to pay greatest attention to experimental adequacy and accuracy, as we are convinced that bad data cannot be improved by any computerized treatment, powerful as it may be. The interpretation of good data can often be improved. Bad data remain bad. An obvious consequence of this creed of ours is a constant effort to optimize the experimental methods. To this crucial point we will return later.

Chemists are generally familiar with the terminology, the fields of pertinence, the investigation, and the interpretation methods adopted in *Equilibrium Analysis* (EA). But what is, in fact, equilibrium analysis? In answering this question, let us confine our interest to aqueous solutions. In water, chemical interactions mainly take place through acid-base, redox, precipitation or complexation reactions. The products may be species capable of individual existence. If so, they can be characterized by means of the classical analytical methods, which generally imply one or more separation steps prior to the analysis. Alternatively, the products formed may be complexes, i.e. species in a dynamic, often reversible chemical equilibrium, which obey the law of mass action and thus cannot be separated. In this case, they are characterized by measuring one or more physico-chemical properties of the solution (electrode potential, absorbance of electromagnetic radiation, solubility of solid phases, distribution between solvents, etc.) without perturbing the state of equilibrium[1]. It is obvious that to identify and to quantify the products of a reaction requires a totally different approach in the two cases. The experimental techniques and the methods of interpretation of the data developed for the study of chemical equilibria in aqueous solutions, particularly by Scandinavian chemists, constitute the basis of the equilibrium analysis.

To investigate a chemical equilibrium implies the determination of the *composition* of the species (complexes) formed by the reaction of one or more co-ordinating reagents with one or more ligands, as well as their *formation constants*. This is the

[1] For the sake of completeness, the kinetic methods, based on the study of the relaxation, following the perturbation of the state of equilibrium, must also be mentioned.

prelude to the *speciation* of the solution, i.e. *its qualitative and quantitative description in terms of the stoichiometry of the complexes* formed and their *concentrations.* It follows that a study of chemical speciation of a metal ion does not only provide its total analytical concentration but also *a detailed knowledge of its partition among all the ligands in the solution.* The speciation represents a fundamental requirement for the comprehension of the chemical and physical behaviour, as well as for the interpretation of the properties of the solution investigated, such as toxicity to living organisms, therapeutic efficiency and biological activity of chemical preparations. It further allows for understanding the role of the complexes in the genesis of minerals, for evaluating the possibility that toxic metals be transported in the biosphere, etc."

We have now introduced a term, speciation, which was practically unknown three decades ago and is incorrectly and overly used today. However for the sake of clarity, we are dealing with a particular form of speciation of solutions: *that which handles species coexisting under the condition of chemical equilibrium, defined by the law of mass action.* Now let us attempt to place the few concepts cited above in a broader framework, without pretending to be exhaustive.

12.2 Equilibrium Analysis Applied to the Modelling of Natural Systems

Natural systems are too complex to be studied directly, thus we divide them into progressively simpler sub-systems, until they ultimately fit with the requisites and the limitations of idealized laboratory experiments, in which, for instance, the reactions of a single metal ion with one or more ligands, are investigated. The procedure is extended to all metals and to all ligands, with the aim of producing what we call a *Thermodynamic Database.* Finally, an approximate picture of the real system is obtained by reassembling the laboratory results, consisting of all the species found and their formation constants. This is *modelling* of which speciation is the first and essential step.

However speciation and consequently modelling are no better than the database on which they are founded. The databases are then the limiting parameter of the reliability of models. This is the crucial point mentioned above, on which a few comments have to be made. To delimit the boundaries of this theme, let us focus our attention on metals, as the central ions.

Vast collections of data have so far been reported by several authors, starting from Sillén and Martell's fundamental work: "Stability constants of metal-ion complexes" (Sillén and Martell 1964). However, these collections are seldom proposed as critical tables of thermodynamic parameters, and if they are, the criteria of selection are not clearly indicated. In a few cases, bad data are easily spotted. But generally the selection is not so straightforward. The reluctance of compilers of tables to recommend "the most reliable data" can be understood. But one must also realize that equilibrium analysis does not generally produce "fingerprint" evidence of the complexes claimed, rather it gives "probable" results. This means that the model (the stoichiometry of the complexes) that we propose to explain the experimental data may be not unique. The philosophy supporting EA recommends explaining the experimental data with the minimum number of complexes, which, for instance, produce the lowest value of the minimized function or the best fit in graphical approaches. Then, most obviously: the more accurate the experimental data, the more probable the results, in terms of the composition and formation constants of the complexes chosen.

12.3 The Thermodynamic Database (TDB) Example

Recent compilations (Grenthe et al. 1992; Silva et al. 1995; Sandino and Östhols 1999; Lemire 2001) of critically reviewed thermodynamic parameters on the chemistry of solutions and solids of uranium and some actinides in their various oxidation states have been promoted by OECD/NEA (Organization for Economic Cooperation and Development/Nuclear Energy Agency), within a project currently referred to as the TDB (Thermodynamic Database) project (Östhols 2000), which started in 1984. It states that "before that date a number of reviews had already been published, but none could reliably be used as a complete source data table to study the behaviour of radioelements in the environment." An international organization was then assembled (see Fig. 12.1).

Task groups were created to review the existing data and to propose a selection of the literature results or to suggest the acquisition of missing parameters, with the ultimate scope of realizing a model of migration to the biosphere of radionuclides from permanent repositories for radioactive wastes of nuclear reactors. The selection of the reliable data has been done with the greatest possible attention, often by re-elaborating the primary data and ignoring the articles that did not report experimental details. For each system that was validated, the referees were asked to justify their choice, which was then submitted to a higher Board for definitive approval. The TDB is still being assembled relatively to five elements, namely: uranium, americium, technetium, neptunium and plutonium and updated with the newly produced values of relevant parameters. To the complex formation equilibria of uranium, in the oxidation states +6, +5 and +4, with carbonate, oxalate and hydroxide ions, most frequently present in ground waters, a substantial contribution in the last two decades has been given by our laboratories and more recently by some of the Italian groups of thermodynamics of complexes.

To be adequate for performance assessment of the permanent repositories of radioactive wastes, the TDB (and thus the reviewers) had to comply with the following requisites:

- To contain data for all the elements of interest in radioactive waste disposal systems;
- To document why and how the data were selected;
- To document the sources of experimental data used;
- To be internally consistent;
- To treat all solids and aqueous species of the elements of interest.

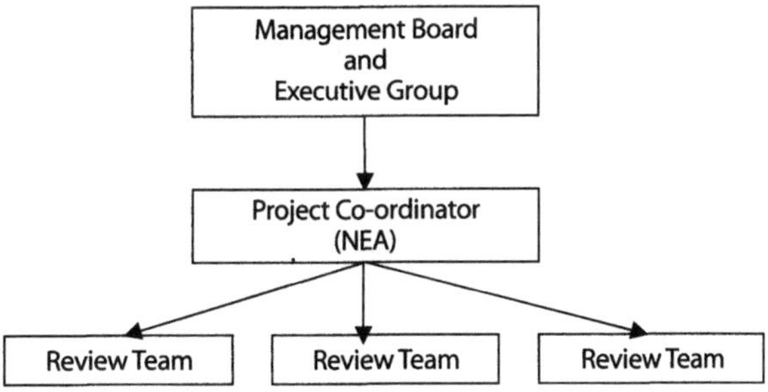

Fig. 12.1. The organization of the NEA-TDB project

At the time (1984) when the TDB project started, none of the existing databases fulfilled all these criteria. The example of the TDB is a clear demonstration of two facts:

a Although the attempts, carried out by diligent authors to give order to the immense mess of the literature on equilibrium studies are greatly appreciated, we must admit that, at the first serious challenge to demonstrate their adequacy to model real systems, the existing compilations have badly failed. Much work is still to be done. The OECD/NEA project has shown that the right way to do the job is through international cooperation. The amount of labour for this enterprise demands for joint ventures, since we do not know the legendary Hercules' present address (read the sarcastic story of the mythical hero cleaning the stables (databases?) of king Augeas (Grenthe and Puigdomenech 1997, pp. 131–152)). While a deeply critical revision of the existing literature is indispensable, missing data must also be provided on mixed complex formation, on complexation of metals by large ligands as humic and fulvic acids, and on thermodynamic quantities, in addition to ΔG. Spectroscopic investigations (to gain structural information on the species found by the methods of equilibrium analysis) must also corroborate, when possible, the results of the equilibrium analysis. Finally, the results obtained in the presence of an ionic medium must be recalculated on other scales of activity or extrapolated to zero ionic strength, possibly by means of the Pitzer (Pitzer 1979) or the SIT (Brøensted 1922; Brøensted 1923; Scatchard 1936; Guggenheim 1966; Biedermann 1975; Ciavatta 1980) equations, which are the most reliable. The first is more precise (see Chapts. 1 and 8) but requests more parameters, which are rarely available. The second, which is simpler to apply, has proved to be the most advantageous choice. At this point it is evident that in our opinion the existing equilibrium data compilations must be considered generally inadequate as sources on which to base modelling. Somebody can suggest to begin amending the databases by eliminating from the literature the infesting data on applying a set of rules, by assigning, for instance, a weight to the experimental method of investigation. Whoever makes use of potentiometry and has experimented with its incomparable power, provided that an adequate electrode exists, would suggest the highest ranking for this method, but probably somebody would disagree. And yet the application of this rule would be only the first step. We should also ascribe different weights to the choice of the ionic medium (if there is one) and its concentration, to the type of electrochemical cell adopted (with or without liquid junction), to the type of glass electrode chosen (single or combined), to the number of free concentrations measured (if more than one is accessible) and so on. Then comes the delicate moment of the interpretation of the data in terms of stoichiometry of the complexes and their formation constants. Recently, May and Murray (2001) discussed the possibility of building up a database that should achieve thermodynamic consistency automatically. But let us go back to the initial statements about the accuracy of the experimental data.

b The advent of high-speed computers and consequently of powerful programmes for the interpretation of the data has somehow favoured a dangerous tendency to use less effort in the acquisition of experimental data, for instance measuring a single free concentration (one electrode, usually the H^+ glass membrane), when a deeper insight in the system could be attained by also measuring other electrodes (amalgam of the metal, second type electrodes for halogenides, oxalate, etc.). The choice of minimizing the time needed for the experimental part is particularly risky in polynuclear or multi-component systems, although extremely versatile computer programmes will suggest a possible interpretation of the data. But the probability that the interpretation might not be unique increases. We, conversely, believe that the utmost care must be reserved to the acquisition of data, bearing a dowry of information on the system investigated that is as rich as possible. As pointed out earlier, good data may always be reinterpreted by more powerful methods of calculation, as the TDB project has shown. Bad data cannot become good data, no matter how sophisticated the method of interpretation is. Bad data, since the first compilation of stability constants have always plagued the literature. However, we continue to produce bad data, whereas the utmost care should be used to design the experimental approach of an investigation, in order to measure as many parameters as possible with the highest accuracy. Time cannot be saved in this stage of the investigation. The amount of labour must instead be decreased by using fully automated acquisition systems that operate around the clock. In our laboratory, we have developed one which is capable of a precision of 10 microvolts, in the course of coulometric or volumetric titrations, performed into stainless steel air boxes, where the temperature is kept constant to within 0.02 °C, with the possibility of holding a free concentration (usually pH or pE) constant.

With the contributions of many solution chemists all over the world, EA has grown into a precise methodology for the study of equilibrium reaction with solid bases of thermodynamics. It is not obsolete because it thus far has not been replaced by any approach of comparable efficiency. Yet its role within the established chemical sectors is not well defined. It is not rare to find people who question the relevance of our work, on the basis of the contradictory and discrepant results reported in the literature, practically on all the systems studied. The project of realizing a thermodynamic database has even incited comments such as the following: "Metal speciation is the analytical chemist's answer for eternal employment" (Grenthe and Puigdomenech 1997, p. 132). However, the discipline EA is not to blame for this distrust. We think instead that to some extent, we the users of EA deserve this sarcastic attitude for the reasons discussed above. As pointed out earlier, most of the methods adopted in EA, because of their indirect nature, only produce probabilistic results that have often been questioned. There have been in the past decades people who did not believe in the existence of complexes in solutions, and they ascribed the deviations from ideal behaviour to variations of the activity coefficients. The development of the X-ray diffraction technique in concentrated solutions demonstrated the existence of complexes whose stoichiometric composition had been obtained by the methods of EA. Later on, other spectroscopic methods, such as NMR, Raman and, more recently, Mass Spectrometry and the possibility to carry out measurements in relatively dilute solutions using light of Synchrotron, offer means to corroborate the results of EA by producing, sometimes, even the structure of the complexes. But one must be careful, because without a prior knowledge of the speciation of the solution, spectroscopic data would be very difficult to interpret, for instance in a polynuclear system. Thus, spectroscopic measurements can provide essential information, particularly on the structure of the complexes, provided the speciation of the solution is known, and they should be carried out every time this is possible. In some way, a few scientific journals are already encouraging this trend and frequently suggest that the authors *complete* the studies submitted with spectroscopic data. We have so far discussed our views vs. the care one uses in following and performing the experimental stage. But the severe approach invoked must often come to a compromise:

a With the many difficulties which characterize the real systems and thus make their modelling problematic. The network of all the interactions within their reagents requests extremely versatile computer programmes able to refine many thermodynamic parameters contemporarily. Computer facilities for the speciation of natural fluids have been reviewed in 1997 by De Stefano et al. (see also references therein).
b With the fact that interactions between the components of natural fluids also give rise to weak complexes, involving alkaline and alkaline earth ions, which are generally neglected in defining chemical models of natural aqueous systems. The interpretation of data concerning the formation of complexes is a delicate and controversial subject, but necessary if we consider the high concentration of alkaline, alkaline earth and chloride ions in sea water or, even worse, in salt brines. An interesting review on weak complex formation has been carried out by Daniele et al. (1994).
c With the use of the available literature. We do not intend to absolve the bad data, but bad data sometimes may tell one how to plan an investigation to collect good data. We mean that there are grades of "bad." In the absence of other resources, partial, inaccurate, or badly designed experiments may give some primary information and help one plan a rigorous study.

As an application of what has been so far discussed, we propose two typical examples of our research, carried out following the guidelines exposed above. The examples are chosen among the studies carried out, either by the Royal Institute of Technology of Stockholm or by the University of Naples, to realize a thermodynamic database, within the Swedish project for the realization of a mathematical model for the migration of radionuclides, from permanent repositories to the biosphere.

12.3.1
1st Example: Uranium-Carbonate System

A series of model studies (Åberg et al. 1983a; Åberg et al. 1983b; Biedermann et al. 1982; Bruno et al. 1986; Ciavatta et al. 1979; Ciavatta et al. 1981; Ciavatta et al. 1983; Ferri et al. 1981; Ferri et al. 1983; Ferri et al. 1993; Ferri et al. 1994; Ferri et al. 1988; Grenthe and Ferri 1981; Grenthe et al. 1984), on the complex formation between U, in the oxidation states +6, +5 and +4, and the carbonate ion was undertaken in the range $2 < pH < 12$, in oxidizing as well as in reducing solutions, using potentiometric and spectrophotometric methods. A number of species were identified, which form the thermodynamic cycle of uranium in carbonate solutions, in function of pH and pE as it is shown in Fig. 12.2.

Particular attention was devoted to a species having a concentration ratio of carbonate to uranyl equal 2 (which was known and had traditionally been assigned the composition $UO_2(CO_3)_2^{2-}$) that proved to have stoichiometry $(UO_2)_3(CO_3)_6^{6-}$. Since this complex had never been reported, although it is the main species at $5.5 < pH < 7$, we characterized it further by X-ray diffraction (Åberg et al. 1983a) in concentrated solu-

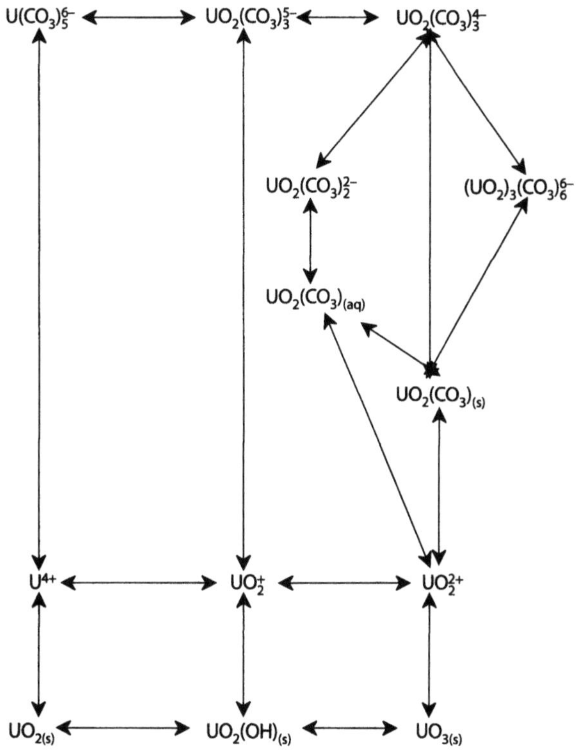

Fig. 12.2. Thermodynamic cycle of U(VI, V and IV) in carbonate solution at $2 < pH < 12$ in oxidizing and reducing environment

tions (up to 1 M). A single distance U-U was found, which implies an equilateral triangle geometry for the three uranium atoms. Further analyses by ^{13}C NMR and by ^{17}O NMR (Ferri et al. 1988; Åberg et al. 1983b) definitively confirmed the structure (Fig. 12.3).

It is interesting to know that rutherfordine, a natural uranyl carbonate, has a layer structure where one may recognize the (3,6) species, by shifting successive layers. The high solubility of uranyl carbonate, at pH around 5.5, can be explained by the small work requested on passing from the structure of the solid to that of the (3,6) complex.

Of course very seldom one meets with such a fortunate coincidence of favourable conditions to fully characterize a complex. However, on the one hand, works like this one add evidence that EA is an incomparable method for the determination of the stoichiometric composition of equilibrium complexes; on the other hand, this is an example of what we should do every time it is possible to validate our results.

12.3.2
2nd Example: Lanthanides Hydrolysis

The second example concerns investigations still in progress on the hydrolysis of lanthanides for two reasons: (*a*) to critically evaluate the relevant literature, (*b*) to obtain data of the quality requested by the TDB. In fact the chemical behaviour of the lan-

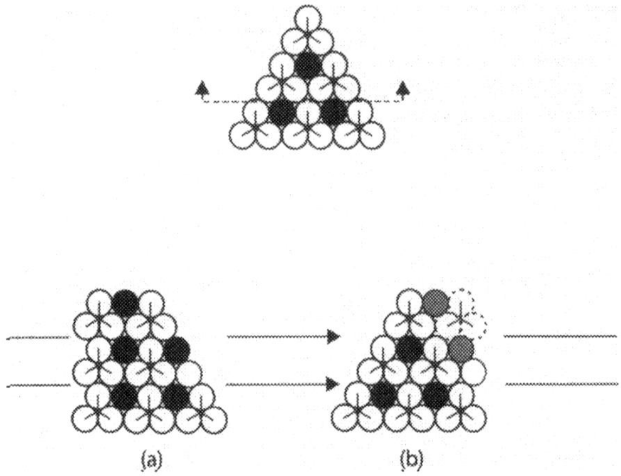

Fig. 12.3. *Top*: Suggested geometry for the trinuclear complex $(UO_2)_3(CO_3)_6^{6-} = (3,6)$; *filled circles* indicate uranyl groups, *open circles* stay for the carbonate groups. *Bottom*: Relation between the structures of the natural carbonate, $UO_2CO_{3\,(s)}$ Rutherfordine (*a*) and the trinuclear species $(UO_2)_3(CO_3)_6^{6-}$ (*b*)

thanides greatly resembles that of the actinides, mostly in the oxidation state +3, which are strongly radioactive. As such they give rise to radiolysis and produce heath, making their direct investigations problematic. The lanthanides are thus used to simulate the actinides in the same state of oxidation. But let us take a look at Table 12.1, showing the results of the studies on the hydrolysis of lanthanides reported in the literature (Pettit and Powell 2001)

As we can see, rarely two lanthanides exhibit the same mechanism of hydrolysis $(-p,q)$. Considering the substantially identical reactions characterizing the trivalent lanthanides (a similarity which for many decades has rendered their separation very laborious), the enormous differences among the literature results look at least suspect. In addition, one must consider that the hydrolysis of the lanthanides seldom exceeds 2% of the total metal concentration. Since this is usually maintained below the 1 M level, it is easy to see that in favourable conditions we can reach a concentration of hydrolyzed species, which at most equals 0.02 M. Some studies report measurements at lanthanide concentrations as low as 0.01 M, where the highest amount of hydrolyzed species hardly reaches the concentration level of 10^{-4} M. At these levels, even using the most precise techniques available, it is impossible to distinguish with certainty between polynuclear species bearing 9 or 10 OH and 5 or 6 lanthanide atoms. We have studied 6 lanthanides (Eu, Gd, Dy, Ho, Er, Yb) (works to be published) by measuring the glass electrode potential with a precision of 0.01 mV, keeping the temperature in a termostatted air box at 25.00 ±0.02 °C and varying the composition of the solutions by using a coulometric technique that allows for the addition or withdrawal of a precise number of micromoles of electrons without introducing protolytic impurities. The interpretation of the data collected indicates that a single mechanism is sufficient to explain the hydrolysis, Eq. 12.1, of all the lanthanides we have so far studied:

$$q\,Ln^{3+} + p\,H_2O = Ln_q(OH)_p{}^{(3q-p)+} + p\,H^+ \tag{12.1}$$

The mechanism corresponds to the following compositions $(-p,q)$: (−1,1), (−2,2), (−9,5) and affords the best fit of the experimental data. But a few other mechanisms

Table 12.1. Survey of the composition $(-p,q)$ (see Eq. 12.1) of the hydrolytic species of the lanthanides

Ce^{3+}	Pr^{3+}	Nd^{3+}	Pm^{3+}	Sm^{3+}	Eu^{3+}	Gd^{3+}	Tb^{3+}	Dy^{3+}	Ho^{3+}	Er^{3+}	Tm^{3+}	Yb^{3+}	Lu^{3+}
(−1,1)	(−1,1)	(−1,1)	(−1,1)	(−1,1)	(−1,1)	(−1,1)	(−1,1)	(−1,1)	(−1,1)	(−1,1)	(−1,1)	(−1,1)	(−1,1)
(−2,1)	(−1,2)	(−2,2)	(−2,1)	(−2,1)	(−2,1)	(−2,1)	(−2,1)			(−2,1)		(−2,1)	(−4,1)
(−3,1)	(−2,2)	(-8,6)		(−2,2)		(−2,2)				(−2,2)		(−2,2)	(−5,1)
(−5,3)	(−3,1)	(−12,6)		(−3,1)						(−3,1)		(−3,1)	(−6,1)
	(−4,1)			(−4,3)						(−4,1)		(−4,1)	
	(−5,3)			(−5,3)						(−5,1)		(−5,1)	
				(−6,4)						(−6,1)		(−5,3)	
				(−9,5)								(−6,1)	
												(−6,4)	

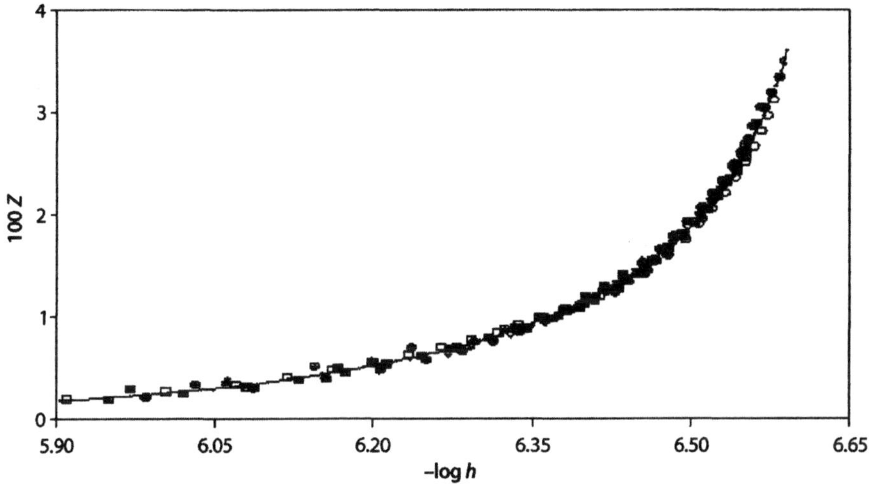

Fig. 12.4. $Z(\log h)_B$ $Z = (h - H) / B$ where $h = [H^+]$, H = Total concentration of acid and B = [Ln(III)] = 0.100 M

cannot be ruled out (due to the small amount of hydrolyzed products), although their fit is poorer. The present accuracy of the measurements does not allow one to distinguish with certainty among a few models that are slightly different. But we are sure that whatever the model, it fits all the lanthanides. This is what one would expect on the basis of the chemical similarity that characterizes the lanthanides. To corroborate this observation, we mixed the data relative to 0.1 M concentration level of the six lanthanides, as if they referred to a unique element. The fit was excellent as can be seen in Fig. 12.4. The next step will be to repeat this test by experimentally mixing up a number of different lanthanides. We have also found linear correlations between the formation constants of the species (2,2), as well as (9,5), and the ionic radius or the atomic number of the various lanthanides. Due to the extreme dilution of the solutions studied, we have not yet succeeded in producing spectroscopic evidence in support of the mechanism proposed, but we shall not give up.

We are presently striving to realize in solutions of oxalate the same cycle obtained for carbonate.

References

Åberg M, Ferri D, Glaser J, Grenthe I (1983a) Structure of the hydrated dioxouranium(VI) ion in aqueous solution. An X-ray diffraction and ^{1}H NMR study. Inorg Chem 22:3986
Åberg M, Ferri D, Glaser J, Grenthe I (1983b) Studies on metal carbonate equilibria. 8. Structure of the exakis(carbonato)tris-dioxouranate(VI) ion in aqueous solution. An X-ray diffraction and ^{13}C NMR study. Inorg Chem 22:3981
Biedermann G (1975) Dahlem Workshop on the Nature of Seawater. Dahlem Konferenzen, Berlin
Biedermann G, Bruno J, Ferri D, Grenthe I, Salvatore F, Spahiu K (1982) Modelling of the migration of lanthanoids and actinoids in ground water: The medium dependence of equilibrium constants. Scientific Basis for Nuclear Waste Management V, Berlin, p 791
Brøensted JN (1922) Studies of solubility: IV. The principle of specific interaction of ions. J Am Chem Soc 44:877–898
Brøensted JN (1923) The individual thermodynamic properties of ions. J Am Chem Soc 45:2898–2910
Bruno J, Ferri D, Grenthe I, Salvatore F (1986) Studies on metal carbonate equilibria. 13. On the solubility of the uranium(IV) dioxide, UO_2(s). Acta Chem Scand A40:428
Ciavatta L (1980) Ann Chim 70:551
Ciavatta L, Ferri D, Grimaldi M, Palombari R, Salvatore F (1979) Dioxouranium(VI) carbonate complexes in acid solution. J Inorg Nucl Chem 41:1175
Ciavatta L, Ferri D, Grenthe I, Salvatore F (1981) On the first acidification step of the triscarbonato-dioxouranate(VI) ion, $UO_2(CO_3)_3^{4-}$. Inorg Chem 20:463
Ciavatta L, Ferri D, Grenthe I, Salvatore F (1983) Studies on metal carbonate equilibria. 4. Reduction of the tris(carbonato)dioxouranate(VI) ion, $UO_2(CO_3)_3^{4-}$, in hydrogen carbonate solutions. Inorg Chem 22:2088
CODATA (1978) CODATA recommended key values for thermodynamics 1977. Committee on Data for Science and Technology (CODATA), Bulletin 28, International council for Scientific Unions, Paris
Daniele PG, De Stefano C, Prenesti E, Sammartano S (1994) Weak complex formation in aqueous solution. Current Topics in Sol Chem 1:95
De Stefano C, Sammartano S, Mineo P, Rigano C (1997) Computer tools for the speciation of natural fluids. In: Gianguzza A, Pellizzetti E, Sammartano S (eds) Marine chemistry – An environmental analytical chemistry approach. Kluwer Academic Publishers, Amsterdam, p 71
Ferri D, Glaser J, Grenthe I (1988) Confirmation of the Structure of $(UO_2)_3(CO_3)_6^{6-}$ by ^{17}O NMR. Inorg Chim Acta 148:133
Ferri D, Grenthe I, Salvatore F (1981) Dioxouranium(VI) carbonate complexes in neutral and alkaline solutions. Acta Chem Scand A35:403
Ferri D, Grenthe I, Salvatore F (1983) Studies on metal carbonate equilibria. 7. Reduction of the tris(carbonato)dioxouranate(VI) ion, $UO_2(CO_3)_3^{4-}$ in carbonate solutions. Inorg Chem 22:3162
Ferri D, Salvatore F, Vasca E, Graser J, Grenthe I (1993) Complex formation in the U(VI)-OH^--F^- system. Acta Chem Scand 47:855
Ferri D, Salvatore F, Vasca E, Miranda R (1994) A combined potentiometric-coulometric method for the analysis of uranium in carbonate media. Ann Chim (Rome) 84:283
Grenthe I, Ferri D (1981), Actinide species in ground water Systems. Near-field phenomena in geologic repositories for radioactive wastes. NEA, Paris, p 21
Grenthe I, Puigdomenech I (eds) (1997) Modelling in aquatic chemistry. OECD-NEA, Paris
Grenthe I, Ferri D, Riccio G, Salvatore F (1984) Studies on metal carbonate equilibria. 10. Solubility study of the complex formation in the U(VI)-water-carbon dioxide system at 25 °C. J Chem Soc Dalton Trans 11:2439
Grenthe I, Fuger J, Konings RJM, Lemire RJ, Muller AB, Nguyen-Trung C, Wanner H (1992) Chemical thermodynamics of uranium. In: Wanner H, Forest I (eds) Elsevier, Amsterdam
Guggenheim EA (1966) Applications of statistical mechanics. Clarendon Press, Oxford
Lemire RJ (2001) Chemical thermodynamics of neptunium and plutonium. Elsevier, Amsterdam
May PM, Murray K (2001) Database of chemical reactions designed to achieve thermodynamic consistency automatically. J Chem Eng Data 46:1035
Östhols E (2000) The NEA thermochemical database project. NEA-TDB, Issy-les-Moulineaux
Pankratz LB (1982) Thermodynamic properties of elements and oxides. Bullettin, US Bureau of Mines

Pettit D, Powell K (2001) IUPAC stability constants database. Academic Software, Otley, UK
Pitzer KS (1979) Theory: Ion interaction Approach. In: Activity coefficients in electrolyte solutions, vol I. CRC Press, Boca Raton
Sandino MC, Osthols E (eds) (1999) Chemical thermodynamics of technetium, vol III. Elsevier, Amsterdam
Scatchard G (1936) Concentrated solutions of strong electrolytes. Chem Rev 19:309–327
Sillén LG, Martell AE (1964) Stability constants of metal-ion complexes. Chemical Society, London
Silva RJ, Bidoglio G, Rand MH, Robouch PB, Wanner H, Puigdomenech I (1995) Chemical thermodynamics of americium. Elsevier, Amsterdam

Part III

Toxicants in Marine Environment

Chapter 13

Endocrine-Disrupting Chemicals in Marine Environment

M. R. Preston

13.1 Introduction

The recognition that certain chemicals had oestrogenic activity dates back to the work of Dodds and co-workers in the 1930s (Dodds and Lawson 1938; Dodds et al. 1938). The use of synthetic oestrogens in medicine began in the 1940s with the synthetic oestrogen diethylstilboestrol (DES) being widely used with the intention of preventing abortions and pregnancy complications in women. Its use was, however, largely counterproductive, and it was eventually found to increase abortions, neonatal deaths, premature births and a certain form of vaginal cancer (Herbst et al. 1971). The major use of synthetic hormones since this time has been in the contraceptive pill in the form of ethynyl derivatives of oestradiol.

During the 1990s, support began to grow for the hypothesis that a number of apparently unrelated changes within the environment and in human populations could be related to exposure to environmental chemicals, which has lead to specific damage to endocrine systems. Such changes included such 'classical' pollution problems as eggshell thinning in top predator birds, imposex in molluscs and reproductive disorders in seals, and more recently discovered phenomena as varying degrees of intersex/hermaphrodite occurrence in fish, reptiles and mammals. In human populations, incidences of a variety of cancers (particularly breast cancer) and disorders of the male reproductive system such as declines in sperm counts, cryptorchidism and hypospadias all have potential origins with environmental chemicals.

The linking factor governing these effects is that they are all related (either totally or in part) to problems within the endocrine system. Whilst there is now a growing body of evidence to suggest that certain types of endocrine-disrupting behaviour can be demonstrated by chemicals within the laboratory, only a few chemicals have been conclusively shown to have such effects in the open environment. The best known is probably the link between tributyltin compounds in antifouling paints and imposex in molluscs. In many cases, links between environmental chemicals and other effects remain fairly circumstantial, and a great deal of research effort is presently being placed into finding evidence for or against such connections.

This chapter reviews the present evidence base for endocrine-disrupting effects and extends beyond the marine environment into some discussion relating to human health effects and environmental chemicals. IEH (1995), Hester and Harrison (1999) and Vos et al. (2000) have recently published reviews of the topic, and the present work extends some aspects of these publications and introduces some more recent findings. It is not intended to be a monograph on the details of the endocrine system but rather a general review of the state of knowledge and uncertainty in the field.

13.2
Definition of Endocrine-Disrupting Chemicals

A considerable number of terms have been used to describe the chemicals that fall within the general category of endocrine-disrupting chemicals, which have largely reflected different aspects of their behaviour as understood at particular times. The terms include oestrogen mimics, oestrogenic chemicals, anti-oestrogens, environmental oestrogens, endocrine moderators, eco-oestrogens, xeno-oestrogens, anti-androgens, hormone-related toxicants, and probably other terms. More recently, the term 'endocrine disrupter' has become favoured, because it allows effects arising from changes in any part of the endocrine system including thyroid, thymic and pituitary hormones (Hester and Harrison 1999).

However, there are still a variety of definitions of endocrine-disrupting chemicals that are given in the literature, and some of these are given in Table 13.1 below. The complexity of these definitions reflects two problems. The first problem involves the difficulty in defining what is meant by 'adverse' in this context (i.e. how far outside the normal range of responses is adverse?). The second problem involves the need to distinguish between direct interference with endocrine systems and those caused as a secondary consequence of toxic effects manifest elsewhere within metabolic systems.

13.3
The Effects of Endocrine-Disrupting Chemicals in Invertebrates

13.3.1
General Effects Excluding Imposex

A considerable number of both inorganic and organic chemicals have been shown to cause endocrine-disrupting effects in invertebrates (Table 13.2).

The concentrations of endocrine-disrupting chemicals capable of causing such effects are low. For example, Billinghurst et al. (2000) found increased (doubled) cypris major protein concentrations in naupilus stage larvae of the barnacle *Balanus amphitrite* at concentrations of 1 $\mu g\ l^{-1}$of 4-n-nonyl phenol and 17β-oestradiol. The range of effects noted in invertebrates is considerable, and examples are given in Table 13.3

13.3.2
Imposex

By far, the most widely studied aspect of the impact of endocrine-disrupting chemicals on invertebrates has been the phenomenon of 'imposex.' Imposex is the chemically induced development of the physical characteristics of the opposite sex. The best example of this is the development of a penis and vas deferens in females of the dogwhelk *Nucella lapillus* as a result of exposure to the antifouling paint ingredient tributyltin (TBT) used on ships, boats and harbour equipment. This effect is widespread, and approximately 72 spp and 49 genera of prosobranchs are affected worldwide (Pinder and Pottinger cited in Depledge and Billinghurst 1999). The sensitivity of the organism to TBT varies considerably between species, but effects in *N. lapillus*

Table 13.1. Different definitions of endocrine disrupters and potential endocrine disrupters

Definition	Source
An endocrine disrupter is an exogenous substance that causes adverse health effects in an intact organism, or its progeny, subsequent to changes in endocrine function	EC (1997)
A potential endocrine disrupter is a substance that posseses properties that might be expected to lead to endocrine disruption in an intact organism	EC (1997)
An endocrine disrupter is an exogenous agent that interferes with the synthesis, secretion, transport, binding, action, or elimination of natural hormones in the body that are responsible for the maintenance of homeostasis, reproduction, development and/or behaviour	U.S. EPA (1997)
An endocrine disrupter is an exogenous substance or mixture that alters function(s) of the endocrine system and consequently causes adverse health effects in an intact organism or its progeny, or (sub)populations	IPCS (1998)
A potential endocrine disrupter is an exogenous substance or mixture that posseses properties that might be expected to lead to endocrine disruption in an intact organism, or its progeny, or (sub)populations	IPCS (1998)
An endocrine disrupter is an exogenous chemical substance or mixture that alters the structure or function(s) of the endocrine system and causes adverse effects at the level of the organism, its progeny, populations, or sub-populations or organisms, based on scientific principles, data, weight-of-evidence, and the precautionary principle	U.S. EPA (1998)

Table 13.2. Chemicals having endocrine-disrupting effects in invertebrates (Depledge and Billingshurst 1999)

Class	Identified endocrine disrupting chemicals
Metals	Cadmium, selenium, zinc, mercury, lead, tributyltin
Herbicides	Diquat bromide, Atrazine, Simazine, Diuron
PCBs	Clophen A50, Aroclor 1254
Alkylphenols	Nonylphenol, pentylphenol
Natural and synthetic vertebrate steroids	Diethylstilbestrol, testosterone
Insecticides	Pyriproxyfen, DDT, MCPA, Endrin, Toxaphene, Piperonyl butoxide, Methoprene, Endosulfan, pentachlorophenol, Diflubenzuron, Kelthane
Mixtures	Tannery effluent, paper and pulp mill effluent, crude oil derivatives, sewage effluent

can be induced at the remarkably low concentration of 1–2 ng l^{-1}, and complete sterilization occurs at concentrations in the 3–5 ng l^{-1} range (Gibbs et al. 1988). The relationship between TBT and imposex in *N. lapillus* has recently been reviewed by Gibbs and Bryan (1996)

Imposex in dogwhelks follows a number of identifiable phases (Alzieu 2000). Phase 1 involves the formation of a vas deferens, a channel between prostate and penis present in males, appearance and growth of a penis (Phases 2–4), and finally ster-

Table 13.3. Examples of chemically induced changes in invertebrate development

Organism	Chemical	Effects	References
Crustaceans	Cd, Se	Moulting changes	Fingerman et al. (1996); Schultz et al. (1980)
Mussels	Cd	Impaired gonadal follicle development	Kluytmans et al. (1988)
Sea urchins *Stronglocentrotus intermedius*	Cd	Oogenesis changes	Khristoforova et al. (1984); Gnezdilova et al. (1985)
Sea star *Asterias rubens*	Cd, Zn, PCBs	Altered steroid metabolism, reproductive anomalies	Voogt et al. (1987); Den Besten et al. (1989)
Estuarine crustaceans	Diflubenzuron	Development and moulting inhibition	Cunningham (1986)
Shrimp *Palaeomenetes pugio*	Endrin (30 $\mu g\ l^{-1}$)	Delayed spawning and reduced embryo viability	Tyler-Schroeder (1979)
Crustaceans	Methoprene	Mimics juvenile hormone	McKenney and Matthews (1990); Celestial and McKenney (1994)

ilization of the subject with blocking of the oviduct and accumulation of eggs within the gland (Phases 5 and 6). In the final stages, the female become sterile, and populations may drastically decline or disappear completely. The mechanism of imposex is still not well-understood but probably involves an increase in testosterone within females, while other hormones such as progesterone and 17β-oestradiol remain constant. The mode of action may be through the inhibition of cytochrome P450-dependent aromatase, causing a breakdown in the conversion of testosterone to 17β-oestradiol and hence a build-up of testosterone (Alzieu 2000).

Imposex is a world-wide phenomenon associated most closely with proximity to ports or marinas (Batley 1996) and not generally extending far outside (Evans 1999). For example, occurrence of imposex has recently been reported in Australia (Reitsema and Spickett 1999), Singapore (Tan 1999) and Thailand (Bech 1999). Concern about the effects of TBT led to the ban on the use of paints containing it in 1982 (in the U.K., though somewhat later in other countries; Stewart 1996). A recommendation has been made by the Marine Environmental Protection Committee of the International Maritime Organization that TBT-based antifoulants should be completely banned from the year 2003. Although widespread, the effects of TBT are localized. This has led Evans (1999) to argue strongly that this ban is misguided and that the lack of suitable alternatives will impose both economic and environmental penalties (TBT antifoulants are estimated to save the shipping industry US$ 5.7 billion per year; Rouhi 1998), and annual fuel savings through TBT usage are estimated to be around 7.2 million tonnes (Bennett 1996)).

Where environmental concentrations of TBT have fallen, there is an increasing body of evidence (see e.g. Evans 1999, Table 2) that, following bans, populations of a range of imposex impacted marine organisms have recovered as TBT concentrations have declined. At present there appear to be no clear successors to TBT antifoulants. Some alternatives such as the herbicide Irgarol 1051 are already coming under suspicion as having non-target species effects (see e.g. Evans 2000; Okamura et al. 2000; OSPAR 1999).

13.4 The Effects of Endocrine Disrupting Chemicals in Vertebrates

13.4.1 Fish

The fact that certain observed changes in fish (including intersex, production of the female yolk precursor lipoprotein vitellogenin and abnormal testicular development in both adult male and juvenile fish) could be related to substances present in sewage and industrial effluent gave the first clear indications of an endocrine-disrupting chemical problem (Purdom et al. 1994; Harries et al. 1993; Jobling and Sumpter 1993; Sumpter and Jobling 1995; Jobling et al. 1996; Kime 1998; Vos et al. 2000). Such effects have been linked to oestrogenic hormones from human waste and from alkylphenols (Jobling et al. 1996). In the U.K., a considerable number of studies have now also demonstrated a high prevalence (locally up to 100%) of intersex occurrence (Jobling et al. 1998). Table 13.4 provides some examples of the types of effects and associated contaminants in marine systems.

In the U.S., other effects have been observed, including delayed sexual development, altered blood sex hormone profiles, impaired reproductive performance, and masculinization in fish exposed to effluents from kraft pulp mill effluents (Hester and Harrison 1999). Tall oil, a major component of kraft mill effluent, contains 3% plant sterols, which are dominated by sitosterol and stigmastanol, but the link between this possible cause and the observed effects has not been established (Howell and Denton 1989), though β-sitosterol is known to be oestrogenically active (Phillips and Harrison 1999).

The endocrine system in fish has similar features to that of mammals (Fig. 13.1; Kime 1999), though there are a number of important ways in which reproductive physiology is different between mammalian and non-mammalian animals (Dawson 1998). External stimuli such as seasonal changes in temperature or day length of the behaviour of a potential mate are processed by the brain, leading to the release of gonadotrophin releasing hormone (GnRH) by the hypothalamus. GnRH induces the pituitary gland to release gonadotrophin (GtH), which stimulates steroid synthesis in the gonads. Some fish are known to produce two gonadotrophins (GtH-1 and GtH-2), which are analogous to the follicle stimulating and luteinising hormones (FSH and LH) that regulate the female cycle in mammals. In fish, GtH-1 stimulates the ovary to produce oestradiol, which in turn induces production of a yolk protein (vitellogenin) by the liver. GtH-2 is most important prior to spawning when it stimulates ovarian synthesis of a progestogen (17,20-β-dihydroxy-4-pregnen-3-one; 17,20βP), which induces maturation of the oocytes prior to ovulation. This progestogen may play a role in the maturation of sperm, but there is no known role for progesterone, which is a key mammalian hormone. A further difference between fish and mammals is that the main hormone produced by the testis is 11-ketotestosterone rather than testosterone. However, both male and female fish produce testosterone, which may be involved in feedback signals to the pituitary. Fish gonads have some of the same functions as mammalian livers insofar as they are able to convert steroid hormones into metabolites that in some fish species may act as pheromones.

The interrelationships between the processes shown in Fig. 13.1 indicate that disruption of a single part of the system can lead to a profound overall disturbance.

Table 13.4. Possible endocrine-disrupting effects in fish (modified from Vos et al. 2000)

Effect/disorder	Location	Species	Contaminants	References
Raised blood vitellogeneis in males; ovotestis up to 20%	U.K. estuaries and coastal waters	Flounder (*Platichthys flesus*)	(Xeno)estrogens	Allen et al. (1999); Matthiessen et al. (1998)
Raised blood vitellogeneis in females; testicular abnormalities	Scotland, estuaries	Flounder (*Platichthys flesus*)	Sewage and industrial waste, (xeno)estrogens	Lye et al. (1997)
Reproductive impairment	U.S. Puget Sound	English sole (*Parophrys vetulus*)	PCBs PAH	Johnson et al. (1988, 1997, 1998)
Reproductive impairment	U.S. San Francisco Bay	Starry flounder (*Platichthys stellatus*)	Organic contaminants	Spies and Rice (1988)
Decreased hatching success	Baltic Sea	Baltic herring (*Clupea hagengus*)	Chlorinated hydrocarbons	Hansen et al. (1985)
Decreased fecundity	Baltic Sea	Cod (*Gadus morhua*)	Lipophilic xenobiotics	Petersen et al. (1997)
Disruption of normal ovarian cycle	France, Brittany	Plaice (*Pleuronectes flesus*)	Amoco Cadiz oil spill	Stott et al. (1983)
Premature vitellogenesis (decreased turnover of steroids)	Netherlands, Texel	Flounder (*Platichthys flesus*)	Harbour dredge contaminants	Janssen et al. (1997)
Elevated vitellogenin in female fish	U.S. Boston Harbour	Winter flounder (*Pleuronectes americanus*)	Unknown	Pereira et al. (1992)
Decreased age at first maturation	North Sea	Plaice, sole	Unknown	Janssen et al. (1997)
Changes in sex ratio	North Sea	Dab (*Limanda limanda*)	Unknown	Lang et al. (1995)
Reproductive disorder M74 syndrome	Baltic Sea	Atlantic salmon (*Salmo salra L.*)	Dietary vit B1 deficiency	Bengtsson et al. (1999)
Immunodysfunction	Puget Sound (estuary)	Chinook salmon (juveniles) (*Oncorhynchus tshawytscha*)	PAH, PCBs	Arkoosh et al. (1998)
Serum vitellogenin	U.K. estuaries	Flounder (*Platichthys flesus*)	nonylphenols + other organic chemicals	Allen et al. (1999)

Changes in the function(s) of the endocrine system in fish can occur as a result of disruption of hypothalmic, pituitary or gonadal processes and also by changes in the steroid hormone deactivation function of the liver. Whilst such effects can be seen in laboratory experiments, identification of causative agents in the open environment is much more difficult. This is due not only to the complex mixture of potential endocrine-disrupting chemicals but also the fact that their behaviour may be antagonistic or synergistic in action. Thus, dose-response relationships are particularly hard to identify unambiguously. Kime (1999) and Vos et al. (2000) have reviewed the potential effects of endocrine-disrupting chemicals under a number of main headings, including changes in testicular structure and hormones, altered pheromone activity and decreased sperm viability in male fish. In female fish vitellogenin production may be reduced, abnormalities in the ovary may be observed (decreases in the numbers of

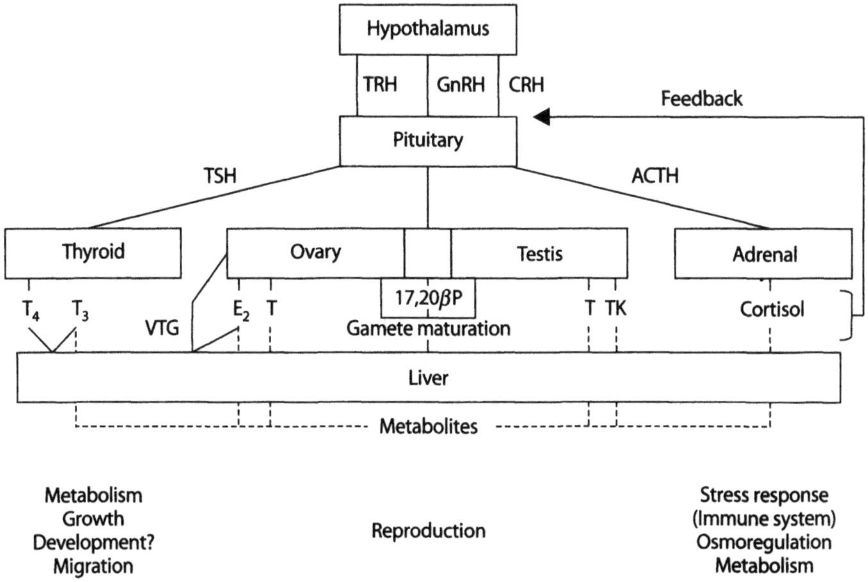

Fig. 13.1. A schematic diagram of the endocrine system of fish (Kime 1999). *TRH* = thyrotrophin releasing hormone; *GnRH* = gonadotrophin releasing hormone; *TSH* = thyroid stimulating hormone; *GtH* = gonadotrophins I and II; *ACTH* = adrenocorticotrophic hormone; T_4 = thyroxine; T_3 = triiodothyronine; E_2 = oestradiol; *T* = testosterone; *17,20βP* = 17,20β-dihydroxy-4-pregnen-3-one; *KT* = 11-ketotestosterone; *VTG* = Vitellogenin

large, yolky eggs, increased numbers of immature oocytes), possible changes in spawning patterns, reduced egg numbers and/or viability and changes in secondary sexual characteristics.

It should be appreciated that the origins of hermaphrodite or intersex fish do not come about because of their adult exposure to environmental chemicals. It is far more likely that exposure at very early life stages is much more important, particularly at the critical points where structural differentiation of the gonads and sexual behavioural patterns are being laid down. Indeed, environmental chemicals may be used deliberately in aquaculture to produce monosex cultures (Foresti 2000). On the other hand, phenomena such as production of plasma vitellogen can be observed in adults (Harries et al. 1999).

13.4.2 Reptiles and Amphibians

The effects of endocrine-disrupting chemicals on reptiles have not been as widely studied as with other organisms, though there is an increasing evidence base that there are problems (Guillette 2000), and reptiles have been used as indicators of bioindicators of environmental contaminants (Crain and Guillette 1998). The most widely cited example is that of alligators in Lake Apopka, Florida, which in 1980 was highly polluted by a spill of the chemical dicofol (contaminated with a variety of chemicals from the

DDT family) (Botham et al. 1999; Vos et al. 2000). This was introduced to a lake which, despite being already contaminated by agricultural runoff and sewage effluent, had a successful alligator population. By 1984 the juvenile alligator population had declined by around 90%. Eggs collected from the lake region had high contaminant concentrations, and abnormalities were also noted in the embryos and young. Males from Lake Apopka eggs had poorly organized testes with aberrant structures while six-month-old females had abnormal ovaries with prominent polyovular follicles and an unusually large number of multinucleated oocytes (Botham et al. 1999). Older animals also showed abnormalities that could be linked to endocrine disruption (Semenza et al. 1997). For example, concentrations of plasma testosterone were higher in females and lower in males than in cleaner environments, whilst 17β-oestradiol were higher than normal in males. Males also showed a link between reduced penis length and androgen doses (Botham et al. 1999).

Less research has been performed on the influence of endocrine-disrupting chemicals and amphibians, though there is increasing concern about world-wide declines in amphibian populations (see e.g. Carey 2000). This is in part because of the complex sex determination factors in these animals that include temperature as a major parameter. Nevertheless, links have been made between the presence of endocrine-disrupting chemicals and various abnormalities in frogs (*Xenopus laevis*, Palmer and Palmer 1995; *Acris crepitans*, Reeder et al. 1998) and salamanders (Clark et al. 1998). Pickford and Morris (1999) have hypothesized that endocrine-disrupting chemicals may disrupt progesterone-induced oocyte maturation in the adult amphibian ovary and have tested this with the African clawed frog (*Xenopus laevis*) and the pro-oestrogenic pesticide methoxychlor. Effects were noted at very low concentrations (mean inhibitive concentration 72 nM), and the effects of the chemical were shown to be dose-dependent, reversible and early acting.

Recently a number of researchers (e.g. Lutz and Kloas 1999; Kloas et al. 1999; Palmer et al. 1998) have examined the use of amphibians as a model to study endocrine-disrupting chemicals. Such models are based around the mechanisms of vitellogenin production, binding to liver receptors and in vivo effects on sexual development caused by larval exposure to chemicals. An overview of a workshop for detecting potential (anti-)oestrogenic/androgenic chemicals in wildlife has also recently been provided by Ankley et al. (1998).

13.4.3 Birds

The damaging effects of organochlorine pesticides (notably DDT/DDE, PCBs and the cyclodiene pesticides) on bird populations were firmly established after the major declines in top predator bird populations in many industrialized countries during the 1950s and 1960s (Botham et al. 1999). Effects include eggshell thinning (Blus et al. 1997; Grasman et al. 1998; Vos et al. 2000), female-female pairing (Hunt and Hunt 1977) and supernormal clutches (Fry and Toone 1981; Fry et al. 1987; Dawson 2000). Species such as peregrine falcons (*Falco perigrinus*), cormorants (e.g. *Phalacrocorax auritus*) and brown pelicans (*Pelecanus occidentalis*) have been shown to be particularly vulnerable though a variety of other species (e.g. dippers, *Cinclus cinclus*; Ormerod et al. 2000) may also be at risk. Whether such effects are due to endocrine-disruption mechanisms seems open to debate with for example Bowerman et al. (2000) citing evidence for population level effects of hormone-disrupting chemicals in bald eagles (*Halineetus leucocephalus*) in the Great Lakes and Leatherland (1998), indicating that polyhalogenated aromatic hydrocarbons represent a significant threat to the thyroid hormone economy in both birds and seals. Conversely, Dawson (2000) argues that there is no evidence that avian wildlife has suffered endocrine disruption and refutes the link between eggshell thinning, supernormal clutches and endocrine disrupters. As part of the efforts to determine the validity of these opposing views, it is possible to identify the work of Berg et al. (1998), Fossi et al. (1999) and Halldin et al. (1999), who have examined methods for determining the effects of endocrine-disrupting chemicals on birds and that of Huet (2000), who has described the areas currently under active investigation by the OECD in this area.

13.4.4 Mammals

Much of the earlier work on the effects of endocrine disrupters on mammals focused on typical laboratory test animals, particularly rats and mice (see e.g. Kelce et al. 1995). Whilst such work is really outside the scope of this chapter, a useful comparative study of the relative potency of a number of natural and synthetic endocrine disrupters has been constructed by Nilsson (2000). Table 13.5 reproduces some key data.

Several groups of marine mammals have come under close scrutiny for evidence of endocrine disruption, including polar bears (*Ursus maritimus*). Polar bears are Arctic top predators and are known to accumulate particularly high concentrations of con-

Table 13.5. Relative potencies in vivo assays of selected synthetic as well as natural endocrine disrupters in comparison with 17β-oestradiol (modified from Nilsson 2000)

Chemical[a]	Approximate potency
17β-oestradiol	1.0
2,6,-cis-Diphenylhexamethyl-cyclotetrasiloxane	0.03
Coumestrol	0.01
Coumestrol	0.01
Genistein	0.001
Nonylphenol	0.0005
Nonylphenol	0.00001
Bisphenol-A	0.0001
Octylphenol	0.00001
Methoychlor	0.0004
Methoxychlor	0.00001
Methoxychlor	0.0001
2',4',6'-Trichlorobiphenol	0.00001
2',3',4',5'-Tetrachloro-4-biphenol	0.00001

[a] Multiple entries are from different studies.

taminants, particularly chlorinated ones. A recently discovered phenomenon is that in some cases (e.g. chlordane and αHCH), bioaccumulation in polar bears, seals and cetaceans is enantioselective, and enantiomer ratios are important variables for sample groupings and for class separation of male/female seals and fat/liver tissues (Minh et al. 2000; Wiberg et al. 2000). The consequences of this observation are likely to be profound, and further research will clearly be necessary.

The distribution of contaminants in high latitudes is not uniform, and several researchers have reported temporal and geographical trends (see e.g. Norstrom et al. 1998; Muir and Norstrom 2000; Muir et al. 1999, 2000). The main source of these contaminants to polar bears is probably their food and in particular ringed seals (*Phoca hispida*) and harp seals (*Phoca groenlandica*) (Kleivane et al. 2000). Very high concentrations of PCBs have been reported in the literature in other top predators such as glaucus gulls (Henriksen et al. 2000). However, whilst there is growing evidence that immunotoxic effects such as changes in immunoglobulin levels are arising as a result, there is no evidence that there is a relationship between organochlorine concentrations and fertility (Bernhoft et al. 1997, 2000). Nevertheless, of considerable concern is the observation made by Wiig et al. (1998) that 4 of 269 animals examined exhibited female hermaphroditism. Two possible explanations for this phenomenon are (*i*) excessive androgen excretion by the mother (possibly as a result of a tumour) and (*ii*) that there is endocrine disruption as a result of the presence of environmental pollutants.

Polar bears are thought to be particularly vulnerable to contaminant exposure, because they undergo one of the most extreme fasts known for any mammal (Polischuk et al. 1995). As organochlorine (and other lipophilic contaminants) are very largely stored in body fat, it is not surprising that contaminant concentration in both adipose tissue and milk increase as the fast progresses (Polischuk et al. 1995). The transfer of contaminants to cubs therefore also increases as the fast continues.

A second group of marine mammals that have been studied extensively are seals. Declines in populations of seals notably in the Baltic and North Sea areas have been evident over many years. As far back as 1976 Helle et al. (1976a,b) reported correlations between pathological changes in seals and PCB concentrations. A number of diseases have been linked to organochlorine contaminants, but the mechanisms are incompletely understood, and Vos et al. (2000) observed that it is conceivable that the disease syndrome is caused by several classes of persistent organohalogens and several independent mechanisms of action. It is also possible that the problems lie at least in part with metabolic or degradation products of the original compounds. For example, $MeSO_2$-DDE apparently has a role in the etiology of adrenocortical hyperplasia in Baltic seals (Brandt et al. 1992). Significant cytochrome P450-catalysed irreversible binding of $MeSO_2$-DDE has also been observed in seals (Lund 1994). Vos et al. (2000) and Van Loveren et al. (2000) have concluded that both reproduction and immune functions in Baltic grey and ringed seals and Wadden Sea harbour seals are impaired by PCBs in the food chain. Damage to reproductive systems has led to population declines, and impairment of immune function has probably contributed to mass mortalities due to morbillivirus infections.

Experimental evidence relating to mammals such as dolphins suggests that there is the potential for damage as a result of exposure to contaminants (Wilson et al. 1999). However, in the wild such conclusions are hard to validate because of the natural variability. So, for example, Wilson et al. (op. cit.) determined that there were no links between epidermal diseases and contaminants in four populations of dolphins but that there were highly significant correlations with oceanographic variables such as low salinity and temperature.

13.4.5 Humans

Although the focus of this chapter is on endocrine disruption in the marine environment, it is also pertinent to consider the effects on humans, particularly those who rely on the marine ecosystem for significant parts of their diet. That such effects may be relevant has recently been demonstrated by Fournier et al. (2000), who demonstrated that immunosuppression can be observed in mice fed on beluga whale meat from both the St. Lawrence Estuary and Arctic Sea. Recently, concerns have been expressed about the effects of environmental contaminants on general populations and particularly in high latitude indigenous peoples either directly in terms of their health or indirectly in terms of changes in cultural habits because of increased environmental risks (Ayotte et al. 1996; Bard 1999; Dewailly et al. 2000;Hansen 2000; Kuhnlein et al. 1995; Kuhnlein and Chan 2000; Van Oostdam et al. 1999).

The types of human effects that have been attributed to endocrine disruption include decreasing sperm counts, cryptorchidism, malformations of the male genital tract, testicular cancer (Turner 1999) and breast cancer and effects on the cardiovascular system (IEH 1995). Much of the evidence available is circumstantial and subject to considerable disagreements as to methods or interpretation. For example, a very widely cited paper presenting evidence for declining semen quality during the past 50 years (Carlsen et al. 1992) has been widely criticized for both methodological and statistical inadequacies (e.g. Marmor et al. 1998; Gandini et al. 2000). Nevertheless, reports are still appearing, indicating both temporal and regional differences in sperm quality (Auger et al. 1995; Auger and Jouannet 1997). The debate on the relationship between environmental chemicals and health can be followed on a number of web sites including that of the Center for BioEnvironmental Research, Tulane and Xavier Universities (http://www.tmc.tulane.edu/ecme/eehome/) and the Introduction to Hormone Disrupting Chemicals run by Dr Michael Warhurst (http://website.lineone.net/~mwarhurst/).

References

Allen Y, Matthiessen P, Haworth S, Thain JE, Feist S (1999) Survey of estrogenic activity in United Kingdom estuarine and coastal waters and its effects on gonadal development of the flounder *Platichthys flesus*. Environ Toxicol Chem 18:1791–1800

Alzieu C (2000) Impact of tributyltin on marine invertebrates. Ecotoxicology 9:71–76

Ankley G, Mihaich E, Stahl R, Tillitt D, Colborn T, McMaster S, Miller R, Bantle J, Campbell P, Denslow N, Dickerson R, Folmar L, Fry M, Giesy J, Gray LE, Guiney P, Hutchinson T, Kennedy S, Kramer V, LeBlanc G, Mayes M, Nimrod A, Patino, R, Peterson R, Purdy R, Ringer R, Thomas P, Touart L, Kraak G Van der, Zacharewski T (1998) Overview of a workshop on screening methods for detecting potential (anti-)estrogenic/androgenic chemicals in wildlife. Environ Toxicol Chem 17:68–87

Arkoosh MR, Casillas E, Huffman P, Clemons E, Evered J, Stein JE, Varanasi U (1998) Increased susceptibility of juvenile chinook salmon from a contaminated estuary to *Vibrio anguillarum*. Trans Am Fish Soc 127:360–374

Auger J, Jouannet P (1997) Evidence for regional differences of semen quality among fertile French men. Human Reproduction 12:740–745

Auger J, Kuntsmann JM, Czyglik F, Jouannet P (1995) Decline in semen quality among fertile men in Paris during the past 20 years. New Engl J Med 332:281–285
Ayotte P, Carrier G, Dewailly E (1996) Health risk assessment for Inuit newborns exposed to dioxin-like compounds through breast feeding. Chemosphere 32:531–542
Bard SM (1999) Global transport of anthropogenic contaminants and the consequences for the Arctic marine ecosystem. Mar Pollut Bull 38:356–379
Batley G (1996) The distribution and fate of tributyltin in the marine environment. In: Mora SJ (ed) Tributyltin: Case study of an environmental contaminant. Cambridge University Press, Cambridge, pp 139–166
Bech M (1999) Increasing levels of tributyltin-induced imposex in muricid gastropods at Phuket Island, Thailand. Appl Organomet Chem 13:799–804
Bengtsson BE et al. (1999) Reproductive disturbances in Baltic fish: A synopsis of the FiRe project. Ambio 28(1):2–8
Bennett RF (1996) Industrial manufacture and applications of tributyltin compounds In: Mora SJ (ed) Tributyltin: Case study of an environmental contaminant. Cambridge University Press, Cambridge, pp 21–61
Berg C, Halldin K, Brunstrom B, Brandt I (1998) Methods for studying xenoestrogenic effects in birds. Toxicol Lett 103:671–676
Bernhoft A, Wiig O, Skaare JU (1997) Organochlorines in polar bears (*Ursus maritimus*) at Svalbard. Environ Pollut 95:159–175
Bernhoft A, Skaare JU, Wiig O, Derocher, Larsen HJS (2000) Possible immunotoxic effects of organochlorines in polar bears (*Ursus maritimus*) at Svalbard. J Toxicol Env Heal A 59:561–574
Billinghurst Z, Clare AS, Matsumura K, Depledge MH (2000) Induction of cypris major protein in barnacle larvae by exposure to 4-n-nonyl phenol and 17β-oestradiol. Aquat Toxicol 47:203–212
Blus LJ, Wiemeyer SN, Bunck CM (1997) Clarification of effects of DDE on shell thickness, size, mass and shape of avian eggs. Environ Pollut 95:67–74
Botham C, Holmes P, Harrison P (1999) Endocrine disruption in mammals, birds, reptiles and amphibians. In: Hester RE, Harrison RM (eds) Endocrine disrupting chemicals. Royal Society of Chemistry, Cambridge (Issues in Environmental Science and Technology, No. 12, pp 61–82)
Bowerman WW, Best DA, Grubb TG, Sikarskie JG, Giesey JP (2000) Assessment of environmental endocrine disruptors in bald eagles of the Great lakes. Chemosphere 41:1569–1574
Brandt I, Jönsson C-J, Lund B-O (1992) Comparative studies on adrenocorticolytic DDT metabolites. Ambio 21:602–605
Carey C (2000) Infectious disease and world-wide declines of amphibian populations, with comments on emerging diseases in coral reef organisms and in humans. Environ Health Persp 108:143–150
Carlsen E, Giwercman A, Keiding N, Skakkebaek NE (1992) Evidence for decreasing quality of semen during past 50 years. Brit Med J 305:609–613
Celestial DM, McKenney CL (1994) The influence of an insect growth regulator on the larval development of the mud crab *Rhithropanopeus harrisii*. Environ Pollut 85:169–173
Clark EJ, Norris DO, Jones RE (1998) Interactions of gonadal steroids and pesticides (DDT, DDE) on gonaduct growth in laval tiger salamanders, *Ambystoma tigrinum*. Gen Comp Endrocr 109:94–105
Crain DA, Guillette LJ (1998) Reptiles as models of contaminant-induced endocrine disruption. Anim Reprod Sci 53:77–86
Cunningham PA (1986) A review of toxicity testing and degradation studies used to predict the effects of Diflubenzuron (Dimilin) on estuarine crustaceans. Environ Pollut A 40:63–86
Dawson A (1998) Comparative reproductive physiology of nonmammalian species. Pure Appl Chem 70:1657–1669
Dawson A (2000) Mechanisms of endocrine disruption with particular reference to occurrence in avian wildlife: A review. Ecotoxicology 9:59–69
Depledge MH, Billinghurst Z (1999) Ecological significance of endocrine disruption in marine invertebrates. Mar Pollut Bull 39:32–38
Den Besten PJ, Herwig HJ, Zandee, DI, Voogt PA (1989) Effects of cadmium and PCBs on reproduction of the sea star *Asteria rubens*: Aberrations in the early development. Ecotox Environ Safe 18:173–180
Dewailly E, Ayotte P, Bruneau S, Gingras S, Belles-Isles M, Roy R (2000) Susceptibility to infections and immune status in Inuit infants exposed to organochlorines. Environ Health Persp 108:205–211
Dodds EC, Lawson W (1938) Molecular structure in relation to oestrogenic activity: Compounds without a phenanthrene nucleus. Proc Roy Soc 125:222–232
Dodds EC, Goldberg L, Lawson W, Robinson R (1938) Oestrogenic activity of certain synthetic compounds. Nature 141:247–248
EC (1997) European Commission DGXII, Report EUR 17549. European Workshop on the Impact of Endocrine Disrupters on Human Health and Wildlife, Brussels
Evans SM (1999) Tributyltinpollution: the catastrophe that never happened. Mar Pollut Bull 38:629–636

Evans SM (2000) Response to letter by Nichols. Mar Pollut Bull 40:713–715
Fingerman M, Devi M, Reddy PS, Katyayani R (1996) Impact of heavy metal exposure on the nervous system and endocrine-mediated processes in crustaceans. Zool Stud 35:1–8
Fournier M et al. (2000) Immunosuppression in mice fed on diets containing beluga whale blubber from the St. Lawrence Estuary and the Arctic populations. Toxicol Lett 112:311–317
Foresti F (2000) Biotechnology and fish culture. Hydrobiologia 420:45–47
Fossi MC, Casini S, Marsili L (1999) Non-destructive biomarkers of exposure to disrupting chemicals in endangered species. Chemosphere 39:1273–1285
Fry DM, Toone CK (1981) DDT-induced feminisation of gull embryos. Science 213:922–924
Fry DM, Toone CK, Soeich SM, Peard RJ (1987) Sex ratio skew and breeding patterns of gulls: Demographic and toxicological considerations. Stud Avian Biol 10:26–43
Gandini L, Lombardo F, Culasso F, Dondero F, Lenzi A (2000) Myth and reality of the decline in semen quality: An example of the relativity of data interpretation. J Endocrinol Invest 23:402–411
Gibbs PE, Bryan GW (1996) Reproductive failure in the gastropod *N. lapillus* caused by tributyltin pollution. A review. In: Champ MA, Seligman PF (eds) Organotin. Environmental Fate and Effects. Chapman and Hall, London, pp 259–280
Gibbs PE, Pascoe PL, Burt GR (1988) Sex change in the female dog-whelk *Nucella lapillus* induced by tributyltin from antifouling paints. J Mar Biol Assoc UK 68:715–731
Gnezdilova SM, Khristoforova NK, Lipina IG (1985) Gonadotoxic and embryotoxic effects of cadmium in sea urchins. Symposia Biologica Hungarica 29:239–251
Grasman KA, Scanlon PF, Fox GA (1998) Reproductive and physiological effects of environmental contaminants in fish-eating birds of the Great Lakes: A review of historical trends. Environ Monit Assess 53:117–145
Guillette LJ (2000) Contaminant-induced endocrine disruption in wildlife. Growth Horm IGF Res 10:45–50
Halldin K, Berg C, Brandt I, Brunstrom B (1999) Sexual behaviour in Japanese quail as test end point for endocrine disruption: Effects of in ovo exposure to ethinyloestradiol and diethylstilbestrol. Environ Health Persp 107:861–866
Hansen JC (2000) Environmental contaminants and human health in the Arctic. Toxicol Lett 112:119–125
Hansen PD, Vonwesternhagen H, Rosenthal H (1985) Chlorinated hydrocarbons and hatching success in Baltic herring spring spawners. Mar Environ Res 15(1):59–76
Harries JE, Sheaman DA, Jobling S, Matthiessen P, Neall P Routledge, EJ, Rycroft R, Sumpter JP, Tylor T (1996) A survey of estrogenic activity in United Kingdom inland waters. Environ Toxicol Chem 15:1993–2002
Harries JE, Janbakhsh A, Jobling S, Matthiessen P, Sumpter JP, Tyler CR (1999) Estrogenic potency of effluent from two sewage treatment works in the United Kingdom. Environ Toxicol Chem 18:932–937
Helle E, Olsson M, Jensen S (1976a) DDT and PCB levels and reproduction in ringed seals from the Bothnian Bay. Ambio 5:188–189
Helle E, Olsson M, Jensen S (1976b) PCB levels correlated with pathological changes in seal uteri. Ambio 5:261–263
Henriksen EO, Gabrielsen GW, Trudeau S, Wolkers J, Sagerup K, Skaare JU (2000) Organochlorines and possible biochemical effects in glaucous gulls (*Larus hyperboreus*) from Bjornoya, the Barents Sea. Arch Environ Con Tox 38:234–243
Herbst AL Ulfelder H, Poskanzer DC (1971) Adenocarcinoma of the vagina. Association of maternal stilbestrol therapy with tumour appearance in young women. New Engl J Med 284:878–881
Hester RE, Harrison RM (eds.) (1999) Endocrine disrupting chemicals. Royal Society of Chemistry, Cambridge (Issues in Environmental Science and Technology, No. 12)
Howell WM, Denton TE (1989) Gonopodial morphogenesis in female mosquitofish, *Gambusia affinus*, mascilinized by exposure to degradation products from plant sterols. Environ Biol Fish 24:43–51
Huet MC (2000) OECD activity on endocrine disrupters test guidelines development. Ecotoxicology 9:77–84
Hunt GL, Hunt MW (1977) Female-female pairing in western gulls (*Laurus occidentalis*) in southern California. Science 196:1466–1467
IEH (1995) IEH assessment on environmental oestrogens: Consequences to human health and wildlife. Institute for Environment and Health, Leicester
IPCS (1998) International Programme on Chemical Safety, Report of IPCS/OECD Scoping Meeting on Endocrine Disruptors (EDCs), 16–18 March, Washington
Janssen PAH, Lambert JGD, Vethaak AD, Goos HJT (1997) Environmental pollution causes elevated concentrations of estrogens and vitellogenin in the female flounder *Platichthys flesus* L. Aquat Toxicol 39:195–214
Jobling S, Sumpter JP (1993) Detergent components in sewage effluent are weakly oestrogenic to fish: An in vitro study using rainbow trout (*Oncorhynchus mykiss*) hepatocytes. Aquat Toxicol 27:361–372

Jobling S, Sheahan DA, Osborne JA, Matthiessen P, Sumpter JP (1996) Inhibition of testicular growth in rainbow trout (*Oncorhynchus mykiss*) exposed to estrogenic alkylphenolic chemicals. Environ Toxicol Chem 15:194–202
Jobling S, Tyler CR, Nolan M, Sumpter JP (1998) The identification of estrogenic effects in wild fish. R&D Technical Report W119, Environment Bristol UK (cited in Vos et al. 2000)
Johnson LL, Casillas E, Collier TK, McCain BB, Varanasi U (1988) Contaminant effects on ovarian development in English sole (*Parophrys vetulus*) from Puget Sound, Washington. Can J Fish Aquat Sci 45:3133–2146
Johnson LL, Sol SY, Lomax DP, Nelson GM, Sloan CA, Casillas E (1997) Precocious sexual maturation and other reproductive anomalies in English sole from an urban waterway. ICES CM 1997/U:07
Johnson LL, Landahl JT, Kubin LA, Horness BH, Myers MS, Collier TK, Stein JE (1998) Assessing the effects of anthropogenic stressors on Puget Sound flatfish populations. J Sea Res 39:125–137
Kelce WR, Stone CR, Laws SC, Gray LE, Kemppainen JA, Wilson EM (1995) Persistent DDT metabolite pp'DDE is a potent androgen receptor antagonist. Nature 375:581–585
Kime DE (1998) Endocrine disruption in fish. Kluwer, Boston
Kime DE (1999) Environmentally induced endocrine abnormalities in fish. Royal Society of Chemistry, London (Issues in Environmental Science and Technology, 12, pp 1–151)
Kleivane L, Severinsen T, Skaare JU (2000) Biological transport and mammal to mammal transfer of organochlorines in Arctic fauna. Mar Environ Res 49:343–357
Kloas W, Lutz I, Einspanier R (1999) Amphibians as a model to study endocrine disruptors: II. Estrogenic activity of environmental chemicals in vitro and in vivo. Sci Total Environ 225:59–68
Kluytmans JH, Brands F, Jandee DI (1988) Interactions of cadmium with the reproductive cycle of *Mytilus edulis*. Mar Environ Res 24:189–192
Khristoforova NK, Gnezdilova SM, Vlasova GA (1984) Effect of cadmium on gametogenesis and offspring of the sea urchin *Stronglocentrotus intermedius*. Mar Ecol Prog Ser 17:9–14
Kuhnlein HV, Receveur O, Muir DCG, Chan HM, Soueida R (1995) Arctic indigenous women consume greater-than acceptable levels of organochlorines. J Nutr 125:2501–2510
Kuhnlein HV, Chan HM (2000) Environment and contaminants in traditional food systems of northern indigenous peoples. Annu Rev Nutr 20:595–626
Lang T, Damm U, Dethlefsen V (1995) Changes in the sex ratio of North Sea dab (*Limanda limanda*) in the period 1981–1995. ICES CM 1995/G:25 Ref E
Leatherland JF (1998) Changes in thyroid hormone economy following consumption of environmentally contaminated Great Lakes fish. Toxicol Ind Health 14:41–57
Loveren H van, Ross PS, Osterhaus ADME, Vos JG (2000) Contaminant-induced immunosuppression and mass mortalities among harbor seals. Toxicol Lett 112:319–324
Lund B-O (1994) In Vitro adrenal bioactivation and effects on steroid metabolism of DDT, PCBs and their metabolites in the grey seal (*Halichoerus grypus*). Environ Toxicol Chem 13:911–917
Lutz I, Kloas W (1999) Amphibians as a model to study endocrine disruptors: I. Environmental pollution and estrogen receptor binding. Sci Total Environ 225:49–57
Lye CM, Frid CLJ, Gill ME, McCormick D (1997) Abnormalities in the reproductive health of flounder *Platichthys flesus* exposed to effluent from a sewage treatment works. Mar Pollut Bull 34:34–41
Marmor D, Izard V, Schahmaneche D, benoit G, Jardin A (1998) Is male fertility really declining? Presse Medicale 27:1484–1490
Matthiessen P, Allen Y, Allchin CR, Feist SW, Kirby MF, Law RJ, Scott AP, Thain JE, Thomas KV (1998) Estrogenic endocrine disruption in flounder (*Platichthys flesus* L.) from UK estuarine and marine waters. CEFAS, Lowestoft (Science Series Technical Reports 107)
McKenney CL, Matthews E (1990) Influence of an insect growth regulator on the larval development of an estuarine shrimp. Environ Pollut 64:169–178
Minh TB, Nakate H, Watanabe M, Tanabe S, Miyazaki N, Jefferson TA, Prudente M, Subramanian A (2000) Isomer-specific accumulation and toxic assessment of polychlorinated biphenyls, including coplanar congeners, in cetaceans from the North Pacific and Asian coastal waters. Arch Environ Con Tox 39:398–410
Muir DCG, Norstrom RJ (2000) Geographical differences and time trends of persistent organic pollutants in the Arctic. Toxicol Lett 112:93–101
Muir DCG, Braune B, DeMarch B, Norstrom R, Wagemann R, Lockhart L, Hargrave B, Bright D, Addison R, Payne J, Reimer K (1999) Spatial and temporal trends and effects of contaminants in the Canadian Arctic marine ecosystem: A review. Sci Total Environ 230:83–144.
Muir DCG, Born EW, Koczansky K, Stern GA (2000) Temporal and spatial trends of persistent organochlorines in Greenland walrus (*Odobenus rosmarus rosmarus*). Sci Total Environ 245:73–86
Nilsson R (2000) Endocrine modulators in the food chain and environment. Toxicol Pathol 28:420–431
Norstrom RJ, Belikov, SE, Born EW, Garner GW, Malone B, Olpinski S, Ramsay MA, Schliebe S, Stirling I, Stishov MS, Taylor MK, Wiig O (1998) Chlorinated hydrocarbon contaminants in polar bears from

eastern Russia, North America, Greenland and Svalbard: Biomonitoring of Arctic pollution. Arch Environ Con Tox 35:354–367
Okamura H, Aoyama I, Takami T, Maruyama T, Suzuki Y, Matsumoto M, Katsuyama I, Hamada J, Beppu T, Tanaka O, Maguire RJ, Liu D, Lau YL, Pacepavicius GJ (2000) Phytotoxicity of the new antifouling compound Irgarol 1051 and a major degradation product. Mar Pollut Bull 40:754–763
Oostdam J van, Gilman A, Dewailly E, Usher P, Wheatley B, Kuhnlein H, Neve S, Walker J, Tracy B, Feeley M, Jerome V, Kwavnick B (1999) Human health implications of environmental contaminants in Arctic Canada: A review. Sci Total Environ 230:1–82
Ormerod SJ, Tyler SJ, Juttner I (2000) Effects of point-source PCB contamination on breeding performance and post-fledging survival in the dipper *Cinclus cinclus*. Environ Pollut 110:505–513
OSPAR (1999) Organic tin compounds. OSPAR Consortium for the Protection of the Marine Environment of the North-East Atlantic. Report of the Working Group on Diffuse Sources (DIFF), Berne, 18–22 October 1999 (cited in Evans 2000)
Palmer BD, Palmer SK (1995) Vitellogenin induction by xenobiotic estrogens in the red-eared turtle and african clawed frog. Environ Health Persp 103:19–25
Palmer BD, Huth LK, Pieto DL, Selcer KW (1998) Vitellogenin as a biomarker for xenobiotic estrogens in an amphibian model system. Environ Toxicol Chem 17:30–36
Pereira JJ, Ziskowski J, Mercaldo-Allen R, Kuropat C, Luedke D, Gould E (1992) Vitellogenin in winter flounder (*Pleuronectes americanus*) from Long Island Sound and Boston Harbor. Estuaries 15:289–297
Petersen GI, Gerup J, Nilsson L, Larsen JR, Schneider R (1997) Body burdens of lipophilic xenobiotics and reproductive success in Baltic cod (*Gadus morhua* L.) ICES CM 1997/U:10
Phillips B, Harrison P (1999) Overview of the endocrine disrupters issues. In: Hester RE, Harrison RM (eds) Endocrine disrupting chemicals. Royal Society of Chemistry, Cambridge (Issues in Environmental Science and Technology, No. 12, pp 1–26)
Pickford DB, Morris ID (1999) Effects of endocrine-disrupting contaminants on amphibian oogenesis: Methoxychlor inhibits progesterone-induces maturation of *Xenopus laevis* oocytes in vitro. Environ Health Persp 107:285–292
Polischuk SC, Letcher RJ, Norstrom RJ, Ramsay MA (1995) Preliminary results of fasting on the kinetics of organochlorines in polar bears (*Ursus maritimus*). Sci Total Environ 161:465–472
Purdom CE, Hardiman PA, Bye VJ, Eno NC, Tyler CR, Sumpter JP (1994) Estrogenic effects of effluents from sewage treatment works. Chem Ecol 8:275–285
Reeder AL, Foley GL, Nichols, DK, Hanssen LG, Wikoff B, Faeh S, Eisold J, Wheeler MB, Warner R, Murphy JE, Beasley VR (1998) Forms and prevalence of intersexuality and effects of environmental contaminants on sexuality in cricket frogs (*Acris crepitans*). Environ Health Persp 106:261–266
Reitsema TJ, Spickett JT (1999) Imposex in *Morula granulata* as bioindicator of tributyltin (TBT) contamination in the Dampier Archipelago, Western Australia. Mar Pollut Bull 39:280–284
Rouhi AM (1998) The squeeze of tributyltins. Chem Eng News 76:41–42
Schultz TW, Freeman SR, Dumont JN (1980) Uptake, depuration and distribution of selenium in *Daphnia* and its effect on survival and ultrastructure. Arch Environ Con Tox 9:23–40
Semenza JC, Tolbert PE, Rubin CH, Guillette LJ, Jackson RJ (1997) Reproductive toxins and alligator abnormalities at Lake Apopka, Florida. Environ Health Persp 105:1030–1032
Spies RB, Rice DW (1988) Effects of organic contaminants on reproduction of starry flounder (*Platichthys stellatus*) in San Francisco Bay. Mar Biol 98:191–200
Stewart C (1996) The efficacy of legislation in controlling tributyltin in the marine environment. In: Mora SJ (ed) Tributyltins: Case study of an environmental contaminant. Cambridge University Press, Cambridge, pp 264–297
Stott GG, Haensly WE, Neff JM, Sharp JR (1983) Histopathologic survey of ovaries of plaice, *Pleuronectes platess* L. from Aber wrac'h and Aber benoit, Brittany, France: Long-term effects of the Amoco Cadiz crude oil spill. J Fish Dis 6:429–437
Sumpter JP, Jobling S (1995) Vitellogenesis as a biomarker for estrogenic contaminantion of the aquatic environment. Environ Health Persp 103:173–178
Tan KS (1999) Imposex in *Thais gradata* and *Chicoreus capucinus* (Mollusca, Neogastropoda, Muricidae) from the Straits of Johor. A case study using penis length, area and weight as measures of imposex severity. Mar Pollut Bull 39:295–303
Turner KJ (1999) Oestrogens, environmental oestrogens and male reproduction In: Hester RE, Harrison RM (eds) Endocrine disrupting chemicals. Royal Society of Chemistry, Cambridge (Issues in Environmental Science and Technology, No. 12, pp 83–108)
Tyler-Schroeder DB (1979) Use of grass shrimp *Palaeomenetes pugio* in a life-cycle toxicity test. UAS American Society for Testing and Materials, Philadelphia
US EPA (1997) Special Report on Environmental Endocrine Disruption: An effects assessment and analysis. EPA, Washington (Report No EPA/630/R-96/012)

US EPA (1998) Endocrine Disrupter Screening and Testing Advisory Committee, Final Report. EPA, Washington

Voogt PA, Den Besten PJ, Kusters GCM, Messing MWJ (1987) Effects of cadmium and zinc on steroid metabolism and steroid level in the sea star *Asteria rubens* L. Comp Biochem Phys C 86:83–89

Vos JG, Dybing E, Greim HA, Ladefoged O, Lambre C, Tarazona JV, Brandt I, Vethaak AD (2000) Health effects of endocrine-disrupting chemicals on wildlife, with special reference to the European situation. Crit Rev Toxicol 30:71–133

Wiberg K, Letcher RJ, Dandau CD, Norstrom RJ, Tysklind M, Bidleman TF (2000) The enantioselective bioaccumulation of chiral chlordane and alpha-HCH contaminants in the polar bear food chain. Environ Sci Technol 34:2668–2674

Wiig O, Derocher AE, Cronin, Skaare JU (1998) Female hermaphrodite polar bears at Svalbard. J Wildlife Dis 34:792–796

Wilson, B, Arnold H, Bearzi G, Fortuna CM, Gaspar R, Ingram S, Liret C, Pribanic S, Read AJ, Ridoux V, Schneider K, Urian KW, Wells RS, Wood C, Thompson PM, Hammond PS (1999) Epidermal diseases in bottlenose dolphins: Impacts of natural and anthropogenic factors. Proc Roy Soc Lond B Bio 266:1077–1083

Chapter 14

Chemistry of Organic Toxicants in Marine Environment

V. S. Petrosyan

The problems of environmental pollution with the persistent organic compounds are of great importance both on national and international levels. Many of the persistent organic pollutants (POPs) are toxic, and this is why their chemistry and toxicology in different environmental media (aquatic ecosystems, soils, atmosphere, biota, etc.) are under the detailed investigations.

In this paper, some of the important POPs entering the marine environment, their properties and behaviour in various compartments of the marine ecosystems (water, sediments), the problems of qualitative and quantitative assessments, and the ways of detoxification of POPs by means of naturally occurring humic substances will be considered.

First of all, it is important to mention the most toxic POPs, the polychlorinated dibenzodioxins (PCDDs, *I*) and dibenzofuranes (PCDFs, *II*), which are today widely known as "dioxins" (Fig. 14.1).

The congeners (isomers) of these important ecotoxicants occupy the first thirteen positions in the World Health Organization's (WHO) list of the most toxic organic compounds. Four of these congeners (*III–VI*) are given in Fig. 14.2.

Unfortunately, these days there are many sources of PCDDs and PCDFs in the environment: incineration of wastes, containing various types of organochlorine compounds, polyvinylchloride (PVC, *VII*, Fig. 14.3) and related products, organochlorine pesticides (like timber preservative 2,4,5-trichlorophenol (*VIII*) and defoliant 2,4,5-T (*IX*)) (see Fig. 14.4).

In the following discussion, we will see that substantial amounts of "dioxins" appear in the marine environment, negatively affecting the aquatic biota.

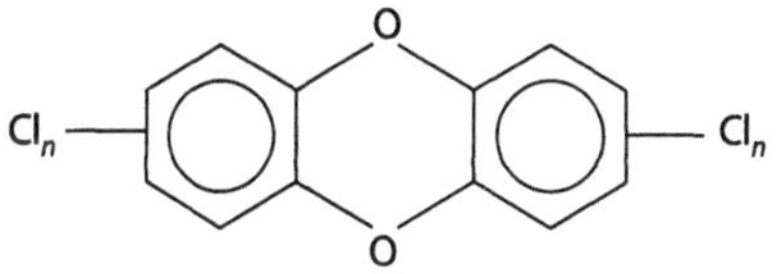

PCDDs (I)

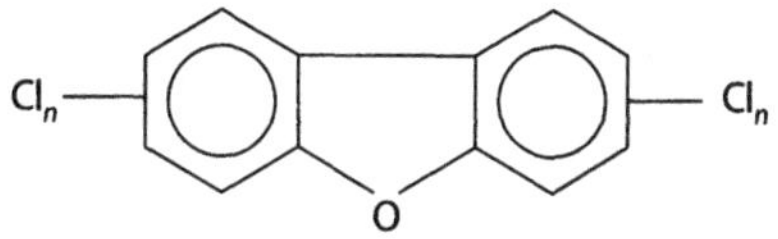

PCDFs (II)

Fig. 14.1. Persistant organic pollutants (POPs). *(I)* Polychlorinated dibenzodioxins (PCDDs), *(II)* polychlorinated dibenzofuranes (PCDFs)

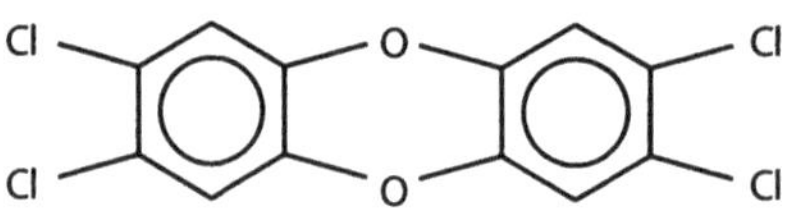

2,3,7,8-TCDD (III)

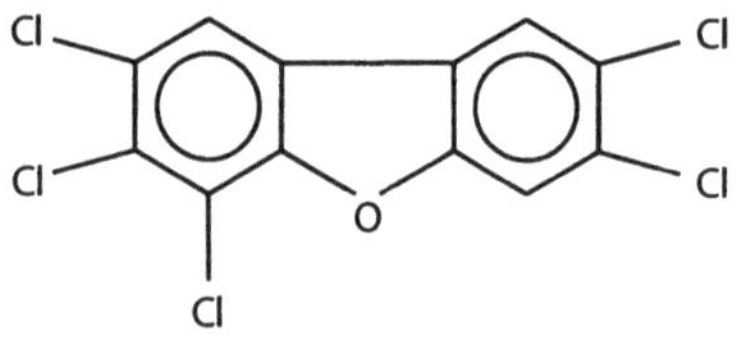

2,3,4,7,8-PeCDF (IV)

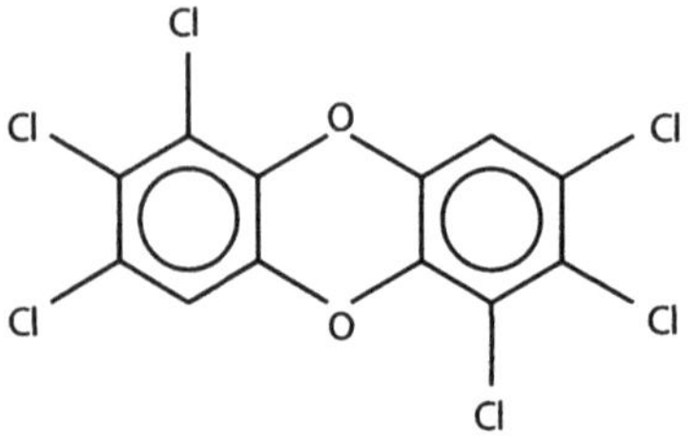

1,2,3,6,7,8-HxCDD (V)

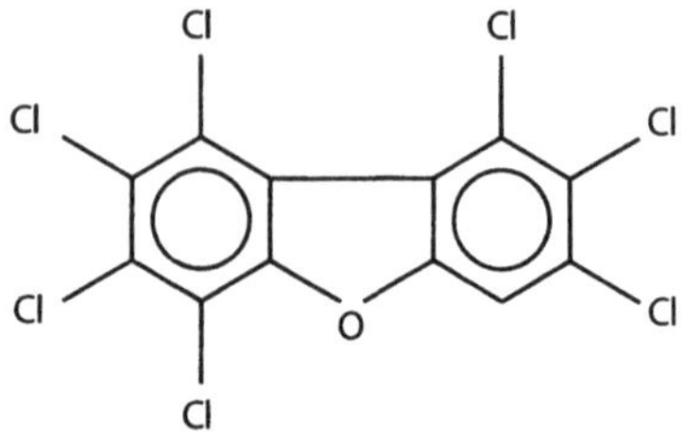

1,2,3,4,7,8,9-HpCDF (VI)

Fig. 14.2. Congeners (isomers) of PCDDs and PCDFs

Fig. 14.3. Polyvinylchloride (PVC)

$$-(-CH_2-CHCl-)_n-$$

(VII)

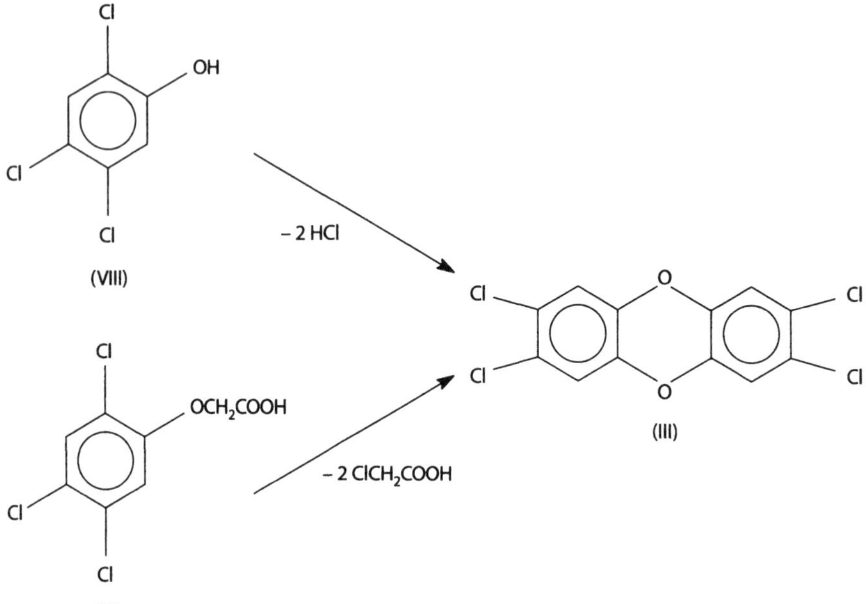

Fig. 14.4. Organochlorine pesticides: 2,4,5-trichlorophenol *(VIII)* and 2,4,5-T *(IX)*

Another important class of POPs are polychlorinated benzenes (*X*) and biphenyls (PCBs, *XI*) (Fig. 14.5), which are widely used as the liquid dielectrics, lubricants, high temperature heat transferrals and hydraulic liquids. Some of the PCBs are toxic (for example, for 2,3',4,4',5-Cl_5 biphenyl, the toxicity equivalent according to the WHO's international system is 0.0001), but most of them are dangerous for the environment, mainly as the precursors for the relevant dioxins and furans.

Organochlorine pesticides like DDT (*XII*) (plus its metabolite DDE, *XIII*) and lindane (HCH, *XIV*) (Fig. 14.6) are mostly dangerous for the environment mainly as precursors for the relevant dioxins and furans.

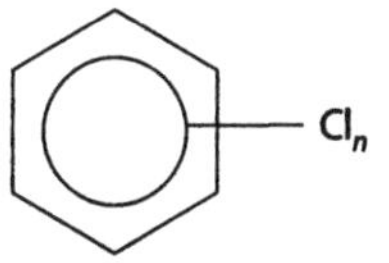

(X)

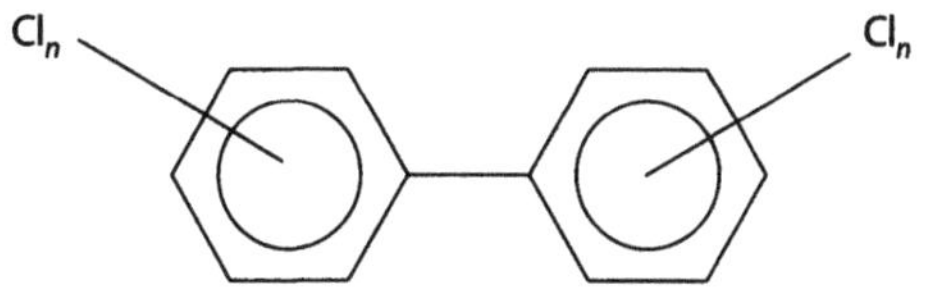

(XI)

Fig. 14.5. Polychlorinated benzenes *(X)* and biphenyls *(XI)*

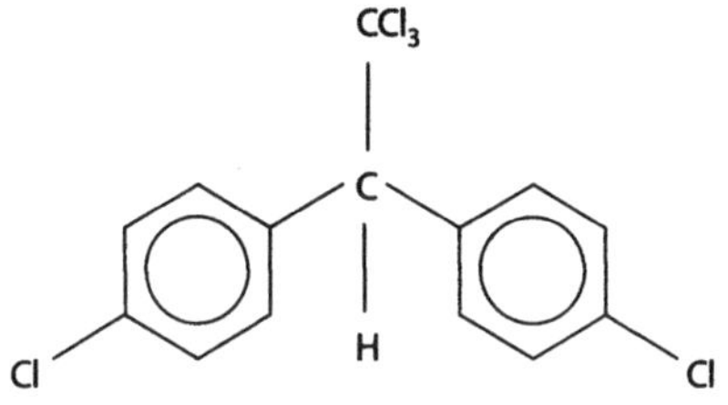

(XII)

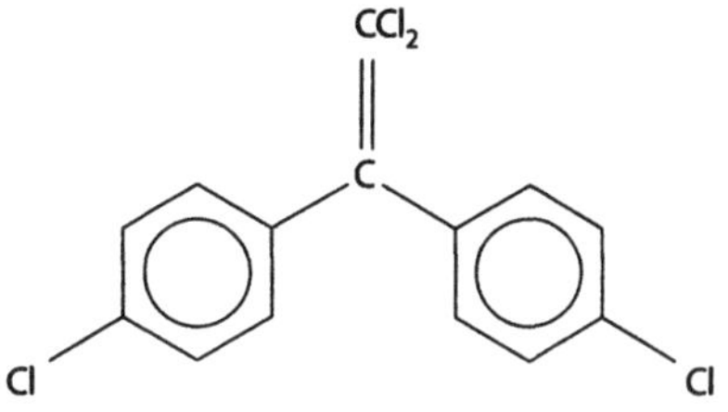

(XIII)

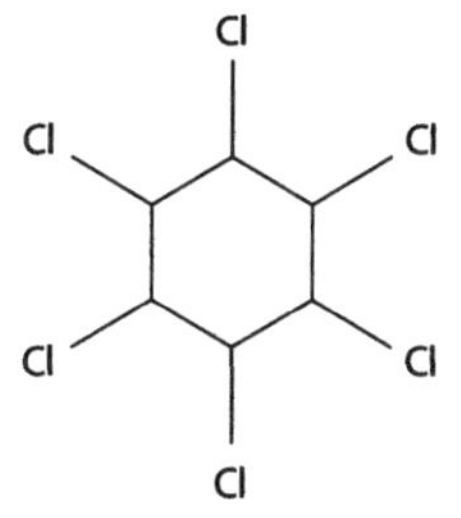

(XIV)

Fig. 14.6. DDT *(XII)*, its metabolite DDE *(XIII)* and lindane *(XIV)*

Usually the models evaluating transport and accumulation of POPs consider the following compartments: air, soil, sea, vegetation and litterfall (Pekar et al. 1998). For example, in the recent joint report (Pekar et al. 1999) of the Meteorological Synthesizing Centre-East and Norwegian Institute for Air Research, it is assumed that initially emitted POPs enter the atmospheric air. The exchange of POPs takes place between gaseous and aerosol phases (Fig. 14.7), while the atmospheric transport is accomplished. Then in the course of wet and dry depositions of gaseous and aerosol phases, POPs enter soil, sea and vegetation. The dry deposition process of the pollutant gas-phase is considered to be reversible, directly allowing one in the modelling course to consider the reemission process, which is important for some pollutants. From vegetation with fallen leaves, POPs enter the litterfall where they sometimes find their way to the soil. The accumulation in compartments decreases due to POPs degradation in the course of chemical, photochemical and biochemical reactions.

The total depositions and deposition densities of PCBs (*XI*), benzo(a)pyrene (*XV*), 2,3,4,7,8-PeCDF (*IV*) and lindane (*XIV*) have been estimated for some of the regional seas using the above-mentioned model (Pekar et al. 1999), and the calculated values for the Mediterranean, Baltic and North seas are given in Tables 14.1 and 14.2.

According to the calculation results, the highest depositions of the POPs considered fell in the area of the Mediterranean Sea. If we consider mean deposition density on the sea basins, then the highest intensity is accounted for on the Baltic Sea. Comparative diagrams of total depositions and their densities on the Mediterranean, Baltic and North Seas are presented in Figs. 14.8–14.11.

The recent analysis (Shatalov et al. 2000) of calculated spatial distribution of selected POPs in 1997 gave the preliminary estimates of contamination levels in different media. The data obtained for the European seas are given in Table 14.3.

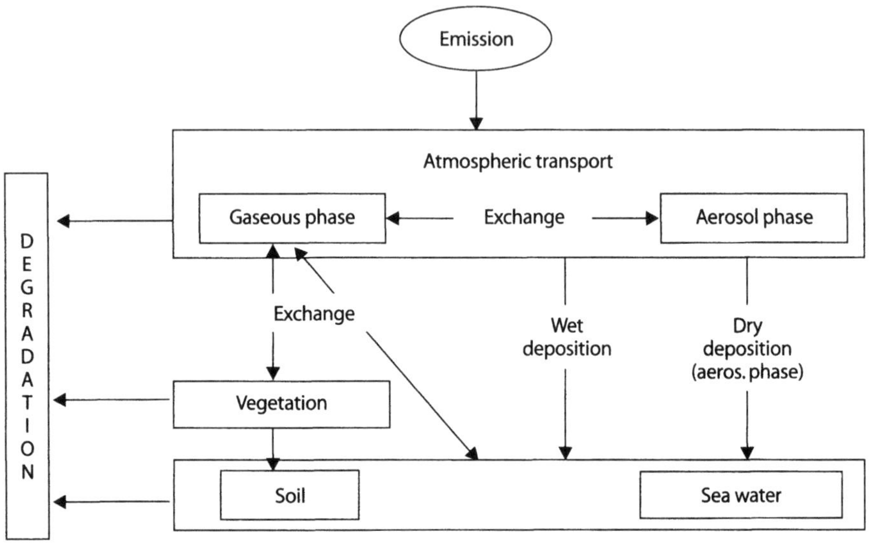

Fig. 14.7. Scheme of POPs exchange between the various environmental compartments

Table 14.1. Total depositions of PCBs, benzo(a)pyrene, 2,3,4,7,8-PeCDF and lindane for 1996

Compounds	Unit	Mediterranean Sea	Baltic Sea	North Sea
PCBs	kg yr^{-1}	1623	715	594
B[a]P	t yr^{-1}	30	7	12
2,3,4,7,8-PeCDF	g I-Teq yr^{-1}	492	65	101
Lindane	t yr^{-1}	56	18	26

Table 14.2. Deposition densities of PCBs, benzo(a)pyrene, 2,3,4,7,8-PeCDF and lindane for 1996

Compounds	Unit	Mediterranean Sea	Baltic Sea	North Sea
PCBs	ng m^{-2} yr^{-1}	472	1843	886
B[a]P	µg m^{-2} yr^{-1}	9	18	18
2,3,4,7,8-PeCDF	pg I-Teq m^{-2} yr^{-1}	143	167	151
Lindane	µg m^{-2} yr^{-1}	16	46	38

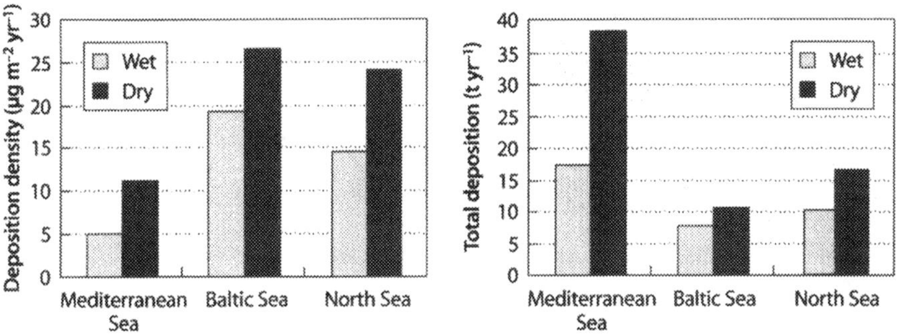

Fig. 14.8. Total depositions (*right*) and deposition densities (*left*) of lindane on regional seas for 1996

Back to 1990–91 during the oceanography expeditions on the research vessel *Moskovsky Universitet*, we were measuring in particular the contamination of the Adriatic, Ionic, Mediterranean, Tirrenian and Black seas with organic pollutants both in waters and sediments (Petrosyan and Gianguzza 1999). In the spring of 1990, the expedition, headed by the author of this paper, started in Casablanca (Marocco) and ended in Sebastopol (U.S.S.R.), and there were 19 stations in the Mediterranean, Ionic, Adriatic and Black Seas (Fig. 14.12).

The data on the contamination of waters with hydrocarbons and phthalates have been obtained by Drs. P. I. Demyanov and A. N. Fedotov (M. V. Lomonosov University, Moscow) by means of high-resolution gas chromatography and presented in Table 14.4. These results show that the three most contaminated aquatic zones were evidently situated next to the estuaries of rivers Rhone (Station *3*), Po (Station *11*) and Danube (Station *19*).

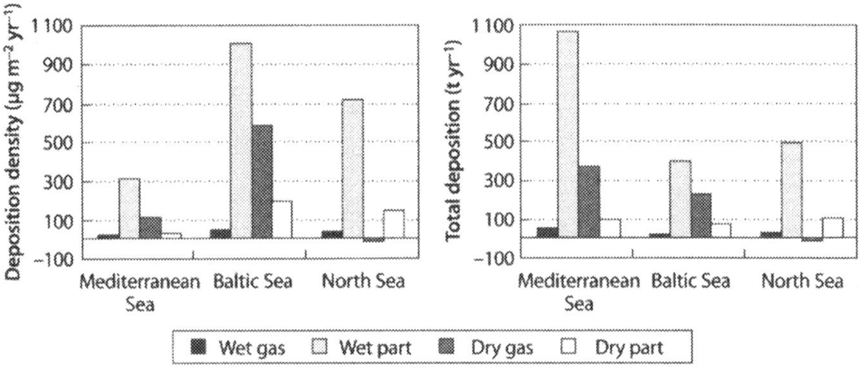

Fig. 14.9. Total depositions (*right*) and deposition densities (*left*) of PCB on regional seas for 1996

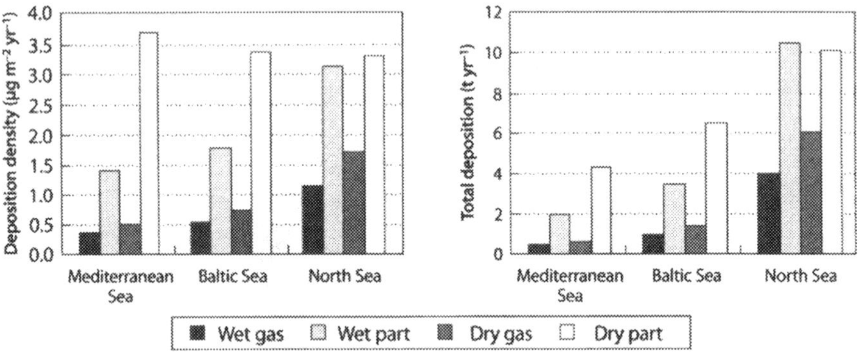

Fig. 14.10. Total depositions (*right*) and deposition densities (*left*) of benzo(a)pyrene on regional seas for 1996

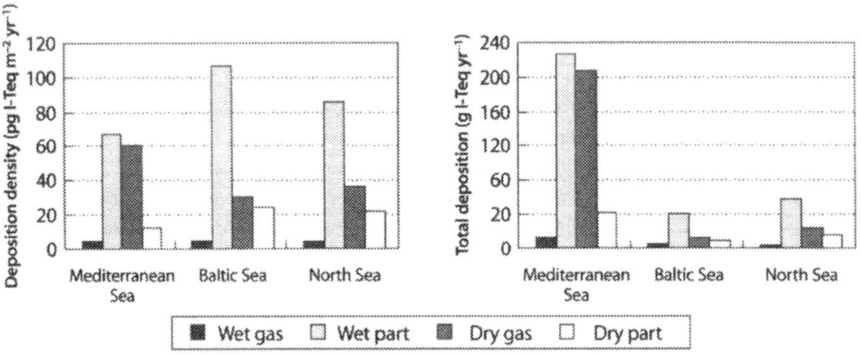

Fig. 14.11. Total depositions (*right*) and deposition densities (*left*) of 2,3,4,7,8-PeCDF on regional seas for 1996

Table 14.3. Concentrations of organic toxicants in European countries (sea water, ng m^{-3}, 1997)

Country	PCB	B[a]P	HCH	Country	PCB	B[a]P	HCH
Albania	32.9	574.7	2668.5	Latvia	66.3	693.8	2487.7
Azerbaidjan	8.3	907.2	123.8	Lithuania	74.8	985.7	2742.7
Belgium	157.4	1228.8	24220.0	Malta	15.4	137.2	2275.8
Bosnia and Herzegovina	46.8	1075.2	3024.1	Macedonia (The FYR)	34.4	651.0	2535.4
Bulgaria	28.3	714.8	1565.8	Moldova	37.7	1191.0	2101.1
Croatia	54.9	870.4	4621.0	Netherlands	150.9	891.9	12927.0
Cyprus	5.0	101.7	389.4	Norway	36.1	147.3	906.3
Denmark	122.7	738.3	3931.5	Poland	116.5	1539.3	3868.4
Estonia	74.2	627.6	2139.3	Portugal	15.3	121.2	2239.7
Finland	59.9	418.3	1497.7	Romania	29.3	773.0	1646.3
France	73.4	472.4	18485.0	Russia	30.9	674.7	853.3
Georgia	24.4	1332.4	368.7	Slovenia	98.0	1004.1	8372.5
Germany	181.9	1149.1	5577.7	Spain	27.5	205.3	5587.2
Greece	14.6	333.9	1405.3	Sweden	62.6	401.4	2183.8
Iceland	8.3	24.5	132.6	Turkey	14.4	370.7	680.8
Ireland	20.7	110.0	1251.7	Ukraine	31.5	1182.4	1459.3
Italy	45.5	462.2	6080.4	U.K.	43.7	356.1	4605.0
Kazakhstan	7.7	296.3	241.6	Yugoslavia	39.0	731.6	2673.9

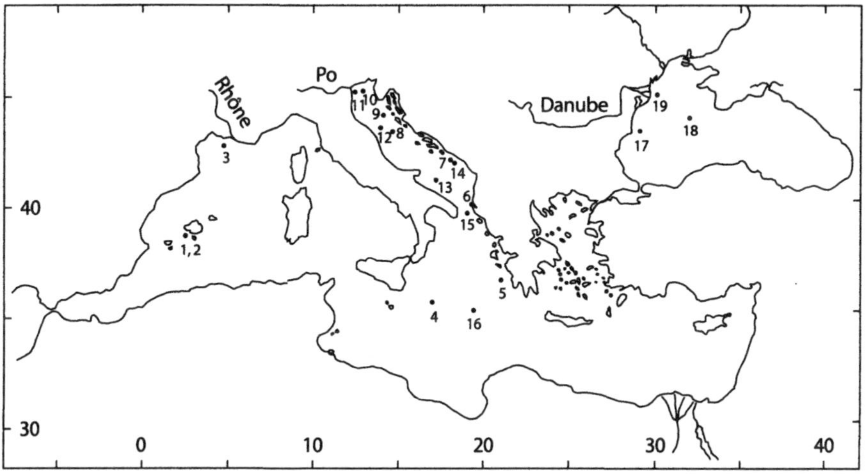

Fig. 14.12. Sampling stations within the 1990 Mediterranean, Ionic, Adriatic, and Black Sea expedition

Table 14.4. Contamination of Mediterranean, Ionic, Adriatic and Black Sea waters with hydrocarbons and phthalates (ng l^{-1})

Organics	No. of stations											
	2	3	4	5	7	10	11	12	13	15	16	19
C-15	10	4	6	5	11	2	18	7		7	7	8
C-16	26	18	12	14	19	13	27	25	12	23	16	10
C-17	12	7	11	8	16	4	298	13		12	10	96
C-18	23	18	15	14	21	16	21	24	13	11	16	7
C-19	9	4	12	5	14	4	28	7		11	5	6
C-20	15	10	18	6	13	10	24	13	12	18	6	3
C-21	11	6	14	5	12	4	18	8	11	11		1
C-22	18	11	18	7	9	12	50	17	25	17	5	2
C-23	15	10	16	8	16	12	61	18	36	17	5	1
C-24	24	16	19	11	19	17	93	36	51	23	6	2
C-25	121	33	37	13	19	19	114	30	59	23	6	8
C-26	58	40	29	12	24	30	136	39	66	23	8	2
C-27	73	51	22	11	23	30	113	39	62	21	8	2
C-28	77	58	52	10	23	41	136	41	60	17	9	2
C-29	89	77	31	10	21	47	113	80	57	16	10	3
C-30	117	76	27	7	23	45	116	39	41	12	10	2
C-31	119	74	21	6	15	46	87	37	33	11	9	1
C-32	110	67	15	4	11	47	77	32	36	8	9	1
C-33	106	56	11		10	43	61	29	23	5	6	1
C-34	89	37	6		6	29	36	20	12	3	6	2
C-35	62	29	4		4	23	32	18	11	3	6	
C-36	49	21	3		5	20	26	12	9	2	4	
C-37	39	14	3		16	16	20	11	8	3	5	
C-38	28	16	4			19	25	14	13		4	
C-39	30	15	4			16	20	11	8	6	5	
C-40	30					13	19	10	6			
Sum of alkenes	1 360	768	410	156	334	578	1 764	620	635	313	181	160
BuPht	66	47	132	62	63	30	7 367	70	56	136	171	2 870
OctPht	773	190	412	130	146	250	554	318	179	231	94	24

The data demonstrate particularly that the petroleum hydrocarbons were mostly in the estuary of River Po and in the area south of Balearic Islands (Station 2). It was interesting to see that the highest concentrations of the individual alkanes have been obtained for the hydrocarbons with 25–33 carbon atoms. In regards to the phthalates, it was very interesting to see that butyl phthalate had been contaminating the waters in the Adriatic and Black seas next to the estuaries of rivers Po and Danube. Contrary to this, octyl phthalate was not found in substantial amounts at these stations, but the highest concentration of this compound has been found in the vicinity of Balearic Islands.

In the spring of 1991, the joint Russian-Italian expedition headed by Professor A. Gianguzza (University of Palermo) and the author of this paper started in Palermo, and all 33 stations were within the Tirrenian Sea (Fig. 14.13).

The concentrations of organic contaminants in the Tirrenian Sea waters were measured by means of high-resolution gas chromatography (Drs. P. I. Demyanov, A. N. Fedotov, M. V. Lomonosov University, Moscow), and are presented in Table 14.5.

Sea waters with hydrocarbons and phthalates at most of the stations were at the same levels, but the contents of alkanes, dibutyl phthalate and other dialkyl phthalates were much higher at Station *1*. The changes in concentrations of all the compounds that had been found in the Tirrenian Sea waters are shown graphically in Fig. 14.14.

Drs. L. de Domenico and G. Magazzu (Oceanography Institute, CNR, Messina) obtained the data on the contamination of the Tirrenian Sea sediments with polycyclic aromatic hydrocarbons (PAHs), using two different standards (chrysene and crude oil). The results obtained are presented in Table 14.6.

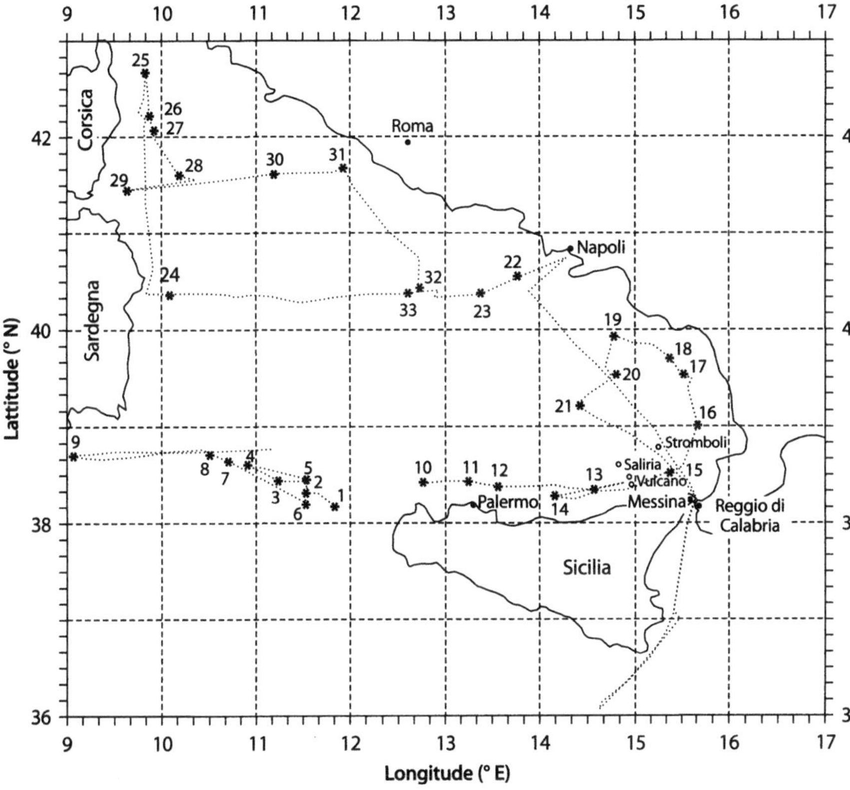

Fig. 14.13. Sampling stations within the 1991 Tirrenian Sea expedition

Table 14.5. Pollution of the Tirrenian sea waters with some of the organic contaminants ($\mu g\, l^{-1}$)

No. of stations	Organics				
	Alkanes	Squalene	DBP[a]	DEHP[b]	DAP[c]
1	2.60	0.40	3.80	0.32	4.70
3	0.84	0.41	0.45	0.31	0.83
13	0.61	0.12	0.08	0.22	0.33
15	0.69	0.43	0.14	0.51	0.73
19	0.74	0.16	0.09	0.15	0.26
22	0.23	0.11	0.05	0.07	0.13
24	0.93	0.13	0.14	0.13	0.38
31	0.49	0.13	0.09	0.09	0.18
33	0.49	0.08	0.09	0.09	0.18

[a] *DBP*: dibutylphthalate.
[b] *DEHP*: di(ethyl)hexylphthalate.
[c] *DAP*: dialkylphthalates.

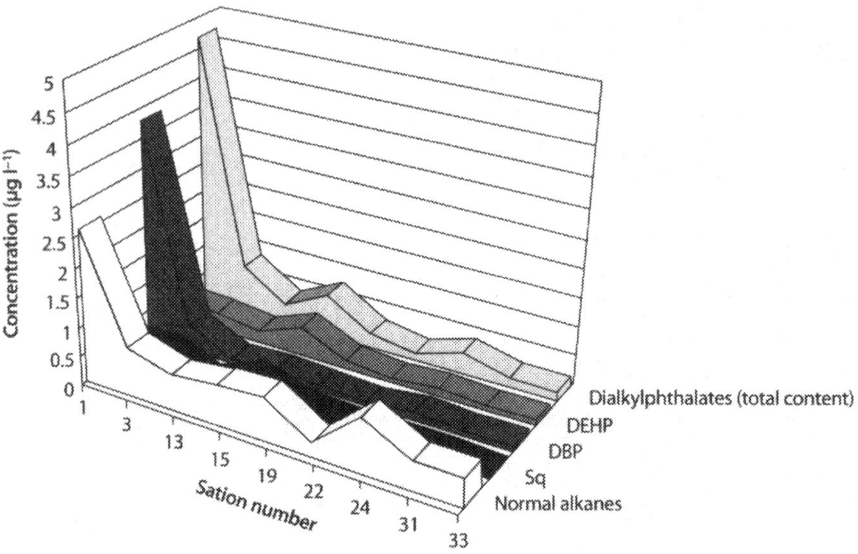

Fig. 14.14. The concentrations of organic contaminants in the Tirrenian Sea waters

It is evident from the results of Table 14.6 that the data obtained with two different standards are quite different: the values in the last column (which are in our opinion more reliable) are 7–8 times higher than the figures from the middle column.

Table 14.6. Contamination of the Tirrenian sea sediments with polycyclic aromatic compounds (μg l^{-1})

No. of stations	Content of PACs	
	With chrysene (as a standard)	With crude oil (as a standard)
10	0.05	0.38
11	0.15	1.10
12	0.22	1.64
13	0.06	0.43
14	0.10	0.78
15	0.06	0.45
16	0.13	0.96
17(1s)	0.08	0.54
17(2s)	4.87	36.13
18	0.69	4.93
19(1s)	0.13	0.86
19(2s)	0.08	0.65
20(1s)	0.07	0.47
20(2s)	0.05	0.40
20(3s)	0.13	1.04
30	1.06	7.22

We have shown (Petrosyan et al. 1994) that the ecotoxicological effects of various types of contaminants are diminished by means of their binding by humic substances (humic and fulvic acids) contained in the aquatic ecosystems. Particularly, the humic substances detoxify different PAHs (Perminova et al. 1999), increasing the binding and detoxification constants in the same series (e.g. anthracene < fluoranthene < pyrene).

References

Pekar M, Gusev A, Pavlova N, Strukov B, Erdman L, Ilyin I, Dutchak S (1998) Long-range transport of selected POPs: Development of transport models. EMEP/MSC-E Report 2/1998, Part 1

Pekar M, Pavlova N, Gusev A, Shatalov V, Vulikh N, Ioannisian D, Dutchak S, Berg T, Hjellbrekke A-G (1999) Long-range transport of selected POPs: Development of transport models. EMEP/MSC-E and CCC Report 4/1999

Perminova IV, Grechishcheva NY, Petrosyan VS (1999) Relationships between structure and binding affinity of humic substances for polycyclic aromatic hydrocarbons: Relevance of molecular descriptors. Environ Sci Technol 33:3781–3789

Petrosyan V, Gianguzza A (1999) Contamination of Adriatic, Ionic, Mediterranean, Tirrenian and Black Seas with heavy metals and organic pollutants. Proceedings of 2nd Symposium on the "Protection of groundwater from pollution and seawater intrusion". Bari, September 27–October 1, (1999)

Petrosyan VS, Perminova IV, Danchenko NN, Yashchenko NY, Lebedeva GF, Kovalevsky DV, Kulikova NA, Filippova OI, Venedictov PS, Polynov VA (1994) Detoxification of heavy metals, polyaromatic hydrocarbons and pesticides by humic substances in waters and soils. Proceedings of the International Congress "Water: Ecology and technology". Moscow, pp 1136–1142

Shatalov V, Malanichev A, Berg T, Larsen R (2000) Investigation and assessment of POP transboundary transport and accumulation in different media. EMEP/MSC-E and CCC Report 4/2000, Part I

Chapter 15

Toxic Effects of Organometallic Compounds towards Marine Biota

L. Pellerito · R. Barbieri · R. Di Stefano · M. Scopelliti · C. Pellerito · T. Fiore · F. Triolo

15.1
Organometallic Derivatives

Organometallic derivatives are compounds containing a direct σ or π carbon metal linkage. Furthermore, the concept of the metallic atom must be extended to all the elements that are less negative than the carbon atom. As a consequence, taking into account all elements that are less negative than carbon and the number of existing organic compounds, it is possible to synthesize millions of organometallic derivatives. Several of these are extensively used in organic syntheses; others may find application in agriculture and in many other fields as pesticides, fire retardants, wood preservatives, antifouling agents, etc. In general, the organic derivatives of the metals are more toxic than the parent inorganic metal, with the alkyl derivatives bearing greater toxicity than the aryl ones. Furthermore, the mechanism of toxicity depends on the co-ordination atoms present in the attached biological molecule.

This review will focus on a limited number of organometallic compounds, i.e. those whose presence and toxicity have been evidenced in marine biota.

15.2
Organoarsenic

15.2.1
Organoarsenic Derivatives

Arsenic occurs mainly as inorganic arsenate in sea water (2–3 μg dm^{-3}) and in marine biota (up to 100 μg kg^{-1} of wet weight), as arsenobetaine in marine animals (Fig. 15.1) and as dimethylarsinylribosides (Fig. 15.1) in marine algae, in which arsenobetaine is absent. The arsenical derivatives are mainly used as industrial chemicals (wood preservatives), agricultural chemicals (herbicides and desiccants), fining agents in glass manufacturing, metallic arsenic in non-ferrous alloys, high-purity metallic arsenic for the electronics industry, and finally a small amount of arsenic is used as additive in animal food (Ishiguro 1992).

15.2.2
Biotransformation of Arsenic

The biotransformation of arsenic in marine ecosystems has been reported in several recent papers (Kaise and Fukui 1992; Francesconi et al. 1992; Garman et al. 1993). Owing to the fact that the phosphate/arsenate concentrations may be close to equimolar

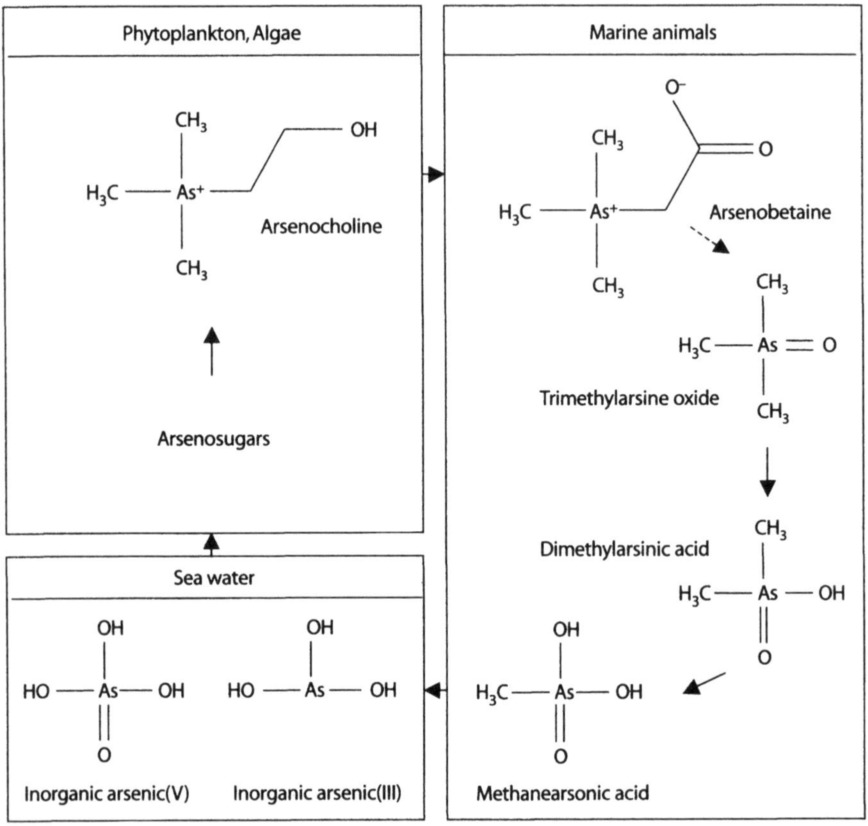

Fig. 15.1. A tentative arsenic cycle in marine ecosystems

in the coastal boundary zone, the discrimination between arsenate and phosphate by phytoplankton is relatively poor, the concentrations differing only by a factor two to ten. As a consequence, arsenic is accumulated in marine organisms mainly in the form of organoarsenic compounds (Hanaoka et al. 1992a), in particular arsenobetaine, dimethylarsinic acid, methanearsonic acid, etc. (Fig. 15.1), which may be interconverted according to the following tentative scheme.

15.2.3 Organoarsenic in Marine Biota

Water-soluble arsenic compounds in several bivalves (*Mytilus coruscum, Meretrix lusoria, Tapes japonica, Tresus keenae,* etc.) (Shibata and Morita 1992) were extracted with 1:1 methanol/water and analysed with a high performance liquid chromatograph, combined with an inductively coupled argon plasma mass spectrometer. The results obtained are reported in Table 15.1.

The study evidenced that *Meretrix lusoria, Tapes japonica* and *Tresus keenae* contain not only arsenobetaine, but also the arsenosugar derivative reported in Fig. 15.2a, while *Mytilus edulis, Mytilus coruscum, Cassostrea gigas, Anadara broughtonii,* and *Corbicula japonica* contain either one or both of the arsenosugars of Fig. 15.2a,b. Since these bivalves are plankton-feeders, one can hypothesize that algae are one of the possible sources of the above-mentioned arsenosugars, while arsenobetaine could be er-

Table 15.1. Arsenic species in bivalves, adapted from Shibata and Morita (1992)

Sample (Japanese name)	Arsenic concentration (μg As g^{-1} fresh tissue)[a] Arsenobetaine	Tetramethyl-arsonium	Arsenosugar Fig. 15.2a	Arsenosugar Fig. 15.2b	Others[e]
Meretrix lusoria (Hamaguri)					
Whole 1	0.78	0.17	0.92	0.1	–
Whole 2	0.33	0.25	0.17	0.03	0.57
Adductor muscle	2.06	0.24	0.57	–	0.34
Foot	1.82	0.46	0.65	–	0.16
Digestive gland[b]	1.39	2.07	1.58	–	–
Mantle	1.03	1.35	1.12	0.1	–
Mantle edge	0.26	1.34	1.14	0.1	–
Gill	0.14	6.07	2.44	–	–
Tapes japonica (Asari)					
Whole 1	0.63	–	0.67	0.07	0.57
Whole 2	0.75	–	0.70	0.07	0.59
Whole 3	0.49	–	0.46	0.05	0.48
Whole 4	0.73	–	1.48	0.15	0.73
Corbicula japonica (Yamatosijimi)					
Mix 1[c]	–	–	0.54	0.14	0.68
Mix 2[d]	–	–	0.53	0.22	0.06
Anadara broughtonii (Akagai)					
Whole	1.03	–	0.11	–	0.05
Tresus keenae (Mirukui)					
Whole	0.57	0.03	0.16	–	0.26
Spisula sachalinensis (Hokkigai)					
Whole	0.66	–	–	–	0.15

Table 15.1. *Continued*

Sample (Japanese name)	Arsenic concentration (μg As g^{-1} fresh tissue)[a]				
	Arsenobetaine	Tetramethyl-arsonium	Arsenosugar		Others[e]
			Fig. 15.2a	Fig. 15.2b	
Mytilus coruscum (Igai)					
Adductor muscle	2.57	–	0.01	–	0.02
Foot	0.81	–	0.05	–	0.11
Digestive gland	1.35	0.06	0.13	0.03	0.18
Remaining part of the body	1.41	0.02	0.06	–	0.12
Mantle	0.90	–	0.03	–	0.15
Mantle edge	1.36	0.04	0.03	–	0.09
Gill	0.93	0.06	0.08	–	0.05

[a] Column, Asahipak GS220; buffer, 25 mmol dm^{-3} tetramethylammonium, 25 mmol dm^{-3} malonic acid (pH 6.8 adjusted by NH_3).
[b] –, not detected (detection limit is less than 0.01 μg As g^{-1} fresh tissue for any species).
[c] Including soft tissues surrounding digestive gland.
[d] Including whole tissues of three bivalves.
[e] Others: I: AsO_4^{3-}; II: AsO_3^{3-}; III: $CH_3AsO_2^{2-}$; IV: $(CH_3)_2AsO_2^-$; V: $(CH_3)_3AsO$; VI: $(CH_3)_4As^+CH_2CH_2OH$; VII: $(CH_3)_2As(O)CH_2CH_2OH$.

roneously accumulated by mussels because of its chemical similarity with the glycinebetaine used by the same organisms for osmo-regulation. The absence of arsenobetaine in *Corbicula japonica* could be due either to a lower amount of arsenobetaine in the food or to a lower amount of osmo-regulator necessary for *Corbicula japonica*, a mussel living in low-salinity regions. It has been demonstrated in vitro that microorganisms occurring in sediments induce the formation of arsenobetaine from arsenocholine (Hanaoka et al. 1992b). In particular, two or three metabolites, Fig. 15.3a,b, have been isolated from two different culture media (1/5 ZoBell 2216E and an aqueous solution of inorganic salts, respectively), after the addition of synthetic arsenocholine to 1 g of the sediment and incubation at 25 °C in the dark.

The metabolites have been identified by high performance liquid chromatography, thin layer chromatography, FAB mass spectrometry, and a combination of gas chromatography and selected-ion monitoring mass spectrometry. The metabolites were structurally identified as arsenobetaine, trimethylarsine and dimethylarsinic acid, which led the AA to conclude that the cycle of Fig. 15.1 could only be carried out by the microorganisms.

An in vitro investigation of the chemical form and acute toxicity of arsenic compounds in sixty specimens of marine organisms has been carried out by Kaise and Fukui (1992). The chemical form of arsenic compounds in Demospongia, Coelenterata, Echinodermata, Mollusca, herbivorous and carnivorous Conches, plankton feeder Bivalvia, herbivorous, carnivorous and plankton feeder fish (*Squalus brevirostris* and *Mustelus manazo*), Crustacea and seaweed (Phaeophyceae (*Laminaria japonica*,

Fig. 15.2. Arsenic-containing ribofuranosides, adapted from Shibata and Morita (1992); Edmonds et al. (1997); Francesconi et al. (1999)

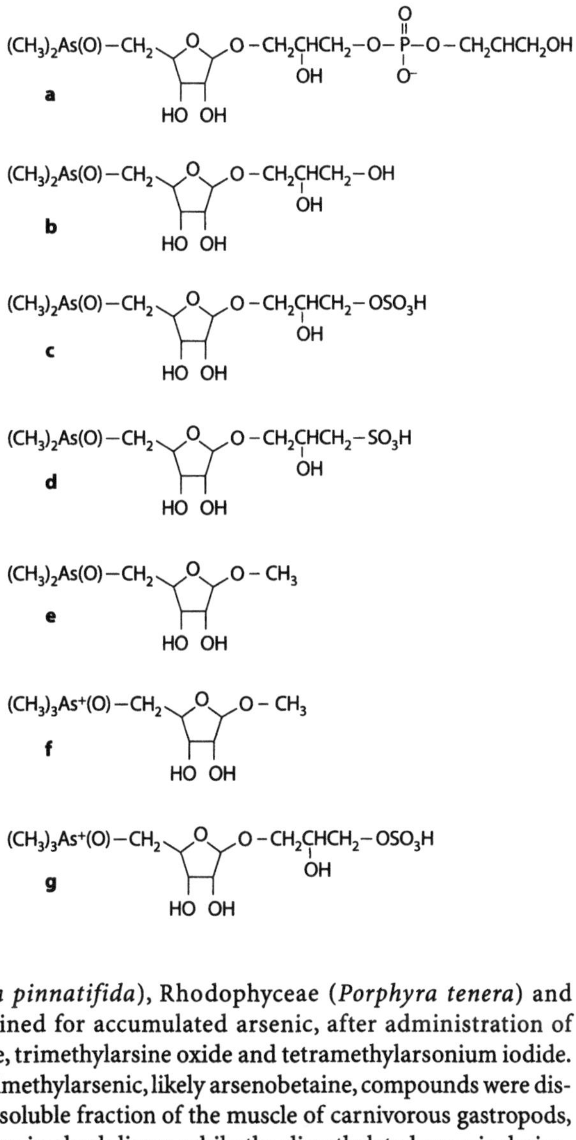

Hizikia fusiforme, *Undaria pinnatifida*), Rhodophyceae (*Porphyra tenera*) and Chlorophyceae) were examined for accumulated arsenic, after administration of arsenobetaine, arsenocholine, trimethylarsine oxide and tetramethylarsonium iodide.

The results showed that trimethylarsenic, likely arsenobetaine, compounds were distributed mainly in the water-soluble fraction of the muscle of carnivorous gastropods, crustaceans and fish as well as in shark livers, while the dimethylated arsenic derivatives were present in the water-soluble fraction of Phaeophyceae (Kaise and Fukui 1992).

The appearance of methylated arsenic derivatives, in particular monomethylarsenic and dimethylarsenic, in the overlying waters from macroalgae, *Ascophyllum nodosum*, was monitored using a coupled hydride generation/GC AA analytical technique (Millward et al. 1993).

In Table 15.2 the seasonal variation of the methylated arsenic species in *Ascophyllum nodosum*, in phytoplankton and in sediment pore waters are reported.

The horizontal distribution of As(V), As(III) monomethylarsonic acid and dimethylarsinic acid have been determined (Santosa et al. 1994) in order to investi-

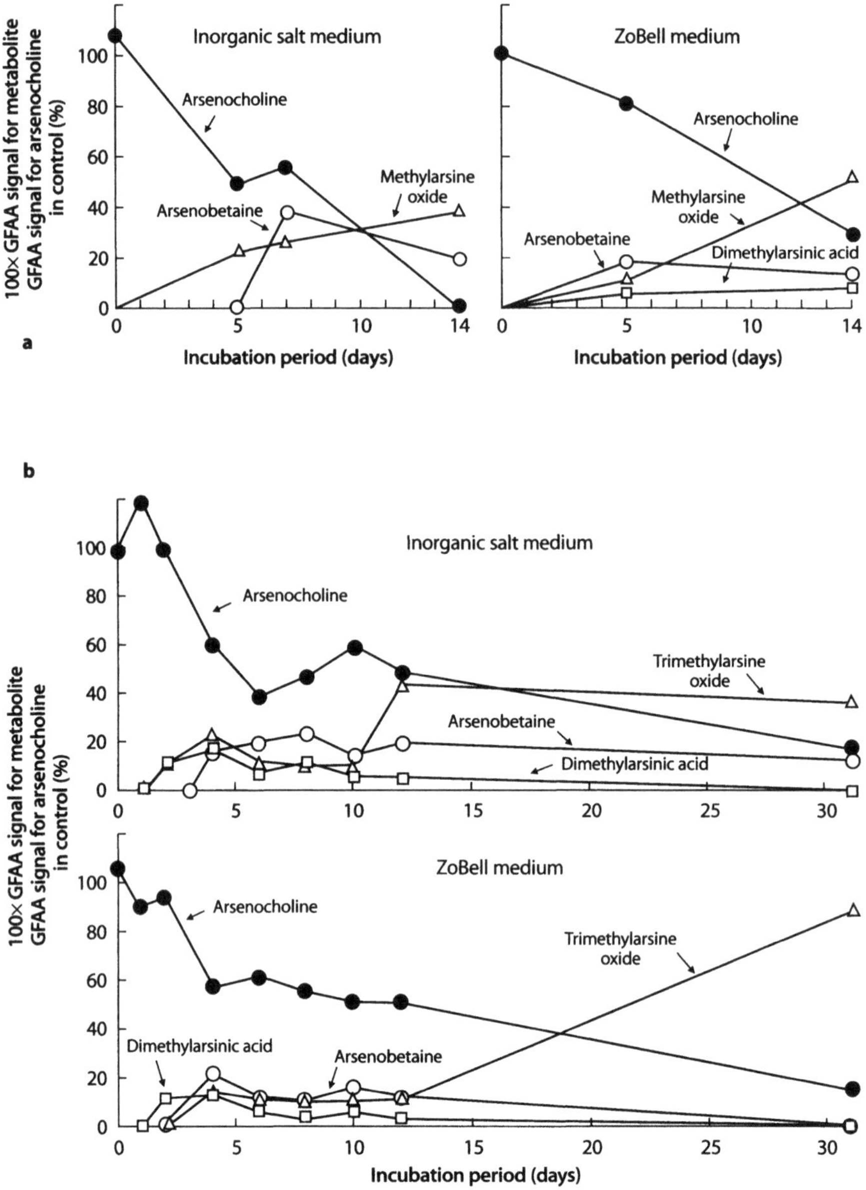

Fig. 15.3. Conversion of arsenocholine into metabolites during aerobic incubation at 25 °C in an inorganic medium and a ZoBell medium added to the sediment collected; **a** in May 1990; **b** in July 1990, adapted from Hanaoka et al. (1992b)

gate their distribution and the cycle of the arsenic derivatives in the marine environment. In particular, during the cruises of the *Shirase* and *Hakuho-maru* ships as well as of the tanker *Japan Violet*, the Indian Pacific Oceanic (Santosa et al. 1994) surface was being investigated, while vertical distribution was being determined in the west Pacific Ocean. The concentrations of the biologically produced species, namely monomethylarsonic acid and dimethylarsinic acid, were extremely low in the Antarctic and southwest Pacific waters (8 and 22 ng dm^{-3}, and 12 and 25 ng dm^{-3}, respectively). In all the other regions (East Indian Ocean, northwestern Pacific Ocean), the concentrations of monomethylarsonic acid (MMAA) and dimethylarsinic acid (DMAA) were very similar to each other (Fig. 15.4).

Three more cruises were performed by the same authors in the 1993–95 period, extending the determinations to inorganic and organometallic germanium derivatives (Table 15.3) (Santosa et al. 1997).

Polyphysa peniculus, a marine unicellular alga, is capable of metabolize arsenate, arsenite and monomethylarsonate, yielding as excreted metabolites, respectively, arsenite and dimethylarsenite, dimethylarsenite, trace amounts of monomethylarsonate,

Table 15.2. Seasonal variation of methylated arsenic species in waters (lower Tamar Estuary), adapted from Millward et al. (1993)

Source	DMA (µg dm^{-3})		MMA (µg dm^{-3})	
	Summer	Winter	Summer	Winter
Macroalgae	0.08	0.01	0.01	<0.01
Phytoplankton	0.04	0	0	0
Sediment porewaters	<0.01	<0.01	<0.01	0.03

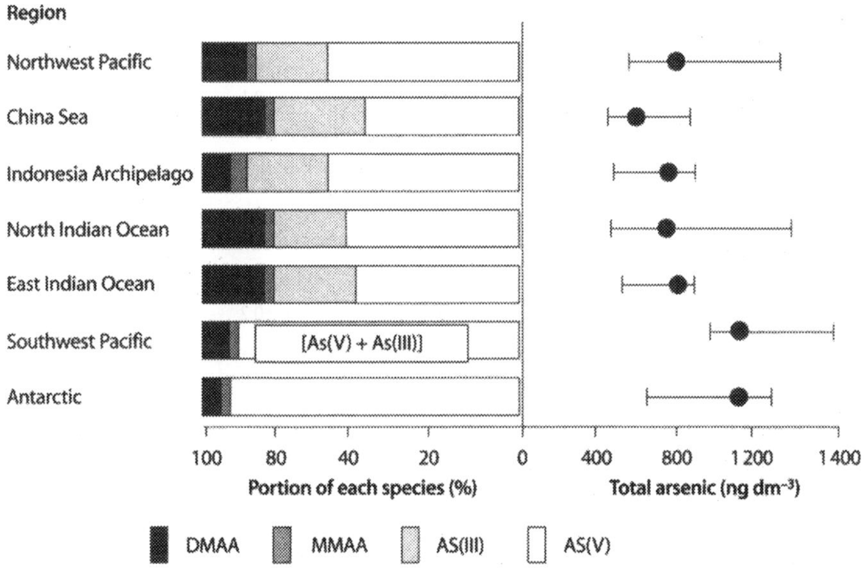

Fig. 15.4. Total dissolved arsenic and the percentage of each species in a variety of ocean surface waters, adapted from Santosa et al. (1994)

Table 15.3. Average concentration of monomethylarsonic acid (MMAA), dimethylarsinic acid (DMAA), monomethylgermanium (MMGe) in the surface zone (< 100 m) of three sampling stations in the Pacific Ocean, adapted from Santosa et al. (1997)

Sampling station	Arsenic Species ($ng\ dm^{-3}$)			Germanium species ($ng\ dm^{-3}$)	
	Inorg. As	MMAA	DMAA	Inorg. Ge	MMGe
North-west Pacific (LM-6)	1483	11.8	47.7	2.5	15.9
Central north Pacific gyre (C)	1110	15.9	184.5	2.0	16.5
Central Pacific equatorial Region (D2)	1450	12.4	53.3	<0.7	15.8

Table 15.4. Arsenic distribution in aqueous extracts of cells determined by HG GC AA ($\mu g\ g^{-1}$, dry weight), adapted from Cullen et al. (1994)

Arsenic exposed	Arsenic species founds in cells	Cells treated with 10 ppm arsenicals		Cell treated with 0.9 ppm arsenicals	
		A[a]	B[b]	A[a]	B[b]
Arsenate	Arsenate	12.8 ±0.8	9.7 ±0.7	0	5.1 ±0.3
	Arsenite	25.1 ±1.3	0	26.3 ±1.6	0
	MMAA	0	0	0	0
	DMAA	1.9 ±0.2	0	21.7 ±2.0	0
	Total	39.8	9.7	48.0	5.1
Arsenite	Arsenate	8.0 ±0.5	0	1.8 ±0.1	0
	Arsenite	37.7 ±2.5	2.3 ±0.2	6.5 ±0.4	1.2 ±0.1
	MMAA	3.1 ±0.2	0.4 ±0.03	1.0 ±0.1	0
	DMAA	0.8 ±0.1	0.8 ±0.08	5.4 ±0.5	0.8 ±0.1
	Total	49.6	3.5	14.7	2.0
MMAA	Arsenate	0	0.3 ±0.03	0	0
	Arsenite	0	0.9 ±0.07	0	0
	MMAA	5.3 ±0.3	0.8 ±0.06	1.4 ±0.1	0
	DMAA	6.5 ±0.6	2.7 ±0.3	1.6 ±0.2	0
	Total	11.8	4.7	3.0	0
MMAA	Arsenate	1.6 ±0.1	0	0	0
	Arsenite	0	0	0	0
	MMAA	0.6 ±0.04	0.1 ±0.01	0.7 ±0.04	0
	DMAA	28.1 ±2.5	3.6 ±0.3	20.4 ±1.8	2.5 ±0.3
	Total	30.3	3.7	21.1	2.5

[a] Seven days after the cells were exposed to arsenicals.
[b] Seven days after the cells were transferred to fresh media.

and finally dimethylarsinate. *Polyphysa peniculus* did not metabolize dimethylarsinic acid when it was used as substrate (Cullen et al. 1994) (Table 15.4).

The data reported in Table 15.4 strongly support the biomethylation pathway of Fig. 15.5 for the microbial process.

According to the proposed model (Cullen et al. 1994), the arsenate is taken up from the medium, via the phosphate transport system, into the algal cells. The arsenate is reduced to arsenite inside the cells by thiols and/or dithiols, and the cells excrete most of the arsenite into the growth medium by means of an active transport system.

Subsequently, methylation of the endocellular arsenite to MMAA occurs by using SAM; because of its low passive diffusion coefficient, the endocellular MMAA is not excreted in the growth medium, rather it remains in the cells, where it is reduced and further methylated to DMAA. Possessing a greater diffusion coefficient, DMAA passively diffuses into the growth medium.

According to an in vitro investigation of the blue mussel *Mytilus edulis* carried out by Gailer et al. (1995) exposure to arsenite, arsenate, methylarsonic acid, dimethylarsinic acid, arsenobetaine, arsenocholine, trimethylarsine oxide, tetramethylarsonium iodide or dimethyl-(-hydroxyethyl)arsine oxide (100 μg As dm^{-3} in sea water, for 10 days), results in conversion of arsenobetaine and arsenocholine to trimethylarsine oxide,

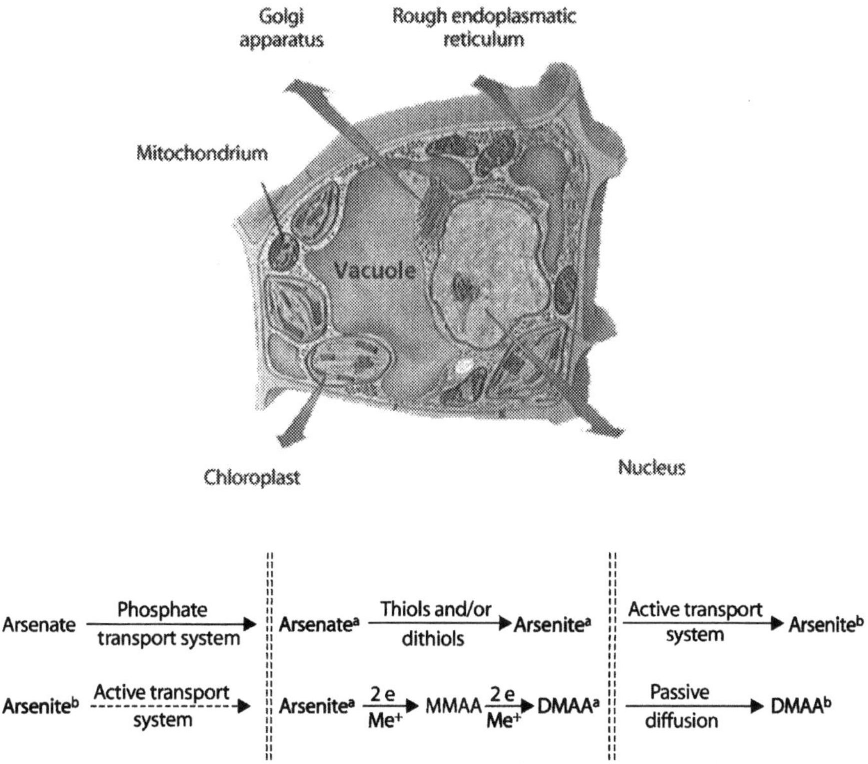

Fig. 15.5. Proposed model for biomethylation of arsenate in marine alga *Polyphysa peniculus*, adapted from Cullen et al. (1994)

whereas trimethylarsine oxide and tetramethylarsonium iodide remain unaltered. None of the other arsenic derivatives are significantly accumulated. Furthermore, arsenobetaine, arsenocholine and tetramethylarsonium iodide are accumulated by the mussel, although arsenobetaine is accumulated more efficiently. Finally, following exposure to arsenobetaine or arsenocholine, the accumulated arsenic is present in the tissue as arsenobetaine (Table 15.5).

Cytotoxicity of arsenobetaine, arsenocholine, trimethylarsine oxide and tetramethylarsonium iodide derivatives, which are contained in fishery products, has been investigated by Kaise et al. (1998) in mammalian cells, see Table 15.6.

Arsenobetaine, arsenocholine and trimethylarsine oxide did not inhibit cell growth (BALB/c 3T3 cells) at a concentration ≤10 mg cm^{-3}. Chromosome aberrations, consist-

Table 15.5. Total arsenic (μg) in 12 mussels at the end of accumulation, adapted from Gailer et al. (1995)

Arsenic compound	After accumulation in whole animal	
	Range	Mean
Control	15.4–38	24.5
Arsenobetaine	456–1201	727
Tetramethylarsonium Iodide	49–113	82
Control	13–41	24
Arsenocholine	99–317	197

Table 15.6. Toxicity of arsenic compounds in experimental animals and cultured cells, adapted from Kaise et al. (1998)

Arsenical	BALB/c T3T cells			Chromosomal aberrations on cultured human fibroblast	
	LD_{50} [a] (g kg^{-1})	IC_{50} [b] (mg cm^{-3})	(%) [c]	Concentration (mg cm^{-3})	SCE [d] (mg cm^{-3} $cell^{-1}$)
Arsenite	0.0345	0.0007	20	0.001	0.001 [e]
Arsenate		0.006	33	0.02	0.02 [e]
Methylarsonic acid	1.8	1.2	37	1	0.5 [e]
Dimethylarsinic acid	1.2	0.32	90	0.5	0.1 [e]
Trimethylarsine oxide	10.6	>10	42	2	1.0 [e]
Arsenobetaine	>10	>10	18	10	1.0 [e]
Arsenocholine	6.5	>10	15	10	1.0 [e]
Tetramethylarsonium	0.9	8	24	10	1.0 [e]
Arsenosugar [f]		2	15	5	1.0 [e]

[a] 50% lethal dose.
[b] 50% growth inhibition.
[c] Percent of aberrant.
[d] Sister chromatid exchange (SCE) metaphase.
[e] Not induced.
[f] See Kaise et al. (1996).

ing in chromatid gaps, chromatid breaks and, rarely, disruption of centromeres, were provoked by arsenite, arsenate, methylarsonic acid and trimethylarsine oxide. Arsenocholine, arsenobetaine and tetramethylarsonium iodide were less toxic than inorganic arsenic (Kaise et al. 1998).

Arsenobetaine may be decomposed into trimethylarsine oxide, dimethylarsinic acid and inorganic arsenic(V) by microorganisms occurring in particles, which are present in deep sea at 1100–3500 m (Hanaoka et al. 1997). The arsenic transformations in short marine food chains were investigated by HPLC-ICP MS (Edmonds et al. 1997). In particular, the organisms studied were the copepod *Gladioferens imparipes* fed on the diatom *Chaetoceros concavicornis* cultured under axenic conditions, the amphipod *Allorchestes compressa*, the Antarctic krill *Euphausia superba*, the abalone *Haliotis roeii*, and finally the teleosts fish silver drummer *Kyphosus sydneyanus*. The experiments were designed to provide information on the ability of marine animals to convert arsenosugars to arsenobetaine, and to determine whether absorption from water is the source of arsenobetaine in herbivorous animals (Table 15.7).

Chaetoceros concavicornis accumulated the arsenosugar of Fig. 15.2c to a concentration dependent on its degree of exposure to inorganic arsenate in sea water; this suggests that the conversion is part of a detoxification process. A high concentration of arsenate was detected in extracts of copepods fed with algal cells that had been grown in water containing 1 mg kg^{-1} of arsenate. Arsenobetaine was absent from the muscle of the silver drummer *Kyphosus sydneyanus*, although trimethylarsine oxide was present (Edmonds et al. 1997).

Arsenobetaine was absent both in the copepods *Gladioferens imparipes* fed only with the diatom *Chaetoceros concavicornis*, grown in axenic culture, and from the muscle of the silver drummer *Kyphosus sydneyanus*, although trimethylarsine oxide was present (Edmonds et al. 1997).

Liver samples of pinnipedes (nine ringed seals (*Phoca hispida*), one bearded seal (*Erginathus barbatus*) and cetaceans (two pilot whales (*Globicephalus melas*), one beluga whale (*Deliphinapterus leucus*)) contained arsenobetaine as the predominant arsenic derivative (0.052–1.67 mg As kg^{-1} wet mass); arsenocholine was also present at 0.005–0.044 mg As kg^{-1} wet mass; dimethylarsinic acid ranged between <0.001 to 0.109 mg As kg^{-1} wet mass and finally methylarsonic acid was below the detection limit (<0.001 to 0.025 mg As kg^{-1} wet mass), (Table 15.8) (Goessler et al. 1998).

As determined by HPLC-GFAA and HPLC-ICP MS, arsenobetaine was the major arsenic compound extracted from the water-soluble fraction of jellyfish *Aurelia aurita* and *Carybdea rastonii* (Fig. 15.6) (Hanaoka et al. 1999a). Arsenocholine and tetramethylarsonium ions were also present.

Animals of higher trophic levels generally do not eat jellyfish. As a consequence, the arsenobetaine that accumulates in the latter may not be passed directly to these higher-level animals. Instead, microorganisms occurring in sediments, suspended substances, macroalgae, etc., could degrade arsenobetaine in a multi-step process to inorganic arsenic, so that the arsenic may circulate in a smaller ecosystem composed of sea water, plankton, small fish and jellyfish rather than in a general ecosystem that includes animals of all trophic levels.

After feeding *Crangon crangon* shrimp with dimethyl- and trimethylarsenosugars, the muscle, midgut gland, gills and the remainder tissue of the crustacean were

Table 15.7. Arsenic compounds detected in various samples and approximate percentage that each compound contributed to the total arsenic in each organism, adapted from Edmonds et al. (1997)

Sample	Arsenic compounds detected[a]
Chaetoceros in normal sea water	Arsenosugar of Fig. 15.2a (2%), arsenosugar of Fig. 15.2c (90%), unknown compounds(5%)
Chaetoceros with elevated arsenic	Arsenosugar of Fig. 15.2c (>99%)
Chaetoceros in artificial (reduced arsenic) sea water	Arsenosugar of Fig. 15.2a (2%), arsenosugar of Fig. 15.2c (60%), unknown compounds(30%)
Copepods fed *Chaetoceros* from normal sea water	Arsenosugar of Fig. 15.2c (70%), trimethylarsine oxide (10%), unknown compounds (>15%)
Copepods fed *Chaetoceros* with elevated arsenic	As(V) (40%), trimethylarsine oxide (25%), arsenosugar of Fig. 15.2c (20%), unknown compounds (>10%)
Copepods fed *Chaetoceros* from artificial sea water	Trimethylarsine oxide (70%), arsenosugar of Fig. 15.2c (20%), unknown compounds (>5%)
Amphipods	Arsenobetaine (60%), arsenosugar of Fig. 15.2a (5%), Fig. 15.2b (7%), Fig. 15.2c (5%), Fig. 15.2d (1%)
Antarctic krill	Arsenobetaine (60%), arsenosugar of Fig. 15.2a (5%), Fig. 15.2b (1%), Fig. 15.2d (1%), DMAA[b] (20%)
Silver drummer muscle	Trimethylarsine oxide (>95%), tetramethylarsonium ion (3%), arsenosugars of Fig. 15.2a and of Fig. 15.2c (1%)
Silver drummer gut 1 (stomach)	Trimethylarsine oxide (65%), arsenosugar of Fig. 15.2a (5%), Fig. 15.2b (2%), Fig. 15.2c (1%), Fig. 15.2d (10%)
Silver drummer gut 2	Trimethylarsine oxide (50%), arsenosugar of Fig. 15.2a (10%), Fig. 15.2b (15%), Fig. 15.2c (1%), Fig. 15.2d (15%)
Silver drummer gut 3	Trimethylarsine oxide (90%), arsenosugar of Fig. 15.2a (2%), Fig. 15.2b (2%), Fig. 15.2d (2%)
Abalone foot muscle	Arsenobetaine (90%), tetramethylarsonium ion(1%), arsenosugar of Fig. 15.2a (3%), Fig. 15.2b (2%), Fig. 15.2c (1%)
Abalone digestive gland	Arsenobetaine (40%), tetramethylarsonium ion (1%), arsenosugar of Fig. 15.2a (10%), Fig. 15.2b (35%), Fig. 15.2c (<1%), Fig. 15.2d (10%)
Abalone fore-gut contents	Unknown compounds (60%), arsenosugar of Fig. 15.2a (5%), Fig. 15.2b (30%), Fig. 15.2d (2%)
Abalone hind-gut contents (faeces)	Unknown compounds (35%), arsenosugar of Fig. 15.2a (5%), Fig. 15.2b (50%), Fig. 15.2d (5%)
Abalone food (algae)	Unknown compounds (30%), arsenosugar Fig. 15.2a(25%), Fig. 15.2b (10%), Fig. 15.2c(5%), Fig. 15.2d (25%)

[a] Approx. percentages in parentheses
[b] DMAA, dimethylarsinic acid.

analysed by HPLC-ICP MS for their arsenical metabolites, see Fig. 15.2a–g (Francesconi et al. 1999). The aim of the investigation was to ascertain the possible role of the arsenosugars as precursors to arsenobetaine. The obtained results showed that while only 0.9% of dimethylarsenosugar was accumulated by the shrimps, giving dimethyl-

Table 15.8. Concentrations of Organoarsenic compounds in liver samples of whales and seals determined by HPLC-HHPN-ICP-MS, adapted from Goessler et al. (1998)

Species	Concentration of arsenic compounds[a] (mg As kg^{-1} wet mass)			
	Arsenobetaine	Arsenocholine	DMAA	MMAA
Pilot whale	0.887	0.005	0.004	0.003
Pilot whale	0.147	0.005	<0.001	<0.001
Ringed seal	0.337	0.005	0.021	<0.001
Ringed seal	1.67	0.020	0.109	<0.001
Ringed seal	1.02	0.012	0.040	<0.001
Ringed seal	1.46	0.015	0.067	<0.001
Ringed seal	0.764	0.008	0.032	<0.001
Ringed seal	1.17	0.011	0.018	<0.001
Ringed seal	0.702	0.025	0.028	<0.001
Ringed seal	0.417	0.021	0.036	0.001
Ringed seal	0.386	0.044	0.021	<0.001
Ringed seal	0.684	0.018	0.004	0.025
Beluga whale	0.0502	0.002	0.016	0.002
Bearded seal	0.0156	0.016	0.021	0.004

[a] Mean of two different extracts.

arsinate and dimethylarsinoylethanol as major metabolites, about half of the accumulated 4.2% of trimethylarsenosugars were converted to arsenobetaine. Shrimps fed with arsenobetaine retained 57% of the dose (Table 15.9). The main conclusion from these results is that the dimethyl and trimethyl arsenosugars do not represent the major source of betaine for wild *Crangon*.

The investigation of the structure of the lipid-soluble arsenic compound is important for the elucidation of arsenic circulation in the marine ecosystem.

Phosphatidyl arsenocholine and phosphatidyl arsenosugar, -OCR^1 and -OCR^2 being fatty acyl groups (Hanaoka et al. 1999b) (Fig. 15.7) were extracted and identified in several tissues of five demersal sharks, in particular, the starspotted shark *Mustelus manazo*.

In particular, the muscle, kidneys and brains contained alkali-labile arsenic derivatives; the livers, stomaches, hearts and gallbladders contained alkali-stable derivatives; finally, the intestines, skin, dark muscle, spleens and bones contained both types of arsenic compounds, as shown in Fig. 15.8.

Further hydrolysis has been carried out on the above-mentioned derivatives obtained from muscle and liver. The HPLC-ICP-MS analysis showed the occurrence of arsenocholine, suggesting that arsenolecithins were present in the tissues, as shown in Fig. 15.9.

Finally, the presence of dimethylated arsenolipids in the liver was evidenced from the detection of dimethylarsinic acid in liver hydrolysates (Fig. 15.10) (Hanaoka et al. 1999b).

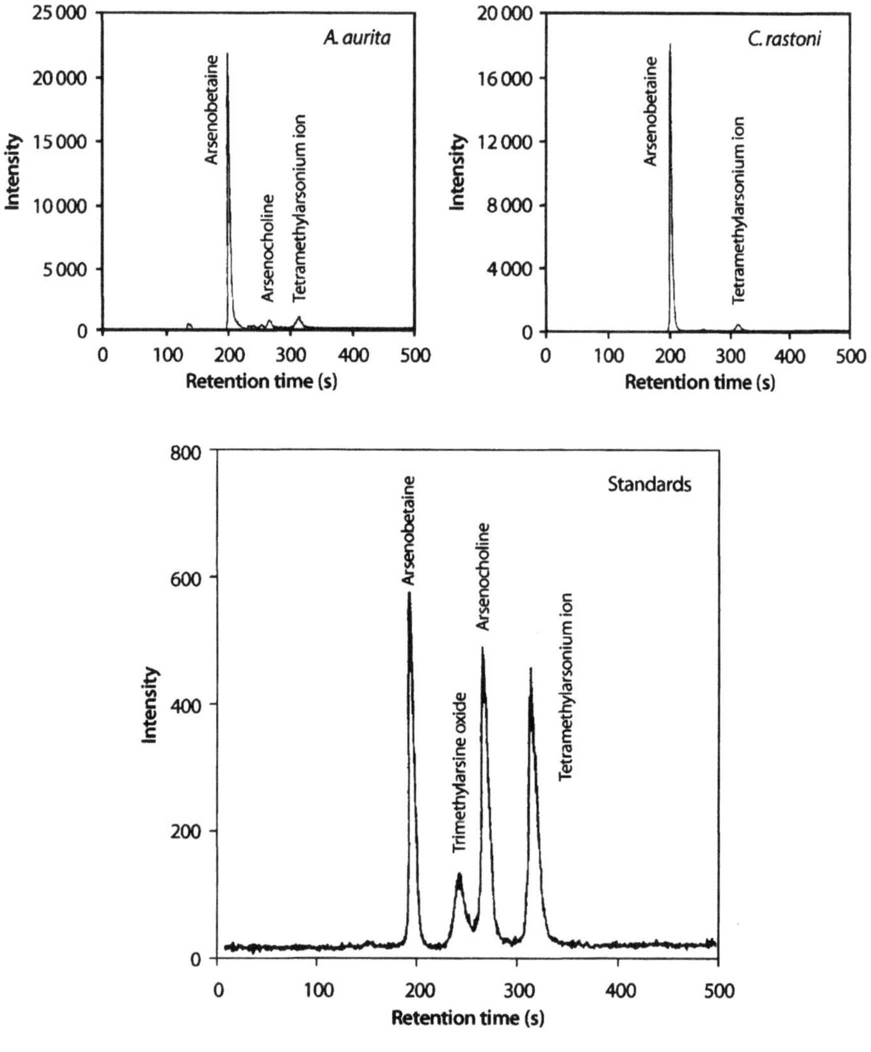

Fig. 15.6. HPLC-ICP MS chromatogram of the water soluble fraction containing arsenic compounds extracted from *Aurelia aurita* and *Carybdea rastonii*, adapted from Hanaoka et al. (1999a)

Table 15.9. Retention (%) of three arsenic compounds in individual tissues and whole shrimp fed 95 µg As (normalized see Francesconi et al. 1999) over 19 days. The proportion of arsenic body burden in each tissue is shown in parentheses. Values for midgut gland and gills are based on arsenic contents of pooled tissues

Tissue	Me_2As-sugar group	Me_3As-sugar group	Arsenobetain group
Tail muscle	0.32 (0.355)	1.08 (0.258)	27.6 (0.488)
Midgut gland	0.02 (0.022)	0.40 (0.096)	2.1 (0.038)
Gills	0.03 (0.035)	0.12 (0.029)	1.1 (0.019)
Remainder[a]	0.53 (0.588)	2.58 (0.617)	25.7 (0.455)
Whole animal	0.9 (1.00)	4.2 (1.00)	57 (1.00)

[a] Not including haemolymph and other fluids lost during dissection. Values have been rounded to two significant figures.

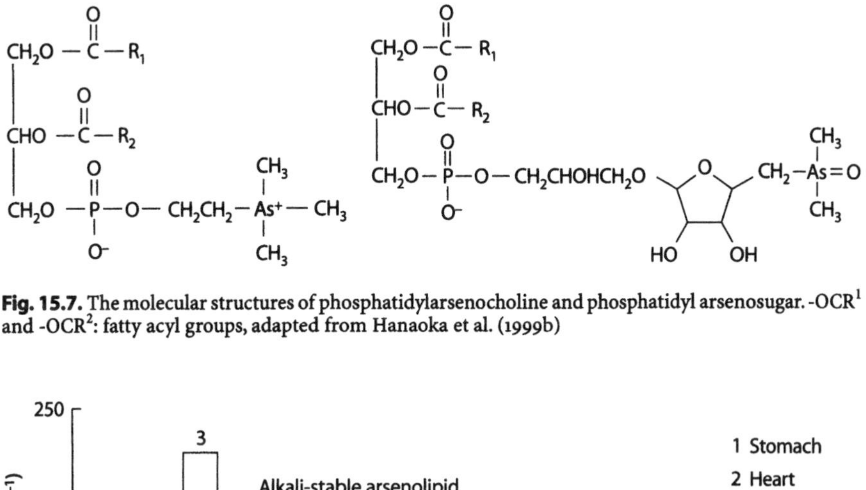

Fig. 15.7. The molecular structures of phosphatidylarsenocholine and phosphatidyl arsenosugar. -OCR[1] and -OCR[2]: fatty acyl groups, adapted from Hanaoka et al. (1999b)

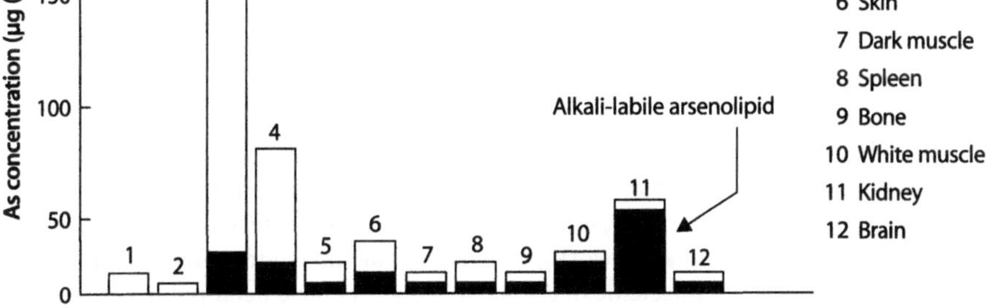

Fig. 15.8. Distribution of alkali-labile and alkali-stable arsenolipids in the tissue and organs of starspotted shark, adapted from Hanaoka et al. (1999b)

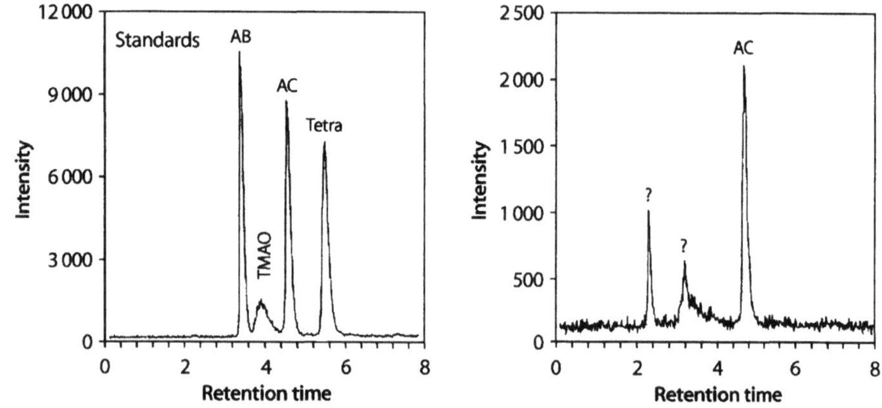

Fig. 15.9. HPLC-ICP-MS chromatograms of standards and the water soluble arsenic residues obtained after severe acid hydrolysis of alkali-labile arsenic fraction prepared from ordinary muscle, adapted from Hanaoka et al. (1999b)

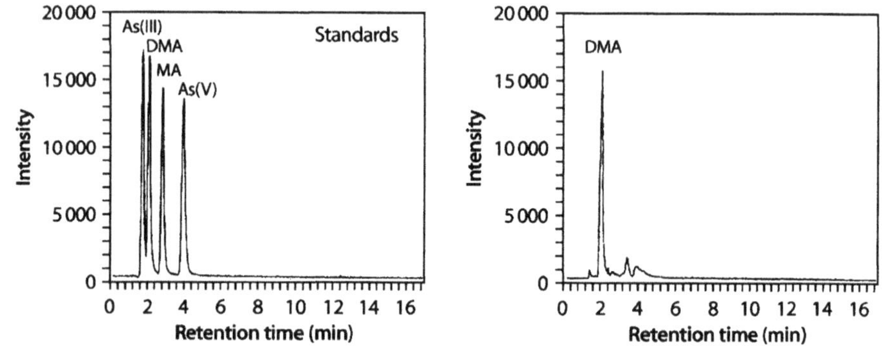

Fig. 15.10. HPLC-ICP-MS chromatograms of standards and the water soluble arsenic residues obtained after severe alkaline hydrolysis of alkali-stable arsenic fraction prepared from the liver, adapted from Hanaoka et al. (1999b)

15.3 Organotin

15.3.1 Organotin Derivatives

Organotin(IV) derivatives have been widely studied for several reasons. Their extensive world-wide use is evident in industry as PVC stabilizers; in agriculture as agrochemicals (fungicides, acaricides), disinfectants (biocides, bactericides) and preservatives (cellulose, wood and stonework); and in the marine environment as antifouling agents (algaecides, molluscicides). Books and reviews have been published in the last few years, covering almost all applications and chemical and biological properties of these compounds (Patai 1995; Abel et al. 1995; Champ and Seligman 1996; De Mora 1996; Smith 1998; Barbieri et al. 2000). As a consequence, organotin(IV) compounds are extensively present in natural and marine waters, marine biota, and sediments. Several good books covering the toxic effects of organometallics and comprising organotin(IV) derivatives have been published recently (Arakawa and Wada 1993; Mennie and Craig 1993; Crompton 1998). Again, for the scope of this review, we will describe the toxic effects of organotin(IV) derivatives on marine biota reported in the last ten years.

15.3.2 Organotin in the Marine Biota

The level of tributyltin(IV)oxide (TBTO) in sediments and in *Mytilus galloprovincialis* sampled in the harbour area of Taranto, where mussel cultures are highly developed, has been detected by electrothermal atomic absorption after *n*-hexane/methylene chloride mixture extraction in 4 different sampling stations and purification of the extract with a sodium hydroxide wash. The TBTO levels obtained in sediments and mussels were in the range of 15–47 ng g^{-1} (wet weight) and 11–30 ng g^{-1} (wet weight), respectively (Cardellicchio et al. 1992) (Fig. 15.11).

When exposed to 10^{-5} and 10^{-7} mol dm^{-3} diorganotin(IV)dichloride and triorganotin(IV)chloride solutions for different times, the gastropod *Truncatella subcylindrica* suffered the following chromosomal structural lesions: (1) breakages, (2) bridging, (3) irregular outline and (4) light areas after staining with acetic orcein, see Figs. 15.12 and 15.13, Table 15.10 (Vitturi et al. 1992).

The experimental evidence shows that tributyltin(IV)chloride is more toxic to this organism than diorganotin(IV)dichloride.

Effects of the exposure to 10^{-4} and 10^{-5} mol dm^{-3} dibutyltin(IV) derivatives of carbohydrates solutions as D(–)sorbitol, D(+)glucose, D(–)fructose and D(+)glyceraldehyde (Mansueto et al. 1993b), Fig. 15.14, on different stages of *Ciona intestinalis* development, larval movements and metamorphosis have been tested in vivo.

The results are summarized in Table 15.11, where it is possible to note that the two-cell stage embryos, if incubated for 1 hour in the organotin(IV) solutions, stopped cleavage, which was restored when the embryos were transferred in organotin-free sea water.

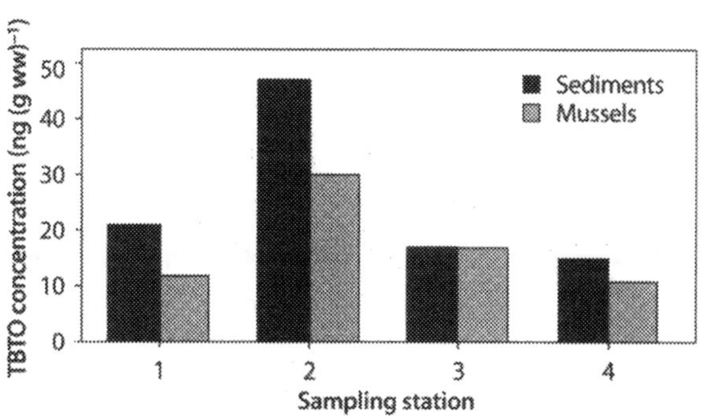

Fig. 15.11. Distribution of TBTO average concentrations in mussels and sediments in the four sampling stations, adapted from Cardellicchio et al. (1992)

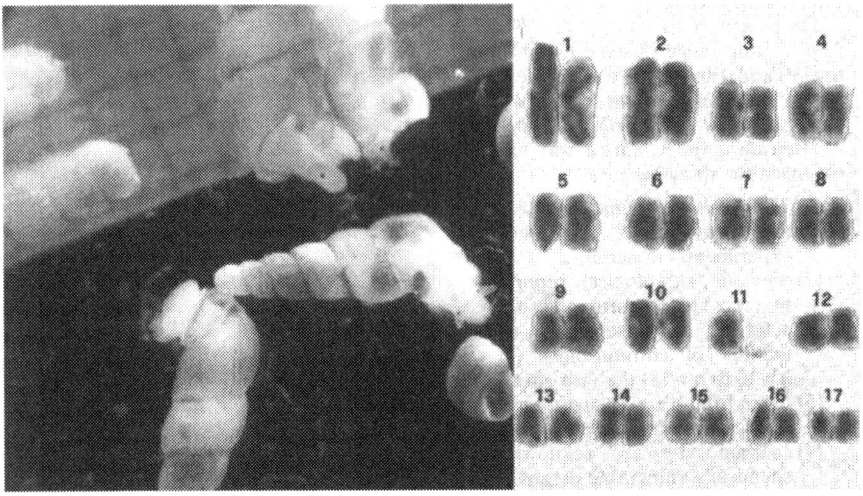

Fig. 15.12. *Truncatella subcylindrica* and its representative karyotype, adapted from Vitturi et al. (1992)

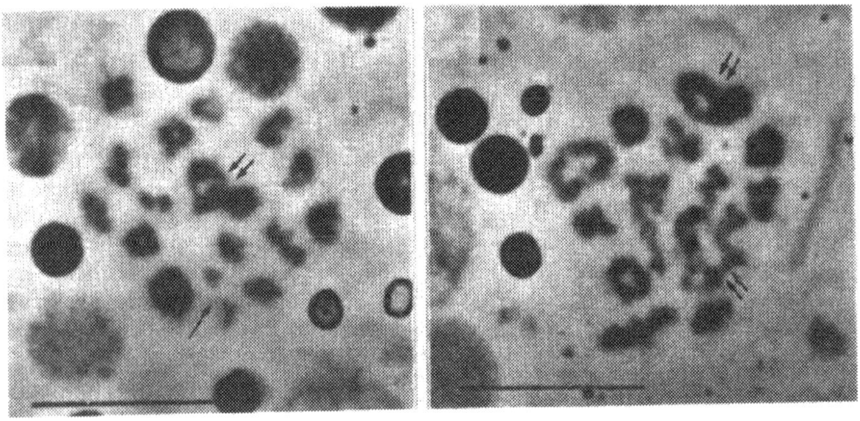

Fig. 15.13. Diakinetic bivalents of *Truncatella subcylindrica* treated with Bu_3SnCl for 72 h (*single arrow* indicates a breakage and *double arrow* indicates the bridging), adapted from Vitturi et al. (1992)

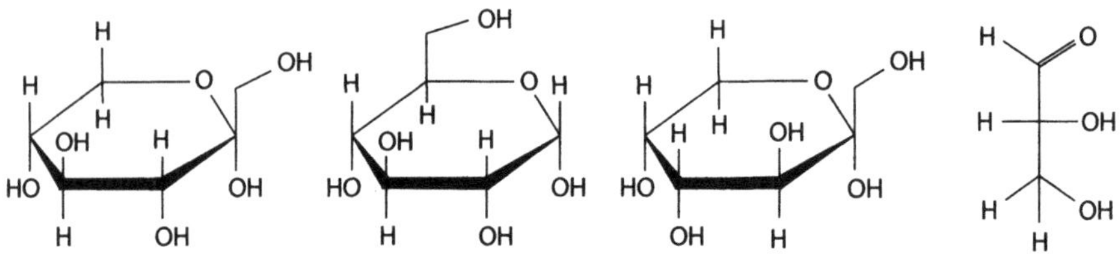

a **b** **c** **d**

Fig. 15.14. Structures of; **a** D(–)sorbitol; **b** D(+)glucose; **c** D (–)fructose; **d** D(+)glyceral-dehydes

Table 15.10. Spermatocyte chromosome alterations in *Truncatella subcylindrica*. Dibutyltin(IV) and tributyltin(IV) chloride concentrations, incubation times and percentages of different categories of anomalous spreads observed during analysis of 100 spreads per specimen, adapted from Vitturi et al. (1992)

Anomaly	**Time interval**	**3 h**		**24 h**		**48 h**		**144 h**
	Organometal	**DBTD**	**TBTC**	**DBTD**	**TBTC**	**DBTD**	**TBTC**	**DBTD**
	No. of specimens	3		2		4		2
	Conc. (mol dm^{-3})	10^{-4}	10^{-4}	10^{-4}	10^{-4}	10^{-4}	10^{-4}	10^{-4}
Irregular outlines		Norm.	45	45	86	82	91	86
Breakages		Norm.	19	20	37	40	35	28
Chromosome bridging		Norm.	24	26	40	32	36	35
Light stained areas		Norm.	3	43	78	67	86	84
	No. of specimens	3		6		6		2
	Conc. (mol dm^{-3})	10^{-7}	10^{-7}	10^{-7}	10^{-7}	10^{-7}	10^{-7}	10^{-7}
Irregular outlines		Norm.	Norm.	7	55	42	85	78
Breakages		Norm.	Norm.	3	21	15	38	45
Chromosome bridging		Norm.	Norm.	3	19	15	39	32
Light stained areas		Norm.	Norm.	5	50	33	80	65
	No. of specimens	3		6		6		2
	Conc. (mol dm^{-3})	10^{-9}	10^{-9}	10^{-9}	10^{-9}	10^{-9}	10^{-9}	10^{-9}
Irregular outlines		Norm.	Norm.	Norm.	46	16	60	35
Breakages		Norm.	Norm.	Norm.	14	7	21	15
Chromosome bridging		Norm.	Norm.	Norm.	16	10	26	18
Light stained areas		Norm.	Norm.	Norm.	44	15	48	35
	No. of specimens	3		6		6		2
	Conc. (mol dm^{-3})	10^{-11}	10^{-11}	10^{-11}	10^{-11}	10^{-11}	10^{-11}	10^{-11}
Irregular outlines		Norm.	Norm.	Norm.	12	Norm.	38	25
Breakages		Norm.	Norm.	Norm.	8	Norm.	20	15
Chromosome bridging		Norm.	Norm.	Norm.	10	Norm.	19	16
Light stained areas		Norm.	Norm.	Norm.	11	Norm.	35	22

DBTD = dibutyltin(IV)dichloride ; *TBTC* = tributyltin(IV)chloride; The term *Norm.* (= Normal) indicates no difference from the control. As 100 spreads of each specimen were examined, the percentage values in this table are calculated from *n* observations in each case, where *n* = 100 No. of specimens.

The gastrula stage was seriously affected by exposure to 10^{-4} mol dm^{-3} solutions of the previously-mentioned complexes: 85% of the embryos were anomalous neurulae with open neural folds; 5% were twisted larvae. The gastrulae, if incubated in 10^{-5} mol dm^{-3} solutions, developed twisted larvae in ovular envelopes and immobile larvae with twisted tails. Finally, larvae treated with both 10^{-4} and 10^{-5} mol dm^{-3} solutions stopped swimming, did not metamorphose and underwent cytolysis, as shown in Fig. 15.15.

Table 15.11. Development of *Ciona intestinalis* embryos after incubation in solutions of organometallic complexes in sea water for a limited time (1 h) and after transferring to normal sea water, after Mansueto et al. (1993)[a]

Compound	Conc.	Development stage				
	($mol\ dm^{-3}$)	Two cells	Gastrula	Neurula	Late neurula	Larva
AG1	10^{-4}	Larvae,90	Anomalous embryos, 85	Delayed larvae, 90	Delayed larvae, 90	Immobility
	10^{-5}	Larvae,90	Delayed, twisted larvae, 90	Delayed larvae, 90	Larvae, 90	Immobility
AG2	10^{-4}	Larvae,90	Anomalous embryos, 85 twisted larvae, 5	Larvae, 90	Twisted, 60 immobile, 10 anomalous larvae, 20	Immobility
	10^{-5}	Larvae,90	Delayed larvae, 90	Larvae, 90	Larvae, 90	
AG3	10^{-4}	Delayed larvae,90	Delayed twisted larvae,90	Larvae, 90	Larvae, 90	Immobility
	10^{-5}	Delayed larvae,90	Delayed larvae, 90	Larvae, 90	Larvae, 90	
AG7	10^{-4}	Anomalous neurulae, 80 twisted larvae, 10	Anomalous embryos, 80 twisted larvae, 10	Anomalous larvae, 80 immobile larvae, 10	Twisted larvae, 60 anomalous larvae, 20 immobile larva, 10	immobility
	10^{-5}	Delayed larvae, 90	Delayed, twisted larvae, 90	Delayed larvae, 90	Delayed larvae, 90	Immobility

[a] Data refer to 14 experiments and show the percentage of developed or arrested embryos. Control developed >90% of swimming larvae. *AG1* = Bu_2Sn-D-(-)sorbitol; *AG2* = Bu_2Sn-D-(+)glucose; *AG3* = Bu_2Sn-D-(-)fructose; *AG7* = Bu_2Sn-D-(+)glyceraldehyde.

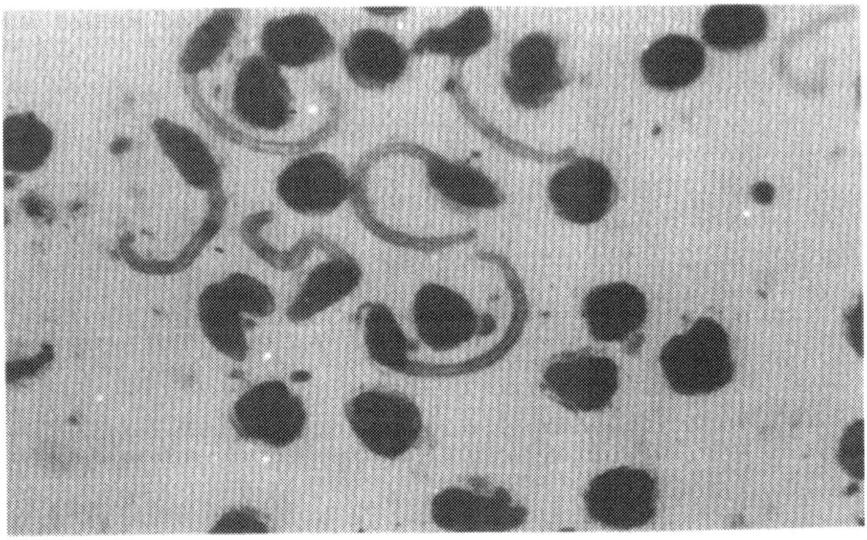

Fig. 15.15. *Ciona intestinalis* gastrulae incubated for 1 h in 10^{-5} mol dm^{-3} Bu_2 Sn-D-(−) sorbitol solution. The embryos developed into twisted larvae in ovular envelopes and immobile larvae with twisted tails (magnification 40×), after Mansueto et al. (1993)

Exposure to 10^{-5} and 10^{-7} mol dm^{-3} bis(dimethyltin(IV)chloro)protoporphyrin(IX), Fig. 15.16, of early developing embryos of *Anilocra physodes* L. (Crustacea, Isopoda), a parasite of various fish, for different exposure times, produced abnormal metaphase and anaphase figures, which depended on the concentration and not the exposure time (Vitturi et al. 1993) (Table 15.12).

According to the reported data, it may be concluded that the bis(dimethyltin(IV)chloro)protoporphyrin(IX) interferes with the formation of the mitotic spindle and acts directly on metaphase chromosome structure (Vitturi et al. 1993).

The above-reported effects are also observed in embryos of the ascidian *Ciona intestinalis* upon exposure to 10^{-5} and 10^{-7} mol dm^{-3} solutions of tributyltin(IV)chloride (Mansueto et al. 1993a).

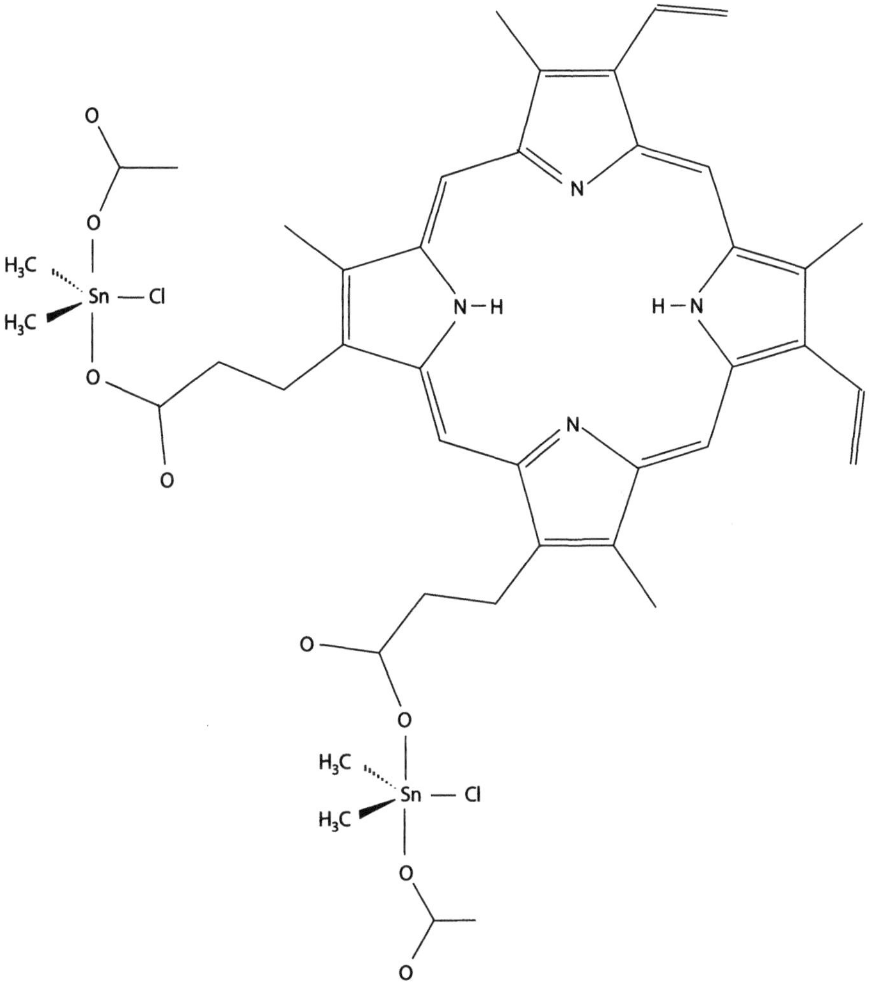

Fig. 15.16. Structure of bis(dimethyltin(IV)chloro)protoporphyrin(IX)

Table 15.12. Genotoxic activity. Metaphase and anaphase chromosomal damage in *Anilocra physodes* early-developing embryos treated with bis(dimethyltin(IV)-chloro)protoporphyrin(IX), adapted from Vitturi et al. (1993)

Compound concentration (mol dm^{-3})	Time interval (h)	No. of embryos employed	Total spreads[a]	Normal[b] metaphases anaphases		Abnormal metaphases anaphases				Total spreads		
						Fragments		Bridges				
				c	d	c	d	c	d	c	d	e
0.956×10^{-5}	24	11	220	10	2	59	3	46	10	104	18	98
	48	9	180	8	1	48	2	36	14	80	16	84
2.07×10^{-7}	24	10	200	20	12	50	1	38	14	87	24	89
	48	8	160	23	10	46	4	14	19	62	33	65
1×10^{-9}	24	12	240	63	59	10	5	9	24	78	78	84
	48	8	160	35	42	13	–	5	10	48	52	60
Control	–	12	270	71	78	–	–	2	1	73	79	88

[a] 20 spreads per embryo were analysed.
[b] The term "normal" indicates no difference from the control.
[c] Number of metaphases.
[d] Number of anaphases.
[e] Number of prophases of which chromosome aberrations were not detected.

In vivo observations showed that cleavage of fertilized eggs was inhibited and embryo blastomeres gave rise to cellular masses not delimited by plasma membranes. Furthermore, electron-dense precipitates, probably due to inorganic tin, were evidenced by transmission electron microscopy (TEM) in the egg cytoplasm of cellular masses. The same type of precipitate was also present in mitochondria, whose structure appeared to be highly modified, indicative of a degenerative process of the embryos after incubation in TBT chloride (Fig. 15.17).

The cytotoxic activity of diorganotin(IV) and triorganotin(IV) derivatives of penicillin G, and of [meso-tetra(4-sulphonatophenyl)porphine], of chloramphenicol, cycloserine and of orotic acid, see Fig. 15.18a–e, has been tested in *Ciona intestinalis* fertilized eggs at different stages of development (Maggio et al. 1994; Pellerito et al. 1997b, 1998; Lencioni et al. 1999) and in *Aphanius fasciatus* (Vitturi et al. 1994).

The genotoxicity of all the organotin(IV)-penicillin G and triorganotin(IV)[meso-tetra(4-sulphonatophenyl)porphinate] derivatives correlates with the blockage of *Ciona intestinalis* fertilized eggs at early stages of development. The results obtained from the in vivo investigation of *Ciona intestinalis* developing embryos, Fig. 15.19a,b and Table 15.13, suggested the following considerations:

1. All the organotin(IV) derivatives are toxic, like the organotin(IV) parents;
2. The triorganotin(IV) derivatives are more toxic than the diorganotin(IV) derivatives;
3. In the triorganotin(IV) series, the tributyltin(IV) is the most toxic.

At least two major nonexclusive hypotheses may explain the results:

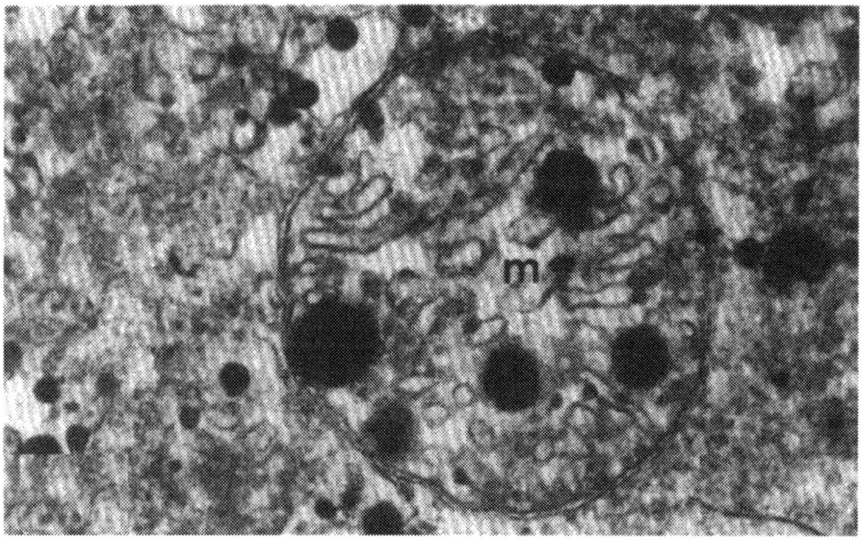

Fig. 15.17. Electron-dense precipitates of TBT chloride in the cell masses. The mitochondria ultrastructure, *m*, is highly modified (magnification 18 500×), adapted from Mansueto, Gianguzza et al. (1993)

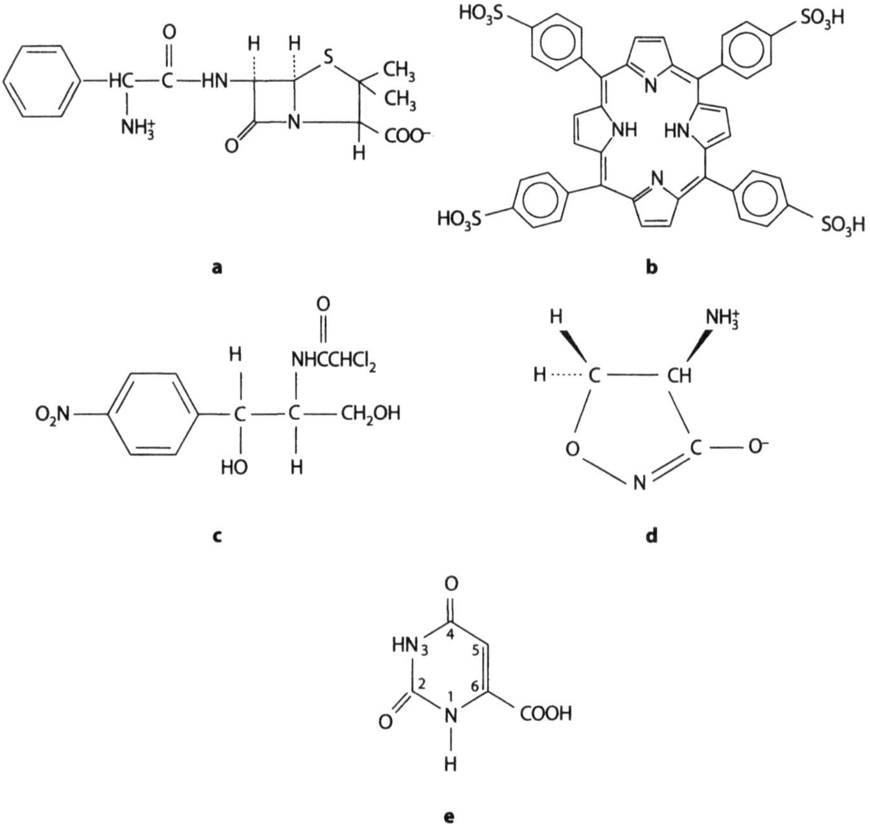

Fig. 15.18. Structures of; **a** penicillin G: **b** [meso-tetra(4-sulphonatophenyl)porphine; **c** chloramphenicol; **d** cycloserine; **e** orotic acid, adapted from Maggio et al. (1994); Pellerito et al. (1997b); Pellerito et al. (1998); Lencioni et al. (1999)

a Consistent ultrastructural damage, including mitochondrial organelles and cytomembranes may be caused by the organotin(IV)-penicillin G derivatives;
b Irregular cleaving processes may reflect damage affecting chromosome structure.

The latter hypothesis is supported by studies that showed that chromosome damages occur in *Aphanius fasciatus* after exposure to the above-mentioned chemicals (Vitturi et al. 1994) and in *Rutilus rubilio* after exposure to diorganotin(IV)amoxicillin derivatives (Vitturi et al. 1995) (Fig. 15.20). The results reported in Tables 15.14 and 15.15 and Figs. 15.21 and 15.22 show that all chromosome abnormalities may be classifiable in *Aphanius fasciatus* as irregular staining, breakages, side-arm bridges or pseudochiasmata of chromosomes, while differentially stained chromosome areas, granular deeply stained zones along the chromosomal body, arm breakage and finally side-arm bridges (pseudochiasmata) were observed in *Rutilus rubilio* chromosomes after exposure to 10^{-5} and 10^{-7} mol dm^{-3} solutions of diorganotin(IV)amoxicillin derivatives (Pellerito et al. 1995).

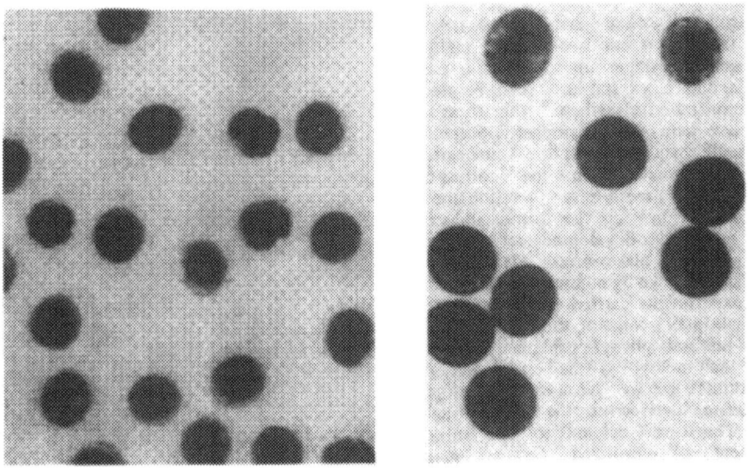

a b

Fig. 15.19. a Anomalous embryos derived from gastrulae incubated for 1 h in 10^{-7} mol dm^{-3} $Bu_3Sn(IV)ClpenGNa$ (magnification 50×); **b** Anomalous two-cell stage incubated in 10^{-5} mol dm^{-3} $(Bu_3Sn)_4TPPS$ solution, adapted from Maggio et al. (1994); Pellerito et al. (1997b)

Table 15.13. Results of development of *Ciona intestinalis* fertilized eggs incubated in the solution of H_4TPPS, $(R_2Sn)_2TPPS$ e $(R_3Sn)_4TPPS$, (H_4TPPS = [meso-tetra (4-sulphonatophenyl)porphine]; R = Me, Bu and Ph) from 2-cell stage. The controls give rise to 90% swimming larvae. Percentage of developed or arrested eggs (average of 5 experiments), adapted from Pellerito et al. (1997b)

Compound	Conc. (mol dm^{-3})	Development stages							
		Blastomere			Early gastrulae	Anomalous neurulae	Undifferentiated embryos	Anomalous larvae	Swimming larvae
		2	4	8–16					
H_4TPPS	10^{-5}						100		
	10^{-7}							100	
$(Me_2Sn)_2TPPS$	10^{-5}						100		
	10^{-7}							100	
$(Bu_2Sn)_2TPPS$	10^{-5}		50	50					
	10^{-7}					50		50	
$(Ph_2Sn)_2TPPS$	10^{-5}		100						
	10^{-7}							100	
$(Me_3Sn)_4TPPS$	10^{-5}				100				
	10^{-7}							50	50
$(Bu_3Sn)_4TPPS$	10^{-5}	100							
	10^{-7}		100						
$(Ph_3Sn)_4TPPS$	10^{-5}		50	50					
	10^{-7}		50	50					

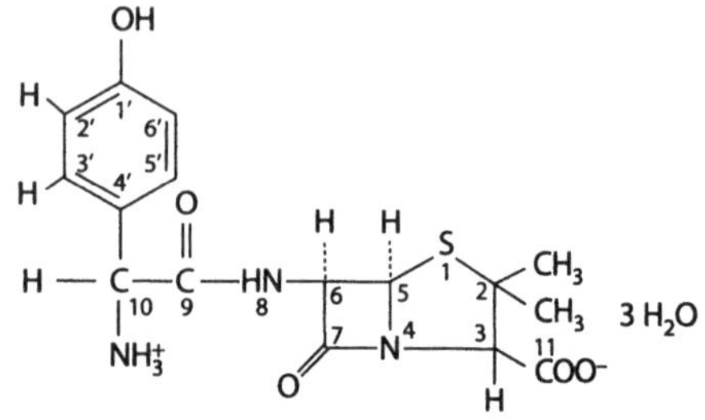

Fig. 15.20. Structure of amoxicillin, adapted from Pellerito et al. (1995)

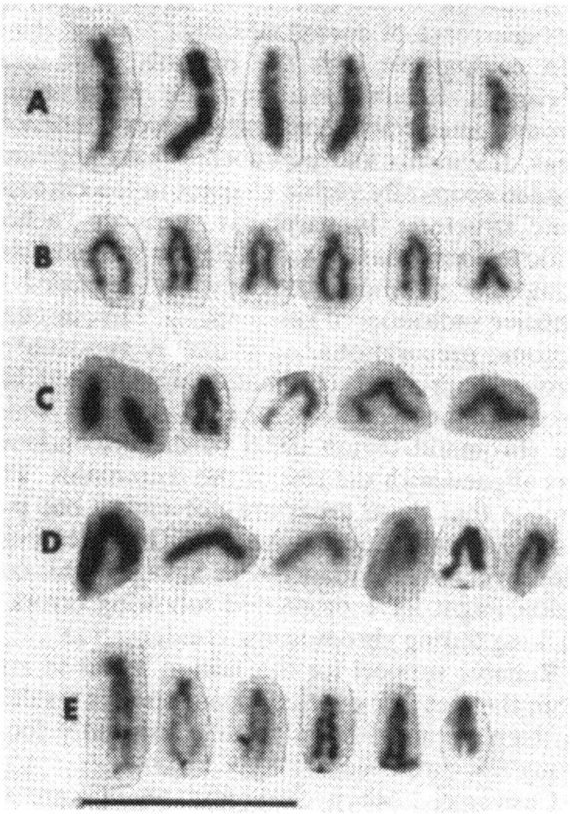

Fig. 15.21. Chromosome aberrations obtained from different giemsa stained spreads of treated *Aphanius fasciatus* specimens; *A* = chromosomes with irregular staining; *B* = chromosomes with black granular regions; *C* = breakages; *D* = chromosomes with arms different in length and *E* = chromosomes with pseudochiasmata, adapted from Vitturi et al. (1994)

Exposure to different concentrations of butyltin(IV)chlorides affected phagocytic activity of *Ciona intestinalis* and of *Botryllus schlosseri* haemocytes, mainly by influencing cellular calcium homeostasis by interacting with calcium pumps (Cima et al. 1995), while dibutyltin(IV)dichloride, triphenyltin(IV)chloride and diphenyltin(IV)-dichloride did not show significant effects on phagocytosis (Cooper et al. 1995) (Figs. 15.23 and 15.24).

On the other hand, tributyltin(IV) derivatives induced apoptosis (Fig. 15.25) in *Botryllus schlosseri* haemocytes (Cima and Ballarin 1999), as evidenced by: (*i*) chro-

Table 15.14. Genotoxic activity: mitotic metaphase chromosomal damage in *Aphanius fasciatus* specimens treated with R_2SnCl_2, $R_2SnClpenG$, R_3SnCl and $R_3SnClpenGNa$ (penG = penicillin G; R = methyl, butyl and phenyl), adapted from Vitturi et al. (1994)

Compound	Conc. (mol dm^{-3})	Time interval (h)	No. of metaphases					
			Normal	Fragments	Irregular outlines	Pseudo-chiasmata	Irregular staining	Total spreads
$Me_2SnClpenG$	10^{-5}	24	4	2	30	28	40	70
		48	1	–		32	8	39
	10^{-7}	24	4	–	52	22	36	60
		48	–	3		26	6	36
Me_2SnCl_2	10^{-5}	24	2	3	18	23	22	30
		48	–	1	16	25	18	40
	10^{-7}	24	12	–	44	32	28	80
		48	8	2	40	39	38	70
$Me_3SnClpenGNa$	10^{-5}		Died after a treatment of 3 hours					
	10^{-7}	24	–	3	15	16	25	36
		48	–	5	22	23	32	42
Me_3SnCl	10^{-5}		Died after a treatment of 3 hours					
	10^{-7}	24	2	2	12	20	18	38
		48	–	5	18	24	20	40
$Bu_2SnClpenG$	10^{-5}	24	4	2	40	16	32	50
		48	4	3	30	8	36	40
	10^{-7}	24	16	5	46	4	40	80
		48	2	7	40	15	36	50
Bu_2SnCl_2	10^{-5}	24	4	–	41	16	32	50
		48	2	2	28	23	18	35
	10^{-7}	24	10	–	50	10	44	64
		48	4	2	35	8	40	50
$Bu_3SnClpenGNa$	10^{-5}		Died after a treatment of 3 hours					
	10^{-7}	24	2	3	12	15	20	30
		48	–	5	14	18	24	28
Bu_3SnCl	10^{-5}		Died after a treatment of 3 hours					
	10^{-7}	24	2	30	18	36	28	46
		48	Died after a treatment of 40 hours					
$Ph_2SnClpenG$	10^{-5}	24	8	4	50	8	30	60
		48	1	6	9	8	8	10
	10^{-7}	24	2	–	42	29	22	50
		48	–	–	48	27	25	58

Table 15.14. *Continued*

Compound	Conc. (mol dm^{-3})	Time interval (h)	No. of metaphases					
			Normal	Fragments	Irregular outlines	Pseudo-chiasmata	Irregular staining	Total spreads
Ph_2SnCl_2	10^{-5}	24	–	3	15	6	15	23
		48	–	–	19	10	18	25
	10^{-7}	24	2	–	10	4	10	20
		48	–	4	12	12	15	22
$Ph_3SnClpenGNa$	10^{-5}		Died after a treatment of 3 hours					
	10^{-7}	24	–	2	8	10	8	12
		48	–	3	10	14	12	18
Ph_3SnCl	10^{-5}		Died after a treatment of 3 hours					
	10^{-7}	24	3	12	15	32	36	50
		48	–	13	18	25	40	45

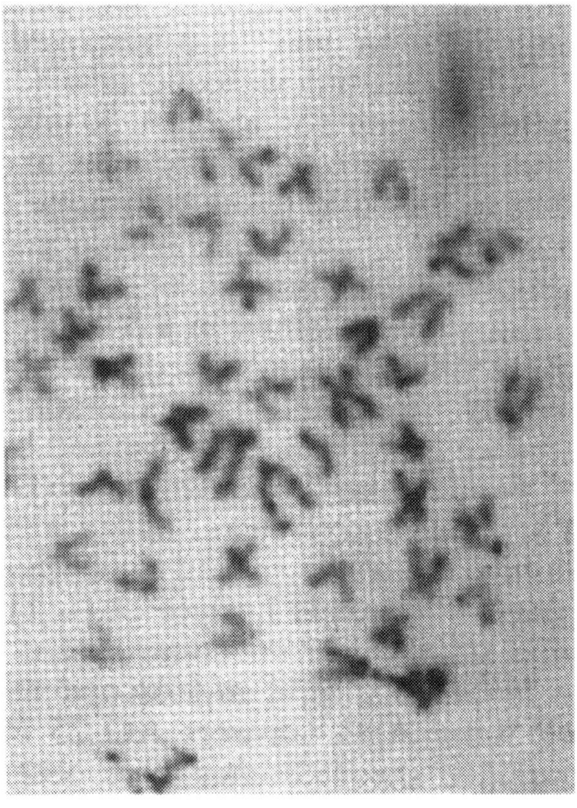

Fig. 15.22. Giemsa-stained metaphase of *Rutilus rubilio* treated with 10^{-7} mol dm^{-3} $Bu_2SnClamox{\cdot}2H_2O$ for 48 h, adapted from Vitturi et al. (1995)

Table 15.15. Genotoxic activity: mitotic metaphase chromosomal damage in *Rutilus rubilio* specimens treated with amox·$3H_2O$, R_2SnClamox·$2H_2O$ and R_2Snamox$_2$·$2H_2O$ (R = methyl, butyl and phenyl), adapted from Vitturi et al. (1995)

Compound	Conc. (mol dm^{-3})	Time interval (h)	No. of metaphases					
			Normal	Irregular staining	Granular zones	Breakages	Side-arm bridges	Total spreads
Controls			326	3	2	1	1	333
Amox·$3H_2O$	10^{-5}	24	5	12	1	1	–	15
		48	7	17	4	2	3	28
	10^{-7}	24	8	22	3	–	2	31
		48	10	26	5	3	6	42
Me_2SnClamox·$2H_2O$	10^{-5}	24	12	10	6	2	3	24
		48	8	6	2	–	3	18
	10^{-7}	24	7	23	8	4	–	36
		48	7	18	7	3	12	40
Me_2Snamox$_2$·$2H_2O$	10^{-5}	24	Died after a treatment of 13–14 hours					
		48						
	10^{-7}	24	22	6	3	4	2	29
		48	13	8	5	4	4	25
Bu_2SnClamox·$2H_2O$	10^{-5}	24	1	.16	5	3	–	21
		48	2	20	7	4	3	32
	10^{-7}	24	3	28	12	6	11	51
		48	2	34	16	7	16	66
Bu_2Snamox$_2$·$2H_2O$	10^{-5}	24	Died after a treatment of 2–3 hours					
		48						
	10^{-7}	24	–	14	8	–	11	27
		48	1	16	13	6	14	43
Ph_2SnClamox·$2H_2O$	10^{-5}	24	2	13	–	–	4	17
		48	1	13	14	1	5	29
	10^{-7}	24	7	23	12	2	–	40
		48	4	25	17	3	–	44
Ph_2Snamox$_2$·$2H_2O$	10^{-5}	24	Died after a treatment of 2–3 hours					
		48						
	10^{-7}	24	3	12	3	2	7	24
		48	1	19	16	10	9	48

matin condensation, with Acridine Orange nuclear staining, (*ii*) DNA fragmentation, with the TUNEL reaction, (*iii*) PS translocation, with the annexin-V assay, and (*iv*) loss of membrane permeability with the Trypan Blue diffusion assay.

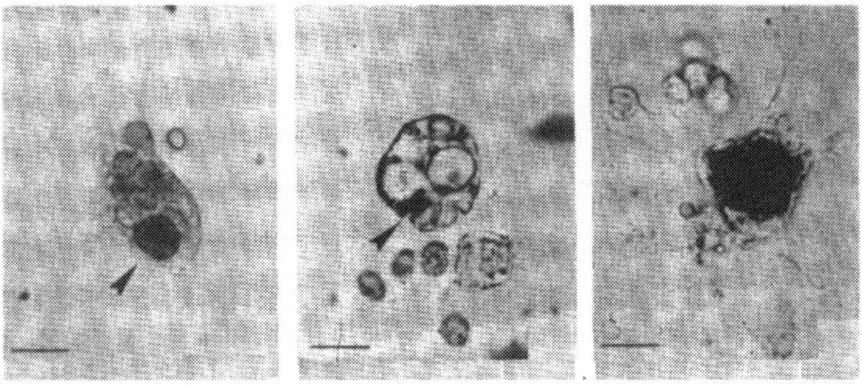

a b c

Fig. 15.23. Phagocytizing haemocytes, showing; **a** positivity for Ca^{2+}-ATPase (*arrowhead*); **b** localized internal calcium rise in normal conditions (*arrowhead*); **c** diffuse internal calcium rise in the presence of TBT. *Bars*: 10 μm, adapted from Cima et al. (1995)

Fig. 15.24. Phagocytosis of *Ciona intestinalis* haemocytes after incubation (90 min) with different concentrations of organotin compounds; (*a*) 1.5 μM, (*b*) 0.15 μM, (*c*) 0.015 μM, (*d*) 0.0015 μM and (*e*) *MS* = marine solution, adapted from Cooper et al. (1995)

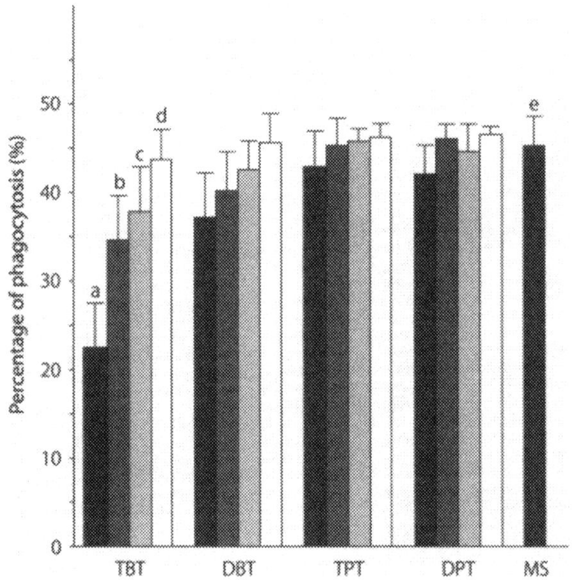

An ultrastructural study carried out by Gianguzza et al. (1996), by exposing embryos of the ascidian *Ciona intestinalis* to tributyltin(IV)chloride solutions, at two different stages of development (1) before hatching (coiled larval stage) and (2) 2 h after hatching (swimming larval stage) led to the following conclusions:

Fig. 15.25. Apoptotic index of *Botryllus schlosseri* exposed to 10 μM TBT for 1 and 2 h as detected with the annexin-V assay (*grey bars*) and the TUNEL reaction (*black bars*). The *white bars* represent the unexposed haemocytes, adapted from Cima et al. (1995)

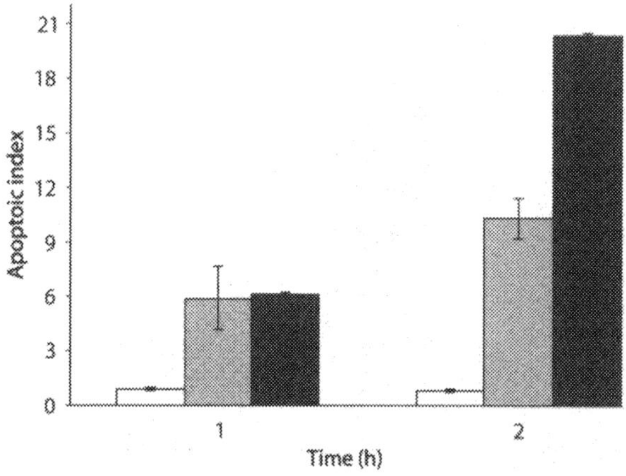

a The embryos at the coiled larval stage do not hatch after exposure to TBT chloride;
b Severe anomalies in the swimming larva occur and include alteration of tail morphology, which appears to be twisted and squatted less than the control.

Furthermore, TEM of tail muscle cells correlated alterations of mitochondrial and myofibrillar ultrastructure with loss of larval mobility (Fig. 15.26a,b).

To correlate these embryonic arrests with cell biochemistry, and especially to understand the molecular details underlying the chemical damages of cellular organelles, fertilized eggs exposed to 10^{-5} and 10^{-7} mol dm^{-3} tributyltin(IV)[meso-tetra-(4-carboxyphenyl)porphinate], were assayed for DNA, RNA, protein, glucose, lipids and ATP contents, comparing the obtained values with those of control fertilized eggs (Fig. 15.27) (Mansueto et al. 2000).

The greater reduction of the content of all tested compounds was induced by 10^{-5} mol dm^{-3} tributyltin(IV)[meso-tetra(4-carboxyphenyl)porphinate] solution, while 10^{-7} mol dm^{-3} solution still inhibited cleavage, but reduced only lipid and ATP content. A possible explanation may be that cleavage did not occur because of lack of energy, or an ATP shortage provoked a cAMP deficiency. Independently of the molecular mechanism underlying the block of the fertilized egg cleavage, the primary effect seems to be the damage to the organellar or cellular membrane, or both.

Cytotoxicity of triorganotin(IV) derivatives towards the sea urchin *Paracentrotus lividus* (Echinodermata) have been gathered by exposure to triorganotin(IV)-L-homocysteate derivatives as well as to the parent triorganotin(IV)chlorides, while the free L-homocysteic acid exerted no significant toxic activity (Pellerito et al. 1997a). Comparative analysis of mitotic chromosomes from untreated embryos and embryos treated with 10^{-5} and 10^{-7} mol dm^{-3} solutions of the above-mentioned organotin(IV) derivatives are reported in Fig. 15.28a–d and Table 15.16.

As expected, the tributyltin- and triphenyltin(IV)-L-homocysteate were the most toxic, while the trimethytin(IV)-L-homocysteate exerted the lowest cytotoxicity. The main defects evidenced were: (1) suppression of stretch between sister chromatids at

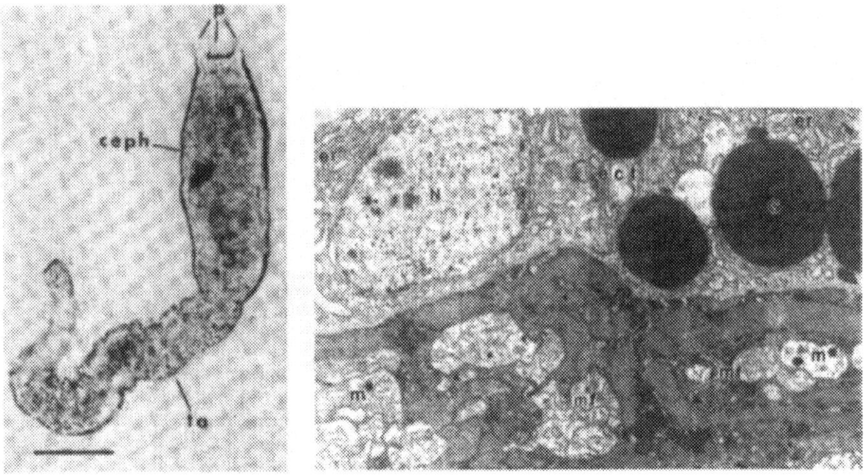

a b

Fig. 15.26. a *Ciona intestinalis* swimming larva incubated for 1 h in 10^{-7} mol dm^{-3} TBT solution and then transferred to sea water. The larva, vitally stained with Nile Blue sulphate, presents severe anomalies in the tail (*ta*) which is strongly twisted; it makes weak and brief movements, and the larva is not able to swim; **b** *Ciona intestinalis* swimming larva incubated for 1 h in 10^{-7} mol dm^{-3} TBT solution. Longitudinal section of the muscle cell. Owing to folding, several sectors of myofibrils (*m*f*) appear as point structures. Some mitochondria (*m**) show a strongly modified ultrastructure. The cells of the ectodermal layer present a large nucleus (*N*), yolk granules (*G*) and are very rich in r.e.r. vesicles (*er*). (*Bar* = 1 μm), adapted from Gianguzza et al. (1996)

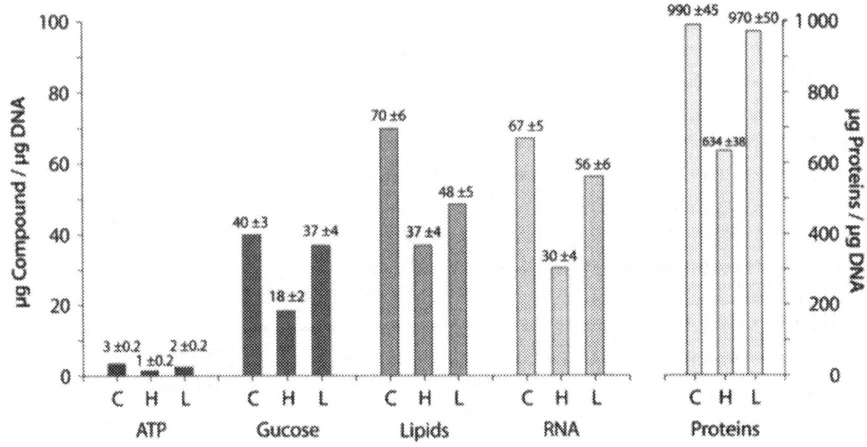

Fig. 15.27. ATP/DNA, glucose/DNA, lipids/DNA, RNA/DNA and proteins/DNA ratios determined at two different concentrations of tributyltin(IV)[meso-tetra(4-carboxyphenyl)porphinate]: $H = 10^{-5}$ mol dm^{-3}; $L = 10^{-7}$ mol dm^{-3}; and C = control, adapted from Mansueto et al. (2000)

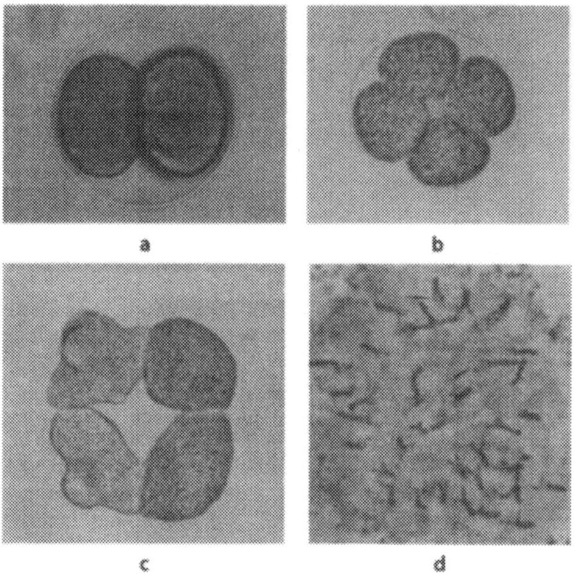

Fig. 15.28. a Early-developing embryos at two blastomere stage; **b** at four blastomere stage; **c** anomalous embryos at four blastomere stage treated with 10^{-5} mol dm^{-3} Bu_3Sn(L-homocysteate) solution after 2 h; **d** metaphase chromosomes treated with 10^{-7} mol dm^{-3} Bu_3Sn(L-homocysteate) solution after 6 h, adapted from Pellerito et al. (1997a)

the beginning of anaphase, (2) deeply stained zones mainly located at the telomeric regions of chromosomes, (3) chromosome arm breakages and (4) chromosome bridges among daughter chromosome at anaphase (Fig. 15.28).

Accumulation of butyltin compounds in marine mammals has been detected by Iwata et al. (1997) in the tissues and organs of three finless porpoises (*Neophocaena phocaenoides*) collected from Japanese coastal waters. The distribution of butyltin(IV) derivatives in the tissues and organs of the porpoises is summarized in Table 15.17 and Fig. 15.29.

The results obtained showed that the distribution of these contaminants was higher in the liver and kidney and lower in the muscle tissue and blubber. Comparison of the hepatic concentrations and compositions with those of water birds, sea turtles and other marine species showed that the levels in the finless porpoises were one order of magnitude higher (Fig. 15.30).

Butyltin compounds have been detected by the HPLC-hydride generation-ICP/AES technique, in water and mussel samples collected in the oil port of Genova, (Rivaro et al. 1999).

The percentage distributions are reported, respectively in Figs. 15.31 and 15.32 and in Table 15.18.

The main conclusions were that owing to the continuous input from vessels and the poor water turnover in the basin, TBT levels in water were high and the TBT level in mussel tissues, particularly gills, reflected the organotin variation found in the sea water during the sampling period. Finally, dibutyltin and monobutyltin(IV) derivatives were present as a result of the degradation processes of the tributyltin derivatives in the water column.

Table 15.16. Genotoxic activity: mitotic metaphase and anaphase chromosomal damage in *Paracentrotus lividus* embryos treated with R_3SnCl and R_3SnL-HCA (L-HCA = L-homocysteic acid; R = methyl, butyl and phenyl), adapted from Pellerito et al. (1997a)

Compound	Concentration (mol dm^{-3})	Time interval (h)	No. of spreads				
			Normal	Irregular staining	Granular zones	Colchicinized-like chromosome	Total spreads
Controls		2	71	2	1	–	74
		6–8	110	–	4	–	114
		18–20	258	3	3	–	264
Me_3Sn(L-HCA)	10^{-5}	2	Arrest				
		6–8					
		18–20					
	10^{-7}	2	6	8	8	20	42
		6–8	4	6	3	25	38
		18–20	–	–	–	5	5
Me_3SnCl	10^{-5}	2	Arrest				
		6–8					
		18–20					
	10^{-7}	2	4	10	12	15	41
		6–8	4	4	5	20	33
		18–20	–	–	3	7	10
nBu_3Sn(L-HCA)	10^{-5}	2	Arrest				
		6–8					
		18–20					
	10^{-7}	2	4	6	6	12	28
		6–8	0	3	5	20	28
		18–20	Arrest				
nBu_3SnCl	10^{-5}	2	Arrest				
		6–8					
		18–20					
	10^{-7}	2	7	12	6	15	40
		6–8	2	4	4	26	36
		18–20	Arrest				
Ph_3Sn(L-HCA)	10^{-5}	2	Arrest				
		6–8					
		18–20					
	10^{-7}	2	Arrest				
		6–8					
		18–20					

Table 15.16. *Continued*

Compound	Concentration (mol dm^{-3})	Time interval (h)	No. of spreads				
			Normal	Irregular staining	Granular zones	Colchicinized-like chromosome	Total spreads
Ph_3SnCl	10^{-5}	2	Arrest				
		6–8					
		18–20					
	10^{-7}	2	Arrest				
		6–8					
		18–20					

Table 15.17. Concentrations (ng of butyltin ion per g, wet weight basis) of butyltins in tissues and organs of higher trophic animals, adapted from Iwata et al. (1997)

Sample	Sampling location	Tissues and organs	MBT	DBT	TBT	ΣBTs
Finless porpoise, 1994	Ise Bay	Muscle	22	38	420	480
		Blubber	73	20	120	213
		Liver	680	1 800	810	3 290
		Kidney	55	340	510	905
		Heart	12	41	640	693
		Lung	90	59	160	309
		Stomach	18	64	130	212
		Brain	9	40	350	399
		Adrenal gland	130	350	250	730
		Spleen	33	98	69	200
		Bone[a]	14	21	33	68
		Esophagus	19	36	49	104
		Pancreas	72	130	220	422
		Intestine	9	70	66	145
		Urinary bladder	26	67	56	149
		Eye	<5	12	35	47
		Blood	82	640	95	817
		Intestine content	6	47	150	203
Ginkgo-toothed beaked whale, 1993	Japan Sea	Muscle	20	8	19	47
		Blubber	20	10	33	63
		Liver	120	130	76	326
		Kidney	66	34	42	132
Largha seal, 1992	Western coast of Hokkaido	Muscle	14	6	20	40
		Blubber	14	1	4	19
		Liver	96	200	32	328
		Kidney	27	55	28	110
Common cormorant, 1993	North Pacific	Liver	150	180	19	349
Loggerhead turtle, 1990	Tosashimizu	Liver	68	60	2.4	130

[a] The bone marrow of a rib was analysed.

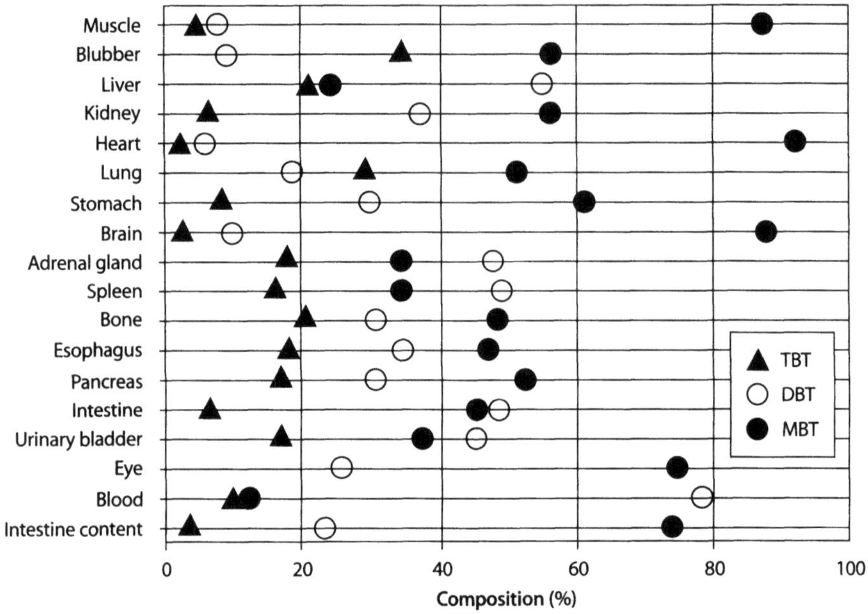

Fig. 15.29. Butyltin compositions in tissues and organs of a finless porpoise collected from Ise Bay, adapted from Iwata et al. (1997)

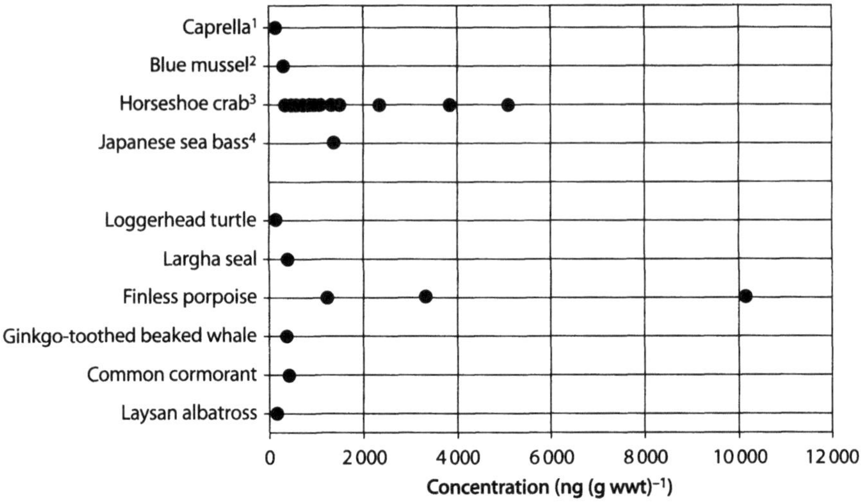

Fig. 15.30. Interspecies comparison of butyltin concentrations in livers, adapted from Iwata et al. (1997); (*1*) Caprella: whole body; (*2*) Blue mussel: soft tissue, (*3*) see Kannan K et al (1995) Arch Environ Toxicol 28:40, (*4*) see Suzuki T et al. (1992) J Agric Food Chem 40:1437

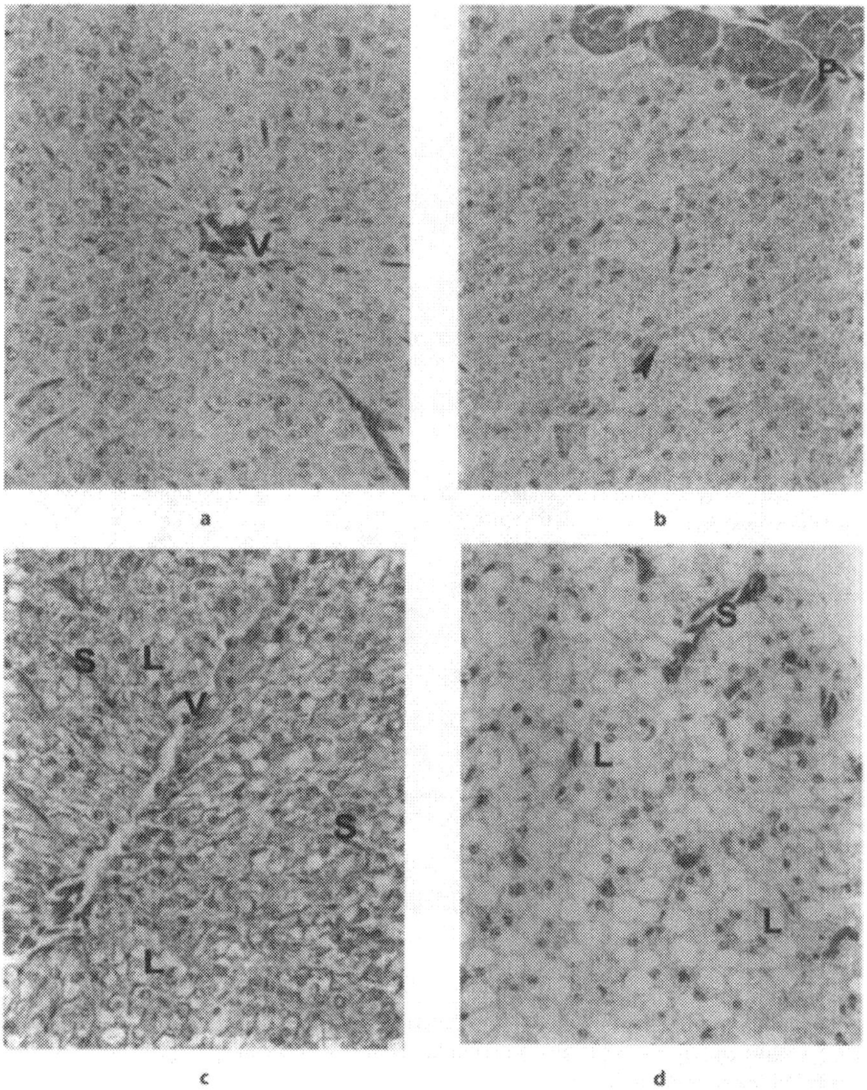

Fig. 15.31. *Nile tilapia* livers of control and TPTH-treated groups after four months of exposure (400×); **a** Control (water) showing normal hepatocytes and sinusoids; **b** Control (0.0015% DMSO), showing small hyaline droplets in hepatocytes. *P*, exocrine pancreas; **c** Treated with 1 mg l^{-1} TPTH, showing congestions of sinusoids (*S*) and central vein. Note marked increase in lipid vacuolation (*L*); **d** Treated with 3 mg l^{-1} TPTH showing more severe sinusoidal congestion (*S*) and lipid vacuolation (*L*), adapted from Visoottiviseth et al. (1999)

Triphenyltin(IV)hydroxide provoked histopathological anomalies in the liver, kidney and gill of *Oreochromis nilotica* (Nile tilapia) (Visoottiviseth et al. 1999). In par-

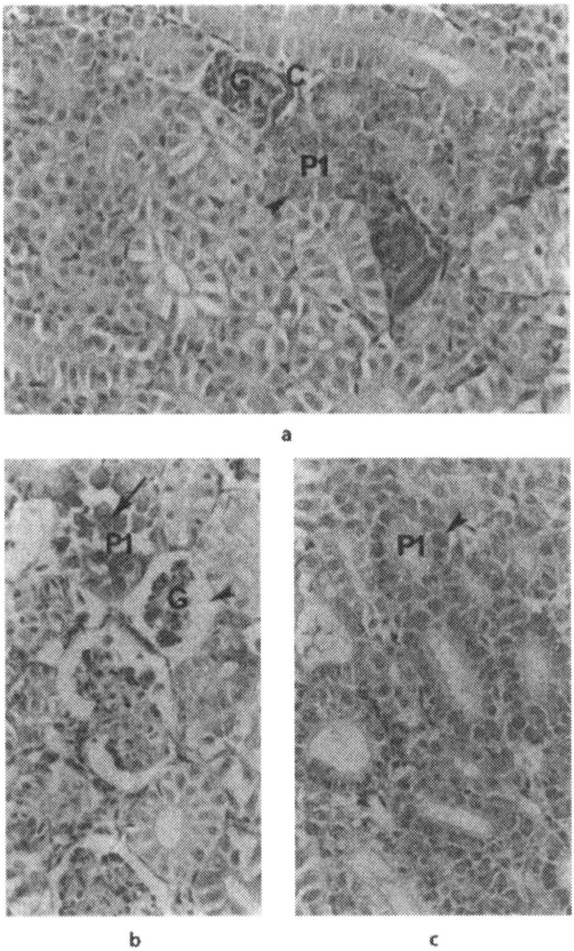

Fig. 15.32. *Nile tilapia* kidneys of control and TPTH-treated groups after four months of exposure (400×); **a** control (water), showing normal appearance of glomerulus (*G*) and renal corpuscle (*C*). Hyaline droplets (*arrowhead*) in the first proximal epithelial cells (*P1*) are observed; **b** treated with 1 mg l^{-1} TPTH, showing more severe glomerulus contraction (*G*) and widening of Bowman's capsules (*arrowhead*). First proximal epithelial cells (*P1*) are filled with large hyaline droplets (*arrow*); **c** treated with 3 mg l^{-1} TPTH, showing numerous large hyaline droplets (*arrowhead*) in epithelial cells of first proximal tubules, adapted from Visoottiviseth et al. (1999)

ticular, the hepatocytes underwent congestion and dilatation of sinusoidal spaces, cytoplasmic pallor, vacuolation and accumulation of hyaline droplets, subcapsular and scattered focal necrosis (Fig. 15.31).

Hydropic degeneration, accumulation of hyaline droplets in the tubular epithelial cells, congestion of peritubular capillaries and detachment of tubular epithelial cells were observed in the kidney, in addition to collapse of the glomerular capillary in more severe cases (Fig. 15.32).

Finally, hyperplasia of the covering gill epithelium, congestion of gill capillaries and vessel and aneurismal formation of gill lamellar capillaries have been noted (Fig. 15.33).

Imposex or masculinization of the muricid *Thais distinguenda* was induced by tributyltin contamination around the major harbours of Phuket Island, Thailand (Bech 1999). 50 specimens of *Thais distinguenda* were collected in 21 stations in the 1996–1998 period and analysed for imposex (Table 15.19).

Table 15.18. Concentration, ng l^{-1} for whole molecule, of butyltin compounds in sea water and mussel tissue samples from Genova oil Port (adapted from Rivaro et al. 1999)

	July 1977	September 1977	October 1977	December 1997	February 1998	May 1998
Unfiltered						
Total butyltin	1 308	1 843	1 237	314	1 906	680
TBT	528 ±15	1 200 ±21	734 ±28	225 ±18	629 ±14	297 ±4
DBT	571 ±12	518 ±29	290 ±7	41 ±4	1 029 ±19	210 ±8
MBT	209 ±21	124 ±14	213 ±13	48 ±3	249 ±2	173 ±15
Filtered						
Total butyltin	603	1 541	1 175	288	1 471	580
TBT	161 ±20	1 041 ±95	724 ±3	200 ±9	394 ±37	222 ±15
DBT	296 ±30	384 ±38	238 ±16	40 ±2	872 ±1	185 ±10
MBT	146 ±15	116 ±10	213 ±11	48 ±4	205 ±6	173 ±15
Whole tissues						
Total butyltin	4.61	11.03	8.46	3.62	6.85	2.18
TBT	2.50 ±0.25	5.79 ±0.28	3.29 ±0.33	0.73 ±0.07	3.36 ±0.02	0.89 ±0.08
DBT	1.57 ±0.14	3.08 ±0.01	3.72 ±0.07	1.04 ±0.11	2.84 ±0.12	0.69 ±0.06
MBT	0.54 ±0.01	2.19 ±0.12	1.45 ±0.03	1.85 ±0.18	0.65 ±0.01	0.60 ±0.01
Gills						
Total butyltin	13.73	16.03	11.40	7.07	12.46	3.63
TBT	6.52 ±0.13	7.59 ±0.56	6.02 ±0.22	2.54 ±0.11	5.94 ±0.03	1.21 ±0.06
DBT	3.70 ±0.03	5.84 ±0.37	3.38 ±0.18	2.32 ±0.24	5.40 ±0.03	0.84 ±0.04
MBT	3.51 ±0.03	2.60 ±0.01	2.00 ±0.06	2.21 ±0.01	1.12 ±0.12	1.58 ±0.08
Digestive glands						
Total butyltin	6.47	4.46	3.82	3.70	6.68	2.50
TBT	2.76 ±0.11	2.46 ±0.08	2.03 ±0.02	1.66 ±0.17	2.75 ±0.02	1.20 ±0.10
DBT	2.05 ±0.16	1.15 ±0.07	0.83 ±0.04	0.19 ±0.04	3.34 ±0.04	0.52 ±0.06
MBT	1.66 ±0.10	0.85 ±0.01	0.95 ±0.01	1.85 ±0.02	0.69 ±0.03	0.78 ±0.03

Table 15.19 clearly shows that both the incidence and distribution of imposex dramatically increased from 1996 to 1998, from 10 stations up to 18 stations. A similar trend was found using *Thais bituberculari*s and *Morula musiva* as indicators.

Butyltin derivatives, namely tributyltin and its degradation products, dibutyltin and monobutyltin, were present in higher concentrations in the kidney, with respect to the liver, of striped dolphins (*Stenella coeruleoalba*), bottlenose dolphins (*Tursiops truncatus*) and in a fetus of the common dolphin (*Delphinus delphi*) found stranded along the western Italian and Greek coasts in the 1992–1994 period (Table 15.20) (Focardi et al. 2000).

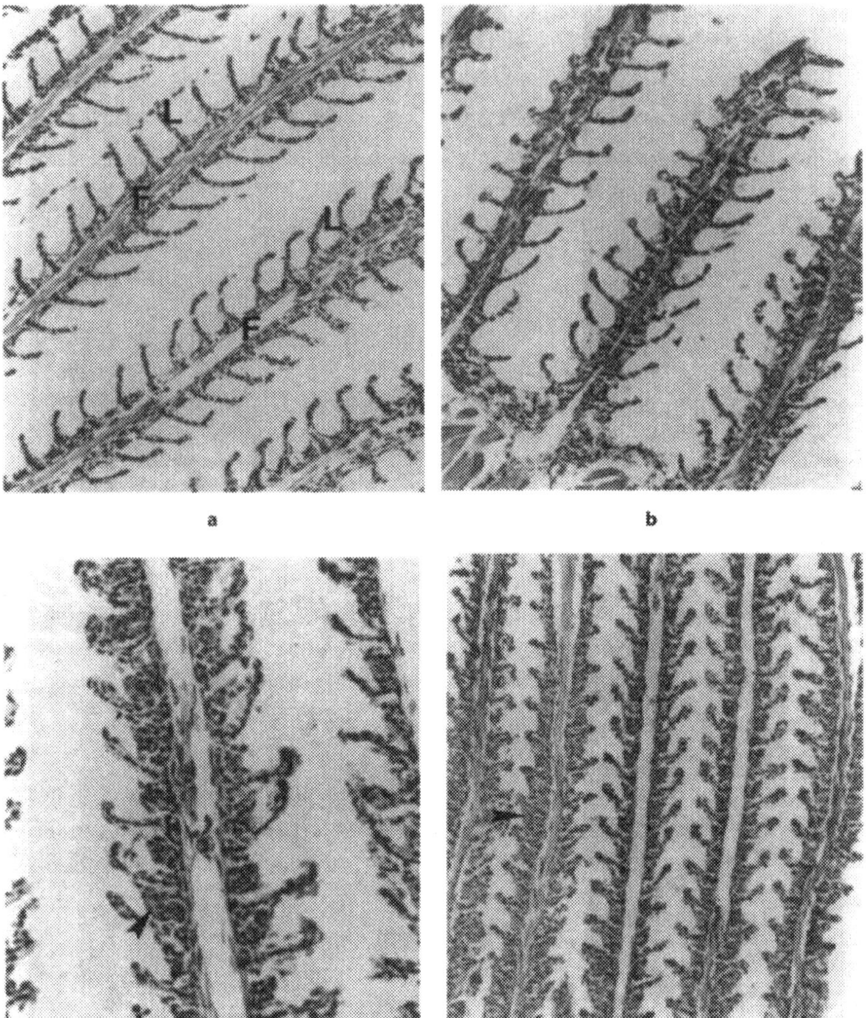

Fig. 15.33. *Nile tilapia* gills of control and TPTH-treated groups after one month of exposure; **a** control (water), showing normal appearance of gill filament (*F*) and lamellae (*L*). Note stratified squamous epithelium covering gill filaments and thin single-layer epithelium of gill lamellae (200×); **b** control (0.0015% DMSO), showing normal gills (200×); **c** treated with 1 mg l^{-1} TPTH, showing slightly thickened epithelium of filament (*arrowhead*) (400×); **d** treated with 3 mg l^{-1} TPTH, showing slightly thickened epithelium of filament (*arrowhead*) (400×), adapted from Visoottiviseth et al. (1999)

From the reported data it was possible to conclude that bottlenose dolphins have lower butyltin concentrations than striped dolphins, while the fetus had a higher concentration in the liver, suggesting that butyltins were transferred from mother to fetus.

Table 15.19. Imposex incidence of *Thais distinguenda* from 21 sites, of *Thais bitubercularis* and *Morula musiva* in 1996 and 1998, adapted from Bech (1999)

Station name	*Thais distinguenda*		*Thais bitubercularis*		*Morula musiva*	
	1996	1998	1996	1998	1996	1998
Laem Nga	0	0	21	79	0	0
Laem Mai Phai	0	62				
Laem Phap Pha	57.4	83	42	100	0	10
Phuket Harbour	89.67	100	100	100	19	62.5
Laem Nam Bor	84	91	41.4	100	0	0
Deep-sea port	65.79	100				
Tin smelting plant	34.38	68				
Taphao Yai (west)	7.1	74				
Taphao Yai (south)	0	62				
Koh Taphao Noi	0	32				
Hotel Cape Panwa	0	63				
Laem Panwa	0	58				
Laem Yam Yen	50	71				
Laem Wing	22.2	62				
Chalong Bay (East)	57	58	100	100	0	5
Koh Thanan	0	34				
Koh Lon (East)	0	14				
Koh Aew	0	2				
Koh Hey	0	0				
Koh Maiton	0	0				

Table 15.20. Concentrations (ng g^{-1} wet wt) of butyltin species in the liver and kidneys of Mediterranean dolphins (Mean and range of concentrations; *N.d.* = not detected, below detection limit of 2 ng g^{-1} wwt), adapted from Focardi et al. (2000)

Species	Tissue	MBT	DBT	TBT	ΣBTs
Stenella coeruleoalba	Liver, 8	97 (4.7–205)	115 (10–434)	67 (N.d.–386)	259 (15–1025)
	Kidney, 7	1 661 (772–6596)	86 (N.d.–468)	200 (8.0–990)	2 230 (783–8055)
Tursiops truncatus	Liver, 2	3.0; 12	36	4.1; 15	27; 43
	Kidney, 2	1 003; 1 968	16	5.3; 46	1 024; 2 014
Delphinus delphi [a]	Liver, 1	5.5	333	4 014	4 352
	Kidney, 1	2 622	400	193	3 215

[a] Fetus.

An analogous investigation has been carried out by using GC-MS on the livers of 21 beluga whales found dead and stranded along the shores of St. Lawrence Estuary and Hudson Strait (northern Quebec), during the 1995–1998 period (St-Louis et al. 2000).

Table 15.21. Concentrations of butyltin compounds in beluga liver samples

Sample[a]	Concentration (ng Sn g^{-1} dwt)				
	MBT	DBT	TBT	ΣBTs	TBT/DBT
St.Lawrence Estuary					
9507, M, 0, 199	1	132	111	244	0.8
9508, F, 0, 157	7	57	67	130	1.2
9509, M, 1.5, 247	26	735	116	877	0.2
9511, M, 26+, 419	399	282	84	764	0.3
9512, M, 11, 400	6	53	8	66	0.2
9601, F, 21+, 427	9	250	22	281	0.1
9602, F, 5, 325	9	45	0	54	
9604, M, 16+, 420	0	96	0	96	
9607, M, 24+, 399	11	100	0	111	
9608, F, 1,2, 216	17	137	0	153	
9609, M, 23++, 393	23	188	0	210	
9701, M, 8, 396	7	229	101	336	0.4
9702, F, 28,5, 338	7	141	58	205	0.4
9703, F, 21+, 363	6	1613	467	2085	0.3
9704[b], M, 0, 139	9	12	15	36	1.2
9705, M, 7.5, 378	14	160	40	214	0.3
9706, F, 31.5+, 385	10	716	308	1034	0.4
9802, M, N/A, 401	9	294	14	316	0.1
9803, F, N/A, 377	27	289	27	342	0.1
9804, F, N/A, 356	16	208	0	224	
9806, F, N/A, 388	16	54	28	98	0.5
Hudson Strait					
98348,M,17, N/A	N.d.	N.d.	N.d.		
98349,F,11, N/A	N.d.	N.d.	N.d.		
98358, M, N/A, N/A	N.d.	N.d.	N.d.		
98359, F, >15, N/A	N.d.	N.d.	N.d.		
98482, F, N/A, N/A	N.d.	N.d.	N.d.		

[a] The first two digits indicate the year when the whale was found stranded.
[b] This beluga whale was a neonate.
N/A = not available.
N.d. = not detected, adapted from St-Louis et al. (2000).

Table 15.21 summarizes the data obtained, from which it was possible to conclude that: (1) the liver was contaminated with concentrations of butyltin ranging from to 0.04 to 2.1 mg Sn kg^{-1} on a dry weight basis, and (2) the contamination with tributyltin derivatives of the mammals of the St. Lawrence did not decrease in the last 10 years after regulating the use of TBT-based antifoulants on small crafts.

Acknowledgements

Financial support by the Ministero dell'Istruzione, dell'Università e della Ricerca (M.I.U.R), Roma, and by Università di Palermo, is gratefully acknowledged.

References

Abel EW, Stone FGA, Wilkinson G (eds) (1995) Comprehensive organometallic chemistry II. Pergamon, Oxford
Arakawa Y, Wada S (1993) Biological properties of alkyltin compounds. In: Sigel H, Sigel A (eds) Metal ions in biological systems, vol XXIX. Marcel Dekker Inc., New York, pp 101–136
Barbieri R, Pellerito L, Ruisi G, Barbieri-Paulsen A, Barone G, Posante S, Rossi M (2000) ^{119}Sn Mössbauer spectroscopy studies on the interaction of organotin(IV) salts and complexes with biological systems and molecules. In: Gianguzza A, Pelizzetti E, Sammartano S (eds) Chemical processes in marine environments. Springer-Verlag, Berlin, pp 229–244
Bech M (1999) Increasing levels of tributyl-induced imposex in muricid gastropods at Phuket Island, Thailand. Appl Organomet Chem 13:799–804
Cardellicchio N, Geraci S, Marra C, Paterno P (1992) Determination of tributyltin oxide in coastal marine sediments and mussels by electrothermal atomic absorption spectrometry. Appl Organomet Chem 6:241–246
Champ MA, Seligman PF (eds) (1996) Organotin: Environmental fate and effects. Chapman and Hall, London
Cima F, Ballarin L (1999) TBT – induced apoptosis in tunicate haemocytes. Appl Organomet Chem 13:697–703
Cima F, Ballarin L, Bressa G, Sabbadin A (1995) Immunotoxicity of butyltins in tunicates. Appl Organomet Chem 9:567–572
Cooper EL, Arizza V, Cammarata M, Pellerito L, Parrinello N (1995) Tributyltin affects phagocytic activity of *Ciona intestinalis* hemocytes. Comp Biochem Physiol 112C:285–289
Crompton TR (1998) Occurrence and analysis of organometallic compounds in the environment. John Wiley and Sons, Chichester
Cullen WR, Harrison LG, Li H, Hewitt G(1994) Bioaccumulation and excretion of arsenic compounds by a marine unicellular alga, *Polyphysa peniculus*. Appl Organomet Chem 8:313–324
De Mora SJ (ed) (1996) Tributyltin: Case history of an environmental contaminant. Cambridge University Press, Cambridge
Edmonds JS, Shibata Y, Francesconi KA, Rippingale RJ, Morita M (1997), Arsenic transformations in short marine food chains studied by HPLC-ICP MS. Appl Organomet Chem 11:281–288
Focardi S, Corsolini S, Aurigi S, Pecetti G, Sanchez-Hernandez JC (2000) Accumulation of butyltin compounds in dolphins stranded along the Mediterranean coasts. Appl Organomet Chem 14:48–56
Francesconi KA, Edmonds JS, Stick RV (1992), Arsenocholine from anaerobic decomposition of a trimethylarsonioriboside. Appl Organomet Chem 6:247–249
Francesconi KA, Hunter DA, Bachmann B, Raber G, Goessler W (1999) Uptake and transformation of arsenosugars in the shrimp crangon. Appl Organomet Chem 13:669–680
Gailer J, Francesconi KA, Edmonds JS, Irgolic KJ (1995) Metabolism of arsenic compounds by the blue mussel *Mytilus edulis* after accumulation from seawater spiked with arsenic compounds. Appl Organomet Chem 9:341–356
Garman GD, Pillai MC, Cherr GN (1993) Inhibition of cellular events during early algal gametophyte development: Effects of select metals and an aqueous petroleum waste. Aquat Toxicol 28(1–2):127–144
Gianguzza M, Dolcemascolo G, Mansueto C, Pellerito L (1996), Effects of tributyltin(IV) chloride exposure on larvae of *Ciona intestinalis* (Urochordata): An ultrastructural study. Appl Organomet Chem 10:405–413
Goessler W, Rudorfer A, Mackey EA, Becker PR, Irgolic KJ (1998) Determination of arsenic compounds in marine mammals with high-performance liquid chromatography and an inductively coupled plasma mass spectrometer as element-specific detector. Appl Organomet Chem 12:491–502

Hanaoka K, Tagawa S, Kaise T (1992a) The fate of organoarsenic compounds in marine ecosystems. Appl Organomet Chem 6:139–146
Hanaoka K, Satow T, Tagawa S, Kaise T (1992b) Formation of arsenobetain from arsenocholine by microorganisms occurring in the sediments. Appl Organomet Chem 6:375–382
Hanaoka K, Kaise T, Kai N, Kawasaki Y, Miyasita H, Kakimoto K, Tagawa S (1997) Arsenobetain-decomposing ability of marine microorganisms occurring in particles collected at depths of 1100 and 3500 meters. Appl Organometal Chem 11:265–272
Hanaoka K, Goessler W, Kaise T, Ohno H, Nakatani Y, Ueno S, Kuehnelt D, Schlagenhaufen C, Irgolic KJ (1999a) Occurrence of a few organo-arsenicals in jellyfish. Appl Organomet Chem 13:95–100
Hanaoka K, Goessler W, Yoshida K, Fujitaka Y, Kaise T, Irgolic KJ (1999b) Arsenocholine- and dimethylated arsenic-containing lipids in starspotted shark mustelus manazo. Appl Organomet Chem 13:765–770
Ishiguro S (1992) Industries using arsenic and arsenic compounds. Appl Organomet Chem 6:323–331
Iwata H, Tanabe S, Mizuno T, Tatsukawa R (1997) Bioaccumulation of butyltin compounds in marine mammals: The specific tissue distribution and composition. Appl Organomet Chem 11:257–264
Kaise T, Fukui S (1992) The chemical form and acute toxicity of arsenic compounds in marine organisms. Appl Organomet Chem 6:155–160
Kaise T, Oya-Ohta Y, Ochi T, Okubo T, Hanaoka K, Irgolic KJ, Sakurai T, Matsubara C (1996) Toxicological study of organic arsenic compound in marine algae using mammalian cell culture technique. J Food Hyg Soc Jap 37:135–141
Kaise T, Ochi T, Oya-Ohta Y, Hanoka K, Sakurai T, Saitoh T, Matsubara C (1998) Cytotoxicological aspects of organic arsenic compounds contained in marine products using the mammalian cell culture technique. Appl Organomet Chem 12:137–144
Lencioni S, Pellerito A, Fiore T, Giuliani AM, Pellerito L, Cambria MT, Mansueto C (1999) Organometallic complexes with biological molecules. X. Dialkyltin(IV) and trialkyltin(IV) orotates: Spectroscopic and *in vivo* investigations. Appl Organomet Chem 13:145–157
Maggio F, Pellerito A, Pellerito L, Grimaudo S, Mansueto C, Vitturi R (1994) Organometallic complexes with biological molecules. II. Synthesis, solid-state characterization and *in vivo* cytotoxicity of diorganotin(IV)chloro and triorganotin(IV)chloro derivatives of penicillin G. Appl Organomet Chem 8:71–85
Mansueto C, Gianguzza M, Dolcemascolo G, Pellerito L (1993a) Effects of tributyltin(IV) chloride exposure on early embryonic stages of ciona intestinalis: *In vivo* and ultrastructural investigations. Appl Organomet Chem 7:391–399
Mansueto C, Lo Valvo M, Pellerito L, Girasolo MA (1993b) Organometallic complexes in ascidian embryonic development: II. Effects on different stages and larvae. Appl Organomet Chem 7:95–107
Mansueto C, Puccia E, Maggio F, Di Stefano R, Fiore T, Pellerito C, Triolo F, Pellerito L (2000) Organometallic complexes with biological molecules. XIV. Biological activity of dialkyl and trialkyltin(IV)-[meso-tetra(4-carboxyphenyl) porphinate] derivatives. Appl Organomet Chem 14:229–235
Mennie D, Craig PJ (1993) Analysis of organometallic compounds in the environment. In: Sigel H, Sigel A (eds) Metal ions in biological systems, vol XXIX. Marcel Dekker Inc., New York, pp 37–78
Millward GE, Ebdon L, Walton AP (1993) Seasonality in estuarine sources of methylated arsenic. Appl Organomet Chem 7:499–512
Patai S (ed) (1995) The chemistry of organic germanium, tin and lead compounds. Wiley, Chichester
Pellerito L, Maggio F, Consiglio M, Pellerito A , Stocco GC, Grimaudo S (1995) Organometallic complexes with biological molecules. IV. Di- and triorganotin(IV)-amoxicillin derivatives: Solid-state and solution-phase spectroscopic investigations. Appl Organomet Chem 9:227–239
Pellerito A, Fiore T, Giuliani AM, Maggio F, Pellerito L, Vitturi R, Colomba MS, Barbieri R (1997a) Organometallic complexes with biological molecules. VIII. Synthesis, solid state and *in vivo* investigation of triorganotin(IV) derivatives of L-homocysteic acid. Appl Organomet Chem 11:601–616
Pellerito A, Fiore T, Giuliani AM, Maggio F, Pellerito L, Mansueto C (1997b) Organometallic complexes with biological molecules: IX. Diorgano- and triorganotin(IV)[meso-tetra(4-sulfonatophenyl)-porphinate] derivatives: Solid-state and in vivo effects. Appl Organomet Chem 11:707–719
Pellerito A, Fiore T, Pellerito C, Fontana A, Di Stefano R, Pellerito L, Cambria MT, Mansueto C (1998) Organometallic complexes with biological molecules. XI. Solid state and in vivo investigations of some diorganotin(IV)-chloramphenicol and cycloserine derivatives. J Inorg Biochem 72:115–125
Rivaro P, Pensiero G, Frache R (1999) Occurrence of butyltin compounds in water and mussel samples collected in an oil port. Appl Organomet Chem 13:727–732
Santosa SJ, Wada S, Tanaka S (1994) Distribution and cycle of arsenic compounds in the ocean. Appl Organomet Chem 8:273–283
Santosa SJ, Wada S, Mokudai H, Tanakas (1997) The contrasting behaviour of arsenic and germanium in seawater. Appl Organomet Chem 11:403–412
Shibata Y, Morita M (1992) Characterization of organic arsenic compounds in bivalves. Appl Organometal Chem 6:343–349

Smith PJ (ed) (1998) Chemistry of tin. Blackie Acad and Prof, London
St-Louis R, Mora S de, Pelletier E, Doidge B, Leclair D, Mikaelian I, Martineau D (2000) Hepatic butyltin concentration in beluga whales (*Delphinapterus leucas*) from the St. Lawrence Estuary and Northern Quebec, Canada. Appl Organomet Chem 14:218–226
Visoottiviseth P, Thamamaruitkun T, Sahaphong S, Riengrojpitak S, Kruatrachue M (1999) Histopathological effects of triphenyl hydroxide on liver, kidney and gill of Nile Tilapia (*Oreochromis nilotica*). Appl Organomet Chem 13:749–763
Vitturi R, Mansueto C, Catalano E, Pellerito L, Girasolo MA (1992) Spermatocyte chromosome alterations in *Truncatella subcylindrica* (L., 1767) (mollusca, mesogastropoda) following exposure to dibutyltin(IV) and tributyltin(IV)chloride. Appl Organomet.Chem 6:525–532
Vitturi R, Catalano E, Lo Conte MR, Pellerito L (1993) Chemically induced chromosome damage in early-developing embryos of *Anilocra physodes* L. (crustacea, isopoda) following exposure to bis-(dimethyl(IV)chloro)protoporphyrin(IX). Appl Organomet Chem 7:295–302
Vitturi R, Mansueto C, Gianguzza A, Maggio F, Pellerito A, Pellerito L (1994) Organometallic complexes with biological molecules, part 3. *In vivo* cytotoxicity of diorganotin(IV)chloro and triorganotin-(IV)chloro derivatives of penicillin G on chromosomes of *Aphanius fasciatus* (Pisces, Cyprinodontiformes). Appl Organomet Chem 8:509–515
Vitturi R, Zava B, Colomba MS, Pellerito A, Maggio F, Pellerito L (1995) Organometallic complexes with biological molecules. V. *In vivo* cytotoxicity of diorganotin(IV)-amoxicillin derivatives in mitotic chromosomes of *Rutilus rubilio* (Pisces, Cyprinidae). Appl Organomet Chem 9:561–566

Part IV

Analytical and Bioanalytical Methodologies for Sea Water

Chapter 16

Flow Injection Techniques for the in situ Monitoring of Marine Processes

P. J. Worsfold · E. P. Achterberg · A. R. Bowie · R. Sandford · V. Cannizzaro · P. Gardolinski

16.1 Introduction

16.1.1 Flow Injection Techniques

Flow injection (FI) analysis has become established as an important tool for sample presentation and on-line treatment in the laboratory environment. It is now being increasingly considered for deployment outside of the laboratory, in both process and environmental locations (Andrew et al. 1994).

FI has been described as an unsegmented flow technique in which a volume of liquid sample is inserted into a moving liquid carrier stream, whereupon it undergoes physical dispersion as it is transported to a flow-through detector for measurement (Ruzicka and Hansen 1981). The transient response is usually in the form of a peak, with a sharp rising edge and a more gradual decay, the shape being due to axial dispersion and radial diffusion of the sample zone as it travels through the FI manifold. The height and area of the peak are usually directly related to analyte concentration, but for convenience, peak height is usually the measured parameter. The degree of sample dispersion is controlled by factors such as sample volume, carrier flow rate, length and diameter of the manifold tubing and manifold geometry and under most conditions is highly reproducible (relative standard deviations are typically less than 5%). The technique is now widely used in analytical laboratories for the automation of wet chemical methods and has considerable potential for use on board ships (Worsfold et al. 2000) and in submersible analysers (David et al. 1998)

A block diagram of a simple single channel FI manifold is shown in Fig. 16.1 and typically consists of a means of propulsion (e.g. a peristaltic pump), a rotary injection valve for sample introduction (similar to HPLC valves but low pressure) and a flow-through detector (e.g. a spectrophotometer). In this manifold, the carrier stream transports the sample to the detector. PTFE tubing (typically 0.8 mm i.d.) is used throughout the manifold for sample and reagent transport, with tightly-wound coils often included to enhance mixing. If the method requires more than one reagent, additional streams can be merged with the carrier stream at suitable points in the manifold. Similarly, if in-line physical treatment of the sample is required, the necessary components can easily be incorporated. These include:

- Solid phase chelating microcolumns for matrix removal and analyte pre-concentration, e.g. 8-hydroxyquinoline for the removal of the major sea water ions and simultaneous preconcentration of trace metals;

- Gas dialysis for the diffusion of a gaseous analyte from a carrier (donor) stream through a microporous membrane into a reagent (acceptor) stream, e.g. for the selective extraction of ammonia from sea water;
- Solid phase reaction columns, in which the injected sample reacts with a solid material, e.g. an immobilized enzyme packed in a column.

Reagent consumption is generally low in FI systems (an important factor for shipboard and submersible applications) and can be reduced still further by using a reagent injection manifold, whereby a discrete volume of reagent is injected into a continuously flowing sample stream. This option is suitable for applications in which the sample is in abundant supply (as in many marine situations) and is particularly beneficial when expensive reagents are required. Simultaneous FI determinations can be performed by designing manifolds in which the sample is injected into more than one flow channel, undergoing different reaction chemistries in each. FI systems are easily automated, using off-the-shelf components and a notebook PC to control the operation of the valves, pumps, data acquisition, and processing. A schematic of an automated FI manifold including the facility for on board calibration is shown in Fig. 16.2.

In the laboratory, FI has been coupled with most detection systems, but this chapter focuses on shipboard (and submersible) deployment of FI instrumentation in order to utilize its ability to acquire high quality analytical data with excellent temporal and/or spatial resolution. The following sections describe examples of FI with chemiluminescence (CL) (Bowie et al. 1996) and spectrophotometric (SPEC) detection, respectively, for the determination of trace metals and nutrients in marine waters.

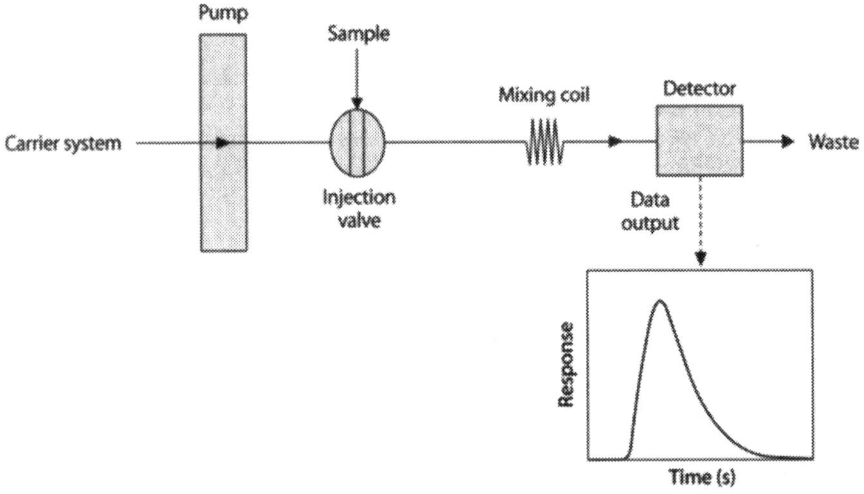

Fig. 16.1. Block diagram of a single channel flow injection manifold

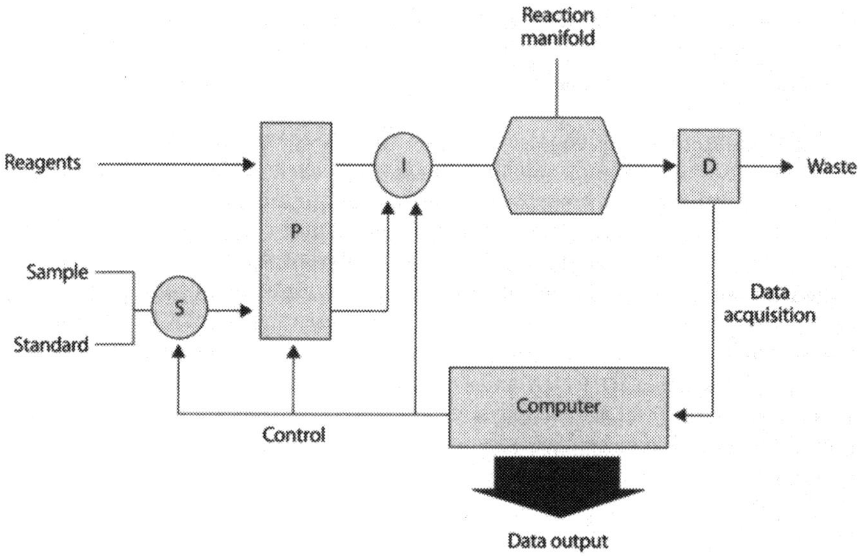

Fig. 16.2. Block diagram of an automated flow injection manifold incorporating a facility for on board calibration

16.1.2 Chemiluminescence Detection

CL is the production of light by a chemical reaction. In solution, CL has many analytical applications including the determination of metal ions in environmental matrices. The advantages of CL include high sensitivity, a wide linear dynamic range and simple instrumentation. For analytical applications, the rapid and transient nature of solution phase CL emission requires rapid and reproducible mixing of sample and reagent, for which FI is well-suited.

The measured CL emission intensity (I_{CL}) is dependent on the rate of reaction and the efficiency of the reaction at generating molecules in an excited state (expressed as the quantum yield). CL reactions commonly used in analysis have quantum yields of 0.001–0.1, but the almost complete absence of background emission (no light source) means that even very inefficient reactions with quantum yields <0.001 can be exploited. Because I_{CL} is proportional to the rate of reaction, any of the reaction components (substrate, oxidant, "catalyst," co-factor or sensitizer) can be determined by adjusting concentrations such that the analyte is the limiting reactant, i.e. all other reagents are present in excess.

For a reaction to emit CL, an excited state molecule must be produced during the course of the reaction. The observed emission arises from the ejection of a photon from this excited state. Three essential features are therefore necessary:

- The reaction must be exothermic in order to generate sufficient energy for formation of the electronically excited state (for emission in the visible region the minimum energy requirement is 180 kJ mol^{-1}).

- A pathway must exist for the formation of the electronically excited state.
- The excited state must be capable of deactivation by the emission of radiation or energy transfer to a fluorophore.

There are empirical guidelines for predicting CL behaviour. For instance, if an analyte itself or an oxidation product is fluorescent, then oxidation of the molecule may produce CL. There are, however, many exceptions to this principle, and in many cases, CL reactions cannot be predicted. Nonetheless, many solution phase CL reactions involve the oxidation of aromatic substances. One of the most commonly used is the reaction involving the oxidation of luminol (5-amino-2,3-dihydrophthalazine-1,4-dione or 5-aminophthalhydrazide), an acyl hydrazide, in basic solution (pH 10–11). The reaction is catalysed by haeme-containing enzymes (e.g. peroxidase) and several metal ions of which Co(II), Cu(II) and Fe(II) are particularly effective. In addition, hexacyanoferrate(III) can act as catalyst/co-oxidant in the reaction. Due to the rapid and transient CL emission from this reaction, it is ideally suited to incorporation within an FI manifold.

16.1.3
Spectrophotometric Detection

Spectrophotometry was the first detection system used in conjunction with FI (Ruzicka and Hansen 1981) and remains the most popular in terms of applications and published papers. The determination of nutrients in natural waters is the most common environmental application of the technique. In recent years, the advent of solid-state detectors has made FI-SPEC a viable system for deployment in the field, including shipboard use. Its compact nature means that it can also be deployed as a submersible device when incorporated in a suitable housing. This provides the possibility of autonomous operation at remote sites for days, or even weeks at a time, which would give high temporal data resolution during transient events such as storms.

16.2
FI-CL Determination of Iron in Sea Water

16.2.1
Marine Chemistry of Iron

Iron (Fe) is the fourth most abundant element in the Earth's crust (~5.6%), but like other reactive trace elements, dissolved Fe levels in open-oceanic waters are often sub-nanomolar. The marine biogeochemistry of Fe is complicated by its redox speciation, low solubility and involvement in biological cycles. Improvements to the understanding of the redox cycling of Fe are required, and chemical nature of Fe associated with various operationally defined parameters (e.g. labile, dissolved, colloidal, organically bound, particulate) needs further clarification.

The major sources of Fe to the world's oceans are atmospheric, fluvial, hydrothermal, and continental shelf regeneration and upwelling of Fe-enriched subsurface waters. In remote areas, the ocean receives the majority of surface water Fe from atmospheric dusts, and the true impact of suspended aerosols on trace element levels can only be made by direct measurements. Iron removal in surface waters is thought to be dominated by biological processes. A schematic of sources, transport and reactivity of Fe in the marine environment is shown in Fig. 16.3.

Fe(III) is the thermodynamically stable form in oxygenated sea water, existing predominantly as insoluble oxyhydroxides or colloidal matter. Fe(II) is a transient species in surface oxic waters, existing via chemical or photochemical reduction or through atmospheric wet or dry deposition and is oxidized rapidly by O_2 and H_2O_2 species at sea water pH. Recently, dissolved Fe has been shown to be strongly complexed to organic material in the marine environment (Gledhill and van den Berg 1994). Laboratory studies have shown that phytoplankton are able to utilize only dissolved Fe(II) or Fe(III) species; uptake of the colloidal or particulate forms is only possible via a thermal or photochemical dissolution pathway. Figure 16.4 illustrates the photoredox cycling of Fe.

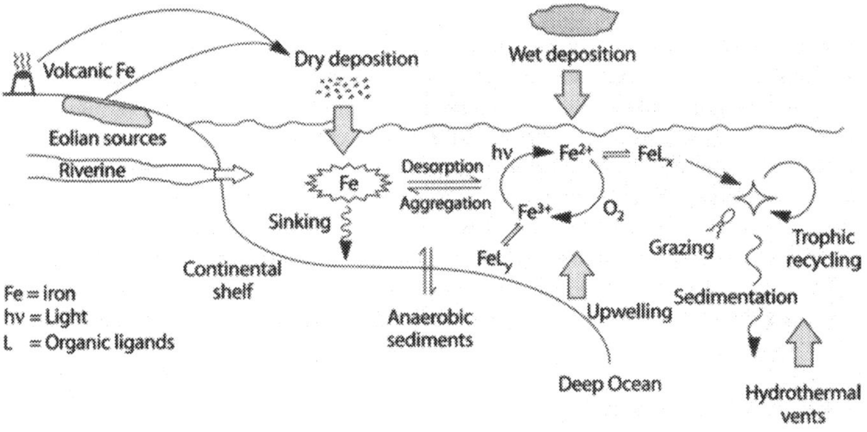

Fig. 16.3. Schematic biogeochemical cycle for iron

Fig. 16.4. Schematic photoredox cycle for iron in the marine environment

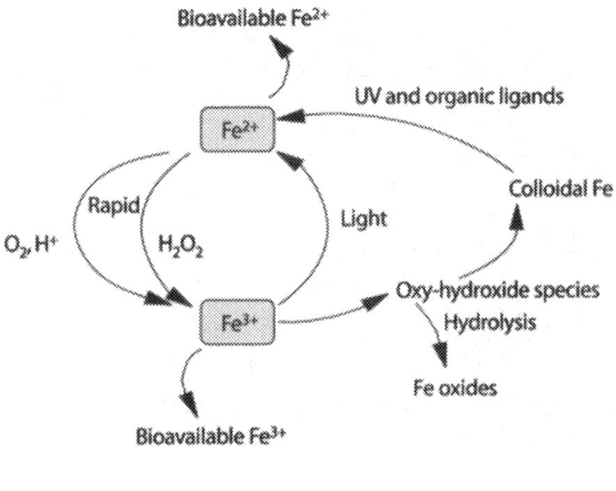

Iron is an essential micronutrient for biological organisms and limits phytoplankton growth in certain high-nutrient, low-chlorophyll (HNLC) areas of the world's oceans, which may have important implications for global carbon cycles. Such hypotheses have recently been tested in the under-productive waters of the equatorial Pacific (Coale et al. 1996) and Southern Ocean (Boyd et al. 2000), where seeding an expanse of surface water with low concentrations of dissolved Fe triggered a massive phytoplankton bloom, and resulted in a significant drawdown of surface water nitrate and atmospheric CO_2. The ongoing debate about the effect of Fe limitation on phytoplankton communities has highlighted how little is actually known about the complex marine speciation of Fe and uptake mechanisms for the growth of microorganisms. Similar mesoscale iron enrichment experiments are planned for different regions of the HNLC Southern Ocean.

16.2.2 FI-CL Manifold for Iron

The manifold for the Fe monitor, shown in Fig. 16.5 (Bowie et al. 1998), utilizes the luminol reaction (see Sect. 16.1.2) and a chelating microcolumn of 8-hydroxyquinoline (8-HQ) immobilized on a commercial hydrophilic vinyl co-polymer (Toyopearl TSK gel, HW 75F, 32-63 micron fine, Toso Haas Co.). After Fe(III) reduction by sulphite to the Fe(II) species, the 8-HQ resin extracts and preconcentrates the Fe from the sea water sample as it flows through the column. Elution is then performed using a HCl stream, which is mixed with the luminol reagent inside a quartz flow cell, emitting light that is detected by a photomultiplier tube and the output sent to a flatbed chart recorder. The analytical figures of merit for this manifold are given in Table 16.1 and show that it is suitable for the determination of Fe(II+III) in coastal and open ocean waters.

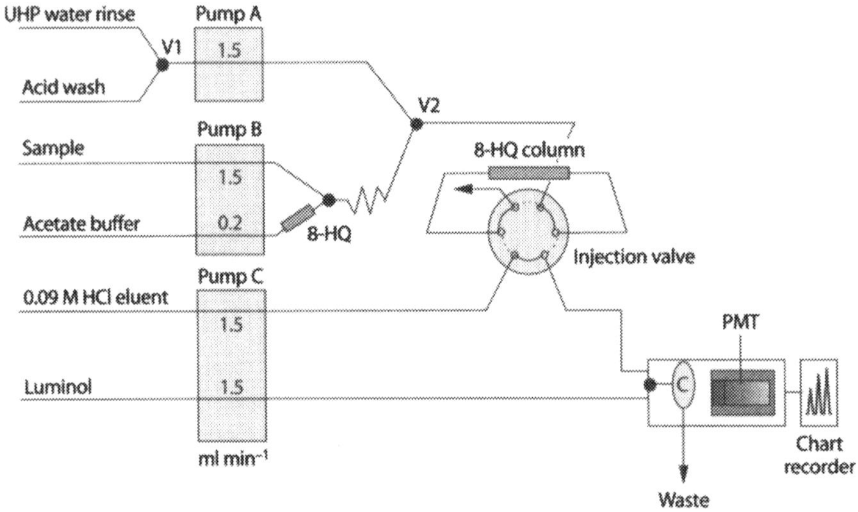

Fig. 16.5. Flow injection manifold with chemiluminescence detection for the determination of Fe(II+III) in the marine environment

Table 16.1. Analytical figures of merit for the Fe(II+III) FI-CL manifold. The time for sample quantification includes two standard additions with analysis of each solution in triplicate

LOD	0.04 nM
RSD ($n = 5$)	3–6% for 1.0 nM Fe
Linear Range	0.04–10 nM ($R^2 = 0.9976$)
Time for one analytical cycle ($n = 3$)	9 min
Time for sample quantification	28 min

16.2.3 Environmental Data

The Atlantic Meridional Transect (AMT)-3 (Aiken et al. 2000) provided an excellent opportunity to map Fe distributions in the upper water column from Portsmouth (U.K.) to Stanley (FI) on board the RRS *James Clark Ross* during September/October 1996. The ship's track crossed a range of ecosystems where oceanic conditions ranged polar to tropical, from productive European shelf seas to oligotrophic mid-Atlantic gyres. The transect from 50° N to 50° S passed through a diverse range biogeochemical provinces with distinct Fe surface water inputs mechanisms. The results from shipboard analysis of the CRM, whilst participating in AMT-3, and a laboratory based analysis were 1.95 ±0.14 nM and 2.01 ±0.12 nM, respectively. This was in good agreement with the NASS 4 certified value of 1.88 ±0.29 nM.

Figures 16.6a and b show the distribution of total dissolvable Fe (II+III), fluorescence, temperature and salinity through the upper water column at Station A308, on September 29, 1996 (24°41' N, 21°24' W). Unfiltered samples were taken at 8 depths through the upper 200 m of the ocean and acidified prior to analysis. Total dissolvable Fe concentrations may include an input from colloidal and labile particulate species present in the sea water. Daily station work showed that Fe concentrations through the upper mixed layer were highly correlated with Fe input mechanisms, hydrography and biological activity. At Station A309, high surface water levels of ~1.2 nM exist due to Fe-laden atmospheric depositions off the West African continent. The Fe content then decreases through the euphotic zone to reach a sharp minimum (~0.4 nM) at the chlorophyll fluorescence maximum. Increased Fe concentrations are then noted below 100 m where biological activity is reduced.

16.3 FI-CL Determination of Copper in Sea Water

16.3.1 Marine Chemistry of Copper

Copper (Cu) is an essential micronutrient (e.g. as an electron donor/acceptor in enzymatic reactions and electron transport mechanisms in photosynthesis), but at enhanced concentrations the dissolved cupric ion can also be toxic to marine biota. At low concentrations, Cu can inhibit growth of dinoflagellates at concentrations $<10^{-13}$ M, causing a decrease in fecundity and even death (Gledhill et al. 1997). Natural inputs (riverine, aeolian and hydrothermal venting) and anthropogenic inputs (e.g. mine

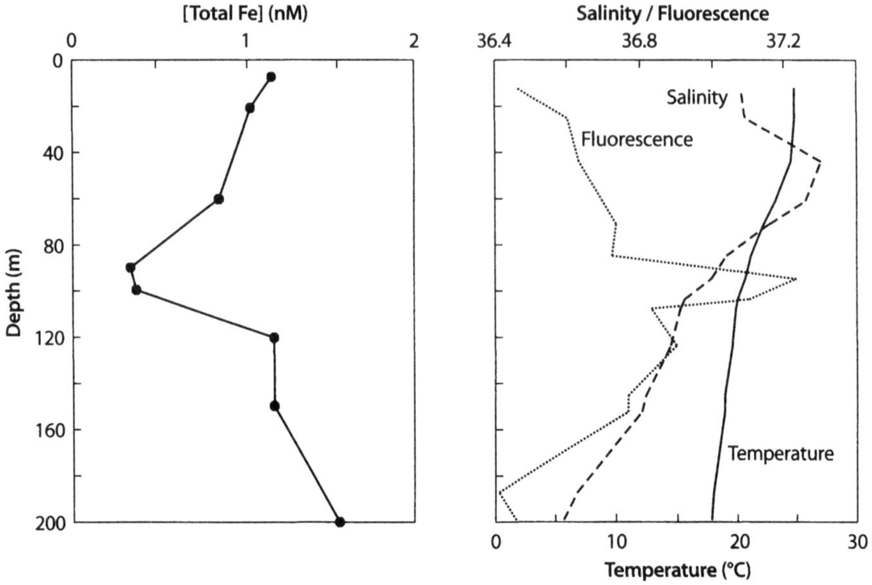

Fig. 16.6. Depth profile and associated hydrographic data for Fe(II+III) in the North East Atlantic

waste, sewage and antifouling paints) can elevate Cu levels in marine waters, particularly in coastal ecosystems.

The speciation of Cu is an important factor, determining its role in marine ecosystems, including its toxicity to marine species. The bioavailable form, and thus the potentially toxic form, is the dissolved free hydrated Cu^{2+} ion (Campbell 1995). Ambient concentrations of dissolved Cu in the open ocean range from 0.5–6 nM, of which >98% is reported to be organically complexed (Coale and Bruland 1988, 1990). The organic ligands have been classified as either weak (conditional stability constants of $K'_{CuL} = 10^7–10^{-11}$), which are usually restricted to the upper or mixed layers of the oceans, or strong (conditional stability constants of $K'_{CuL} = 10^{12}–10^{15}$), usually found throughout the water column (Moffet et al. 1990). Contemporary modelling of the biogeochemical cycling of Cu in marine ecosystems is limited by the ability to accurately measure in situ the concentrations of individual Cu species. Thus, a shipboard analytical method for the determination of Cu(II) at sub-nM levels in sea water would help our understanding of the biogeochemical cycling of Cu and its role in surface redox processes by providing high quality analytical data with good temporal and/or spatial resolution.

16.3.2
FI-CL Manifold for Copper

CL emission is generated in solution by the Cu(II) catalysed oxidation of 1,10-phenanthroline, which produces an excited diformyl species as an intermediate. The 1,10-phenanthroline combines with copper to form a chelate $[\text{Cu-phen}_n]^{2+}$ ($n = 1$ or 2). The bound copper then catalyses the decomposition of hydrogen peroxide to produce superoxide radicals, which oxidize the 1,10-phenanthroline to the excited intermediate state 3,3'-diformyl-2,2 dipyridyl (the primary CL emitter) and eventually to 2'-dipyridyl-3,3'-dicarboxylic acid (Federova et al. 1982). A cationic surfactant, cetyldimethylethylenediamine bromide (CEDAB), creates a less polar micellar environment than the aqueous phase for the uncharged 1,10-phenanthroline, and a positively charged surface for the anionic superoxide radical to migrate to. Thus, a microenvironment is created that allows a higher excitation efficiency and aids decomposition of the excited 1,2-dioxetane so formed. Tetraethylenepentamine (TEPA) is used to reduce the interference (noise) generated from trace metal contamination of the reagents. TEPA is a strong complexing agent for Cu but with slower kinetics than the main CL reaction and thus does not inhibit the primary CL emission.

Many CL reactions are unselective with respect to the metal ions that catalyse them and interfering species alter the quantitative relationship between the CL yield and the analyte concentration in the sample. Significant quenching of the CL signal for Cu(II) from the 1,10-phenanthroline reaction has been observed with some species, e.g. Na(I) (51%), Cr(III) (78%), Mn(II) (83%), and bicarbonate (12.2%) and significant enhancement with others, e.g. K(I) (271%), Mg(II) (143%) and chloride (131%) (all data normalized to 100% for the response for 10 nM Cu). These observations highlight the need for matrix separation when analysing sea water (Coale et al. 1992). A chelating microcolumn of 8-HQ (see Sect. 16.2.2) can be incorporated into the FI manifold (Fig. 16.7) for in-line, solid phase matrix separation and preconcentration. This preconcentrates Cu(II) and removes possible interferents from the sea water matrix, thereby permitting sub-nanomolar determination of Cu(II) in sea water. The analytical figures of merit for this manifold are given in Table 16.2 and show that it is suitable for the determination of Cu(II) in coastal and open ocean waters. The FI-CL analyser can also be easily automated to minimize contamination, improve reproducibility and facilitate continuous monitoring.

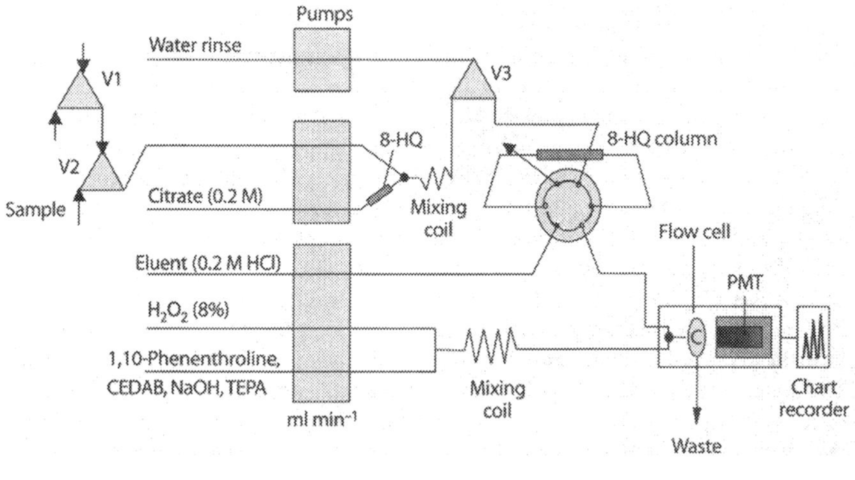

Fig. 16.7. Flow injection manifold with chemiluminescence detection for the determination of Cu(II) in the marine environment

Table 16.2. Analytical figures of merit for the Cu(II) FI-CL manifold. The time for sample quantification includes two standard additions with analysis of each solution in triplicate

LOD	0.01 nM
RSD ($n = 5$)	<5%
Linear Range	0.01–50 nM ($R^2 = 0.9548$) 0.01–10 nM ($R^2 = 0.9941$)
Time for one analytical cycle ($n = 3$)	8 min
Time for sample quantification	26 min

16.3.3 Environmental Data

The FI-CL analyser was validated by analysis of the open ocean CRM NASS 5 and by comparison with voltammetric analysis of an Irish Sea sample. The results from shipboard analysis of the CRM, whilst participating in AMT-9 (September 1999) in the North East Atlantic on RRS *James Clark Ross* as well as a laboratory based analysis were 4.70 ±0.28 nM and 4.37 ±0.17 nM, respectively. This was in good agreement with the NASS 5 certified value of 4.68 ±0.7 nM. A value of 11.5 nM Cu(II) was obtained in the Irish sea water sample using FI-CL, which compared well with the 10.9 nM Cu(II) that was determined by using cathodic stripping voltammetry. A depth profile for Cu(II) from AMT Station A905 in the North East Atlantic (38.78° N, 19.99° W) is shown in Fig. 16.8 as an example of shipboard measurements. The sample was collected using clean sampling protocols with a trace metal clean rosette system. The profile shows good agreement with previously reported data for this region (Danielsson et al. 1985).

16.4 FI-CL Determination of Cobalt in Sea Water

16.4.1 Marine Chemistry of Cobalt

Cobalt (Co) is an essential micronutrient for aquatic organisms. For example, it acts as a co-factor in the vitamin B_{12} complex and is an essential element in some metalloproteins. Co is only toxic to plants and mammals at relatively high concentrations (>17 µM), which are rarely observed in the aquatic environment (Schrauzer 1991). The oceanic concentrations of Co are extremely low (pM), and processes that control Co geochemistry in sea water are not yet well-understood. Depth profiles of Co in oceanic waters do not display the nutrient-like features seen for other micronutrient trace elements such as Cu, Ni or Zn. Instead, they show uniform or decreasing concentrations with depth (Jickells and Burton 1988; Johnson et al. 1988). Donat and Bruland (1995) reported vertical distributions of Co in the open ocean with maxima in surface waters (4–50 pM) and depleted concentrations to less than 20 pM at depth. The surface water maxima have been explained by atmospheric Co inputs (Jickells and Burton 1988). The decrease in Co concentration with depth is has been attributed to redox processes analogous or related to the geochemistry of Mn, leading to enhanced scavenging in mid and deep oceanic waters (Jickells and Burton 1988). Co scavenging in the water column may be due to the oxidation of soluble Co(II) to particle reactive

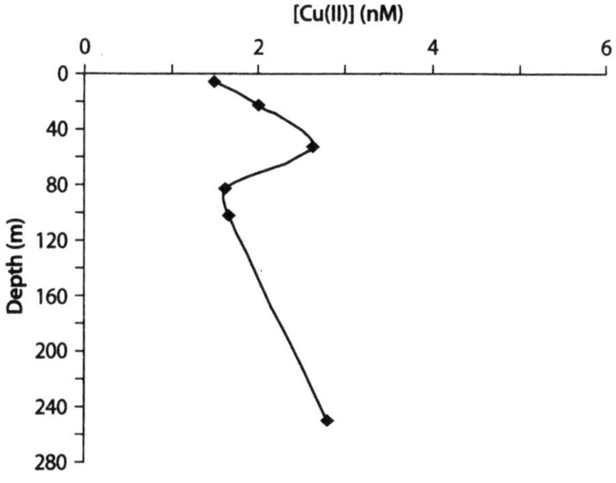

Fig. 16.8. Depth profile for Cu(II) in the Northeast Atlantic

or inert Co(III) on surfaces, probably in association with MnO_2. Co concentrations in estuarine and coastal waters are significantly higher, e.g. 140–310 pM in North Sea coastal waters near the Humber and Wash estuaries (Achterberg et al. 1999). The enhanced Co concentrations in coastal waters have been attributed to atmospheric, fluvial and sedimentary inputs (Achterberg et al. 1999). The predominant inorganic species of Co in sea water are Co^{2+} and its chloride complexes. There is evidence that Co in sea water occurs strongly complexed by organic ligands (Donat and Bruland 1988).

It has been suggested that Co may act as a (co-)limiting nutrient for marine phytoplankton (Knauer et al. 1982). Furthermore, it has been hypothesized (Price and Morel 1990) that Co may promote the growth of Zn-limited phytoplankton by substitution for Zn in some metalloenzymes. These researchers reported that in oceanic surface waters with low Zn concentrations, Co stimulated the growth of the marine diatom *Thalassiosira weissflogii*, indicating that it could be an important nutrient for algal growth for reasons other than its role in vitamin B_{12} (Babior 1975).

16.4.2 FI-CL Manifold for Cobalt

Slawinska and Slawinski (1975) reported a modification of the Trautz-Schorigin reaction (Trautz and Shorigin 1905) for the CL determination of formaldehyde and other organic compounds. This reaction, based on the oxidation of gallic acid (3,4,5-trihydroxybenzoic acid), emits light in two regions: an intense band at 643 nm and a weaker band at 478 nm. The reaction is catalysed by some trace metals, with Co(II) being the most efficient. The sensitivity for Co(II) is significantly enhanced by substituting pyrogallol for gallic acid, allowing a detection limit of 5 pM, using the FI-CL manifold shown in Fig. 16.9 (Cannizzaro et al. 2000).

The manifold incorporates an 8-HQ chelating microcolumn for in-line preconcentration and matrix removal. The sample is passed through the microcolumn for 30 seconds at a flow rate of 1.2 ml min^{-1} after in-line buffering to pH 5.1 with ammonium acetate. Ultra pure water is passed through the column for 30 seconds to remove re-

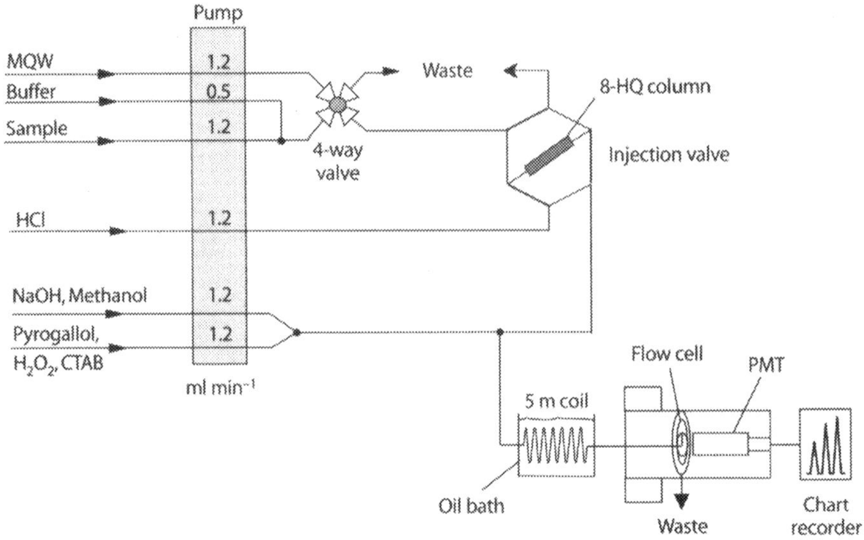

Fig. 16.9. Flow injection manifold with chemiluminescence detection for the determination of Co(II) in the marine environment

sidual alkali and alkali-earth metals (e.g. Mg and Ca) from the column. The eluant (HCl) is then passed through the column at a flow rate of 1.2 ml min^{-1} in the reverse direction to the sample loading, mixed with the reagent solutions (pumped at 1.2 ml min^{-1}) and the resulting mixture introduced into the CL flowcell after heating to 80 °C in a 5 m reaction coil. The analytical figures of merit for this manifold are given in Table 16.3 and show that it is suitable for the determination of Co(II) in coastal and open ocean waters. The accuracy of the method is shown for three marine CRMs and an Irish Sea water sample in Table 16.4.

16.4.3 Environmental Data

The FI-CL Co(II) system has been deployed at sea during the IMPACT cruise (CH 148) on board the RRS *Challenger* in the North Sea (September 16–27, 1999). Samples were collected using ultra-clean sampling protocols with modified Teflon-coated Niskin samplers. Sea water samples were filtered (0.2 µm polycarbonate membrane filters) and acidified (pH 2, quartz-distilled HCl) prior to analysis on board ship. A typical depth profile for Co in the coastal waters of the North Sea (Station 28; 54.02° N, 01.52° E) is shown in Fig. 16.10. Co concentrations ranged between ca. 140 and 190 pM at Station 28, with highest concentrations in the bottom waters. The observed concentrations are in good agreement with values observed previously in this area (between ca. 100 and 200 pM; Achterberg et al. 1999). The salinity profile shows a surface minimum, with a well-mixed water column below ca. 10 m depth. The surface minimum is possi-

Table 16.3. Analytical figures of merit for the Co(II) FI-CL manifold. The time for sample quantification includes two standard additions with analysis of each solution in triplicate

LOD	0.005 nM
RSD ($n = 5$)	<5%
Linear Range	0.005–1 nM ($R^2 = 0.9968$)
Time for one analytical cycle ($n = 3$)	8 min
Time for sample quantification	26 min

Table 16.4. Laboratory based determination of Co in three marine certified reference materials and an Irish Sea water sample. *Error bars* represent ±3 *s*

Sample	FI-CL (nM)	Certified value (nM)	Cathodic stripping voltammetry (nM)
NASS-4	0.16 ±0.01	0.15 ±0.02	–
CASS-3	0.60 ±0.09	0.68 ±0.11	–
SLEW-2	0.93 ±0.13	0.87 ±0.21	–
Irish SW	0.35 ±0.02	–	0.34 ±0.01

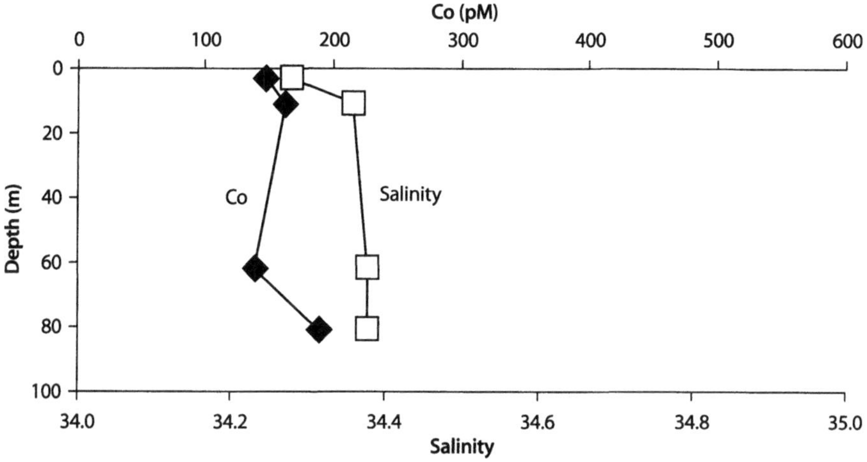

Fig. 16.10. Depth profile for Co(II) in the North Sea

bly due to rain inputs. The Co profile shows a surface minimum (150 pM) coinciding with the salinity minimum and a maximum near the bottom of the water column. The deep water maximum can probably be explained by a sedimentary flux of Co into the overlaying waters. Remobilization of Co from sediments of the North Sea was reported by Burton et al. (1993) and Tappin et al. (1995), following a reduction of the redox potential at the sediment-water interface as a result of mineralization of biogenic material.

16.5 FI-SPEC Determination of Nitrate in Sea Water

16.5.1 Marine Chemistry of Nitrate

The elemental gas (N_2) is the most abundant form of nitrogen in estuarine and coastal waters. For biogeochemical processes, however, the most important species are the dissolved inorganic forms (nitrate, nitrite and ammonia) and organic nitrogen compounds (in dissolved and particulate forms), of which the predominant species in oxic waters is nitrate.

Nitrate plays an important role in primary production in euphotic surface waters, and as a growth-limiting nutrient, it can directly affect the quality of freshwater and marine environments. Therefore, an accurate and reliable technique for determining nutrient concentrations is essential, and a field deployable instrument is desirable to minimize sample degradation and provide high temporal and spatial resolution monitoring. A submersible FI monitor incorporating spectrophotometric (SPEC) detection is therefore a potentially attractive approach.

16.5.2 Submersible FI Monitor

A prototype instrument has been described (David et al. 1998, 1999). The current design is contained within a pressure housing engineered from a single block of PVC in which the FI manifold (Fig. 16.11) is mounted, together with control and processing boards. The FI monitor consists of two battery powered 12 V ISMATEC peristaltic pumps, two switching valves, one acrylic "T" piece for mixing reagents, a sample loop, reduction column, flow cell and detector. Figure 16.12 shows a photograph of the complete monitor, including pressure housing.

For total oxidized nitrogen (*TON*) determination, the FI chemistry is based on the reduction of nitrate to nitrite in a copperized cadmium column followed by diazotisation

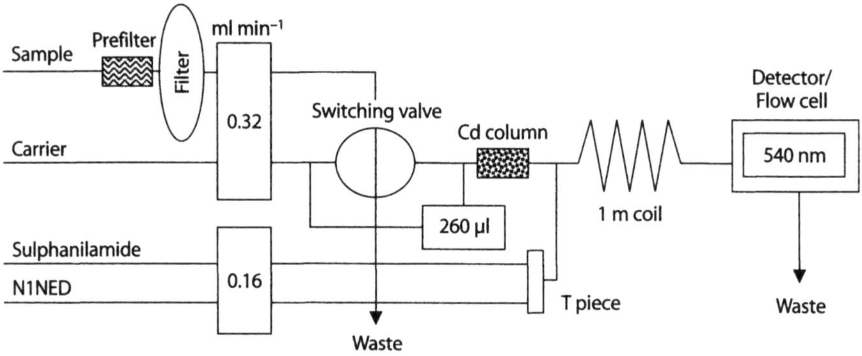

Fig. 16.11. Flow injection manifold with spectrophotometric detection for the determination of total oxidized nitrogen in the marine environment

Fig. 16.12. Photograph of the submersible nitrate monitor showing the flow injection manifold (*left*) and the pressure housing (*right*)

and subsequent coupling with sulphanylamide and N-(1-naphthyl)ethylenediamine (N1NED) to produce a pink-purple azo dye (λ_{max} = 540 nm). This is quantified using a flow-through, solid-state detector incorporating an ultrabright green light emitting diode (LED) as the light source and a photodiode to detect the transmitted radiation. When a sample or standard is injected into the FI manifold, the baseline value and transient peak maximum are recorded. The *TON* concentration, sample identification, date and time are then stored in protected memory. The analytical figures of merit achieved for the FI-SPEC monitor in the laboratory (Table 16.5) show that the limit of detection is sufficiently low for deployment in most marine environments.

16.5.3 Environmental Data

In situ application of the submersible FI system for monitoring of *TON* in the Tamar Estuary and the North Sea has been described (David et al. 1998, 1999). In addition, the monitor has been used on board the RRS *Challenger* during the North Sea Impact Cruise in September 1999. The main objective of this cruise was to study contaminant behaviour in the North Sea in order to predict potential toxic effects of contaminants. The monitor was deployed successfully for nine days, operating on a 30 min cycle time, and provided high quality results for more than 100 samples. The analytical figures of merit of the FI-SPEC monitor during the cruise are summarized in Table 16.6 and show that the limit of detection and linear range can be easily adjusted during the cruise to suit local conditions by, e.g. varying the optical path length. A typical moni-

Table 16.5. Analytical figures of merit for the FI-SPEC submersible nutrient monitor in the laboratory with a 20 mm path length flow cell. The time for sample quantification includes analysis of a sample and an on board standard in triplicate and data processing time

LOD	0.5 µM
RSD ($n = 3$)	<5%
Linear Range	0–10 µM ($R^2 = 0.993$)
Time for one analytical cycle ($n = 3$)	9 min
Time for sample quantification	30 min

Table 16.6. Shipboard analytical figures of merit for the FI-SPEC nutrient monitor during the Impact cruise. The time for sample quantification includes analysis of a sample and an on board standard in triplicate and data processing time

LOD (20 mm path length)	0.6 µM
LOD (10 mm path length)	3.6 µM
RSD ($n = 3$)	<5%
Linear Range (20 cm path length)	0–7.1 µM ($R^2 = 0.994$)
Linear Range (10 cm path length)	3.6–140 µM ($R^2 = 0.997$)
Time for one analytical cycle (n=3)	9 min
Time for sample quantification	30 min

tor profile for *TON* and salinity during a tidal cycle deployment in the mouth of Humber Estuary (53.32° N, 00.05° E) on 25th September 1999 is shown in Fig. 16.13.

An intercomparison exercise for nutrients in sea water was performed by the National Oceanic and Atmospheric Administration – National Research Council of Canada, when sea water samples (MOOS-1) were analysed by 32 laboratories for nutrients (*TON*, nitrite, phosphate and silicate). The consensus mean and standard deviation for *TON* were 22.5 and 3.1 µM, respectively. These samples were also analysed using the submersible FI monitor and the results, 25.1 and 2.0 µM, respectively for the mean and standard deviation were in good agreement with the consensus data when using the z-scoring system for assessing bias ($z < 2$). The results obtained for the same samples using a laboratory segmented continuous flow analyser were also in good agreement with the FI-SPEC monitor ($P = 0.05$), demonstrating that the submersible FI data can be directly compared with historical laboratory data for *TON*.

16.6 Conclusions

Flow injection techniques are well-suited to marine analytical chemistry applications, particularly as shipboard systems for high temporal and/or spatial resolution monitoring. Chemiluminescence is a robust and sensitive means of detection for certain trace metals (e.g. Fe, Cu, Co) and can be made selective by the incorporation of solid phase microcolumns containing immobilized chelating reagents such as 8-hydroxyquinoline. Solid state spectrophotometric detectors can also be incorporated within flow injection systems for the in situ determination of a wide range of analytes, including total oxidized nitrogen. With suitable engineering, such systems can be configured for reliable, long-term submersible deployment. Flow injection in its various forms therefore provides a valuable tool for marine analytical chemists and has the

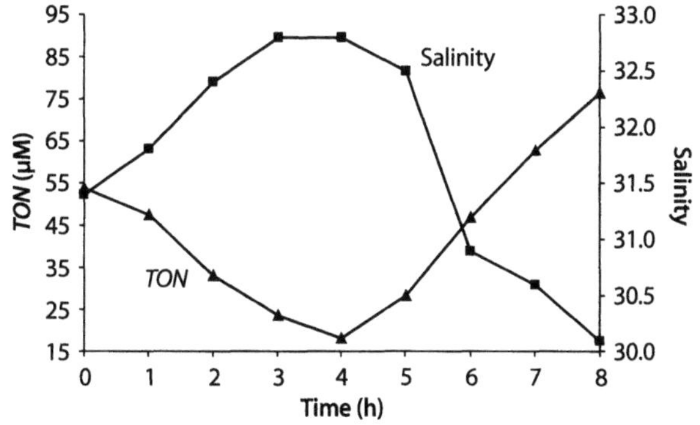

Fig. 16.13. Tidal cycle for nitrate and salinity in the North Sea

potential to provide high quality analytical data to support developments in chemical oceanography.

Acknowledgements

PJW, EPA, ARB, VC and RS acknowledge the European Union for funding (MEMOSEA contract MAS3-CT97-0143 and IRONAGES contract EVK2-CT-1999-00031). PG would like to thank CNPq, Brazil, for financial support. RS would like to thank the University of Plymouth and Plymouth Marine Laboratory for financial support.

References

Achterberg EP, Colombo C, Berg CMG van den (1999) The distribution of dissolved Cu, Zn, Ni, Co and Cr in English coastal surface waters. Cont Shelf Res 19:537–558

Aiken J, Rees N, Hooker S, Holligan P, Bale A, Robins D, Moore G, Harris R, Pilgrim D (2000) The Atlantic meridional transect: Overview and synthesis of data. Prog Oceanogr 45:257–312

Andrew KN, Blundell NJ, Price D, Worsfold PJ (1994) Flow injection techniques for water monitoring. Anal Chem 66:916A–922A

Babior BM (ed) (1975) Cobalamin: Biochemistry and pathophysiology. Wiley, New York

Bowie AR, Achterberg EP, Mantoura RFC, Worsfold PJ (1998) Determination of sub-nanomolar levels of iron in seawater using flow injection with chemiluminescence detection. Anal Chim Acta 361:189–200

Bowie AR, Sanders MG, Worsfold PJ (1996) Analytical applications of liquid phase chemiluminescence reactions – a review. J Biolumin Chemilum 11:61–90

Boyd PW, Watson AJ, Law CS et al. (2000) Mesoscale iron fertilisation elevates phytoplankton stocks in the polar Southern Ocean. Nature 407:695–702

Burton JD, Althaus M, Millward GE, Morris AW, Statham PJ, Tappin AD, Turner A (1993) Processes influencing the fate of trace metals in the North Sea. Philos T Roy Soc 343:557–568

Campbell PGC (1995) Interactions between trace metals and aquatic organisms: A critique of the free ion activity model. In: Tessier A, Turner DR (eds) Metal speciation and bioavailability in aquatic systems. John Wiley and Sons, New York, pp 45–103

Cannizzaro V, Bowie AR, Sax A, Achterberg EP, Worsfold PJ (2000) Determination of cobalt and iron in estuarine and coastal waters using flow injection with chemiluminescence detection. Analyst 125:51–57

Coale KH, Bruland KW (1988) Copper complexation in the Northeast Pacific. Limnol Oceanogr 33:1084–1101

Coale KH, Bruland KW (1990) Spatial and temporal variability in copper complexation in the North Pacific. Deep-Sea Res 47:317–336

Coale KH, Johnson KS, Stout PM, Sakamoto CM (1992) Determination of copper in sea water using a flow-injection method with chemiluminescence detection. Anal Chim Acta 266:345–351

Coale KH, Johnson KS, Fitzwater SE et al. (1996) A massive phytoplankton bloom induced by an ecosystem-scale iron fertilisation experiment in the equatorial Pacific Ocean. Nature 383:495–501

Danielsson L-G, Magnusson B, Westerlund S (1985) Cadmium, copper, iron, nickel and zinc in the Northeast Atlantic Ocean. Mar Chem 17:23–41
David ARJ, McCormack T, Morris AW, Worsfold PJ (1998) Submersible flow injection-based sensor for the determination of total oxidised nitrogen in coastal waters. Anal Chim Acta 361:63–72
David ARJ, McCormack T, Worsfold PJ (1999) A submersible battery-powered flow injection (FI) sensor for the determination of nitrate in estuarine and coastal waters. J Auto Meth Man Chem 21:1–9
Donat JR, Bruland KW (1988) Direct determination of dissolved cobalt and nickel in seawater by differential pulse cathodic stripping voltammetry preceded by adsorptive collection of their nioxime complexes. Anal Chem 60:240–244
Donat JR, Bruland KW (1995) Trace elements in the oceans. In: Salbu B, Steinnes E (eds) Trace elements in natural waters. CRC Press, Boca Raton, pp 247–281
Federova OS, Olkin SE, Berdnikov VM (1982) The chemiluminescence mechanism in 1,10-phenanthroline during catalytic decomposition of hydrogen peroxide. Z Phys Chemie 263:529–549
Gledhill M, Berg CMG van den (1994) Determination of complexation of iron(III) with natural organic complexing ligands in seawater using cathodic stripping voltammetry. Mar Chem 47:41–54
Gledhill M, Nimmo M, Hill SJ, Brown MT (1997) Toxicity of Cu(II) species to marine algae, with particular reference to macroalgae. J Phycol 33:2–11
Jickells TD, Burton JD (1988) Cobalt, copper, manganese and nickel in the Sargasso Sea. Mar Chem 23:131–144
Johnson KS, Stout PM, Berelson WM, Sakamoto-Arnold CM (1988) Cobalt and copper distributions in the water of Santa Monica basin, California. Nature 332:527–530
Knauer GA, Martin JH, Gordon RM (1982) Cobalt in Northeast Pacific waters. Nature 297:49–51
Moffet JW, Brand LE, Zika RG. (1990) Distribution and potential sources and sinks of copper chelators in the Sargasso Sea. Deep Sea Res I 37:27–36
Price NM, Morel FMM (1990) Cadmium and cobalt substitution for zinc in a marine diatom. Nature 344:658–660
Ruzicka J, Hansen EH (1981) Flow injection analysis. Wiley Interscience, New York
Schrauzer GN (1991) Cobalt. In: Merian E (ed) Metals and their compounds in the environment: Occurrence, analysis, and biological relevance. VCH, Heidelberg, pp 879–892
Slawinska D, Slawinski I (1975) Chemiluminescent flow method for the determination of formaldehyde. Anal Chem 47:2101–2109
Tappin AD, Millward GE, Statham PJ, Burton JD, Morris AW (1995) Trace metals in the Central and Southern North Sea. Estuar Coast Shelf S 41:275–323
Trautz M, Schorigin P (1905) Z Wiss Photogr Photochem 121:3
Worsfold PJ, Achterberg EP, Bowie AR, Sandford RC, Mantoura RFC (2000) Flow injection with chemiluminescence detection for the shipboard monitoring of trace metals. In: Varney MS (ed) Chemical sensors in oceanography. Gordon and Breech, Amsterdam, pp 71–94

Chapter 17

Luminescence for the Analysis of Organic Compounds in Natural Waters

A. Roda · P. Pasini · M. Guardigli

17.1 Introduction

Many widely used synthetic chemicals, as well as industrial by-products, are continuously emitted into the environment. Even if present in low quantities, some of these substances or their degradation products represent a threat for the health of humans and wildlife. Sometimes they accumulate in living organisms and target organs (bioaccumulation) so that their concentrations increase along the food chain. Natural waters (rivers, lakes, seas, and ground waters) are among the most endangered environments, since in many cases these substances are directly released in them. Moreover, natural waters represent an important way by which pollutants travel along long distances. Molecules with low water solubility, i.e. with high lipophilicity (pesticides, aromatic hydrocarbons and their derivatives, surfactants, oestrogen-like compounds, etc.) could be present in the particulate matter of fresh and marine water. The analysis of organic compounds in natural waters is therefore of crucial importance in the monitoring of pollution and in risk assessing for humans and environment. This requires rapid and sensitive analytical procedures for the detection of organic pollutants that should be suitable for performing screening analysis of a large number of samples.

Separative chromatographic and hyphenated techniques (such as GC-MS and HPLC-MS) are the most diffused conventional analytical procedures used for the analysis of organic compounds in environmental samples. However, these analytical procedures always require sample preparations, even in the case of environmental matrices as simple as drinking water. Sample preparations are frequently time-consuming, and due to the numerous and sometimes complicated processes involved, errors and sample losses are most likely to occur in this step. In addition, extraction of samples for chromatographic analysis usually requires the use of organic solvents, which frequently poses problems in handling, storage and disposal. A further disadvantage of these techniques is that they are not generally suitable for performing parallel analysis of a large number of samples.

17.2 Immunoassays in Environmental Analysis

In recent years, the expansion of immunochemical detection methods moved beyond the clinical diagnostic field to applications in environmental monitoring. Nowadays, immunoassay is a widely-used analytical method for the detection and quantitation of environmental pollutants. Immunochemical analytical methods have gained favour in environmental analytical processes beginning in the 1980s, when they were recognized as useful screening techniques for the detection of compounds of environmental regulatory concern in many countries (Van Emon 1987; Van Emon and Gerlach 1995). The U.S. Food and Drug Administration (FDA) joined the U.S. Environmental Protection Agency (EPA) in researching immunochemical detection methods for a variety of regulatory and monitoring applications. The high specificity of antibodies used as biospecific recognition elements in immunochemical methods allows for the reduction in the number of pre-analytical steps, and little or no nonaqueous solvents are used. When complex matrix samples have to be analyzed, a separation step of the analyte from the matrix is still required before performing the quantitative step. On the other hand, on samples with a relatively simple and stable matrix with low component variability, such as natural waters, the factors affecting the quantitative procedure can be minimized and kept under control, so that immunoassay can be carried out directly on the sample without any clean-up procedure. Moreover, immunological methods can be easily developed or converted in high throughput screening (HTS) formats (for example, in 96- or 384-well microtitre plates), and optimized to provide quick results in field applications.

17.2.1 Luminescent Immunoassays

The most common immunoassay format in environmental analysis is the ELISA (Enzyme-Linked ImmunoSorbent Assay). Enzyme immunoassays, in which a suitable enzyme is used as a label, allow signal amplification due to the enzyme-catalysed reaction, thus obtaining a high sensitivity. Quantitation by ELISA has become more accepted in the past decade and is now a laboratory standard for environmental analysis, frequently surpassing traditional chromatographic methods in sensitivity, selectivity and cost (Van Emon and Lopez-Avila 1992). Due to the low molecular weight of most of the organic pollutants, competitive immunoassays are the preferred format in environmental analysis. In a typical competitive indirect heterogeneous ELISA, the analyte contained in the sample and an analyte derivative immobilized on a solid support (e.g. a well of a 96-wells microtitre plate) compete for an enzyme-labelled antibody. After the immunoreaction and the washing steps, the enzyme-labelled antibody bound to the solid support is quantified by means of the enzyme-catalysed reaction; the amount of analyte contained in the sample is inversely proportional to the fraction of bound enzyme-labelled antibody. In an alternative format (direct heterogeneous ELISA), the antibody is immobilized on the solid phase, and an enzyme-labelled analyte derivative is used.

Most of the environmental applications involve the determination of a compound or group of compounds present in trace concentrations. Immunoassays commonly used in environmental analysis rely on colourimetric detection procedures, i.e. quantitation is performed through the use of a chromogenic enzyme substrate. The intrinsic detectability of immunoassays can be further improved by the use of luminescent labels such as enzymes detectable by bio- or chemiluminescent (CL) substrates or lanthanide chelates detectable by time-resolved fluorescence (TRF) techniques.

Chemiluminescent detection is probably the most sensitive detection principle for enzyme labels, and the analytical sensitivity of chemiluminescent immunoassays is usually higher than that of colourimetric immunoassays. In many cases, the sensitivity of chemiluminescent immunoassays is comparable to or even better than that obtained with radioactive labels. This trend was clearly demonstrated by the comparison of the detection limits obtained in the colourimetric, luminescent and radio-immunological determination of steroids in real samples, like plasma and saliva (Roda et al. 1984). It should be noted that saliva has a composition resembling quite closely that of surface waters, thus suggesting the convenient applicability of chemiluminescent immunoassays to natural water samples. In addition, luminescent detection is generally characterized by a rapid response, and its dynamic range can be linear for up to 5 decades of concentration. Electrochemiluminescence represents an alternative way to implement luminescent detection in immunoassays. This technique offers the advantage that the luminescent signal is triggered by the application of a suitable potential rather than by the addition of a chemiluminescent reagent, thus reducing problems related to signal handling or to the background emission shown by some chemiluminescent substrates. Time-resolved fluorescence measurement techniques represent an improvement over the conventional fluorescent detection. Fluorescence potentially allows a high detectability, but its sensitivity is reduced by the presence of a background emission, which is mainly due to the excitation of the biological components of the sample. Time-resolved fluorescence detection, in which the measurement of the emission is delayed with respect to the excitation of the sample and long-lived luminescent labels are used, allows one to efficiently suppress the background emission, thus increasing the detectability of the luminescent probe.

17.2.2 Applications

Most of the luminescent immunoassays developed for analysis of environmental water samples are used for the detection and quantitation of pesticides (Table 17.1). Benzo(a)pyrene was detected in environmental water samples using immunoassay methods. It was found that the immunological method could be used for the direct

Table 17.1. Luminescent immunoassays for the detection of organic compounds in water samples

Analyte	Detection principle	Reference
Benzo(a)pyrene	Time-resolved fluorescence	Ius et al. (1992)
Phthalate esters	Time-resolved fluorescence	Ius et al. (1993)
Paraoxon, Aldicarb	Chemiluminescence	Roda et al. (1994)
Chlortoluron	Chemiluminescence	Kameth et al. (1996)
Triazine, 2,4-Dichlorophenoxyacetic acid, Trinitrotoluene	Chemiluminescence	Weller et al. (1999)
2,4-Dichlorophenoxyacetic acid	Chemiluminescence	Dzgoev et al. (1997)
Atrazine	Electrochemiluminescence	Wilson et al. (1997)
2,4-Dichlorophenoxyacetic acid	Electrochemiluminescence	Marquette and Blum (1998)
Atrazine, Terbuthyilazine, Ametryn	Chemiluminescence	Samsonova et al. (1999)

analysis of both freshwater and sea water samples, since the salt content of the sea water did not interfere in the assay (Roda et al. 1991). Instead, a preliminary desalting procedure was necessary if the sea water sample was preconcentrated before analysis.

A time-resolved fluoroimmunoassay was developed for the detection of benzo(a)-pyrene (Ius et al. 1992) on the basis of the previously reported work. This direct heterogeneous immunoassay was developed using a specific antibody labelled with an isothiocyanatophenyl-EDTA-Eu complex. After the immunoreaction, the lanthanide complex was dissociated, and the amount of metal ion in solution was determined via a time-resolved fluorescence measurement upon addition of an enhancement solution forming a luminescent europium complex. The detection limit (about 0.5 ng ml^{-1}) was low enough for the application of the method on water samples without preliminary extraction or concentration procedures. Phthalate esters, widely used as plastifiers, were determined in a direct competitive immunoassay based on a time-resolved fluorescence measurement using rabbit antibodies against phthalate esters (Ius et al. 1993). The bound antibodies were detected using a biotinylated anti-rabbit antibody and a BCPDA-labelled streptavidin. Metal luminescence intensity was then measured in the time-resolved mode upon addition of an excess of europium ion. This assay was able to detect as low as 0.5 pmol ml^{-1} of the most diffused phthalate esters.

A chemiluminescent flow sensor was developed for the determination of organophosphorus (Paraoxon) and carbamate (Aldicarb) pesticides in water (Roda et al. 1994). The flow sensor relies on the inhibition of acetylcholinesterase (either in solution or immobilized on a solid support) due to pesticides. A series of coupled enzymatic reactions, catalysed by the enzymes choline oxidase and peroxidase immobilized on a solid support, was used in order to allow the chemiluminescent measurement of the acetylcholinesterase activity. Detection limits of 0.75 μg l^{-1} and 4 μg l^{-1} were reported for Paraoxon and Aldicarb, respectively.

The water sample content of the herbicide chlortoluron was determined using an enhanced chemiluminescent immunoassay (Kameth et al. 1996) in which the chemiluminescence signal was detected by means of a camera luminometer, providing a semiquantitative assay based on a photographic record of the luminescent end point. This assay could represent a rapid, simple and portable means of monitoring multiple water samples for the presence of chlortoluron or other pesticides. In the reported example, the assay was able to identify and quantitate the herbicide with good accuracy at or above the European limit for individual pesticides in drinking water.

Detection of atrazine in water was also performed by means of an electrochemiluminescence flow injection immunoassay based on anti-atrazine antibodies labelled with glucose oxidase (Wilson et al. 1997). This competitive electrochemiluminescent immunoassay used transparent aminosilanized indium tin oxide-coated glass electrodes, derivatized with aminodextran covalently linked to a triazine derivative. It was possible to detect less than 0.1 ppb of atrazine, thus below the precautionary limit for pesticides in drinking water recommended by the European Commission.

A flow analysis system coupled with an immunosensor exploiting the electrogenerated chemiluminescence of luminol was used for the detection of 2,4-dichlorophenoxyacetic acid (2,4-D) in drinking water (Marquette and Blum 1998). The immunosensor is based on a competitive immunoreaction, in which 2,4-D contained in the water sample and 2,4-D immobilized on the surface of a glassy carbon electrode compete for a luminol-labelled anti-2,4-D antibody. After the immunoreaction, the amount of bound antibody is detected by applying a suitable potential to the electrode, in order to obtain the luminol label electrochemiluminescence. An innovative aspect of this device with respect to other 2,4-D immunosensors reported in the literature is the novel immobilization procedure used to obtain the 2,4-D antigen-coated glassy carbon electrode, which allowed multiple (up to 50 times) regenerations of the immunosensor. This immunosensor was able to detect as low as 0.2 $\mu g\,l^{-1}$ of free 2,4-D, which is close to the accepted level in drinking water in the European Union (0.1 $\mu g\,l^{-1}$).

A chemiluminescent competitive multianalyte immunoassay based on a combination of mono- and polyclonal antibodies was developed for the detection of three cross-reacting s-triazine pesticides (atrazine, terbuthylazine and ametryn) in water samples (Samsonova et al. 1999). Detection of single s-triazines using immunological methods is a difficult task, because even monoclonal antibodies against such small analytes are often unable to efficiently discriminate between structurally similar molecules. The simultaneous use of three polyclonal and two monoclonal antibodies with different specificities towards s-triazines allowed for the development of a multianalyte assay in which antibody cross-reactivity is exploited to identify and quantify the three analytes. The antibodies were immobilized in separate wells of an eight-well microtitre strip, and horseradish peroxidase, which was detected with the enhanced chemiluminescence reaction, was used as a label for antigens. Measurement of the chemiluminescence intensities was performed by using a small, battery-powered portable luminometer, which is of practical importance from the on-site monitoring point of view. The data obtained with the chemiluminescent immunoassays were processed using a neural network, and the best parameters for the correct identification of an individual s-triazine in a mixture and the estimation of its amount were found. When this assay was applied to environmental samples containing various s-triazine mixtures, it was able to correctly classify the analytes in most cases, suggesting that it can be proposed as an alternative field test for multianalyte environmental monitoring. In all the luminescent immunoassays described above, luminescence measurements were performed using conventional photomultiplier tube-based luminometers or time-resolved fluorometers. Imaging devices, which are able not only to measure the intensity but also to evaluate the spatial distribution of the luminescence emission from a sample, represent a significant improvement in luminescence measurement. These devices allow for the simultaneous measurement of the luminescent signals from a whole 96- or 384-well microtitre plate, thus being more rapid than conventional luminometers that measure light output well by well or strip by strip. They also represent the detectors used for the development of multiarray affinity devices, in which several different biospecific reagents are immobilized in an array arrangement on a suitable surface. This sensor design allows for the parallel processing of many immunoassays, allowing simultaneous determination of different analytes in the same sample, using reagents volumes much smaller than those needed to perform separate immunoassays for each analyte of interest. Quantitation of the analyte content of the sample is performed by using image analysis techniques: the localization of the CL signal on the target surface and its intensity are related to the identity and the concentration of the analyte, respectively.

The development of a parallel affinity sensor array based on chemiluminescent labels for the detection of environmental contaminants in water has been reported (Weller et al. 1999). The required reagents (antibodies or haptens) were immobilized as spots on a glass slide, obtaining a biochip with an array containing up to 1600 spots. The glass slide was inserted in a flow cell in which the immunoreactions took place and, after addition of the chemiluminescent substrate, the CL signal was acquired by means of a CCD imaging device. The biochip was used to develop model systems for the simultaneous detection and quantitation of triazine, trinitrotoluene and 2,4-D in water samples based on immunoassays either in direct (antibody-immobilized) or indirect (hapten-immobilized) formats. Both formats proved to be able to detect and quantitate the analytes, offering the possibility of many regeneration cycles of the biochip. This device could thus represent a potential analytical tool for performing multiple analyses on one sample in parallel, also including the identification of the analytes.

The development and optimization of a chemiluminescent ELISA for the simultaneous determination of 2,4-D in multiple samples, using innovative solid supports represented by gold-coated surfaces and glass capillaries with CCD imaging of the chemiluminescent signal has been reported (Dzgoev et al. 1997). The authors observed that such solid supports offered advantages with respect to other supports used in the development of multiarray devices such as the "flat-wells" obtained by thick-film technology, but the analytical sensitivity was still lower than that of the single sample determination assay.

The perception of immunoassay has changed in the past decade from considering it to be an environmental analytical alternative to accepting it as a reliable quantitative method. Luminescent immunoassays represent a valuable alternative to the conventional radiometric or colourimetry ones. They are more sensitive than colourimetry immunoassays, and can be as sensitive as radioimmunoassays, without all the problems related to storage, handling and disposal of radioactive materials. This feature, combined with the rapidity of response and the possibility of performing simultaneous analysis of several samples in a 96- or 384-well microtitre plate format, makes these assays suitable for the development of HTS analytical methods and for extensive automation. New refinements of immunoassay technology, such as biochip-based methods and multianalyte immunoassays, while still in research, will certainly take their place among other environmental monitoring techniques of the future.

17.3 Luminescent Recombinant Cell-Based Biosensors in Environmental Analysis

A different bioanalytical approach has been made possible by the use of recombinant DNA techniques in analytical chemistry, which led to the development of luminescent whole-cell biosensors. Luminescence-based reporter genes have been used to develop efficient bioassays for the specific detection of several inorganic and organic compounds, which are applied especially in the environmental analysis field (Köhler et al. 2000). These biosensors are based on the ability of genetically engineered microorganisms (bacteria and yeast) or mammalian cells to emit visible light in response to specific substances. Such transformed cells are constructed by recombinant DNA techniques, using a plasmid vector containing the luminescent reporter gene under transcriptional control of a specific gene sequence, whose expression is regulated very precisely. The presence of the analyte induces the expression of the specific gene sequence and consequently of the reporter gene in a quantitative manner, followed by the synthesis of the corresponding protein. *Luc* and *lux* genes encoding firefly and marine bacteria luciferases, respectively, are widely used as reporters. When the corresponding luciferin substrate is either added or produced by the recombinant cells, a luciferin/luciferase-mediated light output occurs. The light emission intensity is proportional to the analyte concentration, which allows quantitative analysis to be performed. The use of enzymes as reporter molecules permits the analytical signal amplification, which combined with the high detectability of bioluminescent reactions, lowers the detection limit of the system. Recently, the genes encoding two naturally occurring and recombinantly produced luminescent proteins from the jellyfish *Aequorea victoria*, i.e. aequorin and green fluorescent protein (GFP), have been used as reporter genes in the development of recombinant cell-based biosensors. Aequorin is naturally bioluminescent upon binding to Ca^{2+} ions and has virtually no associated background signal, thus allowing highly sensitive detection. GFP is an autofluorescent protein that does not require addition of substrates or co-factors. These characteristics made the use of these photoproteins very attractive in diverse analytical applications (Shetty et al. 1999; Lewis and Daunert 2000).

17.3.1 Applications

Luminescent recombinant cell-based biosensors for the detection of organic pollutants in natural waters were reported (Table 17.2).

Bacteria are commonly used as test organisms because of their large population sizes, rapid growth rates and easy maintenance. In addition, many natural occurring bacteria are provided with genetically coded detoxifying systems for a given contaminant, and such specific DNA sequences are quite easy to be transformed in other microorganisms. A bacterial biosensor for naphthalene was developed by fusing the *nahG* promoter from the naphthalene degradation pathway of *Pseudomonas fluorescens* to *Vibrio fischeri*'s *luxCDABE* gene and was applied in waste water and ground water

Table 17.2. Luminescent recombinant cell-based biosensors for the detection of organic compounds in water samples

Analyte	Recombinant cell (reporter gene)	Reference
Naphthalene	*Pseudomonas fluorescens* (*luxCDABE*)	King et al. (1990)
Naphthalene	*Pseudomonas fluorescens* (*luxCDABE*)	Heitzer et al. (1994)
Middle-chain alkanes	*Escherichia coli* (*luxAB*)	Sticher et al. (1997)
Estrogen-like compounds	MCF-7 breast cancer cell line (*luc*) and HeLa cells (*luc*)	Balaguer et al. (1999)
Estrogen-like compounds	T47D breast cancer cell line (*luc*)	Legler et al. (1999)
Estrogen-like compounds	*Saccharomyces cerevisiae* (*lac-Z*)	Pasini et al. (2001)
Polyhalogenated aromatic hydrocarbons	Rat hepatoma cell line (*luc*)	Murk et al. (1996)

monitoring (King et al. 1990). A bioluminescent whole-cell biosensor for specific on-line monitoring of naphthalene and its degradation product salicylate was developed by immobilizing the reporter bacteria in calcium alginate (Heitzer et al. 1994). A microbial sensor for the detection of middle-chain alkanes was constructed by introducing in *Escherichia coli* a plasmid containing a fusion between the promoter of *alkB* and *luxAB* of *V. fischeri* and a separate plasmid bearing the regulatory gene *alkS* from *Pseudomonas oleovorans*. The biosensor was induced by *n*-alkanes from pentane to decane and was applied to the analysis of contaminated ground water samples (Sticher et al. 1997).

As concerns the detection of some organic compounds, their ability to bind specific nuclear receptors is exploited to construct recombinant cell-based biosensors. Upon binding an active ligand, the ligand-occupied receptor interacts with specific DNA sequences (responsive elements) carried by an expression plasmid to modulate gene transcription. This causes the expression of the reporter gene and the synthesis of the reporter protein, which is detected by luminescent techniques.

A very wide range of synthetic chemicals and natural compounds have been found to be oestrogenic. These endocrine-disrupting compounds (EDCs) include pesticides, plasticizers, breakdown products of surfactants, synthetic oestrogens used as drugs and phyto-oestrogens, most of which are often detectable in many surface waters. The majority of these molecules exhibit different molecular structures, while sharing the ability to bind the human oestrogen receptors (hER). A reporter gene bioassay was developed using a breast cancer cell line (MCF-7), which naturally expressed the endogenous oestrogen receptor and was stably transfected with an oestrogen-regulated firefly luciferase gene (Pons et al. 1990). Recently, this cell line along with newly developed ones, with detection limits of 1–5 pM oestradiol, was used to detect oestrogenic activity in wastewater sewage treatment plant influents and effluents. The direct analysis of the water samples, after sterile filtration on 0.2 μm membranes, showed that the oestrogenic activity was equivalent to 60–100 pM oestradiol in the influent and was reduced by a factor 4–10 in the effluent, thus suggesting that effluents from sewage treatment works are an important potential source of oestrogenic compounds contaminating the aquatic environment (Balaguer et al. 1999). An oestrogen receptor-mediated, chemical-activated luciferase reporter gene-expression bioassay (ER-CALUX) was developed by transfection of the firefly luciferase *luc* gene in the T47D breast cancer cell line. The system was tested on single oestrogen-like compounds or combinations and exhibited a detection limit of 0.5 pM oestradiol (Legler et al. 1999).

Other reporter gene bioassays for oestrogenic activity assessment are based on recombinant yeast cells. This cellular model is often used because yeast cells are easily engineered and cultured, and have rapid growth rates, thus providing a very robust model suitable for wide-scale screening. Most yeast-based systems use the β-galactosidase coding sequence as a reporter gene with the enzyme activity detected by colourimetric methods, which are often time-consuming and poorly sensitive, providing detection limits inadequate for the direct analysis of environmental aqueous samples (Rehmann et al. 1999). In our laboratory, a chemiluminescent whole cell biosensor based on recombinant yeast cells was developed and optimized for environmental monitoring of oestrogen-like compounds. A recombinant yeast strain (*Saccharomyces cerevisiae*) was used in which the DNA sequence of the human oestrogen receptor was stably integrated into the main chromosome. The yeast cells also contained an expression plasmid carrying oestrogen-responsive sequences and the reporter gene *lac-Z*, encoding the enzyme *β*-galactosidase (Routledge and Sumpter 1996). The activity of the analyte-induced enzyme was detected by a chemiluminescent 1,2-dioxetane-*β*-D-galactopyranoside substrate. The CL detection proved to be much more rapid than the colourimetric one, allowing for the achievement of a detection limit of 10 pmol l^{-1} oestradiol after incubating the cells with the analyte for 24 hours compared to at least 72 hours required with the colourimetric detection. The applicability of the chemiluminescent biosensor to the analysis of natural water samples was preliminarily checked by assessing samples from influent and effluent of an activated sludge sewage treatment plant (Pasini et al. 2001).

A different receptor-mediated, chemical-activated luciferase reporter gene-expression bioassay (CALUX) was developed for the detection of aryl hydrocarbon receptor-active compounds, such as polyhalogenated aromatic hydrocarbons, which include polychlorodibenzo-*p*-dioxins (PCDDs), polychlorodibenzofuranes (PCDFs) and polychlorobiphenyls (PCBs). These substances either derive from combustion or industrial processes as by-products, or are used in productive processes (e.g. plastics manufacture) and are included in a list of 16 persistent organic pollutants (POPs) identified to be monitored at international level. A rat hepatoma cell line, which naturally expresses the aryl hydrocarbon receptor (AhR), was stably transfected with a plasmid containing the dioxin-responsive elements sequence and the luciferase *luc* reporter gene. The biosensor had a detection limit of 1 pM 2,3,7,8-tetrachlorodibenzo-*p*-dioxin (TCDD) and was applied to the analysis of sediments and pore water. In the case of pore water, only a simple and rapid extraction procedure was needed, without additional clean-up (Murk et al. 1996).

In addition to the previously described specific or group-selective luminescent reporter gene bioassays, a number of total toxicity luminescent tests have been developed. The most widely accepted microbial toxicity test system is based on the wild-type luminescent bacterium *Vibrio fischeri*. Bacterial bioluminescence has proved to be a convenient measure of cellular metabolism and consequently a reliable sensor for measuring the presence of toxic chemicals in aquatic samples (De Zwart and Sloof 1981). Sample toxicity is assessed from the decrease in luminescence following a short exposure to several concentrations of the sample. This system is commercially established and marketed as Microtox. It is commonly used to assess the general toxicity of potentially contaminated water samples, e.g. in wastewater treatment plant effluent testing for protection of receiving waters, in surface water monitoring for identification of pollution sources, and in monitoring raw drinking water for contamination due to spills or pollution sources. A recombinant bacterial sensor, in which the DNA damage-inducible promoter *recN* from *Escherichia coli* was fused to the *lux* operon of *V. fischeri* and introduced into *Salmonella typhimurium*, was developed for genotoxicity assessment (Verschaeve et al. 1999). Also this system is now commercially available as the Vitotox test. A different approach for toxicity assessment was based on a set of *E. coli* strains harbouring different plasmids, each carrying a different stress promoter:*lux* fusion. The panel included an oxidative stress sensor, a DNA damage sensor (genosensor) and two general stress sensors. The system allowed for the evaluation of different toxic effects in industrial wastewater samples (Belkin et al. 1997). These luminescent biosensors, though rapid, well-standardized and often ready-to-use, provide information on the sample total toxicity only, giving no indication on the chemical nature of contaminants.

In conclusion, rapid and sensitive detection of the bio-chemiluminescence deriving from transformed cells, together with the selectivity and specificity of the gene regulation, has allowed efficient bioanalytical methods to be developed for the detection of pollutants in environmental aquatic samples. Such methods are usually carried out in 96-well microtitre plates and can be adapted to a 384-well microtitre format and easily automated, thus allowing high throughput screening. Advantages inherent in biological detection systems lie in their ability to indicate bioavailability and report effects on living organisms, thus providing preliminary information on the sample toxicity. The relatively inexpensive and easy handling of recombinant cell-based biosensors make their development and application even more attractive.

17.4 Conclusions and Future Perspectives

The reported bioanalytical methods could be very useful tools for the first-level monitoring of environmental and biological samples. Only positive samples will undergo further investigations, using more accurate analytical techniques such as GC-MS and HPLC-MS to detect and quantify the individual analytes. Since the percentage of negative samples is usually high, the availability of these rapid, sensitive first-level screening tests should allow considerable savings in terms of instrumentation, personnel and reagents costs, and operator time. A possible application of these first-level screening tests could be the systematic study of large and complex natural water systems (rivers, lakes, seas), which implies the analysis of numerous samples; several representative sampling points have to be selected in order to cover the whole aqueous system and to monitor the known possible sources of pollutants; in addition, the sampling has to be performed over a long period of time to evaluate the effects of meteorological variables.

Luminescent immunoassays and recombinant cell-based biosensors could also be advantageously applied in epidemiologic studies involving the analysis of biological samples to assess human exposure to contaminants and in food analysis to find out possible contamination sources.

Finally, continuous advancements in immunoassay and biosensor technology will lead to the development of miniaturized and biochip-based devices, which could further improve the analytical throughput and allow continuous environmental monitoring by carrying out on-site analysis.

References

Balaguer P, François F, Comunale F, Fenet H, Boussioux AM, Pons M, Nicolas JC, Casellas C (1999) Reporter cell lines to study the estrogenic effects of xenoestrogens. Sci Tot Environ 233:47–56

Belkin S, Smulski DR, Danon S, Vollmer AC, Dyk TK van, LaRossa RA (1997) A panel of stress-responsive luminous bacteria for the detection of selected classes of toxicants. Wat Res 31:3009–3016

De Zwart D, Sloof W (1981) The Microtox as an alternative assay in the acute toxicity assessment of water pollutants. Aquat Toxicol 4:129–138

Dzgoev A, Mecklenburg M, Xie B, Miyabayashi A, Larsson PO, Danielsson B (1997) Optimization of a charge coupled device imaging enzyme linked immuno sorbent assay and supports for the simultaneous determination of multiple 2,4-D samples. Anal Chim Acta 347:87–93

Emon JM Van(1987) Interim report on development and demonstration of immunoassay detection systems for rapid screening at superfund sites. EPA/600/X-87/414

Emon JM Van, Gerlach CL (1995) A status report on field-portable immunoassays. Environ Sci Technol 29:312A–317A

Emon JM Van, Lopez-Avila V (1992) Immunochemical methods for environmental analysis. Anal Chem 64:79A–88A

Heitzer A, Malachowsky K, Thonnard JE, Bienkowski PR, Sayler GS (1994) Optical sensor for environmental on-line monitoring of naphthalene and salicylate bioavailability with an immobilized bioluminescent catabolic reporter bacterium. Appl Environ Microbiol 60:1487–1494

Ius A, Bacigalupo MA, Roda A, Vaccari C (1992) Development of a time-resolved fluorimmunoassay of benzo(a)pyrene in water. Fres J Anal Chem 343:55–56

Ius A, Bacigalupo MA, Meroni G, Pistillo A, Roda A (1993) Development of a time-resolved fluoroimmunoassay for phthalate esters in water. Fres J Anal Chem 345:589–591

Kameth MF, Aherne GW, Stevenson D (1996) Development and evaluation of a chemiluminescent immunoassay for chlortoluron using a camera luminometer. Analyst 121:329–332

King JMH, DiGrazia PM, Applegate B, Burlage R, Sanseverino J, Dunbar P, Larimer F, Sayler GS (1990) Rapid, sensitive bioluminescent reporter technology for naphthalene exposure and biodegradation. Science 249:778–781

Köhler S, Belkin S, Schmid RD (2000) Reporter gene bioassays in environmental analysis. Fres J Anal Chem 366:769–779

Legler J, Brink CE van den, Brouwer A, Murk AJ, Saag PT van der, Vethaak AD, Burg B van der (1999) Development of a stably transfected estrogen receptor-mediated luciferase reporter gene assay in the human T47D breast cancer line. Toxicol Sci 48:55–66

Lewis JC, Daunert S (2000) Photoproteins as luminescent labels in binding assays. Fres J Anal Chem 366:760–768

Marquette CA, Blum LJ (1998) Electrochemiluminescence of luminol for 2,4-D optical immunosensing in a flow injection analysis system. Sensor Actuat B 51:100-106

Murk AJ, Legler J, Denison MS, Giesy JP, Guchte C van de, Brouwer A (1996) Chemical-activated luciferase gene expression (CALUX): A novel in vitro bioassay for Ah receptor active compounds in sediments and pore water. Fund Appl Toxicol 33:149–160

Pasini P, Gentilomi G, Guardigli M, Baraldini M, Musiani M, Roda A (2001) A chemiluminescent whole cell biosensor for assessing estrogenic activity. In: Case JF, Herring PJ, Robison BH, Haddock SHD, Kricka LJ, Stanley PE (eds) Bioluminescence and chemiluminescence 2000. World Scientific Publishing Company, Singapore, pp 327–330

Pons M, Gagne D, Nicolas JC, Mehtali M (1990) A new cellular model of response to estrogens: A bioluminescent test to characterize (anti)estrogen molecules. BioTechniques 9:450–459

Rehmann K, Schramm K-W, Kettrup AA (1999) Applicability of a yeast oestrogen screen for the detection of oestrogen-like activities in environmental samples. Chemosphere 38:3303–3312

Roda A, Girotti S, Lodi S, Preti S (1984) Development of a sensitive immunoassay for plasma and salivary steroids. Talanta 31:895–900

Roda A, Bacigalupo MA, Ius A, Minutello A (1991) Development and applications of an ultrasensitive quantitative enzyme immunoassay for benzo(a)pyrene in environmental samples. Environ Technol 12:1027–1035

Roda A, Rauch P, Ferri E, Girotti S, Ghini S, Carrea G, Bovara R (1994) Chemiluminescent flow sensor for the determination of Paraoxon and Aldicarb pesticides. Anal Chim Acta 294:35–42

Routledge EJ, Sumpter JP (1996) Estrogenic activity of surfactants and some of their degradation products assessed using a recombinant yeast screen. Environ Toxicol Chem 15:241–248

Samsonova JV, Rubtsova MY, Kiseleva AV, Ezhov AA, Egorov AM (1999) Chemiluminescent multiassay of pesticides with horseradish peroxidase as a label. Biosens Bioelectron 14:273–281

Shetty RS, Ramanathan S, Badr IH, Wolford JL, Daunert S (1999) Green fluorescent protein in the design of a living biosensing system for L-arabinose. Anal Chem 71:763–768

Sticher P, Jaspers MCM, Stemmler K, Harms H, Zehneder AJB, Meer JR van der (1997) Development and characterization of a whole-cell bioluminescent sensor for bioavailable middle-chain alkanes in contaminated groundwater samples. Appl Environ Microbiol 63:4053–4060

Verschaeve L, Gompel J Van, Thilemans L, Regniers L, Vanparys P, Lelie D van der (1999) VITOTOX bacterial genotoxicity and toxicity test for the rapid screening of chemicals. Environ Mol Mutagen 33:240–248

Weller MG, Schuetz AJ, Winklmair M, Niessner R (1999) Highly parallel affinity sensor for the detection of environmental contaminants in water. Anal Chim Acta 393:29–41

Wilson R, Barker MH, Schiffrin DJ, Abuknesha R (1997) Electrochemiluminescence flow injection immunoassay for atrazine. Biosens Bioelectron 12:277–286

Chapter 18

Affinity Electrochemical Biosensors for Pollution Control

M. Mascini

18.1 Introduction

DNA electrochemical biosensors, realized by immobilizing an oligonucleotide sequence of the calf thymus DNA on a suitable electrode surface, are simple to assemble and can provide reliable results; such DNA biosensors hold enormous potential for environmental monitoring (Wang et al. 1997a,b).

A major application of a DNA biosensor will be the testing of water, food, soil, and plant samples for the presence of analytes (carcinogens, drugs, mutagenic pollutants, etc.) with binding affinities for the structure of DNA. Binding of small molecules to DNA and general DNA damage by ionizing radiation, dimethyl sulphate etc. has been described through the variation of the electrochemical signal of guanine (Mecklenburg et al. 1997; Wang et al. 1996a,b,c; Jelen et al. 1997a,b).

The objective of our work was to develop a disposable electrochemical DNA sensor to evaluate the presence of small DNA binding compounds by measuring changes of the electrochemical signal of guanine in calf thymus DNA extract. Single-use sensors have several advantages, such as avoidance of contamination among samples, constant sensitivity and reproducibility, and ease of use (Del Carlo et al. 1997).

This biosensor was realized by immobilizing calf thymus DNA onto the electrode surface (Wang et al. 1997b; Marrazza et al. 1999a,b). The DNA biosensor was then immersed in the sample solution containing the analyte. After two minutes of interaction, the DNA sensor was washed, immersed in a suitable clean buffer and a chronopotentiometric analysis (PSA) was carried out to evaluate the oxidation of guanine residues on the electrode surface. We report some preliminary experiments showing clear electrochemical effects due to the presence of genotoxic compounds. We can extrapolate and evaluate such electrochemical signals as resulting from potentially genotoxic compounds present in real water samples.

18.2 Procedures

18.2.1 Electrochemical Measurements

All electrochemical measurements were carried out at room temperature in 2 ml Teflon beakers. Potentiometric stripping analysis at a constant current (PSA) was performed with the following parameters; the potentials were sampled at a frequency of 33 kHz, and the derivative signal (dt/dE) was recorded vs. the potential.

The electrodes modified by calf thymus ssDNA immobilization were examined in 0.2 M acetate buffer pH 5.0, using an initial potential of +0.5 V and a constant current of +6 μA. The guanine peak area following baseline fitting was used as the analytical signal.

18.2.2 DNA Sensor for Binding Compounds with an Affinity for DNA

The procedure consisted of the following steps: calf thymus DNA immobilization on the electrode surface, dipping the electrode in the sample/blank solution, and electrochemical interrogation of the surface.

Calf thymus immobilization consisted of an electrochemical oxidation (+1.6 V vs. SCE for 1 min in 0.2 M acetate buffer pH 5.0); then the electrode was immersed in a stirred buffer solution containing 10–20 mg l^{-1} of single stranded or double stranded calf thymus DNA. This immobilization step lasted for 2 min, holding the electrode surface at a potential of +0.5 V vs. SCE. The electrode was then washed with buffer solution for 10 s. The modified electrode was placed in a sample solution for 2 min. A chronopotentiogram was carried out in 0.2 M acetate buffer pH 5.0 by using an initial potential of +0.5 V and a constant current of +6 μA; the guanine peak area was obtained at around +1 V (see Fig. 18.1).

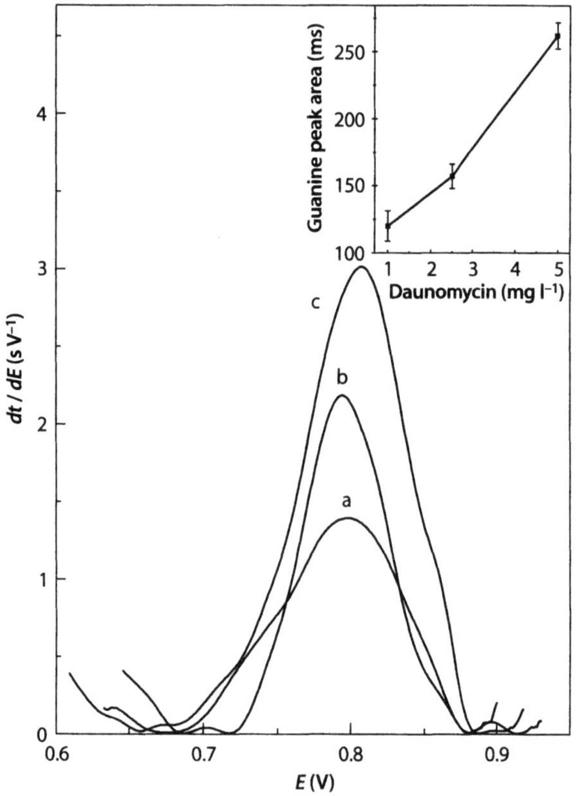

Fig. 18.1. Chronopotentiograms for modified screen printed electrodes calf thymus dsDNA obtained increasing the daunomycin concentrations (*a*) 1 mg l^{-1}, (*b*) 2.5 mg l^{-1}, (*c*) 5 mg l^{-1}. Calf thymus dsDNA immobilization: 10 mg l^{-1} dsDNA for 2 min at +0.5 V vs. Ag/AgCl. PSA transduction: in 0.2 M acetate buffer pH 5.0 with a stripping constant current of +1 μA and an initial potential of +0.5 V

This area was compared with the area obtained when the analyte concentration in the sample solution was zero.

In some cases, the sample solvent was up to 10% methanol to allow analyte dissolution.

18.2.3 Analysis of River Water Sample

A preconcentration step of river water was found necessary to obtain some signals through this technique. The water samples were prefiltered through a 0.45 μm filter to clarify the water. Then 1 l of sample water was passed through an Isolute column SPE. The organic compounds extracted were eluted using 500 μl ethyl acetate. The obtained samples were dried, then 10 ml of phosphate buffer containing 3% v/v methanol were added. The analysis was carried out as described before. The extracts of the water samples were also analyzed using a standard method with a Varian 3400 gas chromatograph coupled to a Finnigan Mat 800 ion trap detector mass spectrometer (GC-ITDMS).

18.3 Results

18.3.1 DNA Sensor for Binding Compounds with an Affinity for DNA

Preliminary studies were performed to identify general assay conditions that affected the electrochemical signal of the guanine oxidation peak, like ionic strength, pH, buffer composition, DNA concentration and form (single-stranded and double-stranded). Figure 18.2 reports the guanine peak area obtained as a function of single-stranded calf thymus concentration. The area increases linearly with concentrations up to 20 mg l^{-1}, then levels off. This value was generally used.

Table 18.1 reports the values of the guanine peak area as a function of the stripping oxidation current value. The area is smaller with larger currents, but the peak appeared very broad and not reproducible (a large RSD). 6 μA was then generally used.

The guanine peak appears very sharp in acetate buffer; the baseline at this high potential value was lower than in other buffer. The modified electrode performance was tested in buffer solution containing methanol up to 10% v/v, and we observed any variation of the electrochemical signal of guanine peak. Methanol is useful for dissolving some organic compounds.

We performed several preliminary experiments for evaluating the variation of the area of the guanine peak obtained using single-stranded or double-stranded DNA immobilized when the sample contained different compounds of environmental interest.

Table 18.2 summarizes these experiments, showing that the guanine peak is higher for single-stranded DNA (ssDNA); the guanine base in single-stranded DNA is obviously more readily available for oxidation than in double-stranded DNA (dsDNA).

Buffer solution: 0.2 M acetate buffer solution pH 5.0 with 10% v/v of methanol.

The area is reported with each measurement repeated four times.

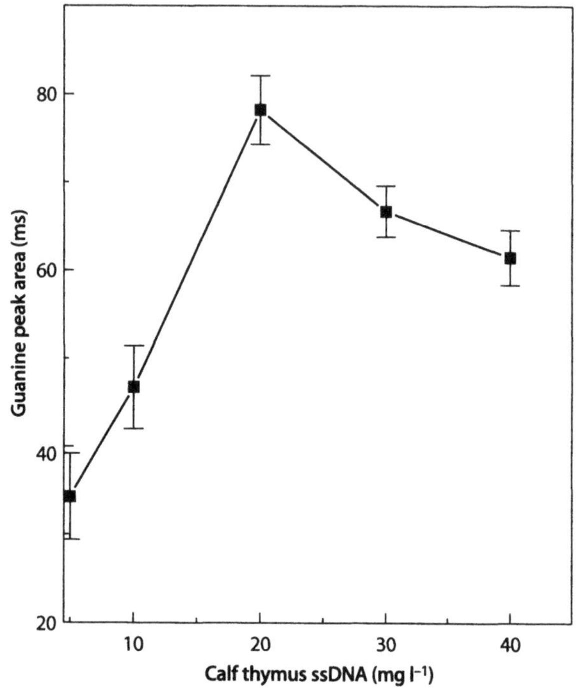

Fig. 18.2. Anodic peak of the guanine obtained increasing the concentration of calf thymus ssDNA for immobilization. Calf thymus ssDNA immobilization: 2 min at +0.5 V vs. SCE in 0.2 M acetate buffer pH 5.0. PSA transduction: in 0.2 M acetate buffer pH 5.0 with a stripping constant current of +6 μA and an initial potential of +0.5 V

Table 18.1. Effect of stripping current on guanine peak area using single stranded calf thymus DNA. Single stranded calf thymus DNA immobilization: 20 mg l^{-1} in 0.2 M acetate buffer solution pH 5.0 for 2 min. at +0.5 V vs. SCE. PSA conditions: in 0.2 M acetate buffer solution pH 5.0 with an initial potential of +0.5 V. The area is reported with each measurement repeated four times

Stripping oxidation current (μA)	Guanine peak area (ms)
1	339 ±60
2	268 ±51
4	103 ±33
6	78 ±5
8	32 ±7

Compounds behaving as intercalators, like daunomycin or cisplatin, showed an increase in the guanine peak area when using dsDNA (Table 18.2). In this case, we observe a linear increase with the daunomycin concentration in the range 1–10 mg l^{-1}.

Aflatoxins are metabolites produced by some strains of the mould *Aspergillus flavus*. They are among the most potent environmental mutagens and are implicated as liver carcinogens. The binding of the aflatoxin B1 to both native and denatured DNA has been demonstrated (Gopalakrishnan et al. 1989).

Figure 18.3a–b demonstrates the applicability of the DNA sensor to the analysis of these pollutants. We obtained the gradual decrease of the guanine peak in the presence of increasing levels of the aflatoxin B1. Such suppression of the guanine response

Table 18.2. Compounds tested with single-stranded or double-stranded calf thymus DNA immobilized on screen printed electrodes. Calf thymus DNA immobilization: 20 mg l^{-1} of single stranded or double stranded calf thymus DNA in 0.2 M acetate buffer solution pH 5.0 for 2 min. at +0.5 V vs. SCE. PSA conditions: in 0.2 M acetate buffer solution pH 5.0 with a stripping current of +6 μA and an initial potential of +0.5 V

Compounds tested	Guanine peak area (ms) using calf thymus dsDNA immobilized	Guanine peak area (ms) using calf thymus ssDNA immobilized
Buffer solution	36 ±7	89 ±13
Daunomycin	52 ±6	86 ±3
Phthalates mixture (20 mg l^{-1})	39 ±16	80 ±26
Atrazine (50 mg l^{-1})	66 ±39	85 ±11
Bisphenol (100 mg l^{-1})	76 ±13	85 ±17
PCB 105 (0.4 mg l^{-1})	54 ±8	68 ±7
PCB mixture (Aroclor 1260) (20 mg l^{-1})	39 ±5	99 ±7
PCB mixture (Aroclor 1016) (20 mg l^{-1})	43 ±8	95 ±17
Aflatoxin B1 (10 mg l^{-1})	36 ±8	77 ±7
Cisplatin (30 mg l^{-1})	58 ±14	78 ±6
Hydrazine (20 mg l^{-1})	22 ±4	73 ±6

results in a well-defined concentration dependence and offers convenient quantification of low levels of aflatoxin. The signal was observed within 10–30 mg l^{-1} (Fig. 18.3b).

In the most of the compounds, ssDNA gave greater effects. The peak area of guanine decreased even when low concentrations were present (PCB 0.2 mg l^{-1}). This can be explained by the binding of the compounds with guanine in a short time (2 min) and a lower availability of guanine for oxidation at the electrode surface.

PCBs have been recognized for several years as ubiquitous environmental pollutants. The high toxicity of some of the PCB congeners represents a public health risk, as these molecules are still present in the environment, even though the production of PCB has been banned. The screen-printed electrodes modified with single-stranded calf thymus DNA were used to detect PCB 105. Figure 18.4a–b displays the chronopotentiometric response of the calf thymus ssDNA modified electrode followed by increasing PCB 105 concentrations.

A decrease of the guanine peak area could be detected with some water samples (Fig. 18.5, Curves *b*, *c*, *d*). By HPLC analysis, we obtained the results reported in Table 18.3.

It seems that there is an approximate relationship between the two sets of analysis. The sensor is not able to distinguish between compounds of environmental concern but could be conveniently used as a screening tool of toxicity.

18.4 Conclusions

The potential of DNA biosensors for detection of toxic compounds has been demonstrated. This kind of procedure offers a sensitive, rapid and portable tool for field monitoring of several environmentally and toxicologically significant compounds.

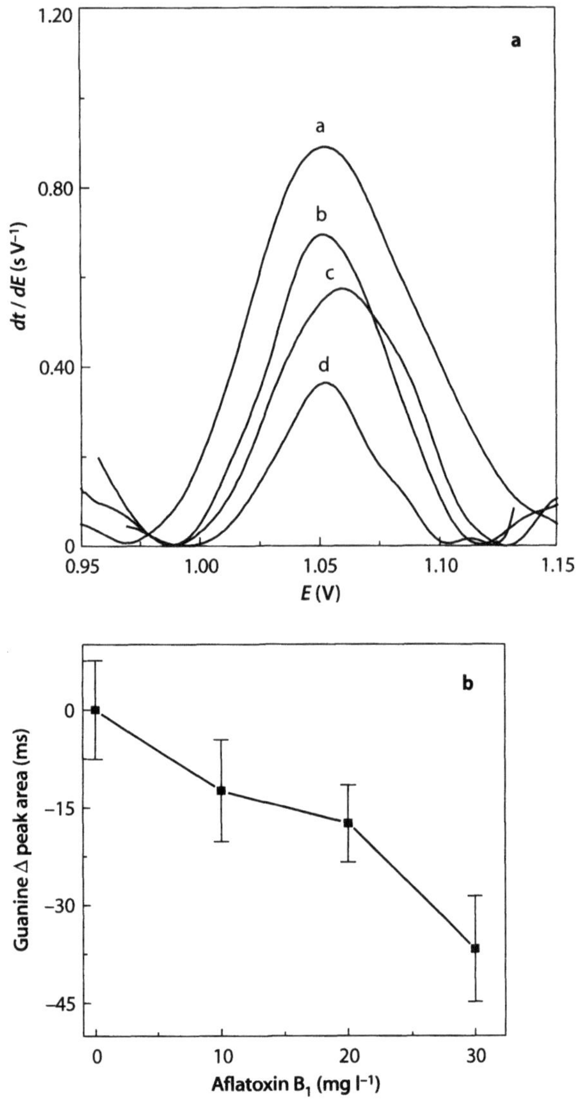

Fig. 18.3. **a** Chronopotentiograms for (*a*) calf thymus ssDNA modified screen printed electrodes and followed increasing aflatoxin B_1 concentrations; (*b*) 10 mg l^{-1}; (*c*) 20 mg l^{-1}; (*c*) 30 mg l^{-1}; **b** Calibration curves obtained increasing the concentration of aflatoxin B_1. The results correspond to the difference between the guanine peak area after interaction with aflatoxin B_1 minus that obtained for the buffer solution. Calf thymus ssDNA immobilization: 20 mg l^{-1} calf thymus ssDNA for 2 min at +0.5 V vs. SCE. PSA transduction: in 0.2 M acetate buffer pH 5.0 with a stripping constant current of +6 μA and an initial potential of +0.5 V

Moreover, we found that the calf thymus ssDNA biosensor interacts with low-molecular mass substances of environmental concern, and it can be used as a general indicator.

The DNA biosensor, realized by immobilization of the calf thymus extract on suitable electrode surface, is a simple tool to assemble and obtains reliable results. This therefore opens the possibility of rapid evaluation as a screening tool of toxicity of water samples.

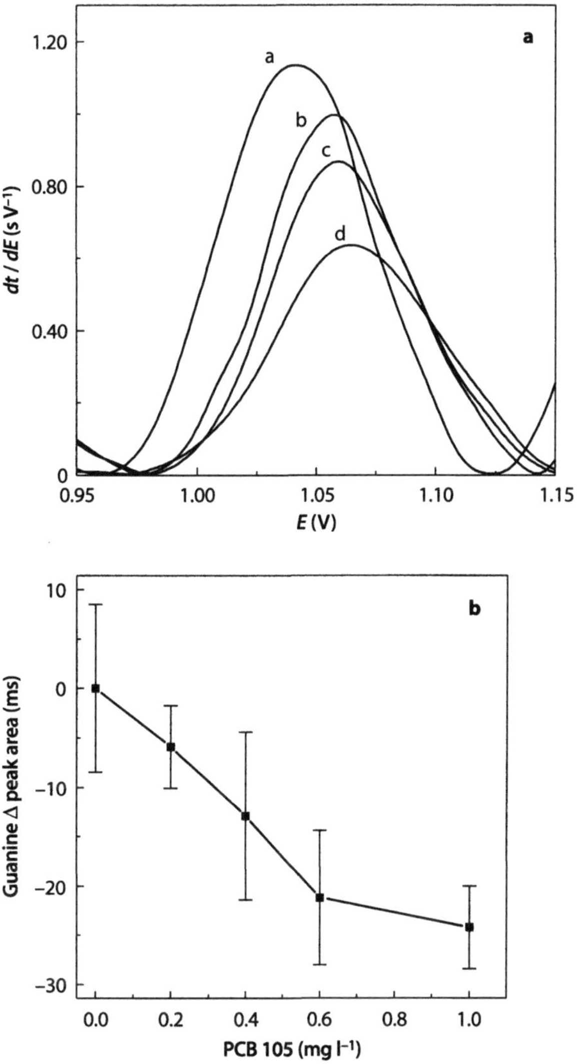

Fig. 18.4. a Chronopotentiograms for (*a*) calf thymus ssDNA modified screen printed electrodes and followed by an increasing PCB 105 concentration, (*b*) 0.2 $mg\,l^{-1}$, (*c*) 0.6 $mg\,l^{-1}$, (*d*) 1 $mg\,l^{-1}$; **b** Calibration curves obtained increasing the concentration of PCB 105. The results correspond to the difference between the guanine peak area after interaction with PCB 105 minus that obtained for the buffer solution. Calf thymus ssDNA immobilization: 20 $mg\,l^{-1}$, calf thymus ssDNA for 2 min at +0.5 V vs. SCE. PSA transduction: in 0.2 M acetate buffer pH 5.0 with a stripping constant current of +6 μA and an initial potential of +0.5 V

Table 18.3. Results of the water samples obtained using a standard method with a Varian 3400 gas chromatograph coupled to a Finningan Mat 800 ion trap detector mass spectrometer (GC-ITDMS)

	b (ng l^{-1})	*c* (ng l^{-1})	*d* (ng l^{-1})
Desetil-terbutilazine	0	12	21
Carbofuran	0	210	101
Simazine	0	0	23
Terbutilazine	79	28	81
Etofumesate	6	184	83
Alachlor	0	0	27
Metolachlor	7	14	225

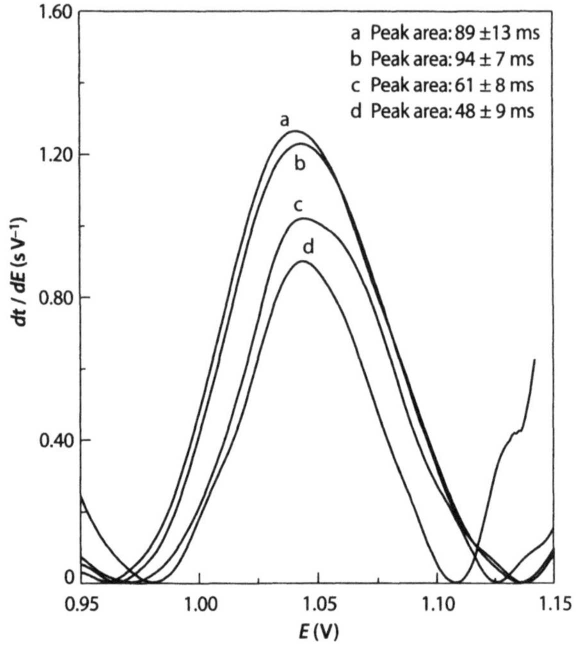

Fig. 18.5. Chronopotentiograms obtained using calf thymus ssDNA modified screen printed electrode with: (*a*) blank solution, (*b*), (*c*) and (*d*) river water (100 times concentrated). Calf thymus ssDNA immobilization: 20 mg l^{-1} calf thymus ssDNA for 2 min at +0.5 V vs. SCE. PSA transduction: in 0.2 M acetate buffer pH 5.0 with a stripping constant current of +6 μA and an initial potential of +0.5 V

References

Del Carlo M, Lionti I, Taccini M, Cagnini A, Mascini M (1997) Disposable screen-printed electrodes for the immunochemical detection of polychlorinated biphenyls. Anal Chim Acta 342:189

Gopalakrishnan S, Byrd S, Stone MP, Harris TM (1989) Carcinogen-nucleic acid interactions: Equilibrium binding studies of aflatoxin B_1 with the oligodeoxynucleotide d(ATGCAT)$_2$ and with plasmid pBR322 support intercalative association with the B-DNA helix. Biochemistry 28:726

Marrazza G, Chianella I, Mascini M (1999a) Disposable DNA electrochemical sensor for hybridization detection. Biosensors and Bioelectronics 14:43

Marrazza G, Chianella I, Mascini M (1999b) Disposable DNA electrochemical biosensors for environmental monitoring. Anal Chim Acta 387:297

Mecklenburg M, Grauers A, Rees Jonsson B, Weber A, Danielsson B (1997) A strategy for the broad range detection of compounds with affinity for nucleic acids. Anal Chim Acta 347:79

Jelen F, Fojta M, Palecek E (1997a) Voltammetry of native double-stranded, denatured and degraded DNAs. J Electroanal Chem 427:49

Jelen F, Tomoschik M, Palecek E (1997b) Adsorptive stripping square-wave voltammetry of DNA. J Electroanal Chem 423:141

Wang J, Rivas G, Cai X, Shiraishi H, Farias PAM, Dontha N, Luo D (1996a) Accumulation and trace measurements of phenothiazine drugs at DNA-modified electrodes. Anal Chim Acta 332:139

Wang J, Chicharro M, Rivas G, Cai X, Dontha N, Farias PAM, Shiraishi H (1996b) DNA biosensor for the detection of hydrazines. Anal Chem 68(13):2251

Wang J, Rivas G, Luo D, Cai X, Valera FS, Dontha N (1996c) DNA-modified electrode for the detection of aromatic amines. Anal Chem 68(24):4365

Wang J, Cai X, Rivas G, Shiraishi H, Dontha N (1997a) Nucleic-acid immobilization, recognition and detection at chronopotentiometric DNA chips. Biosensors and Bioelectronics 12(7):587

Wang J, Rivas G, Palecek E, Nielsen P, Shiraishi H, Dontha N, Luo D, Parrado C, Chicharro M, Farias PAM, Valera FS, Grant DH, Ozsoz M (1997b) DNA electrochemical biosensors for environmental monitoring. A review. Anal Chim Acta 347:1

Palaeoenvironmental Reconstructions Using Stable Carbon Isotopes and Organic Biomarkers

S. Wakeham

19.1 Introduction

Aquatic sediments carry the potential of preserving a historical record, or some remnant thereof, of past environmental conditions. The stratigraphic record has long been a tool for elucidating the character of the Earth and the forces that have altered its face and determined the nature, distributions and fate of living organisms. There have been dramatic changes in the Earth's climate in the past. Increasing levels of atmospheric CO_2 attributed to combustion of biomass and fossil fuels coupled with deforestation have heightened societal and scientific concerns about global warming, a reduction in ice volume and a rise in sea level. These issues have generated a greater interest in understanding past natural and present anthropogenic processes that influence global climate change and in developing predictive capabilities about future climate.

Studies of palaeoenvironment have varied in scope. They can range from qualitative descriptions and interpretation of conditions at a given time in the geological past to quantitative investigations concerned with the nature and the rate of change during specific time intervals. Organic biomarkers, molecules whose molecular structures point to specific biological sources, are proving to be quite valuable for palaeoenvironmental reconstructions. The utility of biomarkers derives from two features. First, organisms that inhabit specific ecological niches may biosynthesize unique biomarker compounds that can be used as indicators of the source organisms and their habitat. Second, organisms often possess an ability to regulate patterns of constituent biochemicals in order to remain viable under changing environmental conditions, so that environmental characteristics may be inferred from these patterns. To be a robust palaeoenvironmental proxy, the signal carried by biomarkers must survive during deposition and burial in sediments over the time frame of interest.

This chapter discusses several ways that palaeoenvironmental records may be obtained from organic geochemical studies, in particular from biomarker records in the sediment. The case studies represent only a few of the many applications that are presently accumulating in the literature and are by no means an exhaustive survey. Readers are referred to the references cited in the respective examples for further reading.

19.2 Stable Carbon Isotopes to Identify Organic Matter Sources

The biological source of sedimentary organic matter may be deduced from chemical studies on the structure of individual compounds (biomarkers) extracted from sediments. A second means of inferring source and ecological setting is through light (C, N, H, S) stable isotope tracers. Carbon fixation during photosynthesis results in a differential partitioning (fractionation) of isotopes of different atomic masses, yielding different ratios of isotope abundances (for definitions and background, see Degens 1969; Fogel and Cifuentes 1993; Hayes 1993). Isotope ratios can be indicative of physiological processes occurring during photosynthetic fixation of inorganic carbon into organic matter and during assimilation of organic compounds up the food chain. The stable isotopic composition of bulk organic matter is an integrative signal. It will reflect isotope fractionation resulting from the range of metabolic processes and environmental conditions that occurred over the life of the organism or multiple organisms that biosynthesized the organic matter and over the time frame during which the organism(s) lived and the sediments deposited. Interpreting isotope ratios for bulk organic material can thus be ambiguous. Isotopic compositions of individual compounds, on the other hand, have the potential of isolating specific biosynthetic sources and metabolic events in the life of an organism, although the biochemical mechanisms and environmental factors determining isotopic fractionation are complex and the subject of continuing research. Carbon isotopes have received the widest attention, because potentially all organic matter and compounds carry a carbon isotopic signal. Isotopes of nitrogen, sulphur, oxygen, and hydrogen have generally been less widely used because of difficulties in analytical procedures (Fogel and Cifuentes 1993), but this is rapidly changing.

Early studies of carbon isotopes found that there were metabolic controls on the photosynthetic isotope fraction (ε_P) that could be used to elucidate organic carbon (*OC*) sources in geochemistry. As a consequence of different carbon fixation pathways and fractionations, there is a wide range of isotopic compositions in various types of living organisms (Fig. 19.1) and in different biochemical constituents (Degens 1969). Most plants use the enzyme ribulose-1,5-biphosphate-carboxylase (RuBP-carboxylase) during photosynthesis to produce a 3-carbon compound, hence the C_3-pathway. It is the RuBP-carboxylase reaction that is responsible for the fractionation of carbon isotopes during photosynthesis, which is about 20–30‰. To a first approximation, land plants are isotopically lighter (depleted in ^{13}C relative to ^{12}C with del-values ($\delta^{13}C$) of about −25 to −30‰ according to the nomenclature; Degens 1969) than marine plankton ($\delta^{13}C$ −18 to −22‰). This difference results in part from isotopic fractionation and input from isotope differences for the source carbon, atmospheric CO_2 for land plants vs. dissolved CO_2 for plankton. Recent work has shown that photosynthetic isotope fractionation in the ocean is in fact quite complicated and the range of isotopic compositions considerably wider ($\delta^{13}C$ −18 to −28‰). Photosynthtic isotope fractionation by marine plankton depends of the concentration of dissolved CO_2, the phytoplankton species physiology (do the cells take up CO_2 by passive diffusion across the cell membrane or do they actively pump CO_2 or HCO_3^-), cellular volume, and cellular growth rate, to name a few (see Popp et al. 1999 and references therein).

A few plants (e.g. grasses, sugarcane, corn) utilize the C_4 pathway that employs an alternate enzyme, phosphoenol-pyruvate-carboxylase (PEP carboxylates), and produces a 4-carbon product. This reaction has a smaller isotope fractionation (~2‰). Land plants fixing carbon by the C_4-pathway are isotopically heavier (enriched in ^{13}C; $\delta^{13}C$ −12 to −16‰) than plants employing the C_3-pathway.

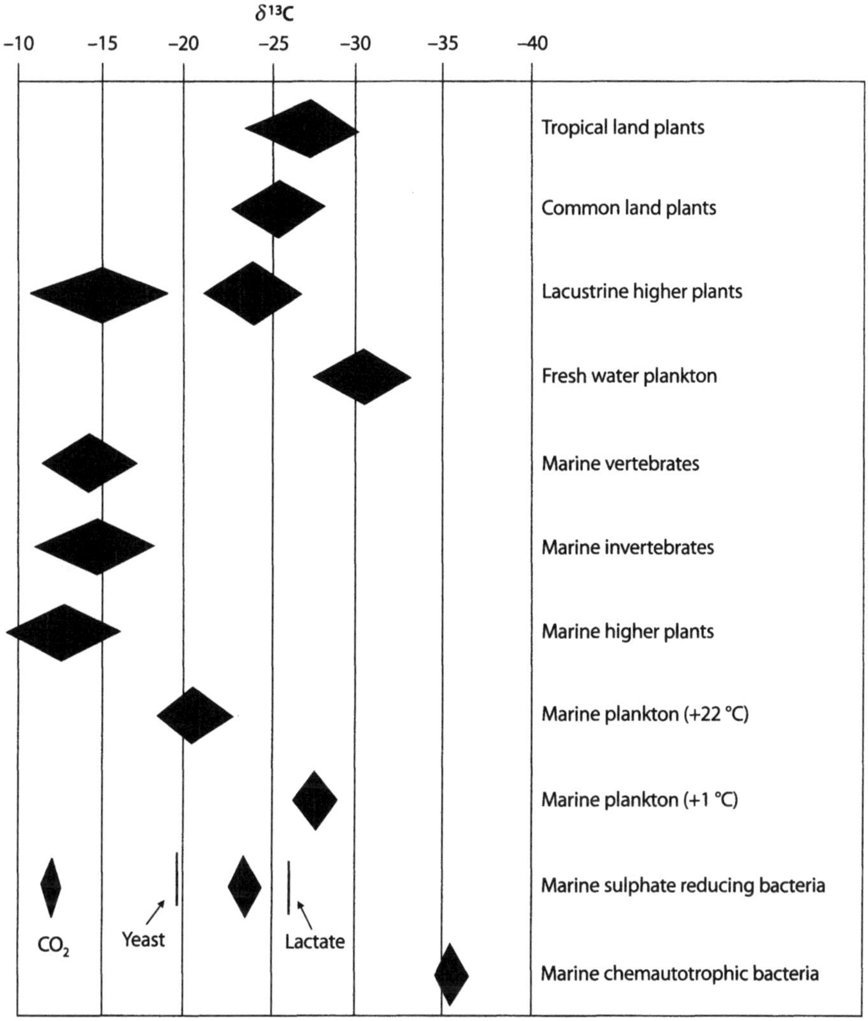

Fig. 19.1. Ranges of stable carbon isotope values for total *OC* in living organisms. Redrawn from Degens (1969)

Stable carbon isotopes are frequently used to estimate contributions from marine and terrigenous sources of organic matter in marine sediments. Early work (e.g. Hedges and Parker 1976; Gearing et al. 1977) showed a trend in δ^{13}C of bulk *OC* as one progresses from continental shelf (δ^{13}C −26‰) to off-shore sediments (δ^{13}C −19 to −21‰) in the Gulf of Mexico. Most terrigenous *OC* would be deposited on the inner shelves, and little is transported to outer shelves and continental slopes, indicating a minimal contribution of land-derived *OC* to surface sediments of the Gulf. Other studies (e.g. Prahl and Muelhausen 1989; Goñi et al. 1998) have suggested that a significant amount of terrigenous *OC* is deposited in off-shore sediments. In part, this difference in interpretation arises from the fact that organic matter delivered by rivers to continental shelves may not necessarily have a uniform carbon isotopic composition, and this composition may change over time as the ecosystem in the catchment basin changes, for example in response to climate change. In the case of the Gulf of Mexico, a continental end-member value of −26‰ assumes input primarily from C_3 vascular plants and neglects C_4 plants (Goñi et al. 1998). However, material carried by the Mississippi River into the Gulf of Mexico includes contributions by both C_3 (trees and shrubs) and C_4 (grasses) plants on the North American continent. Different soil provenances display different $\delta^{13}C$ values, ranging from ~−19 to −23‰ for C_4 grasslands to −26‰ for C_3 forests (Fig. 19.2; Onstad et al. 2000). Particulate matter delivered to the Gulf is relatively enriched in ^{13}C ($\delta^{13}C$ −19 to − 24‰). Thus, $\delta^{13}C$ values of −20 to − 22‰ for *OC* in Gulf of Mexico sediments (Goñi et al. 1998) might in fact result from significant inputs of terrigenous C_4-derived *OC* rather than strictly marine *OC*. If so, then the terrigenous contribution to these sediments would be higher than previously believed, possibly greater than 50% (Goñi et al. 1998).

Part of the uncertainty in interpreting isotopic data for bulk organic carbon arises from the inclusion of organic materials originating from a variety of photosynthetic sources that individually imprint different $\delta^{13}C$ signals depending on their carbon source, degree of photosynthetic isotope fractionation, and differing response to environmental factors. Analyses of individual biomarker distributions in conjunction with bulk isotopic compositions help resolve these complications (Prahl et al. 1994;

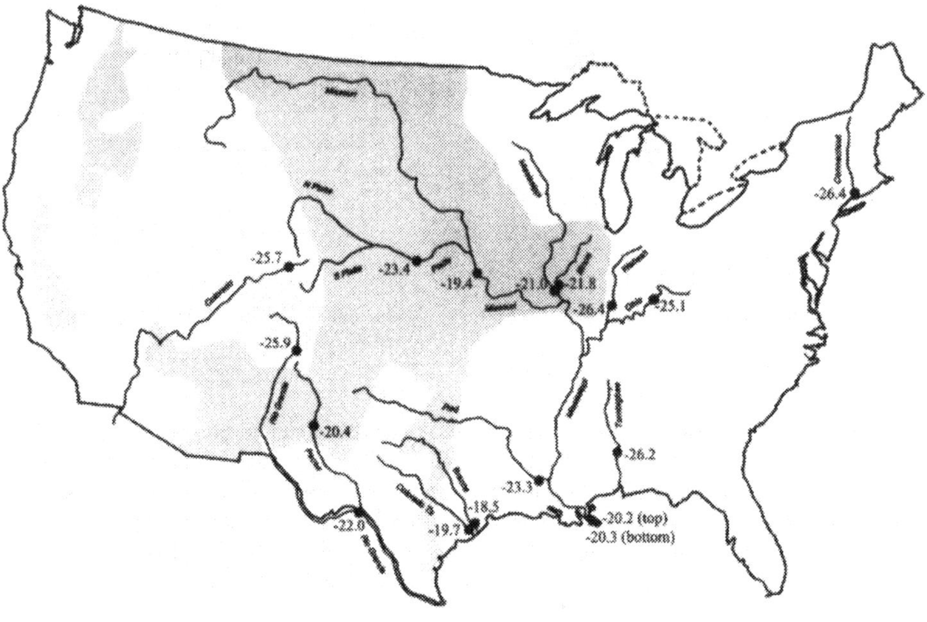

Fig. 19.2. Stable carbon isotope value for riverine suspended particulate matter in the United States. The stippled areas are C_4 grasslands. From Onstadt et al. (2000)

Goñi et al. 1998). However, significantly different results may be obtained depending on the choice of biomarker(s). For example, Prahl et al. (1994) estimated that terrigenous *OC* is a minor component (≤10%) in continental slope sediments on the Washington (USA) margin based on lignin phenol concentrations and ^{13}C of bulk *OC*, but a calculation using *n*-alkane concentrations and ^{13}C of bulk *OC* suggested a significantly higher input (30%).

Better still is measurement of biomarker specific isotopic compositions (Hayes et al. 1990). Compound-specific isotope analyses (CSIA) are now a common tool employed by organic geochemists. A powerful application of CSIA has been in using *n*-alkanes for tracking the anthropogenically-driven shifts in vegetation, for example in northern Australia over the past 6 000 years (Bird et al. 1995). European settlement of the region in the late 19th century resulted in the replacement of the original C_3 forests by a C_4 ecosystem of grasses (and ultimately including cultivated sugarcane). The principle behind the Bird et al. analysis is that *n*-alkanes (1) are robust biomarkers for vascular plants, (2) are stable against diagenesis and hence well-preserved in sediments, and (3) record the stable carbon isotope values of their C_3 ($\delta^{13}C \sim -28‰$) vs. C_4 ($\delta^{13}C \sim -12‰$) plant sources. Figure 19.3 shows carbon isotope values for *n*-alkanes in a set of sediments from the Johnstone River, North Queensland, and a marine sediment core off the mouth of the Johnstone River. Among the river sediments, the sample from the forested area (JR-8B) was the most depleted in ^{13}C ($\delta^{13}C$ −27.9‰), while the sample from a sugarcane cultivating region (JR4) was most enriched ($\delta^{13}C$ −20.9‰).

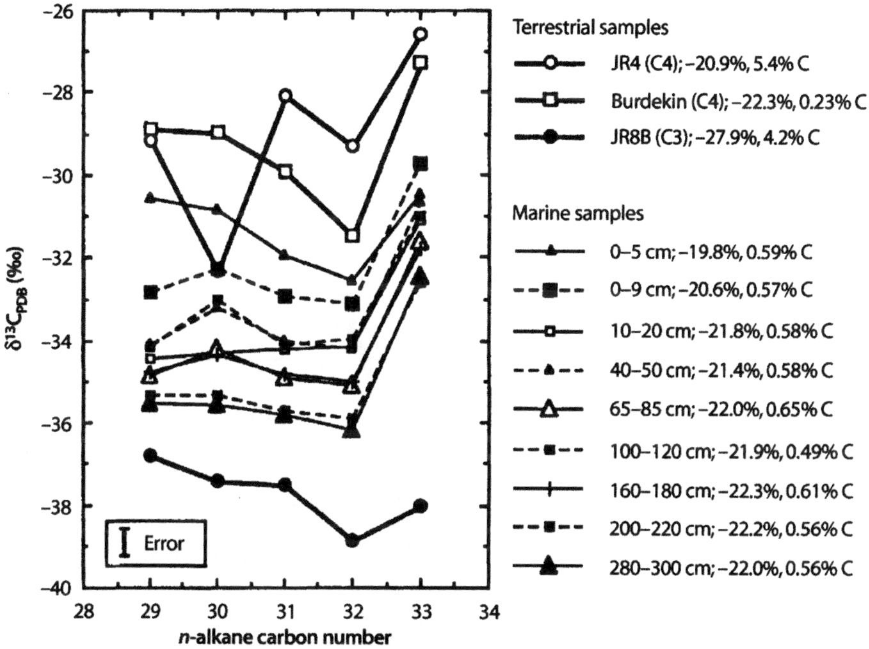

Fig. 19.3. Carbon isotopic compositions of *n*-alkanes from sediments from the Johnstone River, North Queensland, Australia, and a 3-m marine sediment core off the mouth of the Johnstone R. The $\delta^{13}C$ values of bulk organic carbon are given in the legend. Redrawn from Bird et al. (1995)

The spread of about 8‰ between the highest and lowest values for each carbon number for the terrestrial samples agreed with a 7‰ range for bulk *OC*. This suggests that ^{13}C values of the individual *n*-alkanes can be related to vegetation type in the rivers in the same manner that ^{13}C of bulk *OC* is related to vegetation type. The core was studied to evaluate whether the alkane-isotope record could be used to reconstruct the ecosystem history of the region. Bulk *OC* isotope values in core samples ranged from -22.3‰ at the base of the core to -19.8‰ at the top of the core (Fig. 19.3). However, when the alkane δ^{13}C values were plotted as the %-difference relative to the centre of the core that is assigned a 0‰ value, they were remarkably consistent across the range of compounds (Fig. 19.4). There was a steady ~1–1.5‰ increase in the δ^{13}C value of all *n*-alkanes since about 6 000 years B.P. (about 200 cm deep in the core). At the same time, there has been a ~2‰ increase since clearing began in the drainage basin (above 10 cm

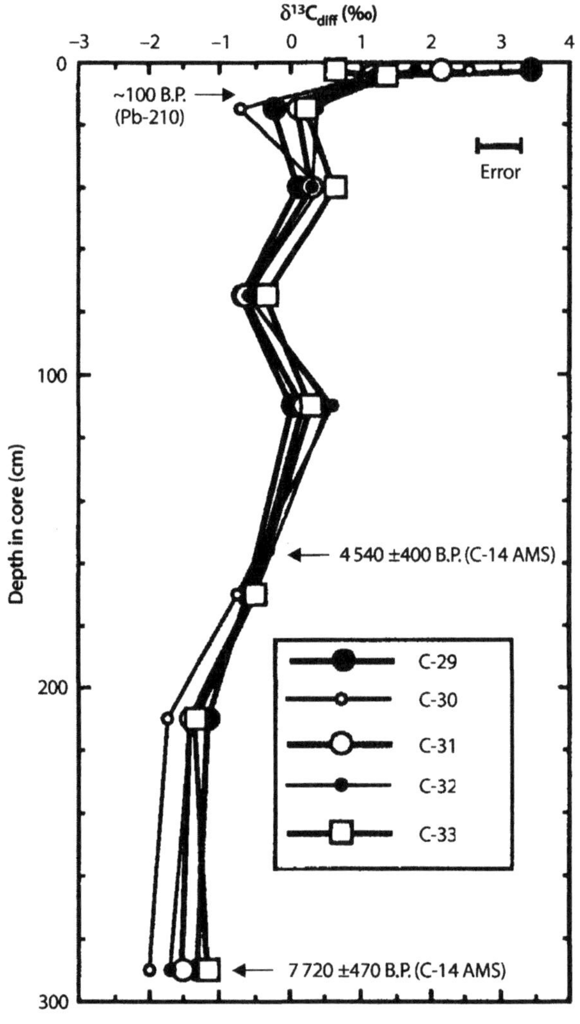

Fig. 19.4. Variations in *n*-alkane δ^{13}C values in a 3-m core off the Johnstone River, North Queensland, Australia. In order to facilitate comparison of trends, an arbitrary amount was added to each *n*-alkane δ^{13}C value so that each *n*-alkane had a $\delta^{13}C_{diff}$ of 0‰. Radiocarbon dates shown were measured by accelerator mass spectrometry. Redrawn from Bird et al. (1995)

depth) about 100 years ago. This isotopic trend is consistent with an increased influx of C_4-derived carbon resulting from clearing of forests in the catchment and replacement with grasses and sugarcane. Alternately, the shift in vegetation could result from aboriginal burning or a decrease in rainfall to the basin, both of which would also change the ecosystem. Overall, the shift in isotopic signal in the past century is twice as large as any other changes over the past 8 000 years.

Organic carbon isotope signatures can provide palaeoenvironmental information over long geologic time frames. Schoell et al. (1994) measured $\delta^{13}C$ values for C_{27} steranes and C_{35} hopanes in the Naples Beach section of the Monterey (California) formation. The Miocene Monterey formation has received considerable attention from petroleum geochemists, since it is an important source bed for the oil fields on the California continental margin. Steranes are fossil biomarkers derived from sterols biosynthesized by many phytoplankton that live in surface waters (0–30 m) in the contemporary ocean and presumably a similar depth range during the Miocene. The hopanes are derived from bacteriohopane polyols biosynthesized exclusively by prokaryotes, including cyanobacteria, the bulk of which generally reside deeper in the photic zone (e.g. 70–100 m). Convergence of $\delta^{13}C$ values of sedimentary steranes and hopanes over the same time period would imply that both phytoplankton and cyanobacteria took up the same dissolved inorganic carbon (*DIC*) pool with a single $\delta^{13}C_{DIC}$, and thus the water column was mixed. Divergence of sterane and hopane $\delta^{13}C$ values would result if the water column had been stratified and there were two *DIC* pools having different $\delta^{13}C_{DIC}$ values. The variation in sterane and hopane $\delta^{13}C$ values over the ~22 million years of the Naples Beach section of the Monterey formation is shown in Fig. 19.5. Isotope values for the steranes were relatively constant (~ -25 to -26‰), while there was a wider range (-31‰ increasing to -26‰) for the hopanes. There was a large difference between sterane and hopane $\delta^{13}C$ values ($\Delta^{13}C_{H\text{-}S}$) of ~6‰ early in the record during a cold climatic period (see the Antarctic ice sheet and $\delta^{18}O$ records in Fig. 19.5) vs. roughly equal values later in the record during a warmer climate. These data suggest that during the colder early Miocene, the water column was more stratified and there was a greater separation between zones of productivity of phytoplankton (shallow photic zone) and cyanobacteria (deep photic zone). However, later the water column was relatively well-mixed.

19.3
Depositional Environment – Anoxygenic Photosynthesis

Preservation of organic matter in sediments is a prerequisite for the development of a record that can be studied. It was traditionally accepted that anoxic depositional environments favour preservation of organic matter (Demaison and Moore 1980), and there has been considerable research both supporting and refuting this hypothesis (e.g. Emerson and Hedges 1988; Calvert and Pederson 1992; Lee 1992; Paropkari et al. 1992; Keil and Cowie 1999). Since anoxia and enhanced organic matter preservation might lead to formation of petroleum source rocks, recognition of any biomarkers that might aid in discerning anoxic palaeoenvironments would be of great interest to petroleum geochemistry and organic geochemistry alike.

Anoxic environments that might be prototypes for petroleum source beds have been intensely studied. One such environment is the Black Sea, where circulation in the

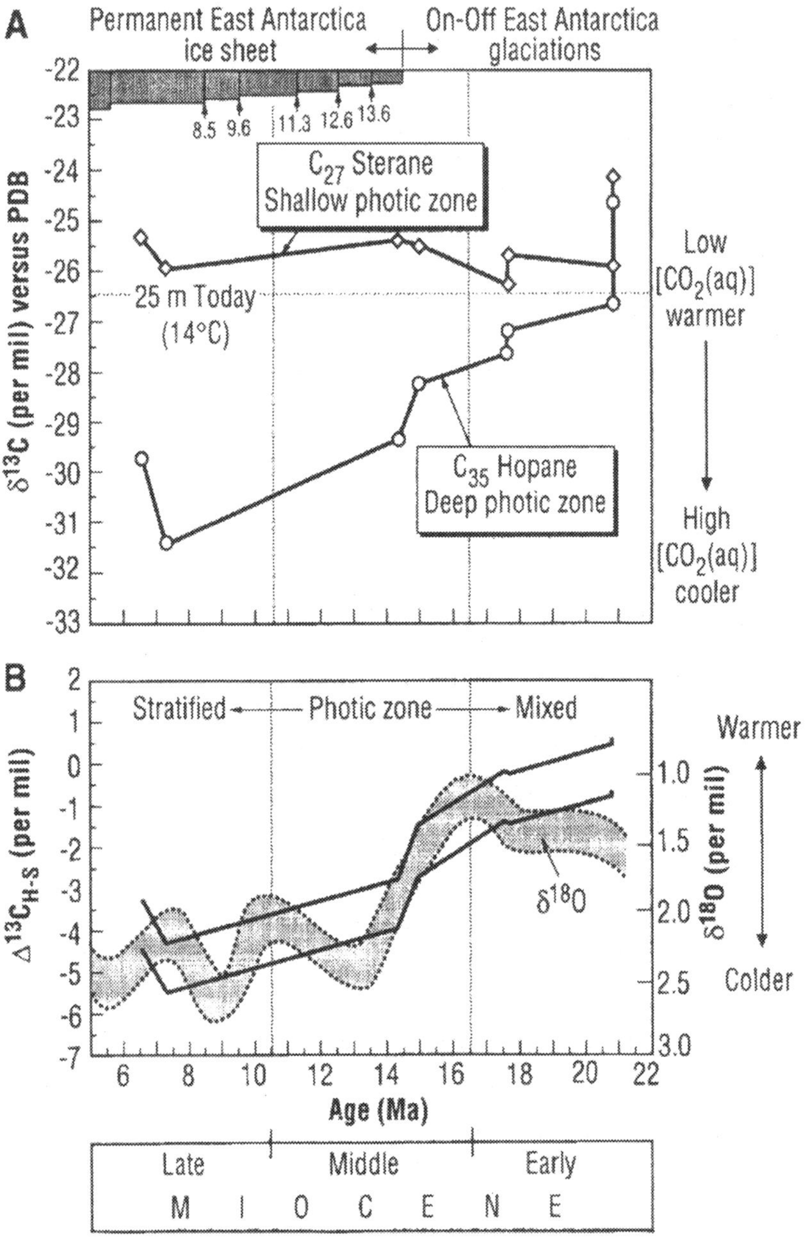

Fig. 19.5. a C_{27} steranes and C_{35} hopanes $\delta^{13}C$ values for the Naples Beach section of the Monterey formation (Naples Beach, California), indicating changing water masses in the shallow and deep euphotic zone. The scope of the Antarctic ice sheet represents relative climatic conditions; **b** Isotopic difference $\Delta^{13}C_{H\text{-}S}$ between C_{27} steranes and C_{35} hopanes. The lower solid line is as measured; the upper solid line is $\Delta^{13}C_{H\text{-}S}$ corrected by 1.2‰ for the difference ($\Delta^{13}C_{DIC}$) in the photic zone of contemporary Santa Monica Bay. The interpretation is that in the late Miocene, the water column was stratified but in the early Miocene it was mixed. The wavy ribbon is the Miocene $\delta^{18}O$ record from benthic foraminifera. From Schoell et al. (1994)

2 200-m deep basin is restricted, and bottom waters (≥120 m) are permanently anoxic. During a cruise to the Black Sea in 1988, organic geochemists and microbiologists were surprised to discover that the oxic-anoxic boundary in the water column (the "chemocline") was located well within in the photic zone (Murray et al. 1989). Associated with the chemocline were concentration maxima for bacteriochlorophyll e and carotenoid pigments, notably isorenieratene (*I*, Fig. 19.6) that are characteristic of anoxygenic primary production by photoautotrophic bacteria, *Chlorobrium* sp. (Repeta and Simpson 1991). These brown sulphur bacteria typically live as strict anaerobes in bacterial plates at shallow oxic/anoxic interfaces. Blooms of *C. phaeobacteriodes* frequently occur in shallow meromictic lakes, where hydrogen sulphide produced in anoxic sediments and bottom waters rises into the euphotic zone (Parkin and Brock 1980); viable of *C. phaeobacteriodes* have also been isolated from surface sediments underlying anoxic water columns (Herbert and Tanner 1977). The high concentrations of bacteriochlorophylls and isorenieratene at the chemocline in the Black Sea generated significant interest among organic geochemists. These pigments and their diagenetic and catagenic products have been found in a wide variety of ancient sediments (Koopmans et al. 1996a), making them potentially excellent biomarkers for anoxic waters reaching into the photic zone.

Distributions of isorenieratene in sediments of the Black Sea have been used to reconstruct the record of anoxygenic primary production over the past 8 200 years (Repeta 1993). Downcore concentrations of isorenieratene ranged from non-detectable to >70 μg g^{-1} or up to 1 μg mg^{-1} *TOC* (Fig. 19.7). There are distinct episodes of intense burial of isorenieratene that suggest temporal variations in the depth of the H_2S chemocline. Maxima in sedimentary isorenieratene should represent periods of a shallow H_2S chemocline, while the absence of isorenieratene should correspond to the chemocline being located below the photic zone. The depth of the chemocline in the water column has significant implications to the balance of freshwater inflow to the Black Sea from adjacent landmasses and through the Bosphorus. The freshwater balance in turn is related to climatic variations throughout the Holocene.

Fig. 19.6. Structures of isorenieratene (*I*), *β,β*-carotane (*II*) and octahydroisorenieratene (*III*)

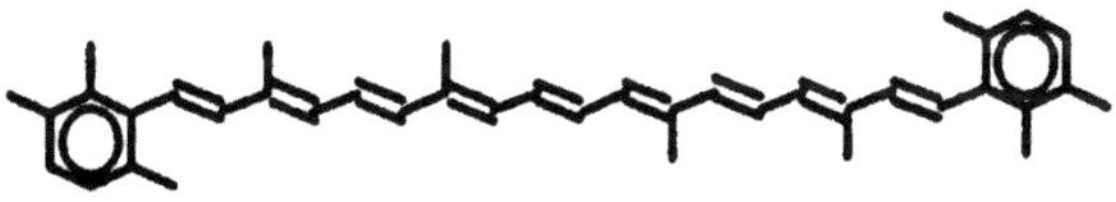

I

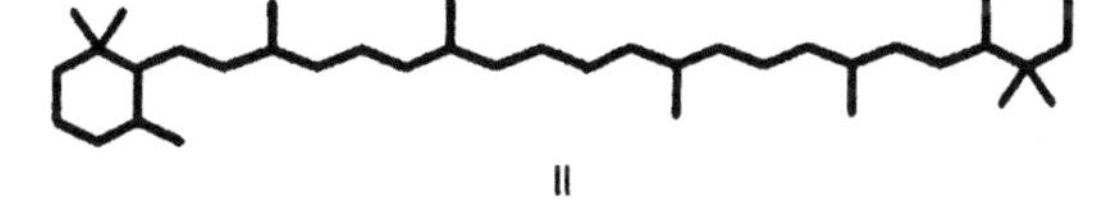

II

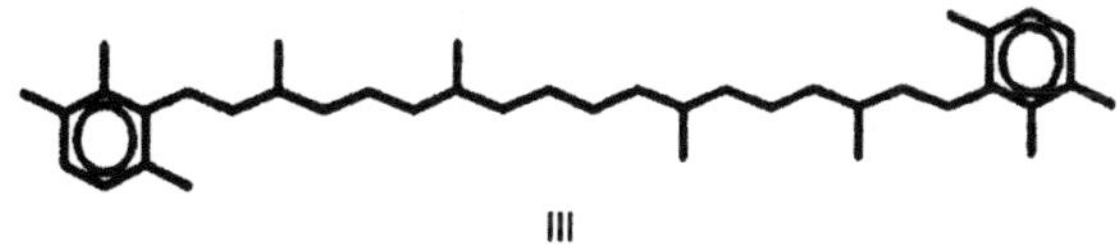

III

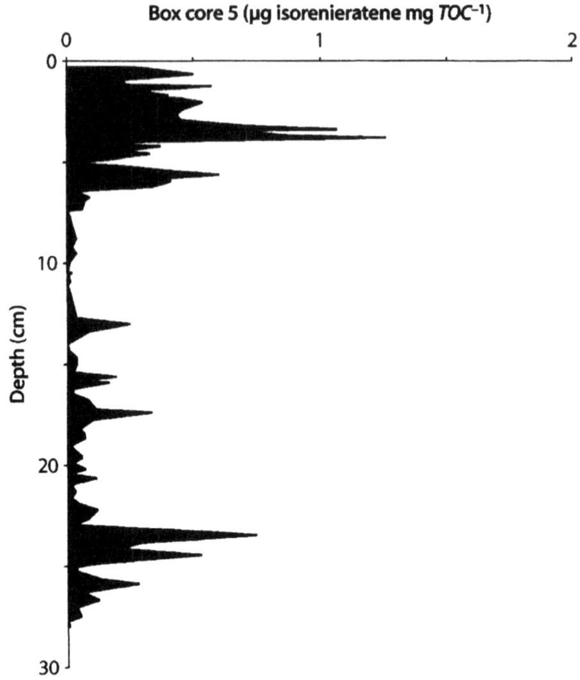

Fig. 19.7. Concentrations of isorenieratene in a Black Sea core covering the past 8000 years. Redrawn from Repeta (1993)

The highly unsaturated nature of carotenoid pigments such as isorenieratene renders these compounds remarkably reactive. In anoxic sediments, they preferentially accumulate in macromolecular fractions, usually through loss of double and incorporation of inorganic sulphur (Sinninghe Damsté et al. 1990; Wakeham et al. 1995). Sulphurization (inter- and intra-molecular cross-linking between S and C) may provide a mechanism that protects otherwise labile molecules from degradation by sequestration into a less reactive polar macromolecule. Chemical release of S-bound *β,β*-carotane and octahydroisorenieratene (*II* and *III*, Fig. 19.6), fully and partially unsaturated derivatives of isorenieratene, by desulphurization of polar macromolecular material extracted from Black Sea sediments produced a historical record (Sinninghe Damsté et al. 1993) similar to that derived from isorenieratene described above. Diagenesis and ultimately catagenesis of isorenieratene generates a wide range of aromatic and S-containing isoprenoids (Koopmans et al. 1996a) that can being applied as indicators of photic zone anoxia in ancient environments. In the Messinian (upper Miocene) evaporitic sequence of the Veno del Gesso Basin (Italy), distributions of S-bound isorenieratane (a derivative of isorenieratene) and the pentacyclic triterpenoid gammacerane have been used to reconstruct the palaeodepositional environment (Fig. 19.8). Gammacerane is thought to derive from anaerobic bacterivorous ciliates living at or just below the chemocline and is an indicator of water column stratification; isotopic evidence suggests that the ciliates were feeding on sulphur bacteria that produced isorenieratene (Sinninghe Damsté et al. 1995b). In sequences where both isorenieratane and gammacerane were present, the water column was stratified and the chemocline was in the photic zone. When gammacerane was present but iso-

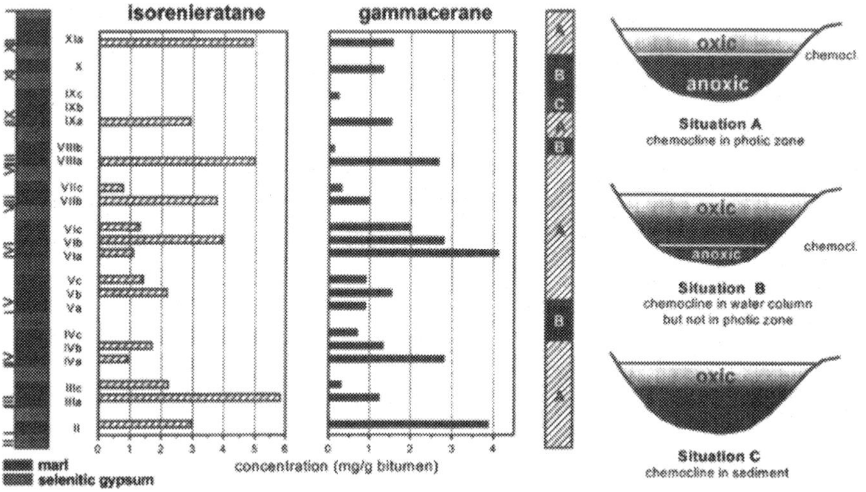

Fig. 19.8. Concentrations of isorenieratane and gammacerane in the Veno del Gesso marl sequence. Also shown are the reconstructed water columns based on these biomarker distributions. From Sinninghe Damsté et al. (1995a)

renieratane absent, the chemocline was below the photic zone but still in the water column, because the ciliates do not live in sediments. When both were absent, the water column was unstratified and there was no anoxic zone. Although carotenoid derivatives are now widely used as indicators of photic zone anoxia, they must be used with caution, however, and should not be made without complementary ^{13}C isotopic analysis (Koopmans et al. 1996b).

19.4
Alkenone Palaeothermometer

Series of long chain C_{37}-, C_{38}-, and C_{39}-di-, tri- and tetra-unsatursated methyl and ethyl ketones (see Fig. 19.9 for the C_{37} alkenones) were first observed nearly two decades ago in Miocene to Pleistocene sediments (Boon et al. 1978). The geographic range of alkenones is now known to encompass sediments from all oceans (Brassell 1993). At the time that the structures of the alkenones were first determined (de Leeuw et al. 1980; see also Rechka and Maxwell 1988a,b), they were also found to be abundant in unicellular algae of the Prymnesiophyceae, namely *Emiliania huxleyi* (Volkman et al. 1980). Alkenones have now been found in a number of prymnesiophytes (now classified as haptophytes), of which the calcifying *E. huxleyi* and *Gephyrocapsa oceanica* and the noncalcifying *Crysotilla* and *Isochrysis* are the most important in the oceans (Conte et al. 1994, 1995; Volkman et al. 1995). No other sources of alkenones are presently known, making them powerful and unusually unambiguous biomarkers. In addition to the limited species distribution of alkenones among phytoplankton, it was also discovered that they might be sensitive to the temperature at which they are biosynthesized and thus have a potential palaeotemperature application (Brassell et al. 1986).

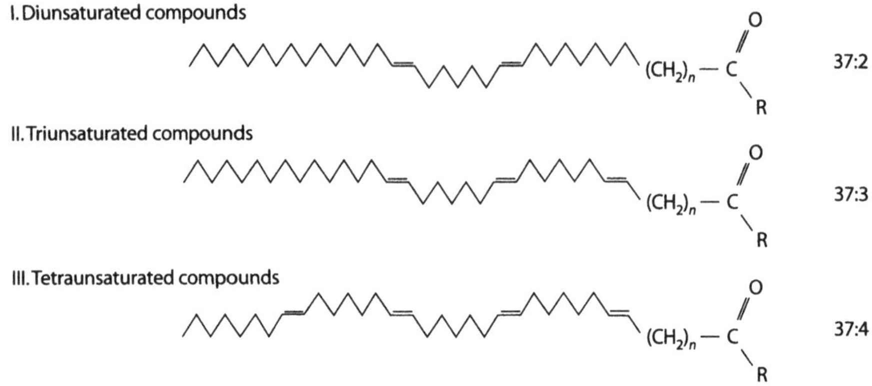

Fig. 19.9. Structures of the 37:2, 37:3 and 37:4 unsaturated alkenones

Variations in the proportions of di-, tri- and tetra-unsaturated alkenones in cultures and sediments (Fig. 19.10) led to an alkenone unsaturation index (Brassell et al. 1986):

$$U^{K}_{37} = ([37{:}2] - [37{:}4]) \,/\, ([37{:}2] + [37{:}3] + [37{:}4])$$

which simplifies to

$$U^{K}_{37}{}' = ([37{:}2] \,/\, ([37{:}2] + [37{:}3])$$

in the absence of 37:4. Culture experiments showed a linear relationship between U^{K}_{37} (or $U^{K}_{37}{}'$) and growth temperature (Brassell et al. 1986; Prahl and Wakeham 1987; Prahl et al. 1988) that could be transformed into a U^{K}_{37} vs. temperature calibration. There have been many calibrations reported in the literature due to apparent variability in alkenone biosynthesis by different strains of haptophytes (Brassell 1993; Conte et al. 1994, 1995). However, analysis of core tops and comparison with sea surface temperatures in many regions of the ocean has produced a relatively robust and "universal" calibration (Fig. 19.11; Müller et al. 1998; see also references cited for numerous recent $U^{K}_{37}{}'$-temperature calibrations), at least for typical open ocean regimes. Two important caveats must be kept in mind when calculating temperatures from alkenone unsaturation patterns. First, there is significant regional variability in $U^{K}_{37}{}'$-temperature calibrations, especially in coastal areas such that a universal open ocean calibration may not always be inappropriate. Second, there remains an issue about whether $U^{K}_{37}{}'$-derived temperatures more accurately reflect sea surface temperatures or mixed layer temperatures depending on the depth at which the alkenone-producers live.

Alkenone thermometry has nonetheless become a widely used tool by palaeoceanographers. Sea surface temperature records have been reconstructed over time scales of years to hundreds of kiloyears. For example, alkenone unsaturation patterns in sediments of the Peruvian margin provide information about El Niño events in the eastern Pacific that can be related to Peruvian palaeoclimate over the past several centuries. McCaffrey et al. (1990) found alkenone unsaturation to be positively correlated with periods of greater or lesser El Niño activity as described by Quinn et al. (1987).

Fig. 19.10. Gas chromatograms of alkenones in cultures of *Emiliania huxleyi* at 25 and 10 °C. Redrawn from Prahl and Wakeham (1987)

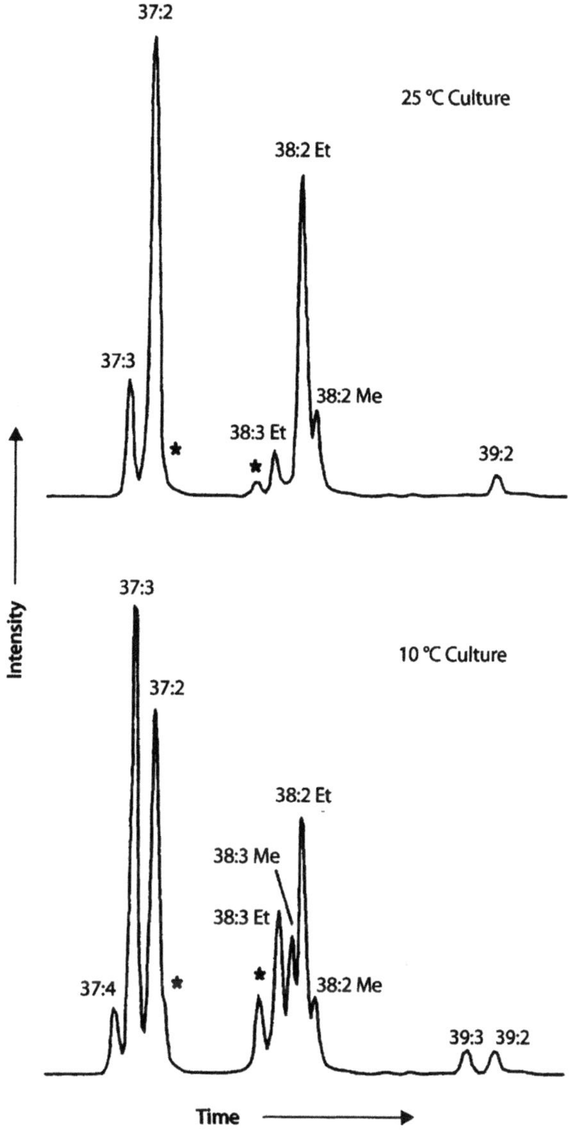

Increased sea surface temperatures are associated with El Niño conditions (Fig. 19.12), because the intensity of coastal upwelling of colder subsurface water in the eastern tropical Pacific decreases. The "very strong" El Niño events of Quinn et al. (1987) had sea surface temperature anomalies of 7–12 °C, considerably higher than the sea surface temperature derived from alkenone analyses (e.g. about 2 °C for the 1982 El Niño). This temperature difference may be related to the fact that the alkenone-producing haptophytes may better record the average temperature of the cooler mixed layer rather than that at the warmer sea surface.

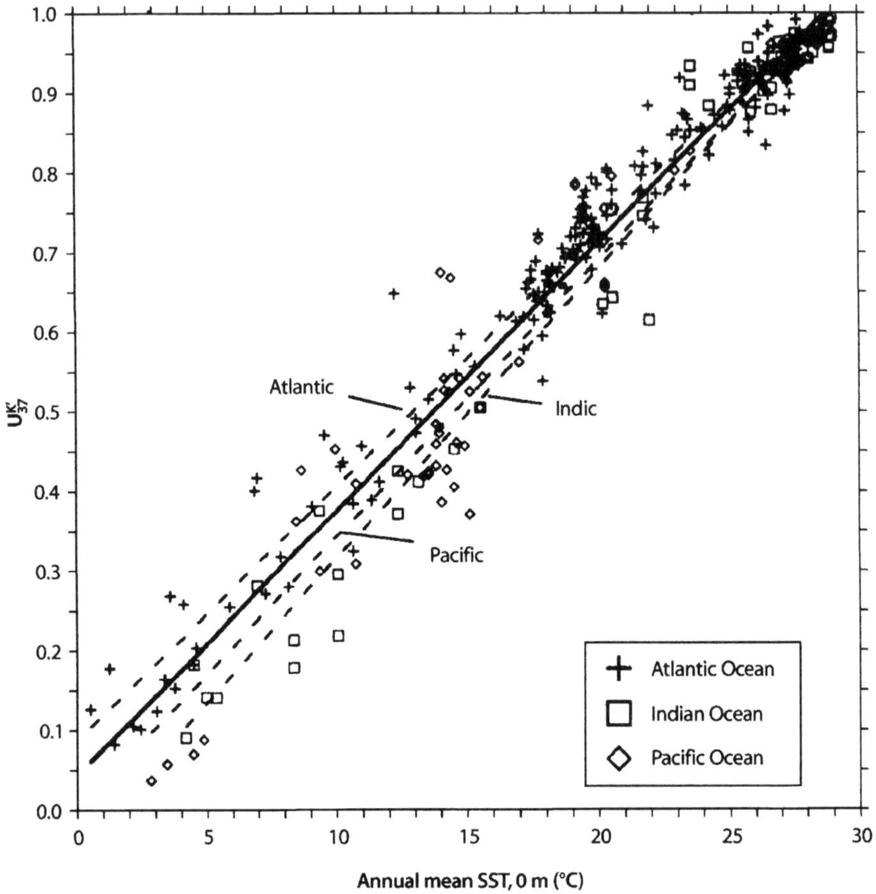

Fig. 19.11. "Global" calibration of U^{K}_{37}' vs. sea surface temperature derived from a compilation of core top data throughout the ocean. Redrawn from Müller et al. (1998)

On a longer scale, for example, the sea surface temperature history over the past 80 ka was determined by Zhao et al. (1995), who used high resolution (~200–500 years) U^{K}_{37}' records from cores off northwest Africa to track effects of glacial-interglacial transitions of sea surface temperatures in the subtropical eastern Atlantic. The derived U^{K}_{37}' sea surface temperatures had a range from ~18 °C during the last glacial (10–75 ka B.P.; Fig. 19.13) to ~21.5 °C during the early Holocene (<10 ka B.P.). In addition to the significant glacial vs. interglacial temperature difference recorded in these cores (and numerous others in the literature) is the rapidity of temperature change. In some cases, temperature changes of 3–4 °C occur in the space of a century, for example after Heinrich events H2 and H4 for core 658C in Fig. 19.13. These apparent rapid alkenone-derived temperature changes are in good agreement with oxygen isotope (^{18}O) records of temperature inferred from ice volume. The very rapid temperature shifts that coincide with Heinrich events are thought to indicate ice sheet surges during periods of

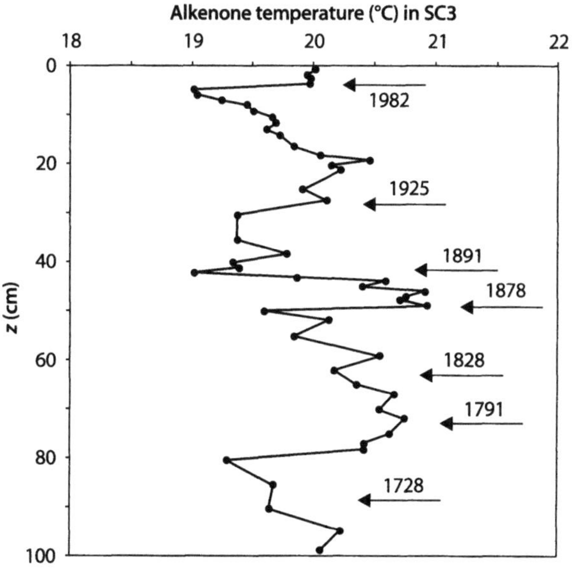

Fig. 19.12. Comparison of the alkenone temperature record for a core from the Peruvian margin with the historical record of "very strong" El Niño events compiled by Quinn et al. (1987). Redrawn from McCaffrey et al. (1990)

intense atmospheric cooling coupled with massive iceberg discharge and subsequent melting in the North Atlantic. These events must have been very large to affect sea surface temperatures as far south as subtropical north Atlantic (20° N).

19.5. Alkenone Palaeobarometer

Increases in atmospheric CO_2 at the end of the 20th century have raised questions about the levels of CO_2 during Earth's past. Estimates of palaeo-CO_2 concentrations have traditionally been made using fluctuations in $^{13}C/^{12}C$ of sedimentary marine carbonates (e.g. Schidlowski and Aharon 1992) or organic matter (e.g. Rau et al. 1991; Freeman and Hayes 1992). The record of isotope fractionation (ε_P) associated with fixation of CO_2 by marine phytoplankton may be useful since the fractionation is a function of the ambient concentration of CO_2(aq), among other things (reviewed by Popp et al. 1999). As described above (Sect. 19.2) for the utility of stable carbon isotopes for source identification, palaeo-CO_2 estimations derived from bulk organic carbon are complicated by the integrative aspects inherent in bulk *OC*, where there may be multiple biosynthetic sources responding to a range of environmental factors. As before, biomarker-specific measurements present a vast improvement in interpretative capability.

An ε_P-based approach using $\delta^{13}C$ values of sedimentary alkenones could provide estimates of palaeo-CO_2 (Jasper and Hayes 1990, Jasper et al. 1994). The principle behind this method (Fig. 19.14) is to compare $\delta^{13}C$ of inorganic carbonate (or dissolved inorganic carbon, *DIC*) with that of photosynthetically-fixed organic carbon (primary photosynthate or biomass) calculated from a compound-specific isotopic analysis of alkenones after applying an appropriate correction for the fractionation between

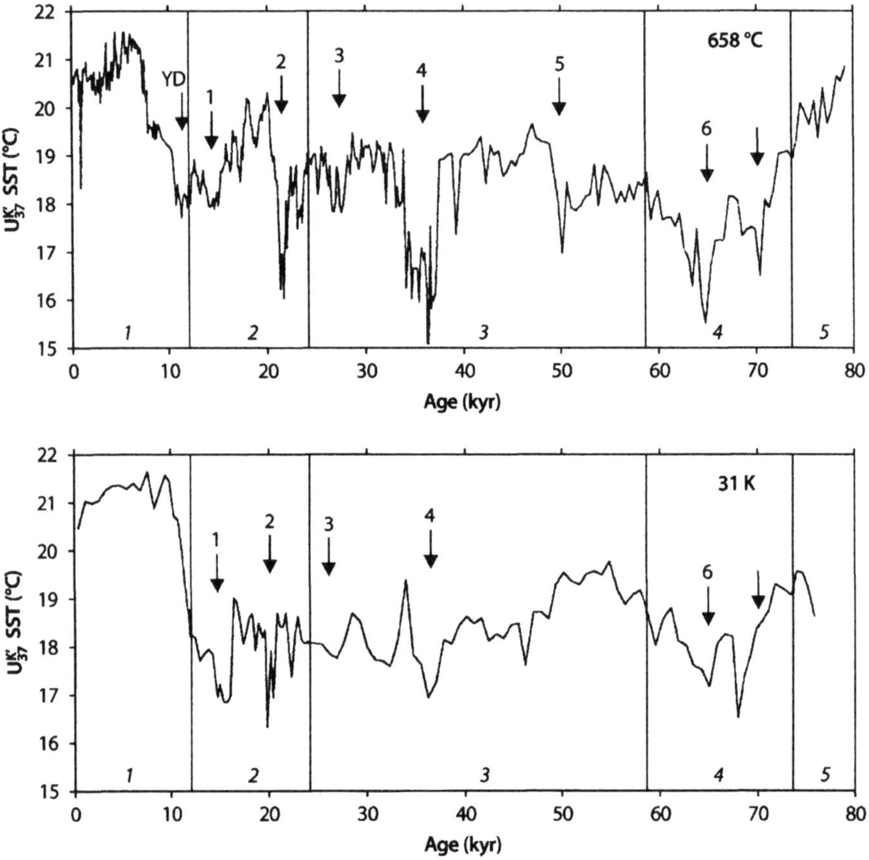

Fig. 19.13. Reconstructed $U^{K'}_{37}$ sea surface temperature records over the past 80 ka derived from cores in the subtropical southeast Atlantic. Arrows mark sea surface temperature maxima that correlate with Heinrich events (see text). Redrawn from Zhao et al. (1995)

$\delta^{13}C_{biomass}$ and $\delta^{13}C_{alkenone}$. This gives an estimate of the isotope fractionation (ε_P, ‰) that occurs during photosynthesis. The resulting ε_P is then placed on a calibration plot relating ε_P and the sea water concentration of dissolved CO_2 [$CO_2(aq)$] to give an estimated p_{CO_2}. Application of Henry's Law allows calculation of the equilibrium atmospheric p_{CO_2}.

Jasper and Hayes (1990) applied their approach to calculate the history of ε_P, surface water p_{CO_2} and atmospheric p_{CO_2} for the northern Gulf of Mexico over the past 100 ka using sediments from the Pigmy Basin (Fig. 19.15). Palaeo-p_{CO_2} trends were compared with the p_{CO_2} record obtained directly by measuring CO_2 in gas bubbles from the Vostok (Antarctica) ice core. The alkenone-based and ice core records are remarkably similar, showing reduced levels of atmospheric CO_2 during the last glacial. In contrast, the p_{CO_2} record that would be obtained from $\delta^{13}C$ of total organic carbon was significantly different from and more poorly tracked the Vostok record. This discrepancy provides a strong argument for using specific biomarkers for palaeo-CO_2 reconstructions rather than bulk *OC*.

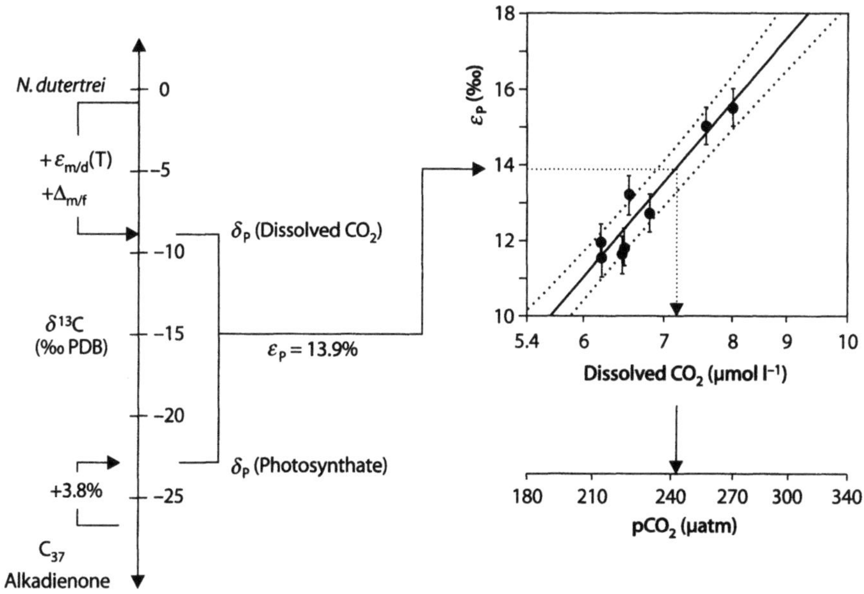

Fig. 19.14. Schematic of the method used to calculate palaeo-p_{CO_2} (sea water) and p_{CO_2} (atmosphere) from $\delta^{13}C$ of carbonate, *OC*, or alkenones. Redrawn from Jasper and Hayes (1990)

Alkenone-based palaeo-CO_2 histories are rapidly accumulating in the literature (e.g. Jasper et al. 1994; Müller et al. 1998; Pagani et al. 1999). In a study of the Angola Current over the last 200 ka, Müller et al. (1998) derived the p_{CO_2} record for surface waters using isotopic compositions of alkenones extracted from cores in the Angola Basin off western Africa. As in the Pigmy Basin in the Gulf of Mexico, there was substantial variability in p_{CO_2} in the Angola Basin over the time scale covered by the cores. p_{CO_2} trends in surface waters were compared to the Vostok ice core record to determine the disequilibrium between the ocean and atmosphere (Fig. 19.16). The magnitude of the difference between sea water and atmosphere (Δp_{CO_2}) determines the flux of CO_2 across the air-sea interface, while the sign of the difference determines the direction of the flux. This exercise suggests that surface waters in the eastern Angola Basin were generally a CO_2 source during the past 200 ka (sea water p_{CO_2} levels were higher than p_{CO_2} in the atmosphere; note that a 70 µatm correction was used in Fig. 19.16 to facilitate comparison of trends). The strength of the flux from the sea surface to atmosphere was reduced during glacial compared to interglacial times ($p_{CO_2} \sim 70$ µatm for glacial times (20–70 ka and 130–180 ka) vs. 100–120 µatm for interglacial periods (15 ka to present and 70–130 ka)). Presumably, this difference was related to lower sea surface temperatures and higher biological productivity during glacial periods that would draw down concentrations of dissolved CO_2 and reduce the flux-driving gradient between the sea surface and atmosphere.

Alkenone palaeobarometry is a relatively new tool, and additional investigation is required to minimize uncertainty that presently exists in this or any other biomarker-based CO_2 proxy. For example, variability in ε_P and $\delta^{13}C$ of inorganic carbon and

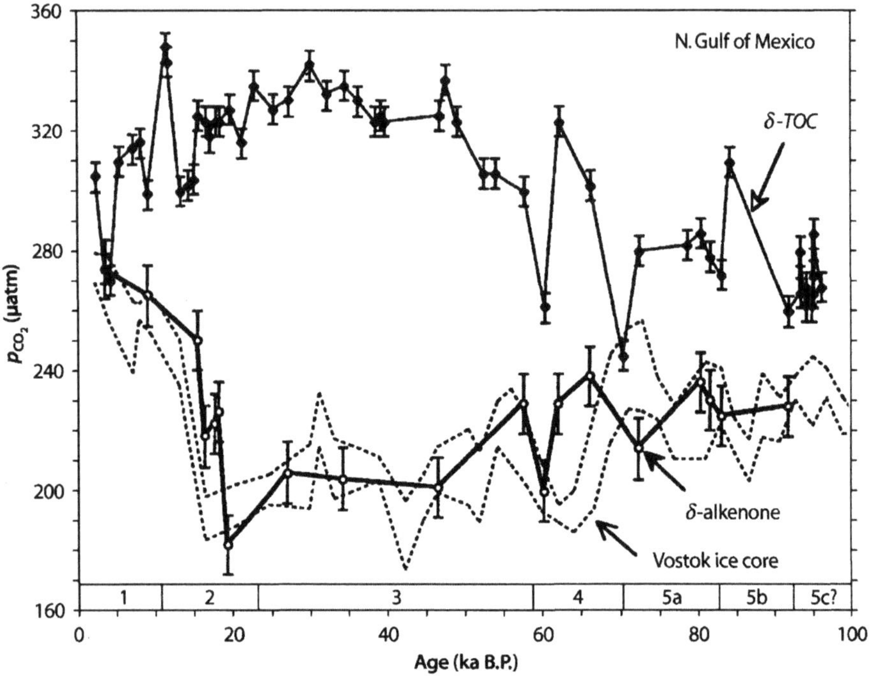

Fig. 19.15. Atmospheric p_{CO_2} over the last 100 ka derived from $\delta^{13}C$ of alkenones extracted from sediments of the Pigmy Basin, compared to the record from the Vostok ice core and a record that would be obtained from bulk *OC*. Redrawn from Jasper and Hayes (1990)

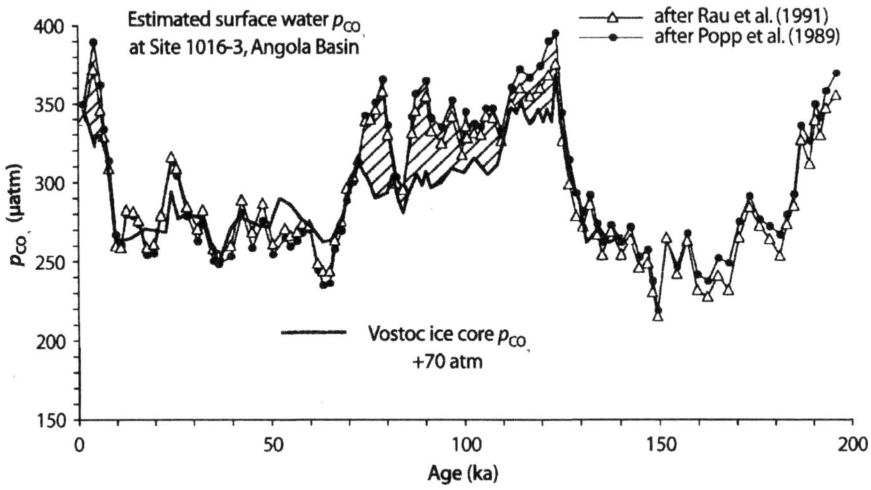

Fig. 19.16. Estimated surface water p_{CO_2} derived from a core from the Angola Basin. The record obtained from the Vostok ice core is shown for comparison and has been adjusted by the average glacial sediment-ice difference of 70 µatm to facilitate comparison. This presentation emphasizes the disequilibrium during the period between 60 and 130 ka. Redrawn from Müller et al. (1998)

biomarkers can be caused by many factors, most of which are poorly constrained in palaeoenvironmental studies but which are being intensely studied in the contemporary ocean (e.g. Popp et al. 1999). Carbon isotope fractionation is strongly influenced by several factors in addition to dissolved CO_2. One must be certain that the source of the specific biomarker being used is as unambiguous as possible. The limited sources of alkenones in open ocean settings – *E. huxleyi* and *G. oceanica* – makes these compounds among the best in the organic geochemists' toolbox to date. The carbon uptake mechanism and cellular geometry of source organisms must be known, because they affect the ε_P that is imprinted into organic matter. Microalgal growth rate is another important factor affecting ε_P that cannot be easily determined in ancient environments. Cd/Ca ratios might be a proxy for surface water phosphate, which could be used to infer primary production and perhaps growth rates. It is thus important to use a multi-proxy approach – e.g. ε_P determined from sedimentary alkenones and inorganic carbonate from planktonic foraminifera along with Cd/Ca measurements – in order to improve confidence in the results.

Acknowledgements

The support of the National Science Foundation during the preparation of this paper is gratefully acknowledged. C. Lee provided helpful comments to improve the manuscript.

References

Bird MI, Summons RE, Gagan MK, Roksandic Z, Dowling L, Head J, Fifield LK, Cresswell RG, Johnson DP (1995) Terrestrial vegetation change inferred from *n*-alkane $\delta^{13}C$ analysis in the marine environment. Geochim Cosmochim Acta 59:2853–2857

Boon JJ, Meer FW van der, Lange F de, Leeuw JW de, Schenck PA, Burlingame AL (1978) In: Initial reports of the Deep-Sea Drilling Project, vols 38, 39, 40 and 41, Supplement, Walvis Ridge DSDP Leg 40. U.S. Government Printing Office, Washington, pp 627–637

Brassell SC (1993) Applications of biomarkers for delineating marine paleoclimate fluctuations during the Pleistocene. In: Engel MH, Macko SA (eds) Organic geochemistry – principles and applications. Plenum Press, New York, pp 699–738

Brassell SC, Eglinton G, Marlowe IT, Pflaumann U, Sarnthein M (1986) Molecular stratigraphy: A new tool for climate assessment. Nature 320:129–133

Calvert SE, Pederson TF (1992) Organic carbon accumulation and preservation in marine sediments: How important is anoxia? In: Whelan J, Farrington JW (eds) Organic matter. Univ. Press, New York, pp 231–263

Conte MH, Volkman JK, Eglinton G (1994) Lipid biomarkers of the Haptophyta. In: Green JC, Leadbeater BSC (eds) The haptophyte algae. Clarendon Press, Oxford, pp 351–377

Conte MH, Thompson A, Eglinton G, Green JC (1995) Lipid biomarkers diversity in the coccolithophoride *Emiliania huxleyi* (Prymnesiophyceae) and the related species *Gephyrocapsa oceanica*. J Phycol 31:272–282

Degens, ET (1969) Biogeochemistry of stable carbon isotopes. In: Eglinton G, Murphy MTJ (eds) Organic geochemistry. Springer-Verlag, New York, pp 304–329

Demaison GJ, Moore GT (1980) Anoxic environments and oil source bed genesis. AAPG Bull 64:1179–1209

Emerson S, Hedges JI (1988) Processes controlling the organic carbon content of open ocean sediments. Paleoceanogr 3:621–634

Freeman KH, Hayes JM (1992) Fractionation of carbon isotopes by phytoplankton and estimates of ancient CO_2 levels. Global Biogeochem Cycles 6:185–198

Fogel ML, Cifuentes LA (1993) Isotope fractionation during primary production. In: Engel MH, Macko SA (eds) Organic geochemistry – principles and applications. Plenum Press, New York, pp 73–98

Gearing P, Plucker FE, Parker PL (1977) Organic carbon stable isotope ratios in continental sediments. Mar Chem 5:251–266

Goñi MA, Ruttenberg KC, Eglington T (1998) A reassessment of the sources and importance of land-derived organic matter in surface sediments from the Gulf of Mexico. Geochim Cosmochim Acta 62:3055–3075
Hayes JM (1993) Factors controlling ^{13}C contents of sedimentary organic compounds: Principles and evidence. Mar Geol 113:111–125
Hayes JM, Freeman KH, Popp BN, Hoham CH (1990) Compound-specific isotopic analyses: A novel tool for reconstruction of ancient biogeochemical processes. Org Geochem 16:1115–1128
Hedges JI, Parker PL (1976) Land-derived organic matter in surface sediments from the Gulf of Mexico. Geochim Cosmochim Acta 40:1019–1029
Herbert R, Tanner A (1977) The isolation and some characteristics of photosynthetic bacteria (Chlorobiacae, Chromatiaceae) from Antarctic marine sediments. J Appl Bacteriol 43:437–447
Jasper JJ, Hayes JM (1990) A carbon isotope record of CO_2 levels during the quaternary. Nature 347:462–464
Jasper JP, Hayes JM, Mix AC, Prahl FG (1994) Photosynthetic fractionation of ^{13}C and concentrations of CO_2 in the central equatorial Pacific during the last 225000 years. Paleoceanogr 9:781–898
Keil RG, Cowie GL (1999) Organic matter preservation through the oxygen deficient zone of the NE Arabian Sea as discerned by organic carbon:mineral surface area ratios. Mar Geol 161:12–22
Koopmans MP, Köster J, Kaam-Peters HME van, Kenig F, Schouten S, Hartgers WA, Leeuw JW de, Sinninghe Damsté JS (1996a) Diagenetic and molecular products of isorenieratene: Molecular indicators for photic zone anoxia. Geochim Cosmochim Acta 60:4467–4496
Koopmans MP, Schouten S, Kohnen S, Sinninghe Damsté JS (1996b) Restricted utility of aryl isoprenoids as indicators for photic zone anoxia. Geochim Cosmochim Acta 60:4873–4876
Lee C (1992) Controls on organic carbon preservation: The use of stratified water bodies to compare intrinsic rates of decomposition in oxic and anoxic systems. Geochim Cosmochim Acta 56:3323–3335
Leeuw JW de, Meer FW van der, Rijpstra WIC (1980) On the occurrence and structural identification of long-chain unsaturated ketones and hydrocarbons in sediments. In: Douglas AG, Maxwell JR (eds) Advances in organic geochemistry. Pergamon Press, New York, pp 211–217
McCaffrey MA, Farrington JW, Repeta DJ (1990) The organic geochemistry of Peru margin sediments. I. A comparison of the C_{37} alkenone and historical El Niño records. Geochim Cosmochim Acta 54:1671–1682
Müller PJ, Kirst G, Ruhland G, Storch I von, Rosell-Melé A (1998) Calibration of the alkenone paleotemperature index $U_{37}^{K'}$ based on core tops from the eastern South Atlantic and the global ocean (60° N–60° S). Geochim Cosmochim Acta 62:1757–1772
Murray JW, Jannasch HW, Honjo S, Anderson RF, Reeburgh WS, Top Z, Frederich GE, Codispoti LA, Izdar E (1989) Unexpected changes in the oxic/anoxic interface in the Black Sea. Nature 338:411–413
Onstad G, Canfield DE, Quay PD, Hedges JI (2000) Sources of particulate organic matter in rivers from the continental USA: Lignin phenol and stable isotope compositions. Geochim Cosmochim Acta 64:3539–3546
Pagani M, Arthur MA, Freeman KH (1999) Miocene evolution of atmospheric carbon dioxide. Paleoceanogr 14:273–292
Parkin TB, Brock TD (1980) The effect of light quality on the growth of phototrophic bacteria in lakes. Arch Microbiol 125:19–27
Paropkari AL, Prakash Babu C, Mascarenhas A (1992) A critical evaluation of depositional parameters controlling the variability of organic matter in Arabian Sea sediments. Mar Geol 107:213–226
Popp, BN, Hanson KL, Dore JE, Bidigare RR, Laws EA, Wakeham SG (1999) Controls on the carbon isotopic composition of phytoplankton: Paleoceanographic perspectives. In: Abrantes F, Mix A, (eds) Reconstructing ocean history: A window into the future. Kluwer Academic, New York, pp 381–398
Prahl FG, Muelhausen LA (1989) Lipid biomarkers as geochemical tools for paleoceanographic study. In: Berger WH et al. (eds) Productivity of the ocean: Present and past. Wiley, New York, pp 271–289
Prahl FG, Wakeham SG (1987). Calibration of unsaturation patterns in long-chain ketone compositions for paleotemperature assessment. Nature 330:367–369
Prahl FG, Muelhausen LA, Zahnle DL (1988) Further evaluation of long-chain alkenones as indicators of paleoceanographic conditions. Geochim Cosmochim Acta 52:2303–2310
Prahl FG, Ertel JR, Goñi MA, Sparrow MA, Eversmeyer B (1994) Terrestrial organic carbon contributions to sediments on the Washington margin. Geochim Cosmochim Acta 58:3035–3048
Quinn WH, Neal VT, Antunez de Mayolo SE (1987) El Niño occurrences over the past four and a half centuries. J Geophys Res 92:14449–14461
Rau GH, Froelich PN, Takahashi T, Des Marais DJ (1991) Does sedimentary organic $\delta^{13}C$ record variations in quaternary ocean $[CO_2(aq)]$? Paleoceanogr 6:339–347
Rechka JA, Maxwell JR (1988a) Unusual long-chain ketones of algal origin. Tetrahedron Lett 29:2599–2600
Rechka JA, Maxwell JR (1988b) Characterization of alkenone temperature indicators in sediments and organisms. In: Mattavelli L, Novelli L (eds) Advances in organic geochemistry 1987. Org Geochem 13:727–734

Repeta DJ (1993) A high resolution record of Holocene anoxygenic primary production in the Black Sea. Geochim Cosmochim Acta 57:4337–4342

Repeta DJ, Simpson DJ (1991) The distribution and recycling of chlorophyll, bacteriochlorophyll and carotenoids in the Black Sea. Deep-Sea Res. 38(Suppl 2):S969–S984

Schidlowski M, Aharon P (1992) Carbon cycle and carbon isotope record: Geochemical impact of life over 3.8 Ga of Earth history. In: Schidlowski M et al. (eds) Earth organic evolution: Implications for mineral and energy resources. Springer-Verlag, Berlin, pp 147–175

Schoell M, Scouten S, Sinninghe Damsté JS, Leeuw JW de, Summons RE (1994) A molecular organic carbon isotope record of Miocene climate changes. Science 263:1122–1125

Sinninghe Damsté JS, Rijpstra WIC, Kock van-Dalen AC, Leeuw JW de, Schenck PA (1990) Characterization of organically bound sulfur in high-molecular-weight, sedimentary organic matter using flash pyrolysis and Raney Ni desulfurization. In: Orr WL, White CM (eds) Geochemistry of sulfur in fossil fuels. American Chemical Society (ACS Symposium 249, pp 486–528)

Sinninghe Damsté JS, Wakeham SG, Kohnen MEL, Hayes JM, Leeuw JW de (1993) A 6000 year molecular record of chemocline excursions in the Black Sea. Nature 362:827–829

Sinninghe Damsté JS, Frewin NL, Kenig F, Leeuw LW de (1995a) Molecular indicators for paleoenvironmental change in a Messinian evaporitic sequence (Vena del Gesso, Italy). I: Variations in extractable organic matter from ten cyclically deposited marl beds. Org Geochem 23:471–483

Sinninghe Damsté JS, Kenig F, Koopmans ME, Köster J, Schouten S, Hayes JM, Leeus JW de (1995b) Evidence for gammacerane as an indicator of water column stratification. Geochim Cosmochim Acta 59:1895–1990

Volkman JK, Eglinton G, Corner EDS, Forsberg TEV (1980) Long-chain alkenes and alkenones in the marine coccolithophoprid *Emiliania huxleyi*. Phytochem 19:2619–2622

Volkman JK, Barrett SK, Blackburn SI, Sikes EL (1995) Alkenones in *Gephyrocapsa oceanica*: Implications for studies of paleoclimate. Geochim Cosmochim Acta 59:513–520

Wakeham SG, Kohnen MEL, Sinninghe Damsté JS, Leeuw JW de (1995) Organic sulfur compounds formed during early diagenesis in Black Sea sediments. Geochim Cosmochim Acta 59:521–533

Zhao M, Beveridge NAS, Shackleton NJ, Sarnthen M, Eglinton G (1995) Molecular stratigraphy of cores off northwest Africa: Sea surface temperature history over the last 80 ka. Paleoceanogr 10:661–675

Studies of Water Masses Mixing in the Ross Sea (Antarctica) Using Chemical Tracers

P. Rivaro · R. Frache

20.1 Introduction

Recent oceanographic observations near the Antarctic continent renewed interest about the role of the high latitudes in climatic changes. In fact, the production of dense waters in the ocean at polar regions has a significant effect on the global deep ocean circulation (Whitworth et al. 1998). These water masses, which have relatively high concentrations of oxygen acquired from the atmosphere, flow away from Antarctic regions and sink. introducing water with near-surface characteristics into deep ocean. This process, called ventilation, is associated with important fluxes of heat, salt, nutrients and gases.

The Antarctic Ross Sea (Fig. 20.1) is an interesting area, where different important aspects such as ice melting, water mass formation, mixing, and spreading can be observed and studied.

The Ross Sea is a wide continental shelf, with an area of about 500 000 km^2, bounded west by the Victoria Land coast and east by Edward VII Land; the northern boundary can be considered the line from Cape Adare to Cape Colbeck, and the southern boundary is the Ross Ice Shelf (RIS). The RIS extends over nearly half the continental shelf and is about 250 m thick on its northernmost side (Jacobs 1989). The Ross Ice Shelf limits only the uppermost waters, while the deeper waters can circulate freely under the ice shelf (Budillon et al.1999).

The topography of the continental shelf is locally modified by canyons and banks, and the mean depth is about 450 m (Picco et al. 1999).

The characteristics of the water masses of the Ross Sea have been described by several authors (Jacobs et al. 1985; Locarnini 1994).

The available evidence indicates that most of the dense deep-water production in high southern latitudes originates on the continental shelf and onto the continental slope, mixing with ambient water. If this mixed fluid becomes equal to the ambient density of the ocean at a given level, it will leave the vicinity of the continental slope and settle at this particular level.

If instead it reaches the bottom of the slope, it may become Antarctic Bottom Water (AABW) (Baines and Condie 1998).

Many of the details of the processes of dense water formation on the continental shelf and its flow down the slope are still obscure, and these details probably vary from location to location.

Dense waters are primarily formed on the Antarctic continental shelf by the formation of sea ice in autumn and winter with the consequent rejection of brine, which produces salty water at ≤ -1.85 °C potential temperature at the freezing point. This pro-

Fig. 20.1. The Ross Sea Map and bathymetry (from Budillon et al.1999, modified)

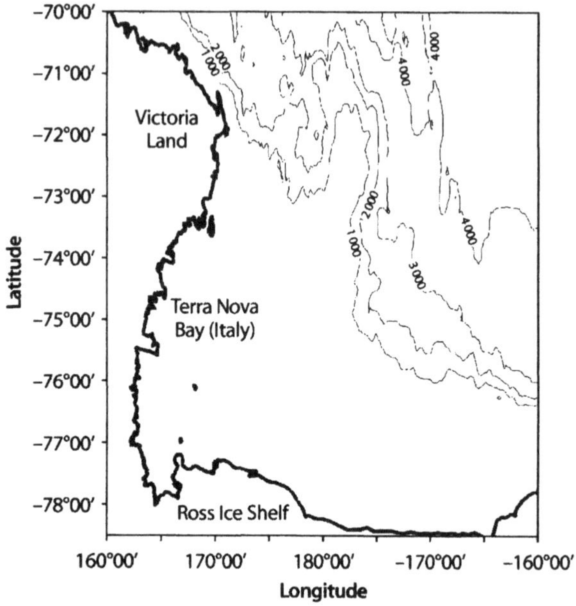

cess is assisted by the presence of coastal polynyas that are formed and maintained by the off-shore; these are directed drainage winds from the continent, which are generally stronger and colder in winter. The rejected brine increases the salinity of the subsurface waters, forming the densest waters in the Southern Ocean. These waters, with salinity values that increase with depth from 4.75 to 35 are named High Salinity Shelf Waters (HSSW).

Some of these dense waters may be modified (made colder and fresher, but generally less dense) by ice melting at the underside of deep ice shelves, which produce water (Ice Shelf Water, ISW) colder than −1.85 °C (to about −2.3 °C), because the pressure at depth lowers the melting point of ice.

All these processes produce a heterogeneous mixture of dense water that may be cold or salty or both, relative to its environment (Baines and Condie 1998).

It may accumulate in the depressions on the shelf and eventually spill over the sill at the shelf break, or find paths down canyons or other openings to the deep sea.

Shelf waters can mix with a warmer (>0 °C) water mass named Circumpolar Deep Water (CDW), the most voluminous water mass of the Southern Ocean, which is carried around Antarctica by the Antarctic Circumpolar Current (ACC). CDW is generally separated from the shelf waters by a front, the Antarctic Slope Front (ASF), a common feature near the continental shelf break, which is supposed to be an important source or sink region of nutrients, particulate matter and heat (Jacobs 1991).

20.2 Chemical Tracers in Oceanography

Traditionally, temperature and salinity have been used as tracers of water masses in order to differentiate between melting, freezing, and mixing processes and to define the water type.

More recently, isotopes and other chemicals such as tritium, ^{3}Helium and chlorofluorocarbons (CFC) have been used as transient tracers of water mass formation processes and circulation.

They do not have any biogeochemical activities, and they enter the marine environment through inputs from atmosphere.

Their time-dependent input into oceanic surface waters can be used in model studies to derive renewal rates of deep water from near-surface waters as well as exchange rates between individual deep-water masses (Mensch et al. 1998).

However, their sampling and measurements are very complicated due to their low concentrations in the water column, atmosphere contamination possibility, and instrumentation not readily available in oceanographic laboratories.

Oxygen and nutrients (nitrate, nitrite, ammonia, phosphate, silicate), which are the most determined oceanographic parameters, are involved in the production and degradation of organic matter, and therefore they are in themselves limited as water mass tracers.

Surface oxygen concentration varies from about 350 μM in surface waters at 0 °C to 200 μM in warm tropical waters due to the change of solubility of oxygen with temperature. However, because a number of effects such as phytoplanktonic activity, the surface concentration is generally supersaturated.

Below the depth of the photic zone, oxygen is consumed by the oxidation of organic matter. The average rate of consumption varies with depth and geographical location (Minster 1989).

For nutrients, the picture is similar. The values at the surface waters are mostly determined by a competition between residence time of the waters, subtraction due to biological activities and regeneration processes, which contribute to their concentrations in the deep layers of the water column.

20.2.1 "NO" and "PO" as Chemical Tracers

Broeker introduced a linear combination of the concentrations of oxygen and nitrate and of oxygen and phosphate, such as the observed changes in concentrations as a result of biological production or degradation, which in principle cancel each other out (Broeker 1974).

Biological studies have shown that for each mole of O_2 consumed, roughly 1/9th of a mole of bound N is released as nitrate ion.

Thus, the combination of $9\ NO_3 + O_2$ (termed NO) and of $135\ PO_4 + O_2$ (termed PO) should be nearly conservative.

The qualification "nearly" is introduced because the coefficient 9 and 135, based on the estimation of Redfield ratio, undoubtedly varies from place to place in the sea.

Evidence that NO is appropriate as a conservative water mass tracer is shown by a plot of NO, NO_3 and O_2 vs. salinity, reported in Fig. 20.2. It can be seen that the concentration of O_2 and NO_3 shows a non linear variation as a function of salinity, and that the curvature in the oxygen and nitrate vs. salinity profile are largely removed in the NO vs. salinity plot (Broecker 1974).

NO and PO have been used in global ocean circulation studies for identifying waters of southern and northern origin in the deep Atlantic (Broecker 1974). More re-

Fig. 20.2. Plot of NO, NO_3 and O_2 values vs. salinity, in the Atlantic Ocean (from Broecker 1974, modified)

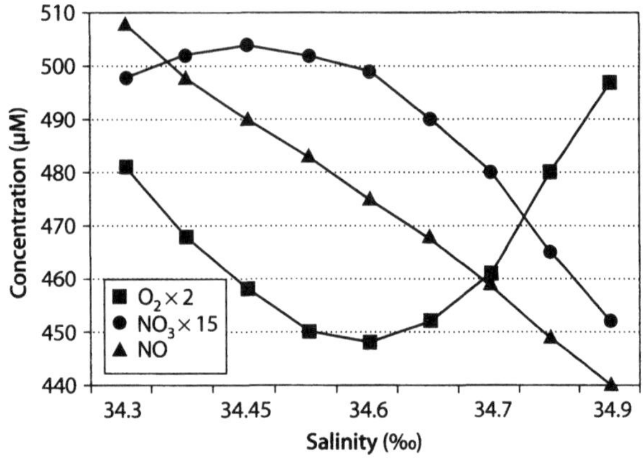

cently, these tracers were applied for studying the mixing of different water masses in the Weddel Sea (Antarctica), involved in the bottom water formation (Lindegren and Anderson 1991; Lindegren and Josefson 1998). NO distribution observed in a section towards the Filchner Ice Shelf in the Weddel Sea showed that the chemical signature of the entrained Warm Deep Water had to be found in the NO minimum (480 µM), which coincided with the temperature maximum (Lindegren and Anderson 1991).

Nevertheless, in the Arctic Ocean, NO and PO and the combined derivative NO/PO have been used to investigate the source of relatively shallow waters (Wilson and Wallace 1990; Anderson and Jones 1992; Cooper et al. 1999).

20.3 The Use of NO and PO as Chemical Tracers in Studying the Mixing of Water Masses in the Ross Sea Shelf Area: A Field Study

As part of an interdisciplinary field project performed during the austral summer 1997–98 devoted to examining the physical and biogeochemical characteristics of the Ross Sea, a study of the interaction between the Shelf Waters and the Circumpolar Deep Water in correspondence to the shelf break was carried out.

These investigations were carried out in the framework of the activities of the Climate Long-Term Interaction for the Mass balance in Antarctica (CLIMA) project of the Italian National Program for Antarctic Research (PNRA) in the Ross Sea.

In this study, the distribution of oxygen and nutrients were considered, together with conservative tracers NO and PO related to the physical variables, to characterize, from a chemical point of view, the mixing processes occurring in correspondence to the shelf break.

20.3.1 Sampling Area and Sea Water Sample Analysis

The sampling area was selected in correspondence to the shelf break in the centre of the Ross Sea basin (latitude 75°54'06" S , longitude 177°44'04" W), on the basis of previous surveys (1995 CLIMA Project Cruise). In this area, a mesoscale experiment was carried out after a large-scale synoptic survey aimed at monitoring the general hydrological characteristics of the Ross Sea.

The map with the bathymetry of the 53 stations sampled in five days of the experiment is shown in Fig. 20.3.

The hydrological casts were carried out using a Sea Bird Electronics SBE 9/11 Plus with double temperature and conductivity sensors coupled with a Carousel water sampler SBE 32 carrying 24 bottles of 12 litres each. Water samples were collected at different depths, according to CTD profiles.

Dissolved oxygen measurements were made aboard ship using automated microtritations, according to the Winkler method (Strickland and Parson 1972).

The subsamples for the determination of nutrients were collected directly from the Niskin bottle, filtered through a 0.7 μm GFF filter and stored at −30 °C in 100 ml LDPE (Low Density PolyEthylene) containers, until analysis.

Measurements of nutrients were performed by an Autoanalyser Technicon II, according to the Hansen and Grasshoff methods (1983).

Estimated detection limits for nitrate, nitrite, ammonium, phosphate and silicate were 0.05, 0.01, 0.05, 0.05, 0.1 μM, respectively.

20.3.2 Distribution of NO and PO in Different Water Masses

On the basis of an accurate θ/S analysis, the stations were clustered, according to the characteristics of the water masses (Bergamasco et al. 1998). In Fig. 20.4, the three-dimensional plot of the minimum temperature, chosen to assess the ISW presence in the studied area, is shown.

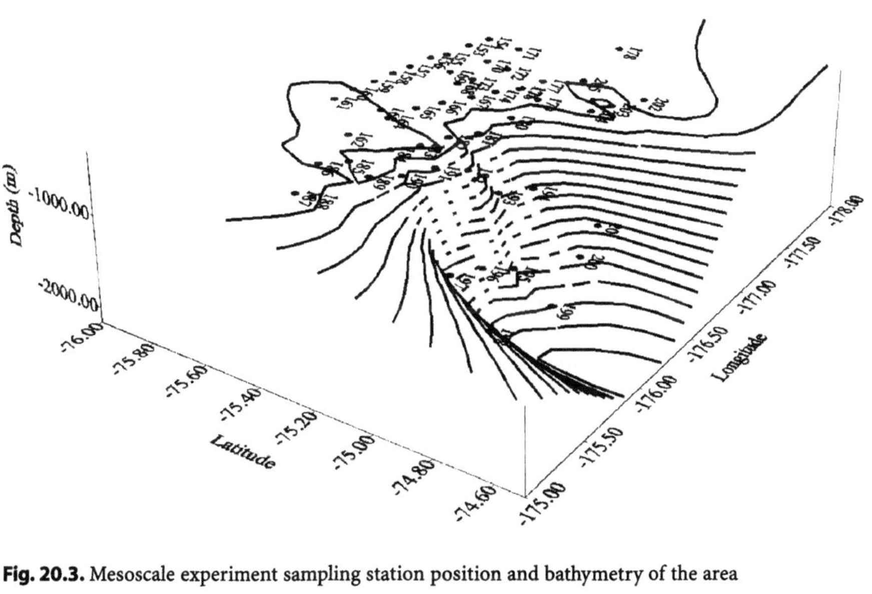

Fig. 20.3. Mesoscale experiment sampling station position and bathymetry of the area

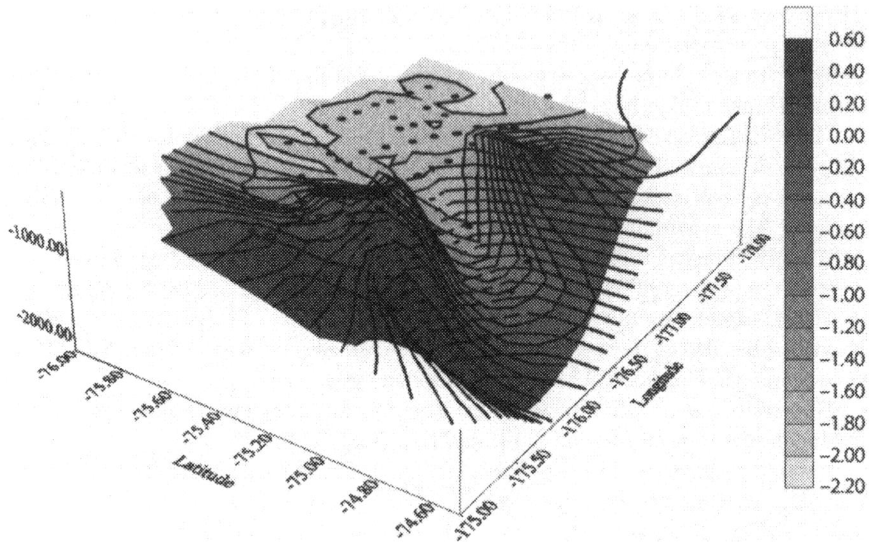

Fig. 20.4. Three dimensional plot of the distribution of the minimum temperature

In particular, in some stations on the shelf the presence of ISW, as indicated by a temperature minimum near the bottom (about 500 metres depth) was assessed; while off-shore CDW was singled out by a temperature maximum between 300 and 600 metres. The minimum temperature trend allowed us to establish the presence of a mass of ISW, which mixes with the southern limits of CDW (Bergamasco et al. 1998).

Between Station 181 and 195, the Antarctic Slope Front is located, characterized by a sharp gradient of temperature from about −0.90 °C until +1.38 °C along the continental slope.

Temperature is the best indicator of the presence on the Antarctic Slope Front, with lateral gradients exceeding 3 °C in 25 km, as reported by Jacobs (1991). At any rate, the gradients can be less pronounced where tongues of warm water intrude onto the continental shelf, as in the case we reported.

The distribution of oxygen is reported in Fig. 20.5.

It can be seen that oxygen minimum values (about 200 μM) were found in stations 195, 194, 181 at 411, 800, 240 m depth, respectively. Moreover, the same depths showed the highest nutrient concentrations in the sections, as shown by the nitrate three-dimensional plot reported in Fig. 20.6. Values fall in the same range as those previously reported in literature for Circumpolar Deep Water (Catalano et al. 1999).

The highest oxygen concentrations, apart from those relative to the surface, were found at bottom depth in some stations in the shelf and in the slope areas, where low values of nutrients (i.e. 25 μM nitrate) were detected. This layer of 100 m thickness presents the characteristics of super-cooled water, and it is colder, lower in nutrients and higher in oxygen than the overlying deep water, features deriving from continental shelf waters. Furthermore, southernmost stations of the mesoscale area, located on the shelf edge, had oxygen concentrations higher than 230 μM.

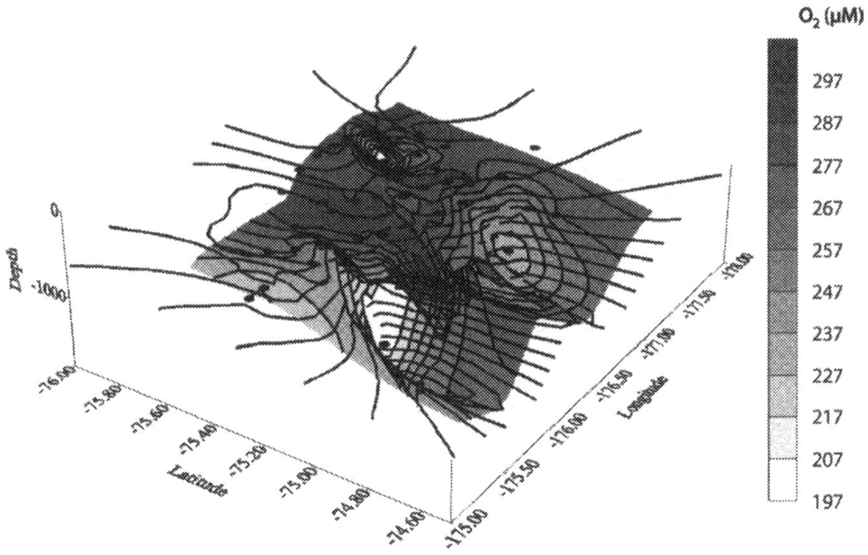

Fig. 20.5. Three dimensional plot of the distribution of oxygen concentration found at the minimum temperature

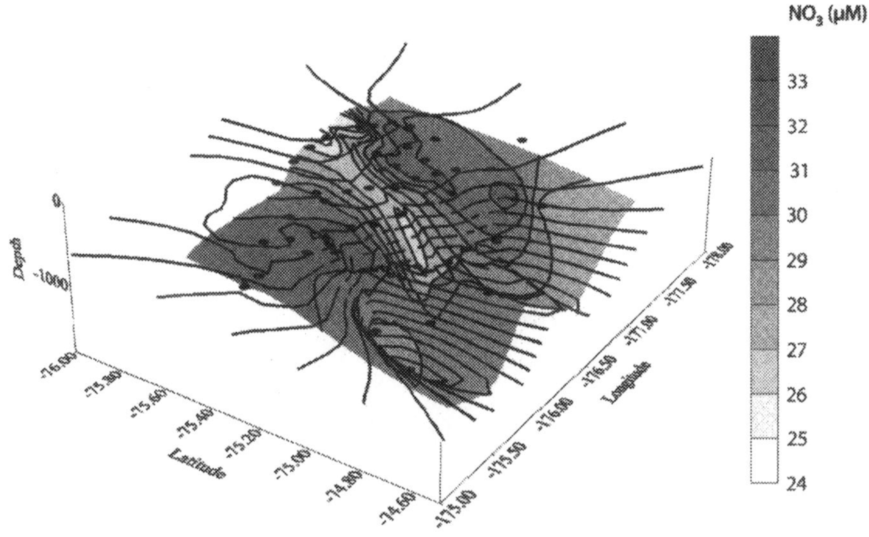

Fig. 20.6. Three dimensional plot of the distribution of nitrate concentration found at the minimum temperature

Comparing the distributions of oxygen and nitrate with a minimum temperature trend, it can be seen that these two parameters are not good chemical tracers of the mixing processes, particularly in the case of nitrate. In fact, chemical gradients are less pronounced than physical ones, where tongues of warmer water intrude onto the continental shelf.

In Fig. 20.7, the distribution of the values of the conservative tracer NO found at the minimum temperature on the shelf break is reported.

It is possible to see that NO maximum levels, higher than 570 μM, were found in those stations close to the Ross Ice Shelf. This maximum coincided with the lowest temperatures found in the studied area, and it reflects the high oxygen concentrations found at the same depths on the shelf. Moreover, the two maxima in the NO section have very nearly the same concentration, and therefore they can be treated as being part of the same water mass.

The lowest NO concentrations (448 and 496 μM) were found respectively in Station 197 and 204, where temperatures higher than 0.5 °C were determined.

The NO values fall in the same range as those calculated for the Weddel Sea (Lindegren and Anderson 1991). In fact, for Ice Shelf Water, NO value was estimated at about 575 μM while a minimum "NO" level of 480 μM was calculated for WDW.

As previously reported by Lindegren for the Weddel Sea, the NO minimum coincided with the temperature maximum.

A NO high level water mass, represented in dark grey, can be observed on the shelf area; it presents the physical features of Deep Ice Shelf Water, which was found at about 500 m depth. Moreover, the presence of two water lenses, which were not put in evidence by the oxygen distribution analysis, seems to be confirmed by the NO distribution.

These masses overflowed along the slope, where mixing processing with the intruding CDW, (bright grey) signed by a lower NO level, occurred.

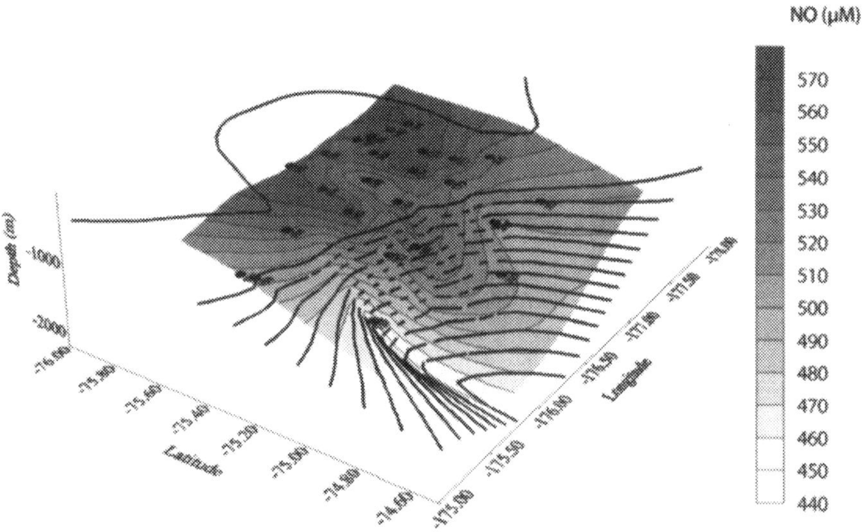

Fig. 20.7. Three dimensional plot of the distribution of NO values found at the minimum temperature

With respect to both oxygen and nitrate, NO distribution points out a patchy distribution of water coming from the Ross Ice Shelf, which modifies its NO content moving along the continental slope. The same trend was observed for PO distribution at the temperature minimum, as shown in Fig. 20.8.

In fact, the PO profile exhibited highest values (>603 µM) in Stations 153–156, while lowest values (<450 µM) were observed near Stations 197 and 204. However, for PO, the patchy distribution observed on the shelf area for NO was not so evident.

We wish to remember that NO distribution is preferred in tracing water masses, because of the better accuracy of the NO_3 measurement with respect to that of PO_4, so that we hypothesize that it could better represent both the outflow of ISW from the shelf and the ascending CDW.

The analysis of the chemical data collected in the mesoscale experiment in shelf break area in the Ross Sea, together with physical variables pointed out some attributes of the Antarctic Slope Front, showing the presence of both ISW and CDW.

In particular, CDW intruded onto the shelf, modifying its chemical, as well as physical properties during its mixing with ISW.

As a result, NO and PO were found to be potentially useful as chemical tracers in outlining the mixing processes and bottom water formation near the continental shelf break of the Ross Sea, in which there seems to be evidence that ISW plays an important role.

The chemical signature of ISW can be recognized in fact in its high NO and PO contents, which decreased along the shelf break. Physical and chemical analysis confirm that the overflow occurred at narrow scales; in fact, NO and to a lesser extent PO revealed a patchy distribution in the shelf area close to the frontal zone. However, they

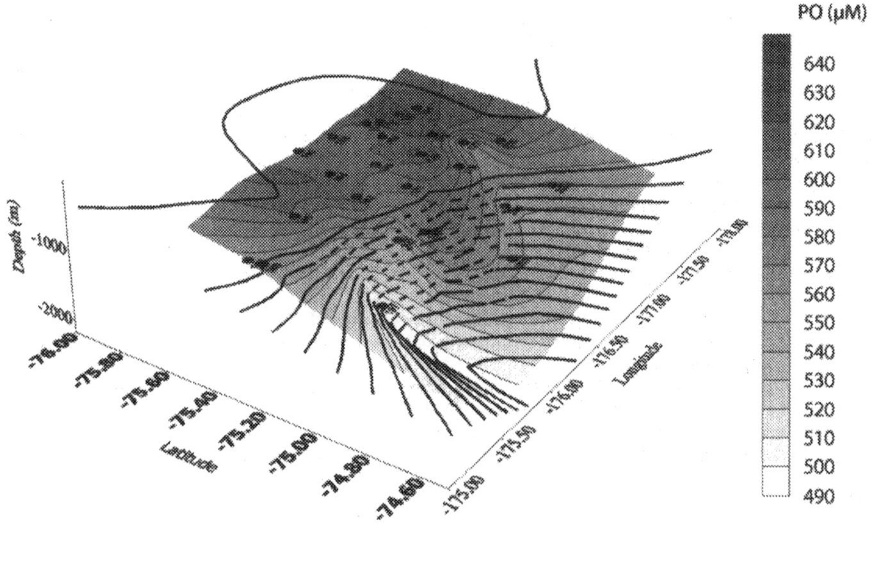

Fig. 20.8. Three dimensional plot of the distribution of PO values found at the minimum temperature

cannot provide additional information on mixing processes without a better understanding of the mixing evolution.

Acknowledgements

This study was performed as a part of the Italian National Program for Research in Antarctica and was financially supported by ENEA through a joint research programme.

The authors are indebted to Prof. Giancarlo Spezie, responsible for the scientific expedition and to Prof. Enrico Zambianchi and Dr. Andrea Bergamasco for their helpful comments and suggestions.

References

Anderson LG, Jones P (1992) Tracing upper waters of the Nansen basin in the Arctic Ocean. Deep Sea Res 39(2):S425–S433

Baines PG, Condie S (1998) Observations and modelling of antarctic downslope flows: A review. Antarctic Res Ser 75:29–49

Bergamasco A, Spezie G, Paschini E, Zambianchi E, Rivaro P, Bottinelli C, Grotti M, (1998). H meso CLIMA 98 experiment: Preliminary analysis and data interpretation. Final Proceedings I Convegno Nazionale delle Scienze del Mare, Ischia, (Italy) 11–14.11.1998

Broecker WS (1974) NO as a conservative water-mass tracer. Earth Planet Sci Lett 23:100–107

Budillon G, Tucci S, Artegiani A, Spezie G (1999) Water masses and suspended matter characteristics of the western Ross Sea. In: Faranda FM, Guglielmo L, Ianora A (eds) Ross Sea ecology. Springer-Verlag, Berlin, pp 63–93

Catalano G, Benedetti F, Predonzani S, Goffart A, Ruffini S, Rivaro P, Falconi C (1999) Spatial and temporal patterns of nutrient distributions in the Ross Sea. In: Faranda FM, Guglielmo L, Ianora A (eds) Ross Sea ecology. Springer-Verlag, Berlin, pp 107–120

Cooper LW, Cota GF, Pomeroy LR (1999) Modification of NO, PO, and NO/PO during flow across the Bering and Chukchi shelves: Implications for use as Arctic water mass tracers. J Geophys Res 104(C4):7827–7836

Hansen HP, Grasshoff H (1983) Automated chemical analysis. In: Grasshoff K, Ehrardt M, Kremling K (eds) Methods of sea water analysis. VCh, Weinheim, pp 347–379

Jacobs SS (1989) Marine controls on modern sedimentation on the Antarctic continental shelf. Mar Geol 85:121–153

Jacobs SS (1991) On the nature and significance of the Antarctic Slope Front. Mar Chem 35:9–24

Jacobs SS, Fairbanks RC, Horibe Y (1985) Origin and evolution of water masses near the Antarctic continental margin: Evidence from $H_2{}^{18}O/H_2{}^{16}O$ ratios in sea water. Antarctic Res Ser 43:59–85

Lindegren R, Anderson LG (1991. "NO" as conservative tracer in the Weddel Sea. Mar Chem 35:179–187

Lindegren R, Josefson M (1998) Bottom water formation in the Weddel Sea resolved by principal components analysis and target estimation. Chemom Intell Lab Sys 44:403–409

Locarnini RA (1994) Water masses and circulation in the Ross Gyre and environs. PhD dissertation, Texas University, College Station, Texas

Mensch M, Simon A, Bayer R (1998) Tritium and CFC input functions for the Weddel Sea. J Geophys Res 103:15923–15937

Minster JF (1989) Introduction to chemical tracers of the ocean circulation. In: Anderson DLT, Willebrand J (eds) Oceanic circulation models: Combining data and dynamics. Kluwer Academic Publisher, Dordrecht, pp 345–376

Picco P, Bergamasco A, Demicheli L, Manzella G, Meloni R, Paschini E (1999) Large scale circulation features in the central and western Ross Sea (Antarctica). In: Faranda FM, Guglielmo L, Ianora A (eds) Ross Sea ecology. Springer-Verlag, Berlin, pp 95–106

Strickland JDH, Parson TR (1972) A practical handbook of sea water analysis. Bull Fish Res Board Can 167–310

Whitworth T III, Orsi AH, Kim SJ, Nowlin WD Jr (1998) Water masses and mixing near the antarctic slope front. Antarctic Res Ser 75:1–27

Wilson C, Wallace DWR (1990) Using the nutrient ratio NO/PO as a tracer of continental shelf waters in the central Arctic ocean. J Geophys Res 95(C12):22193–22208

Chapter 21

Solid Speciation and Selective Extraction Procedures: Trace Metal Distribution and Speciation in Coastal Sediments of the Adriatic Sea

C. Ianni · N. Ruggieri

21.1 Introduction

Studies on distribution and speciation of heavy metals in sediments can not only provide information on the degree of pollution, but also on the effective environmental impact, metal bioavailability, and their origin. Sediments, in fact, are important sinks for various substances including heavy metals, but they can also be a metal source. The extension of the phenomena depends on the association of the metals with the different phases of the sediments, defined as "solid speciation," which is usually determined by utilizing selective sequential extraction procedures.

In this work, some case studies concerning determination of metal speciation in coastal sediments of Adriatic Sea will be shown, in order to better explain the importance of solid speciation in environmental studies better. The data presented are part of different projects: the former is a quasi-synoptic research project for the safeguard of the Adriatic Sea (PRISMA 2), in particular of the sector "Physical, chemical and biological oceanography," undertaken in 1996–98. The latter is an INTERREG II project involving Italy and Greece for the monitoring of the southern Adriatic and Ionic Sea, undertaken in 1999 and still in progress.

21.2 Role of Marine Sediments in the Environment

Marine sediments are caused by unconsolidated accumulations of particles brought to the ocean by rivers, glaciers and winds, mixed with shells and skeletons of marine organisms (Yen and Tang 1977). They are the final destination of trace metals, as a result of adsorption, desorption, precipitation, diffusion, chemical reactions, biological activity and a combination of those phenomena. When some physical disturbance occurs, or because of diagenesis and/or changes in pH or redox potential, sediments can act as a source of metals, releasing them in the overlying water column. This phenomenon can occur even long after the end of direct discharge (Jones and Turki 1997), and its extension depends on the metal association with the different mineralogical fractions of the sediment. Therefore, metal behaviour and availability strictly depends on their chemical form and thus on their speciation. Determination of solid speciation, defined as the identification and quantification of the different species, forms or phases present in a sediment, has become increasingly important. The measurement of total content, in fact, is not sufficient to get information on the potential availability of metals (whether toxic or essential) to biota under various environmental conditions (Davidson et al. 1994).

21.3 Selective Extractions

To date, it has generally been accepted that the most appropriate methods to evaluate solid speciation are selective sequential extraction procedures (Kot and Namiesnik 2000), which are widely used to assess the long-term emission potential of pollutants and to study the distribution of pollutants among the geochemical phases (Rauret 1998). One must keep in mind that selective sequential extractions give, as a result, "operationally defined species" (Tessier et al. 1979), which depend on the different procedures used. It is not possible to obtain specific chemical associations, because the reagents utilized are often insufficiently specific to exclusively dissolve the "target" phase conditions (Davidson et al. 1994) and results obtained can vary widely when different experimental conditions are used (Tipping et al. 1985; Rauret et al. 1989). Nevertheless, useful information can be gained from such studies; in fact, a large number of selective extraction methods have been studied and reported, many of which are variants of the Tessier procedure (Tessier et al. 1979). Selective extractions have also been proven to be adequate for determining the metals associated with source constituents in sedimentary deposits (van der Sloot et al. 1997). Moreover, according to Rubio et al. (1991), metals with an anthropogenic origin are mainly obtained in the first extractions, while in the last stage of the process, the residual fraction is obtained, corresponding to metals with lithogenic origins.

Recently, studies on selective extractions have received new emphasis with the availability of a sediment reference material, certified by the SM&T programme of the European Union (Quevauviller et al. 1997) following a less aggressive procedure. In fact, the increasing performances of the analytical techniques used for element determination in an extract, together with the increasing evidence that exchangeable metals better correlate with plant uptake, has led extraction methods to evolve towards the use of less and less aggressive solutions (Gupta and Aten 1993).

21.3.1 Commonly Used Extraction Procedures

The fractions obtained when applying selective extraction schemes are related to exchangeable metals, metals mainly bound to carbonates, metals released in reducible conditions such as those bound to hydrous oxides of Fe and Mn, metals bound to oxidizable components such as organic matter, and sulphides and residual fraction. The reagents more commonly used in sequential extraction procedures are usually applied according to the following order: unbuffered salts, weak acids, reducing agents, oxidizing agents and strong acids. Exchangeable fraction uses an electrolyte at pH 7 to avoid oxide and carbonate solubilization and to prevent oxyhydroxide precipitation. The carbonate fraction generally uses acetic acid or a buffer acetic acid- sodium acetate at pH 5. These reagents are not able to dissolve all the carbonates, nor can they attack carbonate selectively, as they also remove labile organically bound metals (Rauret 1998; Baffi et al. 1998). The reducible fraction is mainly related to metals bound to Fe and Mn oxides. Hydroxylamine in acid solution is the reducing agent most widely used to solubilize these oxides, although iron oxide is not completely dissolved. Ammonium oxalate seems more effective if used in the dark, even if some precipitation of metals oxalate can occur, also at low pH. Dithionite-citrate reagent dissolves oxides and hydroxides, but can also attack iron rich silicates. Therefore, no reducing agent is either selective or completely efficient. Oxidizing agents destroy organic matter and transform sulphides into sulphates. The reagents most widely used are acidic hydrogen peroxide and sodium hypochlorite, and to a less extent, 8 M nitric acid. Residual fraction is usually extracted using the same reagents utilized for the total solubilization, a mixture of strong acids, with or without fluoridric acid, in closed systems at high temperature (usually microwaves ovens). Alternatively, the amount of metals bonded to the residue can be estimated by the difference between the total content and the sum of the amount extracted in the various selective steps.

21.4 Case Studies

Some case studies will be described and commented on to show how selective extraction can be a useful indication concerning metal solid speciation and how this information completes data on the metal content of marine sediments. In some cases, data on granulometric composition and on organic matter content were also available and supported the discussion of the results. In Fig. 21.1, the location of the three areas studied is reported.

21.4.1 PRISMA 2 Project

This project was undertaken to obtain information and to study the physical and biogeochemical mechanisms that influence and determine the transport and diffusion

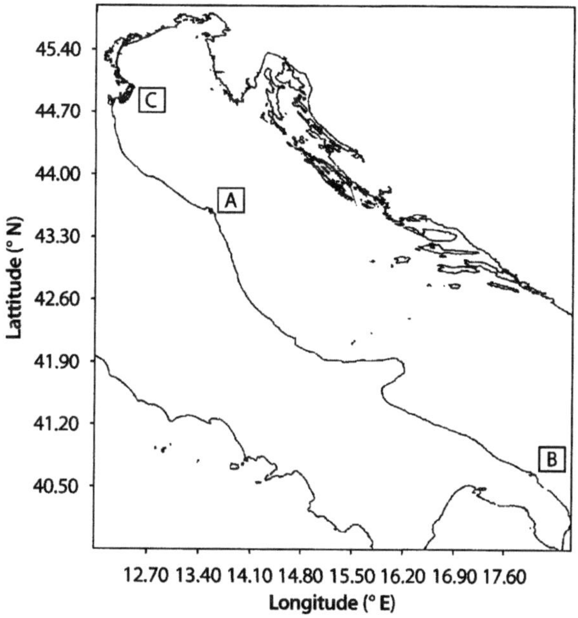

Fig. 21.1. Map of the studied areas

of the substances (pollutants and others) introduced in the marine environment by the freshwaters of the Po River. The aim of the work on sediments was to provide information on the distribution, enrichment and speciation of some trace metals in the surface sediments of two areas of the Adriatic Sea: Chioggia and Ancona. Chioggia was chosen because of the predominance of fluvial inputs, while Ancona was chosen because the freshwaters follow the coast until Conero Mount (near Ancona). The examples reported are of course only a part of the work regarding this project (Ianni et al. 2000).

21.4.1.1
Methods

Total attack. The samples were treated in microwave Teflon vessels with aqua regia and heated in microwave oven.

Selective extraction. *Fraction 1. Exchangeable and carbonates.* Ammonium acetate 1 M (pH 5) was added to the sediment, which was shaken overnight at room temperature. The extracts were separated from the residues by centrifugation.

Fraction 2. Bound to Fe and Mn oxides. Hydroxylamine hydrochloride 1 M and acetic acid 25% (1:1) were added to the residue, and the same procedure of the first extraction was performed.

Fraction 3. Bound to organic matter and sulphides. The residue of the second extraction was put in a microwave Teflon vessel and after the addition of 8 M nitric acid, heated in microwave oven.

The metal content of the residual phase was obtained by the difference between the total content and the sum of the Fractions 1, 2 and 3.

21.4.1.2
Results

In both areas, metals total concentration are comparable with those found in other coastal areas of Mediterranean Sea (Donazzolo et al. 1981, 1984; Martincic et al. 1990; Rapin et al. 1979) and are lower respect to very polluted areas, as for example the Gulf of Naples (De Rosa et al. 1983). Chioggia presents, for some elements, higher concentrations, due to the relevant input of freshwaters from the Po River.

As regard speciation, in both areas for all the samples, the reducible fraction is poor of all the elements, and especially of Fe, which should be a major component of the phase. The negative Eh of these sediments (Colantoni, personal communication) outlines that these sediments are in reducing conditions, particularly rich in sulphides and organic matter (Fabiano, personal communication) and poor of oxides. Generally, all the elements studied, except Cd and Pb, which are found at some extent in the labile fraction (exchangeable, carbonates and/or labile organic matter), are associated prevalently to the oxidizable fraction. This renders the elements in these sediments not immediately available, but can be released if redox conditions change, for example for aeration caused by dredging operations.

In Fig. 21.2, the grain size distribution, together with distribution of Pb total content is shown in the area of Chioggia. Beside Pb, even Cd, Sn and Hg present the same

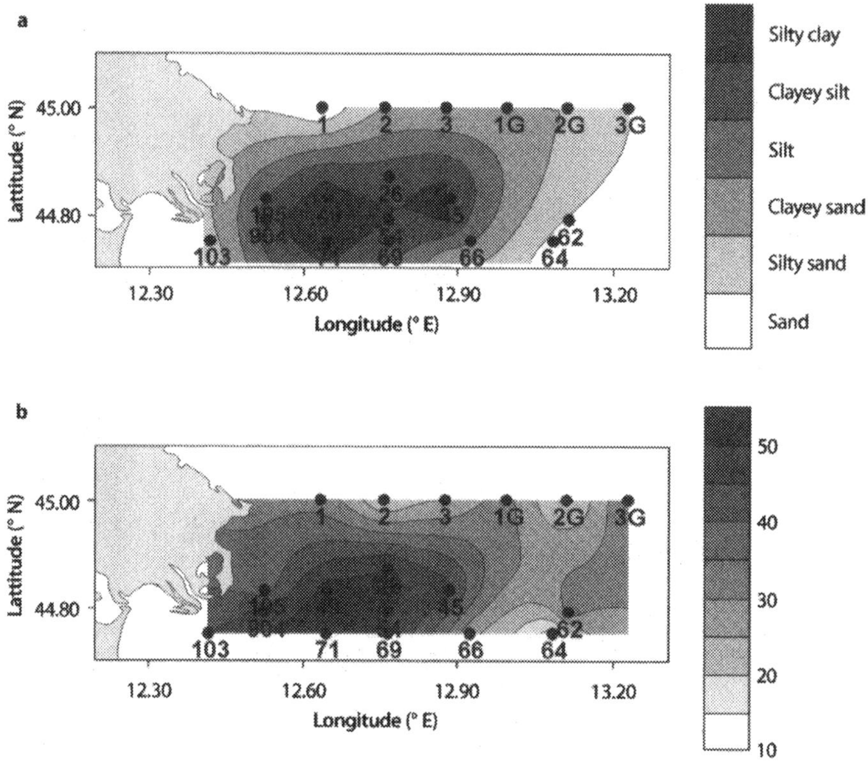

Fig. 21.2. Area of Chioggia, spatial distribution of; **a** grain size; **b** Pb (µg l^{-1})

trend of the grain size, with a higher concentration in the central part of the area, where pelite is predominant. Instead, Cu together with Zn and Fe present a decreasing gradient from the coast towards open sea (Fig. 21.3), which is related to freshwater inputs (Po River).

The same trend has been observed for organic matter, particularly for carbohydrates, proteins and lipids (Fabiano, personal communication), by another group working on the same project. Being that all these elements act as micronutrients, it is probable that they are largely associated to organic phases. Cu, in fact, is principally associated to the organic phase and to the labile biologic matter that can be extracted with the first attack (Baffi et al. 1998; Rauret 1998), and that is highly present in these sediments. In fact, high values of the proteins/carbohydrates ratio have been found (Fabiano, personal communication), and this indicates sediments rich of organic matter of recent origin (excretions, dead organism, etc.), being that bacteria use the proteins more rapidly than the carbohydrates. The association of the copper extracted by the acetate buffer with the labile organic matter is confirmed by the same decreasing trend, as shown in Fig. 21.3. Cd presents also the same trend in step 1, although total Cd follows grain size. Moreover, it is the only element found in great percentage in the acetate buffer extract (20–60%); this is partly explained because this element can also

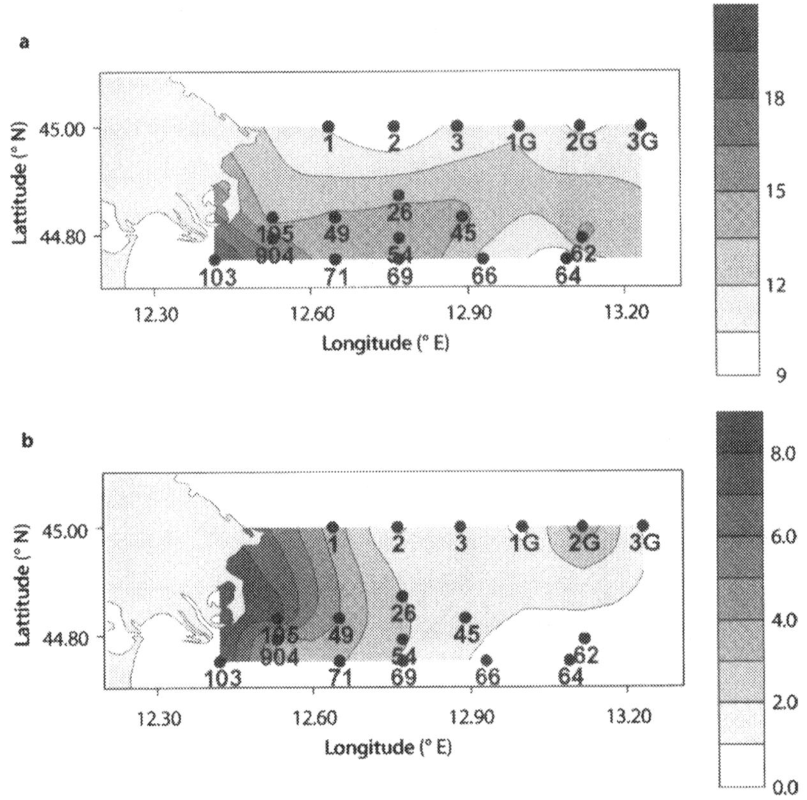

Fig. 21.3. Area of Chioggia, spatial distribution of; **a** Cu (μg l^{-1}); **b** Cu (μg l^{-1}) in Fraction 1

be found as carbonate (Span and Gaillard 1986), but it is also bonded with labile organic matter, being that the higher percentage (50–60%) is found near the coast.

In Fig. 21.4, the speciation of lead in the northern transect and in Station 71 is shown to highlight how using only total content to obtain information on environmental impact can bring one to a careless conclusion. The total content of Station 71 is 55 μg g^{-1}, while the concentration range found in the transect is 15–30 μg g^{-1}. Nevertheless, the percentage of Pb present in the first fraction (more available) is 20% for Station 71 and 35–50% for the transect. Therefore, the availability in the two cases is the same, and the high percentage present in the Station 71 residue (50%) outlines that, according to Rubio et al. (1991), the Pb has prevalently a lithogenic origin, while in the northern transect it has prevalently an anthropogenic origin (5–20% in residue).

As regards the Ancona area, the distribution of Pb in fraction 1 is partially overlapping the total one, which follows grain size (see Fig. 21.5), especially in the two south transects, and presents an area of greater concentration in the central part and in Stations 83 and 85. The grain size distribution (Fig. 21.5) shows that this central part and the mentioned stations are rich of pelites. Therefore, we can suppose that the Pb obtained from the first extraction is prevalently adsorbed on small particles, and so is

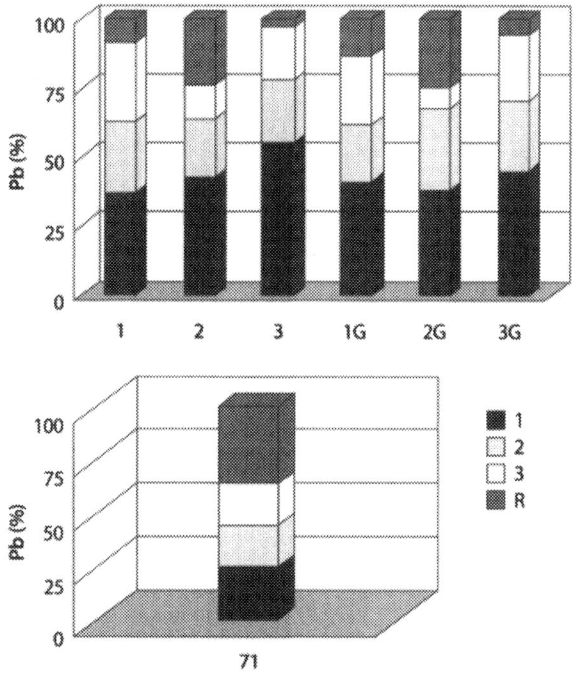

Fig. 21.4. Speciation of Pb in the northern transect and in Station 71 in the area of Chioggia

particularly mobile. Moreover, the amount of Pb in fraction 1 is not negligible, being around 20–30% of the total amount. Fortunately, as was mentioned earlier, the lead content, especially in the Ancona area, is not particularly high, so the bioavailable fraction is low enough to be considered not dangerous.

Even in this area, Cd is present at very high levels in the labile fraction (40–60%), indicating its recent origin with respect to the other metals and its relatively higher bioavailability. Moreover, as shown in Fig. 21.6, the distribution of Cd in this fraction follows the grain size trend, so as Pb, it is probably adsorbed on fine particles, and the same consideration as above can be made.

The distribution of Cu in Fraction 1 (Fig. 21.6) follows a decreasing gradient from the coast towards the open sea, as seen for Chioggia, which indicates a recent coastal input, probably in association with labile organic matter. The maximum concentration in this phase is found in Stations 76 and 80 of the south transect, which are relatively lacking in the other elements, whereas the total content is greater in the other two stations (83, 85). In this transect, the recent coastal input is remarkable if related to the total content.

21.4.2 Interreg Project

The aim of Interreg project is the monitoring and the characterization of the Apulian coast, in order to individuate degradation caused by anthropic activities, and on the other hand, to individuate zones of particular ecological interest, where in the future

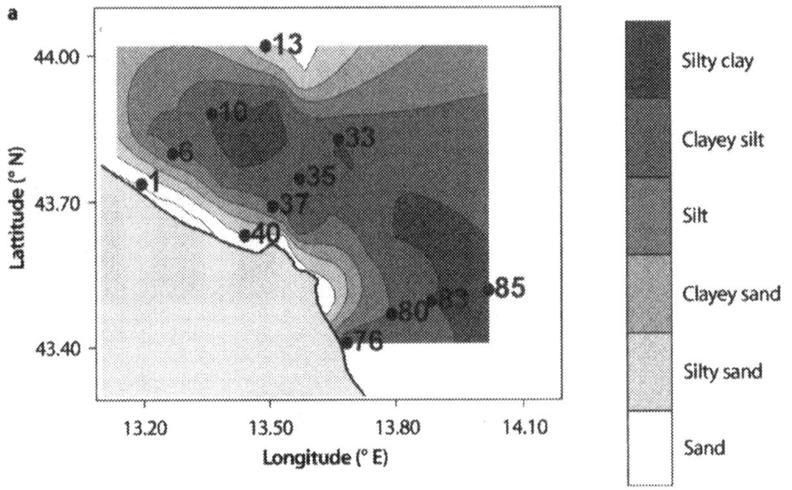

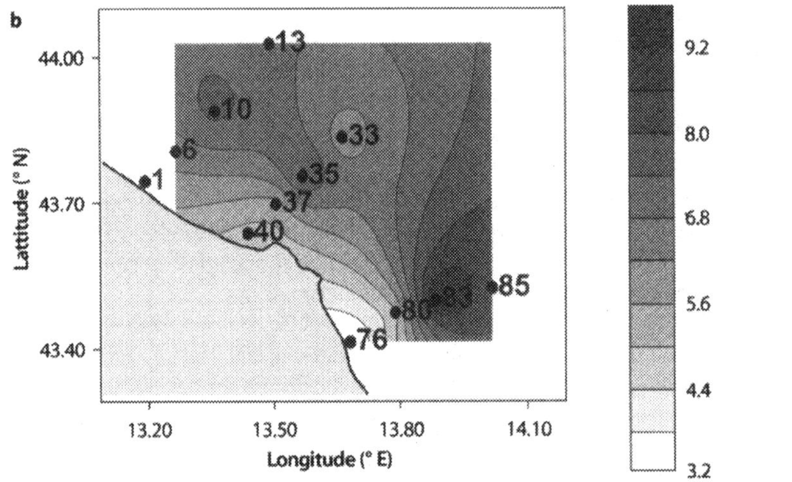

Fig. 21.5. Area of Ancona; **a** grain size; **b** spatial distribution of Pb ($\mu g\ l^{-1}$) in Fraction 1

protected marine areas can be created. The project is still in progress, and it foresees the determination and speciation of heavy metals and some organic compounds in the coastal sediments to evaluate the degree of pollution.

21.4.2.1
Methods

Total attack. As described in Sect. 21.4.1.1.

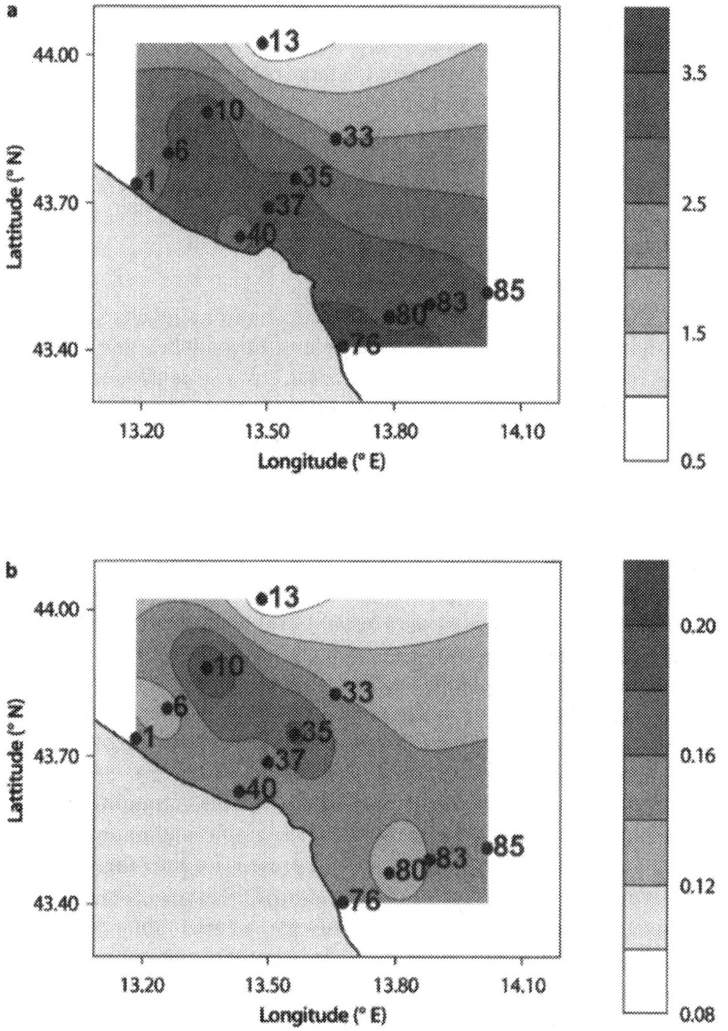

Fig. 21.6. Spatial distribution in Fraction 1 in the area of Ancona; **a** Cu (μg l^{-1}); **b** Cd (μg l^{-1})

Selective extraction. Modified SM&T protocol (accuracy tested on CRM).

Fraction 1. Exchangeable and acid soluble. Acetic acid 0.11 M (pH 2.8) was added to the sediment, which was shaken overnight at room temperature. The extracts were separated from the residues by centrifugation.

Fraction 2. Easily reducible phase. Hydroxylamine hydrochloride 0.1 M adjusted to pH 2 with nitric acid was added to the residue with the same procedure as the first extraction.

Fraction 3. Bound to organic matter and sulphides. The residue was put in a microwave Teflon vessel, and after the addition of 8.8 M hydrogen peroxide, left at room temperature for one hour, then treated at 50 °C in an ultrasound bath for two hours. A further aliquot of hydrogen peroxide was added, and the vessels were heated in a microwave oven. After cooling, 2 M ammonium acetate (pH 2) was added, and the procedure was carried out as for Steps 1 and 2.

The metal content of the residual phase was obtained by the difference between the total content and the sum of the Fractions 1, 2 and 3.

21.4.2.2
Results

In this area, total metal content is generally lower than or similar to natural levels, except for Cr and Zn in all samples, and for Pb and Cd only in some particular stations. No fluvial inputs interest this part of the coast, and as it can be seen by the distribution of the fine fraction (<63 μm) in Fig. 21.7, near the coast we find a high amount of coarse particles, while the fine particles are found off-shore.

Prior to analysis, the sediments were subdivided in two granulometric fractions (< and >63 μm). All the metals in the >63 μm fraction present increasing gradients towards off-shore; in the <63 μm fraction the elements have the same trend (in Fig. 21.7 distribution of Cu is reported as an example), except Cd and Pb, which present an opposite behaviour, as shown in Fig. 21.7 for Pb.

Selective extractions, which were performed on the fine fraction, indicate for Cd and Pb in the area where their concentration is higher (Transects 4 and 6), indicate that both are present, Pb in all stations and Cd in the coastal ones, in the oxidizable fraction (see Fig. 21.8). This phase is not easily mobilizable, but can be oxidized in different environmental conditions. Moreover, Cd is present in the more available fraction, with increasing percentages from the coast towards the open sea. Nevertheless, considering the small percentage of fine fractions in this zone, the concentration of heavy metals in the bulk of the sediment is still low. At any rate, these results outline an evident coastal input, probably associated to organic matter, not present for the other elements.

In all the other cases, selective extractions outlined that metals are present in very low amounts in the three fractions extracted and up to 80% in the residue. As an example, speciation of Cu and Cr in two transects where their total content was particularly high is reported in Fig. 21.9.

As far as this set of data can indicate, these sediments don't seem particularly polluted, even if in some cases variations in physical and chemical conditions can cause negative impacts on the environment. Even in this case, these are some examples drawn from the work, which being still in progress, foresees other sampling and analysis. It will be interesting, for example, to evaluate the results of selective extractions performed on the coarse fraction, which in some cases presents higher contents with respect to the fine fraction.

21.5
Conclusions

The case studies reported confirmed that selective extractions are a very useful and versatile technique to investigate on environmental impact, bioavailability and metal origin in marine sediments. Obviously, for a complete understanding of enrichment,

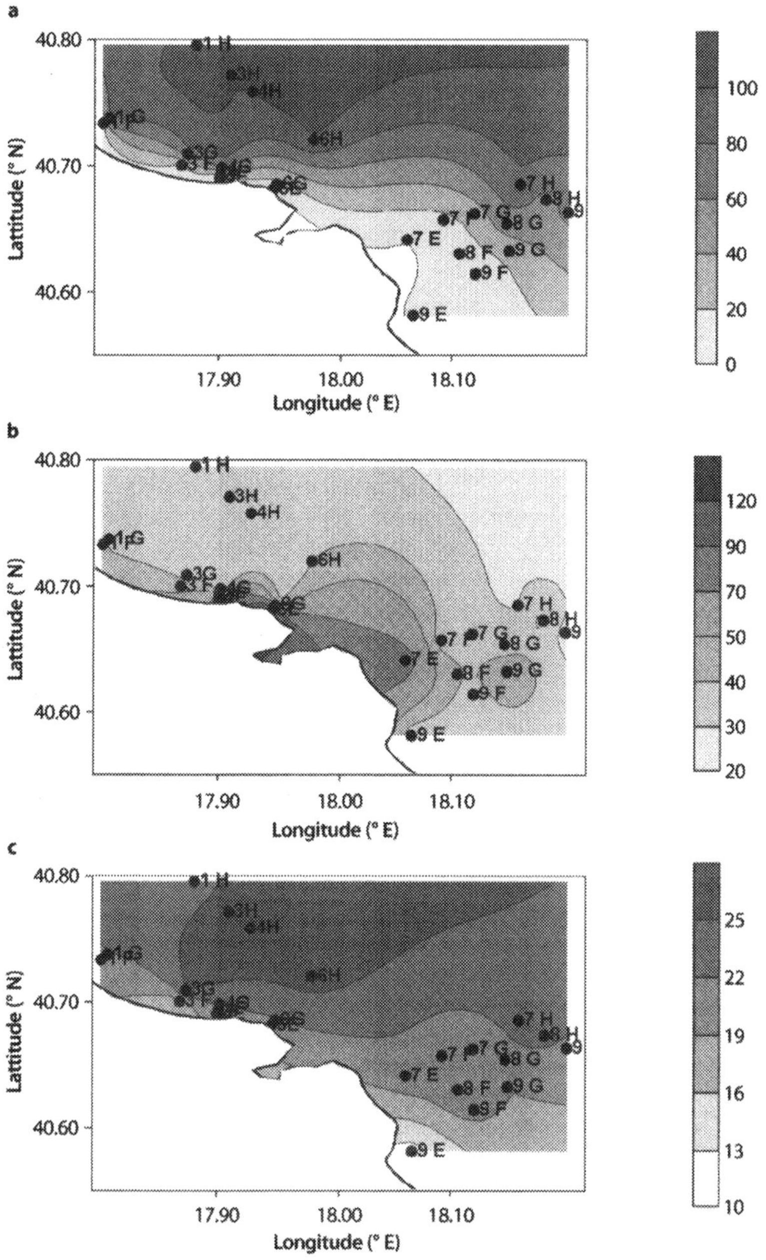

Fig. 21.7. Area of Brindisi; spatial distribution of; **a** fine fraction; **b** Pb ($\mu g\ l^{-1}$); **c** Cu ($\mu g\ l^{-1}$)

diffusion and speciation mechanisms, when possible, knowledge of parameters for grain size, mineralogy, organic matter content, and so on, is of great usefulness. For

Fig. 21.8. Speciation of Pb and Cd in Transect 4 in the area of Brindisi

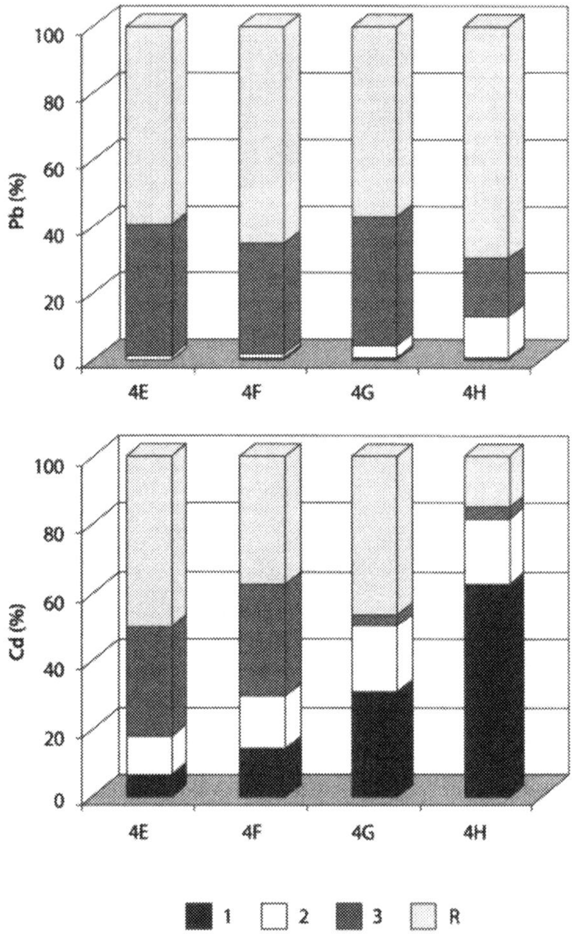

example, the correspondence between a metal distribution in the labile fraction and the fine fraction indicates that the metal is prevalently found in the exchangeable phase. Correlation with organic matter content can indicate that an element in the oxidizable phase is bound to organics rather than to sulphides.

Therefore, as selective extraction procedures are so essential for environmental studies and considering the recent improvements in these analytical methods and the availability of reference material, their utilization is strongly advisable.

Acknowledgements

This work was financially supported by the Italian National "Programma di Ricerca e Sperimentazione per il Mare Adriatico" 2nd phase (PRISMA 2) and by the CE project "INTERREG II Italia – Grecia"

Fig. 21.9. Speciation of Cu and Cr in Transects 4 and 8 in the area of Brindisi

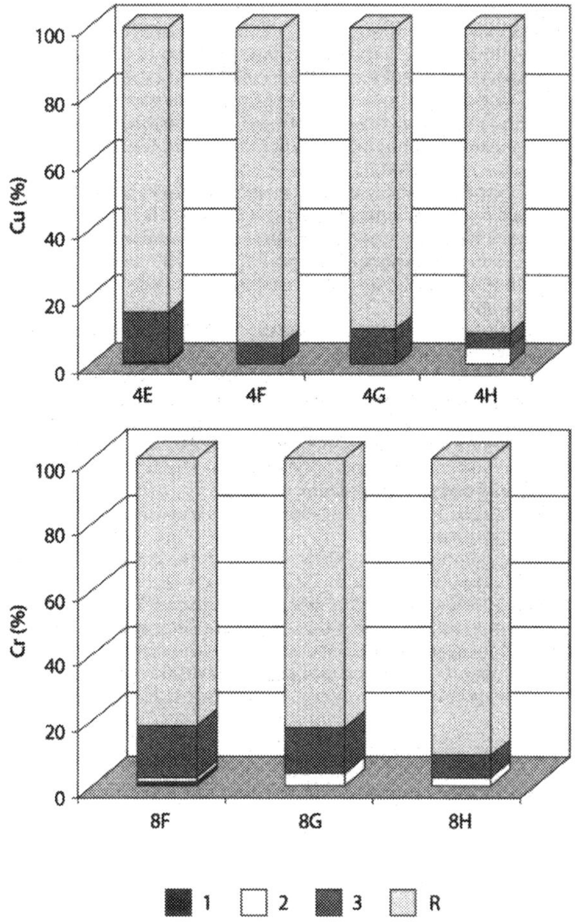

References

Baffi F, Ianni C, Ravera M, Soggia F, Magi E (1998) Evaluation of the acetate buffer attack of a sequential extraction scheme for marine particulate metal speciation studies by scanning electron microscopy with energy dispersive X-ray analysis. Anal Chim Acta 360:27–34

Davidson CM, Thomas RP, McVey SE, Perala R, Littlejohn D, Ure AM (1994) Evaluation of a sequential extraction procedure for the speciation of heavy metals in sediment. Anal Chim Acta 291:277–286

De Rosa S, Damiani V, Serena F (1983) Studio dei sedimenti del Golfo di Pozzuoli: Livelli di contaminazione da metalli pesanti. Atti V Congr Naz AIOL, Stresa, pp 437–447

Donazzolo R, Hieke Merlin O, Menegazzo Vitturi L, Orio AA, Pavoni B (1981) Heavy metal contamination in surface sediments from the Gulf of Venice, Italy. Mar Pollut Bull 12:417–425

Donazzolo R, Hieke Merlin O, Menegazzo Vitturi L (1984) Heavy metal content and lithological properties of recent sediments in the northern Adriatic. Mar Pollut Bull 3:93–101

Gupta SK, Aten C (1993) Comparison and evaluation of extraction media and their suitability in a simple model to predict the biological relevance of heavy metal concentrations in contaminated soils. Int J Environ Anal Chem 51:25–46

Ianni C, Magi E, Rivaro P, Ruggieri N (2000) Trace metals in Adriatic coastal sediments: Distribution and speciation pattern. Toxic Environ Chem 78:73–92
Jones B, Turki A (1997) Distribution and speciation of heavy metals in surficial sediments from the Tees Estuary, north-east England. Mar Pollut Bull 34:768–779
Kot A, Namiesnik J (2000) The role of speciation in analytical chemistry. Trends Anal Chem 19:69–79
Martincic D, Kwokal Z, Branica M (1990) Distribution of zinc, lead, cadmium and copper between different size fractions of sediments. II. The Krka River estuary and the Kornati Islands (central Adriatic Sea). Science Total Environ 95:217–225
Quevauviller P, Rauret G, Lopez-Sanchez JF, Rubio R, Ure AM, Muntau H (1997) The certification of extractable contents (mass fractions) of Cd, Cr, Ni, Pb and Zn in sediment following a three-step sequential extraction procedure. European Commission, Bruxelles (Report EUR 17554 EN)
Rapin F, Fernex F, Favarger PY, Vernet JP, Dievoet E van(1979) Répartition du mercure dans le sediments marins superficiels du plateau continental de la Còte d'Azur (France, Mediterranée). Rev Int Océanog Méd 53:41–49
Rauret G (1998) Extraction procedures for the determination of heavy metals in contaminated soil and sediment. Talanta 46:449–455
Rauret G, Rubio R, Lopez-Sanchez JF, Casassas E (1989) Specific procedure for metal solid speciation in heavily polluted river sediments. Int J Environ Anal Chem 35:89–100
Rubio R, Lopez-Sanchez JF, Rauret G (1991) La especiaciòn sòlida de trazas de metales en sedimentos. Aplicaciòn a sedimentos muy contaminados. Anal De Quim 87:599–605
Sloot HA van der, Heasman L, Quevauviller P (eds) (1997) Harmonization of leaching/extraction test. Elsevier Science, Amsterdam, pp 75–99.
Span D, Gaillard JF (1986) An investigation of a procedure for determining carbonate-bound trace metals. Chem Geol 56:135
Tessier A, Campbell PGC, Bisson M (1979) Sequential extraction procedure for the speciation of particulate trace metals. Anal Chem 51:844–850
Tipping E, Hetherington NB, Hilton J, Thompson DW, Bowles E, Hamilton-Taylor J (1985) Artifacts in the use of selective chemical extraction to determine the distributions of heavy metals between oxides of manganese and iron. Anal Chem 57:1944–1946
Yen TF, Tang JIS (1977) Chemical aspects of marine sediments. In: Yen TF (ed) Chemistry of marine sediments. Ann Arbor Science, Ann Arbor, Mich., pp 1–38

Chapter 22

Organic Matter Sources and Dynamics in northern Adriatic Coastal Waters

M. Pettine · L. Patrolecco · S. Capri

22.1 Introduction

Organic matter in marine environments results from both autochthonous and allochthonous sources. Phytoplanktonic production is responsible for the internal production of organic matter, which according to dominant biological processes, distributes between the particulate and dissolved phases. Photosynthetic activity transforms, in the presence of inorganic nutrients and light, inorganic carbon into particulate organic matter (*POM*), of which more than 30% consists of carbon (*POC*) according to the Redfield stoichiometry and its revisitation by Morel and Hudson (1985). During photosynthetic processes, a variable percentage of photosynthates is released in the surrounding water as dissolved substances (*DOM*), consisting of monomeric and low molecular weight polymeric compounds. Other than being directly released by phytoplankton, *DOM* may be indirectly produced through sloppy feeding (i.e. the release of dissolved compounds following the breaking of large preyed cells that cannot be ingested whole by the zooplankton), dissolution of faecal pellets and marine snow, and virus-induced bacteria cell lysis.

Dissolved organic carbon (*DOC*) is by far the major carbon pool in oceans, outweighing any other marine source of organic carbon by at least a factor of 10 (Kepkay 1994). Furthermore, although algal biomass in oceans is only 0.2% of that of terrestrial plants, annual phytoplankton photosynthesis is roughly equivalent to that of plants, since algae have duplication rates on the order of days compared to month to years for plants (Chisholm 1992). Therefore, the fate of carbon in oceans not only influences the trophic food chain but also affects the exchange of carbon at the sea water-atmosphere interface.

In addition to the autochthonous sources, external inputs may further supply organic matter for marine ecosystems. On a world scale, terrestrial sources contribute about 10% to the total sea water *DOM*, while on a local scale their contribution may be much more important. Riverine inputs either directly discharge organic matter into sea water or indirectly provide new organic matter through the discharge of nutrients that feed primary production.

External inputs have markedly increased the productivity of the shallow and small northern Adriatic over the oligotrophic features of the Mediterranean. External annual contributions of nutrients in this basin have been estimated to be of the same order of magnitude as the regeneration rate, thus giving similar contributions to primary production from new and regenerated material (Degobbis and Gilmartin 1990). This evaluation strengthens the vulnerability of the northern Adriatic to external inputs and therefore to water management of rivers that discharge into this basin. Annual loads of organic matter directly discharged into northern Adriatic waters account for more than 30% of the annual phytoplankton production in its most productive region along the western side (Pettine et al. 1998).

Responses of plankton biomass and productivity to riverine discharged nutrients and organic matter have been studied in many coastal ecosystems including the Hudson River, Narragansett Bay, northern Gulf of Mexico, Chesapeake Bay, Columbia River estuary, and the Baltic Sea (Benner et al. 1992; Maske 1994; Malone 1994; Cloern 1996; Harding et al. 1999; Kemp et al. 1999; Malone et al. 1999; Klinkhammer et al. 2000).

River-induced eutrophication is a general observed phenomenon involving a number of negative consequences such as excess phytoplankton growth, increased frequency of blooms, seasonal decline of oxygen in bottom waters, and changes in the trophic structure (Harding et al. 1999). However, the severity of these problems and in particular the fate of *DOM* resulting from both autochthonous and allochthonous sources may be strongly dependent on seasonal variations of freshwater flow, the geomorphology of the receiving systems and the dominant circulation pattern.

This was clearly disclosed by a recent thorough comparison (Malone et al. 1999) between the Chesapeake Bay (CB) and northern Adriatic (NA). Both of these systems are dominated by a single river, the Susquehanna in CB and the Po in NA; however, despite some similarities in terms of freshwater and nutrient loading, they exhibit significantly different trophic status. CB waters appear to be much more efficient in sequestering nutrients and retaining and recycling phytoplankton biomass than NA waters (Harding et al. 1999); in this latter system the export of nutrients due to water mass exchanges between the northern and middle basins, favoured by the counterclockwise circulation pattern during most of the year, and the confinement of regenerated nutrients, due to the strong vertical stratification during summer, reduce its capacity to retain nutrients on time and scale that promote primary and secondary productivity (Harding et al. 1999).

NA waters exhibit markedly lower average productivity levels than CB waters, although they experience large trophic gradients from coastal eutrophic to off-shore oligotrophic waters. However, the dynamics of *DOM* in this basin has resulted in the formation of a large quantity of sticky mucilaginous masses with serious problems for tourists and fishery activities; in 1989, the damages amounted to billion of dollars according to Italian authorities' estimations. This phenomenon is not reported for CB waters (Malone et al. 1999), while both the CB and NA systems experiment seasonal oxygen decline; anoxic problems are, however, much more severe in the former compared to the latter system.

There is now wide consensus that mucilaginous aggregates consist of organic and inorganic material entrapped in a gelatinous polysaccharides matrix (Degobbis et al. 1995), although timing and evolution of mucilaginous aggregates, trigger mechanisms, and biological species possibly involved in the formation of aggregates are still largely unknown (Funari et al. 1999). The mean percent values and related standard deviation ($\pm x$) of organic carbon, nitrogen, phosphorus, silicon and sulphur in aggregates collected in the summer of 1991 from surface and deep northern Adriatic waters were 24.0 ±7.2, 2.9 ±1.1, 0.26 ±0.12, 6.4 ±4.5 and 1.1 ±0.5, respectively (Pettine et al. 1995).

The scarce information available on the distribution and variability of dissolved organic matter and its important components in the northern Adriatic is one of the restrictions on the understanding of the mucilage occurrences.

In an attempt to improve our knowledge on sources, concentrations and variability of organic matter in coastal waters, we have determined organic matter loads discharged by the Po River in the dissolved and particulate phases, and the concentrations of dissolved (*DOC*) and colloidal (*COC*) organic carbon, together with those of total dissolved carbohydrates (*TDCHO*), free (*DFAA*) and total dissolved (*TDAA*) amino acids in two frontal regions of the northern Adriatic system (see Fig. 22.2). Polymeric organic compounds that are included in the colloidal fraction and in the *TDCHO* and *TDAA* variables are of particular interest in northern Adriatic waters for their involvement in the formation of micro- and macroaggregates.

This paper summarizes the results obtained in previous investigations (Pettine et al. 1998, 1999, 2001) highlighting an interannual variability in seasonal changes of *DOC* concentrations, which is strongly dependent on the hydrological regime of the Po River. Experimental findings also point out contrasting behaviours of mucilage and seasonal changes in *DOC*, *COC* and *TDCHO* concentrations, which strengthen the importance of the qualitative character of *DOM*, rather than its quantitative concentrations in triggering the formation of large mucilaginous aggregates.

22.2
Analytical Methodologies

Dissolved Organic Carbon. Samples were filtered through precombusted (4 hours at 480 °C) and preweighed Whatman GF/F glass fibre filters (0.7 μm nominal pore size). For freshwater samples, filtration was performed in a laboratory within a few hours from collection: in this case about 1 l samples were filtered under negative pressure and the filters, washed with 20 ml milli Q water, were stored at −20 °C for the analysis of *POC*. Filtered freshwater samples for *DOC* analysis were stored in high density polyethylene (HDPE) bottles as in the case of filtered sea water samples (see after) while the filtration procedure was different. For sea water samples, the filtration was performed aboard, and we were not interested in the analysis of *POC*: about 100 ml were filtered in this case by using a disposable polycarbonate syringe and a polypropylene filter holder (Nucleopore). Filtered samples were stored in duplicate into 25 ml HDPE-bottles (previously treated with HNO_3 1.2 M at 50 °C for 1 h), which were quickly frozen in an aluminum block at −20 °C. The suitability of HDPE containers for the storage of *DOC* samples was proved by recent findings (Norrman 1993; Tupas et al. 1994; Yoro et al. 1999), and confirmed by our preliminary tests. For *DOC* analysis, filtered samples were thawed in the laboratory, acidified to pH 2 with ultrapure HCl and purged with N_2 for about 10 minutes to remove inorganic carbon. Dissolved organic carbon (*DOC*) was assayed by high temperature catalytic oxidation (HTCO) and infrared detection using a Shimadzu TOC-5000 A analyser. Carbon concentrations were determined against potassium hydrogen phthalate standards after correction for total blank. This value, which is the analytical system blank plus a Milli Q water blank, was approximately 10–15 μM C under our experimental conditions and was mostly due to the experimental system. Samples were measured in triplicate with a fixed c.v. of 2%; otherwise, further replicates were automatically carried by the instrument.

Particulate Organic Carbon. For *POC* analysis, the filters were dried at 60 °C overnight, re-weighed to determine particle loading and homogenized in an agate mortar mill. Blank filters were processed the same way. Powered, homogenized samples (10–15 mg) were accurately weighed (±0.01 mg) into tin or silver cups (9 × 5 mm) to determine particulate nitrogen (*PN*), total particulate carbon (*PC*) or particulate organic carbon (*POC*). Samples in silver cups were acidified with 20 μl 5 M ultrapure HCl and kept at 50–60 °C for 30 min to remove inorganic carbon. Care was taken to ensure complete saturation of the sample with HCl and to avoid sample loss. Acid treatment was repeated until effervescence was no longer observed (generally three times). *PC*, *POC* and *PN* were determined by high temperature oxidation using a Carlo Erba NA 1500 series 2 C/H/N/O/S analyser. Samples were run in duplicate. The sulphanylamide standard was used to construct the calibration curve. Carbon and nitrogen contents were expressed as percentage of total solid and as concentration (mg l^{-1}) of filtered water sample. Average blank levels were 2.5 ±0.8 μg C and <0.5 μg N; detection limits (calculated as three times the standard deviation of the blank for carbon, and according to sensitivity of the instrument for nitrogen) were 2 and 0.5 μg for carbon and nitrogen, respectively, corresponding to 0.02% C and 0.005% N for a 10 mg sample. Analytical precision was ±2.3 and ±1.8% of the measured values for total and organic carbon and ±1.6% for nitrogen. Particulate inorganic carbon (*PIC*) was determined by the difference between *PC* and *POC*. Only *POC* data will be discussed in this paper.

Colloidal Organic Carbon. A limited number of sea water samples were processed with a tangential flow ultrafiltration (UF) system (Amicon DC10LA) to investigate the molecular weight distribution of the *DOM* pool. For *DOC* fractionation, about 30 l samples were filtered aboard, immediately after collection, through 0.4 μm polycarbonate filters (Nucleopore, ∅ 142 mm) in a closed pressurized system (N_2), using a Sartorius Teflon-covered apparatus. To avoid contamination, the first 1–2 l of the filtrate were discharged. Filtered samples were then passed in cascade through two polysulphone membranes, which have nominal molecular weight cut-offs of 10 kDa (hollow-fibre cartridge, Amicon, H10P10) and 1 kDa (spiral-wound cartridge, Amicon, S10N1), respectively. Operating pressures were 20–25 psi at the inlet and 10–15 psi at the outlet of the system. UF cartridges were thoroughly cleaned, recirculating sequentially 0.1 N NaOH and 0.01 N HCl solutions for at least 30 min before and between samples and rinsing several times with about 5 l Milli-Q water. Finally, after conditioning the UF cartridges with about 1 l filtered sample, about 20 l of sample were processed and the concentrated colloidal fraction was reduced to about 2 l, thus achieving a 10-fold concentration factor.

Following this procedure, we fractionated the *DOC* pool into a low-molecular-weight or "truly dissolved" fraction (*DOC* < 1 kDa) and two colloidal size classes, operationally defined as low-molecular-weight *COC* (1 kDa < *LCOC* < 10 kDa) and high-molecular weight *COC* (10 kDa < *HCOC* < 0.4 μm). Each sample was stored frozen in duplicate into 25 ml high density polyethylene (HDPE) bottles (as previously described for *DOC*). In the laboratory, *DOC* analyses were performed on 0.4 μm polycarbonate filtered samples and their four different molecular weight fractions (>10 kDa; <10 kDa; 1–10 kDa; <1 kDa), in order to calculate partial mass balances on each cartridge. *DOC* recovery (R, as a percentage of the initial *DOC*) was calculated according to Dai et al. (1998). For the 1 kDa cartridge, R % ranged from 87 to 120% (with an average of 101 ±8%), while for the 10 kDa cartridge, R % ranged from 83 to 117% (with an average of 100 ±8%). Overall recovery in our ultrafiltration system was close to 100% (±10%) and was within the ranges reported in the literature (Guo et al. 1994; Benner et al. 1997; Dai et al. 1998), indicating that contamination or loss of organic carbon were minimal. While this fractionation procedure is probably the best one available at the moment and the most widely used, we must note that interactions at membrane surface may lead to conformation changes, irreversible adsorption and /or self-coagulation of small molecules resulting in an overestimation of *COC* (Buffle and Leppard 1995; Guo et al. 2000).

Dissolved Free (*DFAA*) and Total Dissolved (*TDAA*) Amino Acids. Samples for total dissolved amino acids (*TDAA*) were analyzed after liquid hydrolysis in 6 N HCl at 110 °C for 22 h (modified from Parson et al. 1984). 2 ml samples were placed in glass ampoules (acid washed and muffled for 3 hours at 500 °C), sealed under a flow of N_2. Each vial was spiked with α-aminoadipic acid, norvaline and hydroxylysine as recovery standards of the acidic, neutral and basic classes, respectively (Cowie and Hedges 1992), and 2 ml of ultrapure 12 N HCl. Hydrolysates were vacuum dried and then redissolved in milli-Q water to achieve a neutral pH. *DFAA* were analyzed without preliminary hydrolysis of the samples.

DFAA and *TDAA* were determined by high performance liquid chromatography (HPLC-Perkin Elmer Series IV) and fluorimetric detection (Perkin Elmer LS4) with a gradient system of methanol/tetrahydrofuran (90:10 v/v) acetate buffer, following derivatization with o-phthaldialdehyde (OPA) and 2-mercaptoethanol (Lindroth and Mopper 1979, as modified in Pettine et al. 1999); 21 individual amino acids were separated according to this procedure (see Fig. 22.1). Daily standards in aged sea water, samples and blanks (aged sea water) were run in triplicate. The average *TDAA* blank amounted to 62 ±40 nM and was mainly due to glycine, aspartic acid, serine and alanine. Recovery for most of the amino acids was within ±10% except for asparagine, glutamine, tryptophane, histidine and methionine. In the analysis of free amino acids, glutamic acid, glutamine, aspartic acid and asparagine are separated into four distinct peaks, but the acid hydrolysis performed for analyzing total dissolved amino acids causes the deamination of asparagine (asn) and glutamine (gln) to form aspartic acid (asp) and glutamic acid (glu) plus ammonium. Therefore, in the *TDAA* analysis, the amino acids pairs (asn + asp and gln + glu) give only two peaks instead of four. Tryptophane is unstable during acid hydrolysis and may not be detected. Histidine is also partially modified by acid hydrolysis and the recovery is only 20–30% (Keil and Kirchman 1991). *DFAA* and *TDAA* are expressed as individual amino acids (nM) and as the sum of the aminoacidic molar carbon (*DFAA*-C or *TDAA*-C) and nitrogen (*DFAA*-N or *TDAA*-N) units.

Total Dissolved Carbohydrates. Total dissolved carbohydrates (*TDCHO*), including mono-, oligo- and polysaccharides, were analyzed by the MBTH method after a preliminary hydrolysis with 0.09 N HCl (20 h at 100 °C) (Burney and Sieburth 1977). This hydrolysis procedure was preferred to the stronger H_2SO_4 attack more recently proposed (Pakulski and Benner 1992; Mopper et al. 1992), mainly because it has been extensively used in estuarine and coastal waters, thus making the comparison with previous find-

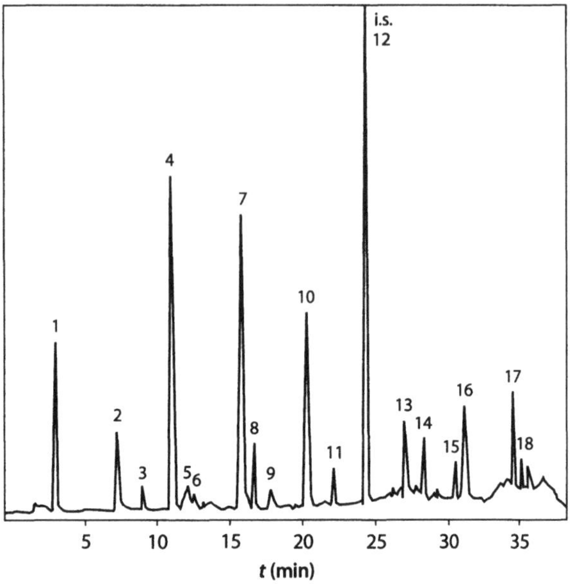

Fig. 22.1. An example of HPLC spectrum of amino acids in a sea water sample; *1*: aspartic acid; *2*: glutamic acid; *3*: asparagine; *4*: serine; *5*: histidine; *6*: glutamine; *7*: glycine; *8*: threonine; *9*: arginine; *10*: alanine; *11*: tyrosine; *12*: γ-aminobutyric acid (internal standard); *13*: valine; *14*: phenylalanine; *15*: isoleucine; *16*: leucine; *17*: ornithine; *18*: lysine; methioninsulphone, methionine, tryptophane and norvaline were not detected in this sample

ings feasible (see Pettine et al. 1999 for more details). Furthermore, polysaccharides determined by this method would more directly reflect changes in the concentration of exuded polymers (Benner 1998), which include transparent exopolymer particles actively involved in the aggregate formation (Passow and Alldredge 1994; Passow 2000). Samples were analyzed in triplicate against glucose standard solutions, and results were expressed in terms of glucose-carbon equivalents (*TDCHO*-C), by assuming 6 moles of carbon per mole of hexose. The relative standard deviation between replicate samples was usually below 5%.

22.3 Role of Organic Matter Dynamics in NA Environmental Problems

The dynamics of the *DOC* pool in the northern Adriatic basin are involved in hypoxic and anoxic crises in bottom waters and massive occurrences of mucilaginous aggregates in the water column.

Anoxic phenomena occur during summer in the western coastal waters, which are influenced by the Po River discharge. Since 1975, when nutrient loads discharged by the Po showed an abrupt increase (Marchetti 1990) as happened in other European rivers (Kempe et al. 1991), large algal blooms have occurred regularly, causing episodes of diffused anoxia at the sea bottom, which resulted in mass mortalities of benthic organisms. The spatial and temporal extensions of areas affected by anoxia have diminished in recent years (Degobbis et al. 1999), probably as a consequence of the reduction and banning of phosphorus in Italian detergents. Toxic algal species on occasion also developed, causing severe damages to mollusc cultures and threatening human health. In addition to these environmental deterioration problems that are typical of eutrophicated systems, the northern Adriatic has experienced a peculiar phenomenon consisting of massive occurrences of mucilaginous aggregates. In 1989, these aggregates covered a large part of the entire surface of the NA basin (Stachowitsch et al. 1990) and drew the attention of public and scientific opinion. Large quantities of sticky mucilaginous masses affected biological, chemical and physical characteristics of the ecosystem (Degobbis et al. 1999). Part of the material floating on the sea surface was deported on beaches by wind and currents, reducing their suitability for bathing and threatening tourism. Suspended and sinking mucilaginous aggregates created serious problems for fisheries and biota. Benthic organisms suffered in many areas from suffocation due to their covering by aggregates.

The impressive mucilage events of 1989–91 stimulated many studies aimed at understanding the possible causes responsible for this phenomenon and its triggering mechanisms; however, in spite of these efforts, this phenomenon remains substantially unknown (Funari et al. 1999). Although it was reported that episodes of massive mucilage formation have been recorded for over a century (Fonda-Umani et al. 1988), the similarities between past and recent phenomena in terms of triggering factors and spatial extensions remain in doubt.

We speculated on the strict temporal parallelism between massive aggregate occurrences and changes in the formulation of detergents (Pettine et al. 1995). In 1988, in fact, phosphorus was banned from domestic detergents and substituted by zeolite A particles and about doubled polycarboxylate concentrations. These changes have further increased the N/P ratio in coastal waters, which was already unbalanced, and probably they were responsible for the shift in the phytoplankton population from dinoflagellate to diatom blooms recently reported (Degobbis et al. 1999).

Therefore, P-substitutes, while contributing to diminish the frequency and extension of anoxia in NA coastal waters, apparently cause an external forcing on biological and chemical processes, which could favour an intensification of macroaggregate occurrences. Possible effects may be on a large scale (changes in algal succession from dinoflagellate to diatoms and on their released chemical compounds; limitation of the bacterial activity due to further increase in N/P ratio and consequently increased residence time for *DOM*) as well as on a micro scale (changes in sorption, partitioning and gelling of polymers, due to the inclusion in colloidal matter of zeolite A particles and polycarboxylates).

22.4
Organic Matter Discharged by the Po River

Concentrations of dissolved and particulate organic carbon (*DOC*, *POC*) were measured in samples taken monthly for a year at two stations in the lower Po River (Figs. 22.2 and 22.3). Samples were collected at each station at two depths (about 0.5 m from surface and 2 m above the bottom) to evaluate possible variations in the water column that may affect the transport. Two major interrelated variables, solid transport and flow conditions, influenced the variability of organic matter concentrations in the water and solid phases, while the water column was reasonably homogeneous (Pettine et al. 1998).

Mass fluxes of *DOC* and *POC* to the northern Adriatic basin were calculated by four different methods. They were based on: (*a*) the mean of products of the instantaneous concentrations and the mean daily discharge of the sampling day; (*b*) the product of

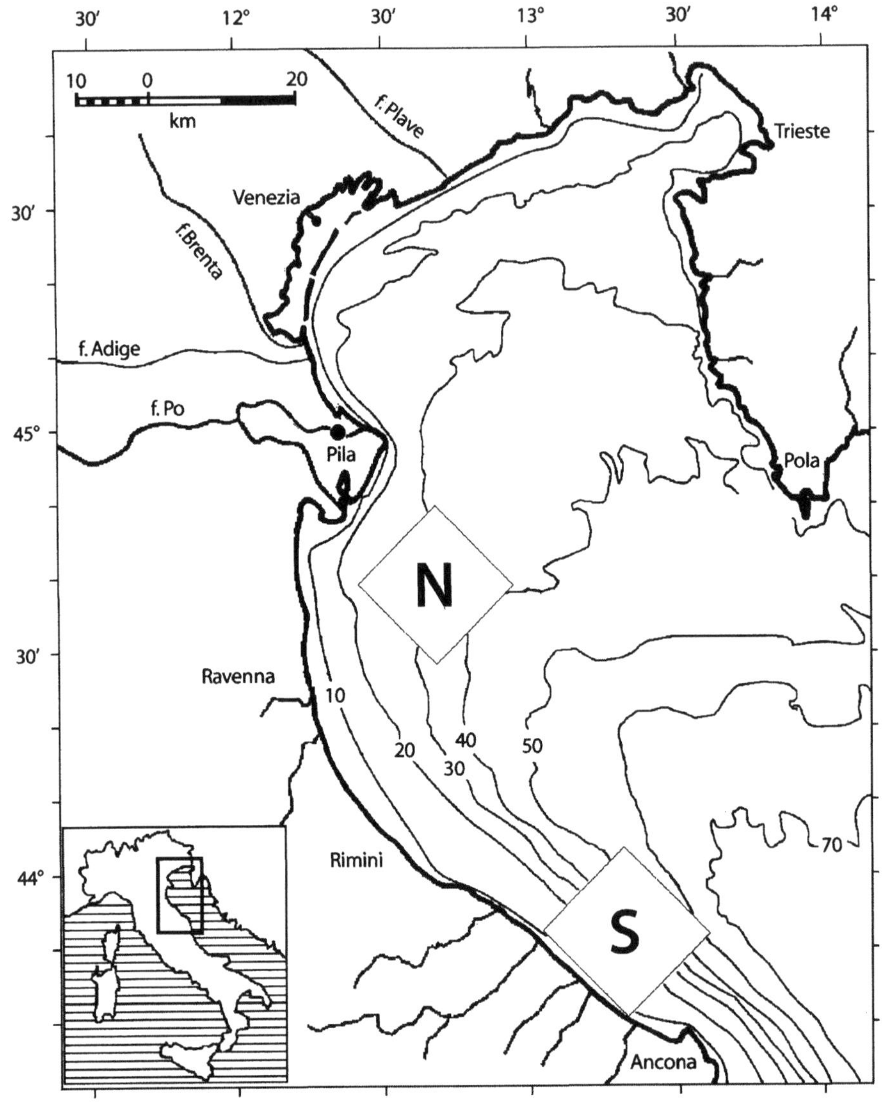

Fig. 22.2. Location of the study area. In the Po River, only the Pila station is shown; the other (Pontelagoscuro) is upstream of the river stretch plotted. In the NA system, the two investigated frontal regions, one in the north (*N*) and the other in the south (*S*), are shown

the arithmetic means of the concentrations and discharges; (*c*) the product of discharge-weighted concentrations and the mean annual discharge; and (*d*) the sum of the products of daily discharge and concentration resulting from the concentration vs. flow equation, extended to a whole year. Methods *a* and *b*, which use the arithmetic mean of instantaneous loads or arithmetic means of concentrations and discharges, give equal weight to each pair of variables. This may introduce a major bias in calcu-

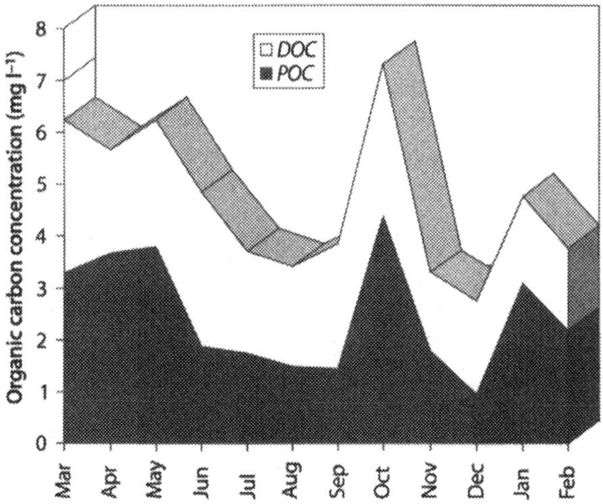

Fig. 22.3. Monthly average organic carbon concentration in the lower stretch of the Po River

lations, since concentrations measured at low discharge have the same importance as those at peak discharge. Method *c* reduces this bias and improves the evaluation of mass transport by use of discharge-weighted means; while Method *d*, based on the total discharge curve, is the more accurate has a better correlation between concentration and discharge. In our case, these different methods provided reasonably close estimations; however, results by Method *c* are more appropriate and gave fluxes of 13.4×10^4 and 12.1×10^4 tonnes $year^{-1}$ for *POC* and *DOC*, respectively.

These loads may be compared with those discharged by the Rhone River, which shows some similarities with the Po River. Both these rivers originate from the Alps, drain large catchment areas (Rhone 96 000 km^2 and Po 70 091 km^2), have reasonably similar solid transport (about 5×10^6 tonnes $year^{-1}$) and water discharge (55 km^3 $year^{-1}$ in the Rhone vs. 47 km^3 $year^{-1}$ in the Po). The *TOC* load for the Rhone was estimated to be 15×10^4 tonnes $year^{-1}$, one third of which is *POC* (Kempe et al. 1991). Thus, in terms of *TOC*, the Po export rate (25.5×10^4 tonnes $year^{-1}$) is about 1.7 times higher than that of the Rhone, making the former river the largest contributor of organic matter to the Mediterranean. The Nile River, which is another large river discharging into the Mediterranean, is extensively exploited for irrigation, hence its discharge into the sea is strongly reduced.

Organic carbon appears to be preferentially transported in the solid phase in the Po compared with the Rhone (*DOC* discharges are in fact similar, while *POC* discharges are markedly different), and this different partitioning reflects a high carbon content in the particulate matter transported by the Po, since solid transport in the two rivers is similar. These differences in the particulate and total transport of organic carbon are strongly related to marked differences in population densities in the watersheds of these rivers (232 and 84 inhabitants km^{-2} for the Po and the Rhone, respectively).

However, the *TOC* export rate by the Po (3.57 tonnes km^{-2} $year^{-1}$), which is given by the ratio between the calculated load and the surface extension of the river catchment, is slightly lower than the average export rate estimated for major temperate rivers

(4 tonnes km^{-2} year^{-1}; Meybeck 1982). Therefore, although the Po exhibits a higher *TOC* discharge compared to other Mediterranean rivers, its *TOC* export rate does not suggest a marked alteration by anthropogenic activity, pointing out that the self-purification capacity of the river is able to maintain organic matter concentrations at a low level in the lower stretch.

22.5 Interannual Variability of *DOC* Concentrations in NA Coastal Waters

DOC concentrations have been measured on more than 400 samples from surface and deep waters at different distances from the Po mouth in two frontal regions, (Pettine et al. 1999 and 2001). The location of these frontal regions, which could move depending on the seasonal period, is exemplified in Fig. 22.2 for the June 1996 cruise. The overall ranges of *DOC* concentrations and salinity resulting from all the four cruises were 70 to 280 µM and 30 to 38, respectively. Average *DOC* values in the four cruises carried out in the northern region, under the direct influence of the Po river, are shown in Fig. 22.4. In winter, deep waters assume a typical background value of 76 ±10 µM, averaged from all the available data in both the northern and southern region. This average value is well within the range 50–92 µM characteristic of deep and surface western Mediterranean (Copin-Montegut and Avril 1993), and is also very close to concentrations typical of surface oceanic waters (70–80 µM; Guo et al. 1995). These low winter concentrations suggest that the Adriatic system has a good capacity to react to external forces through degradation processes and water mass exchange between the northern and middle basins.

Surface values in winter show only a slight increase with respect to the background value, while during summer, changes appear much more marked. However, seasonal changes were found to strongly depend on the hydrological regime of the Po River, which is responsible for over 50% of the total riverine discharge into the northern Adriatic system.

In June 1996, average surface values were a factor of 2.6 and 2.1 higher than the background value (76 ±10 µM) in the northern and southern regions, respectively. In June 1997 the corresponding increasing factors were 1.7 and 1.3. This interannual vari-

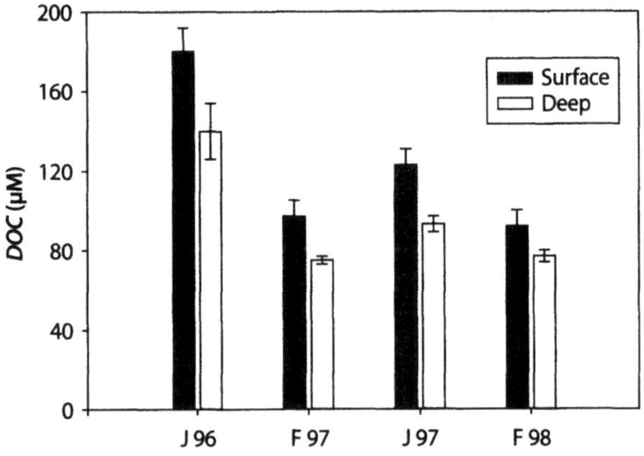

Fig. 22.4. Average *DOC* concentrations in the northern frontal region of the four cruises

ability reflects changes in the Po discharge curve during these two years: in 1996, many peaks preceded the June cruise, while in 1997 a long low flow period prevailed before the June cruise (Fig. 22.5). Accordingly, temporal patterns of chl *a* concentrations in the receiving coastal waters pointed out a much higher productivity in the spring of 1996 than in 1997, which was responsible for the higher increasing factors observed for *DOC* concentrations in the former year compared to the latter (Pettine et al. 2001).

Contrary to the increase in *DOC* concentrations that were stronger in 1996 than in 1997 (Fig. 22.6), mucous aggregates showed a much stronger occurrence in June 1997

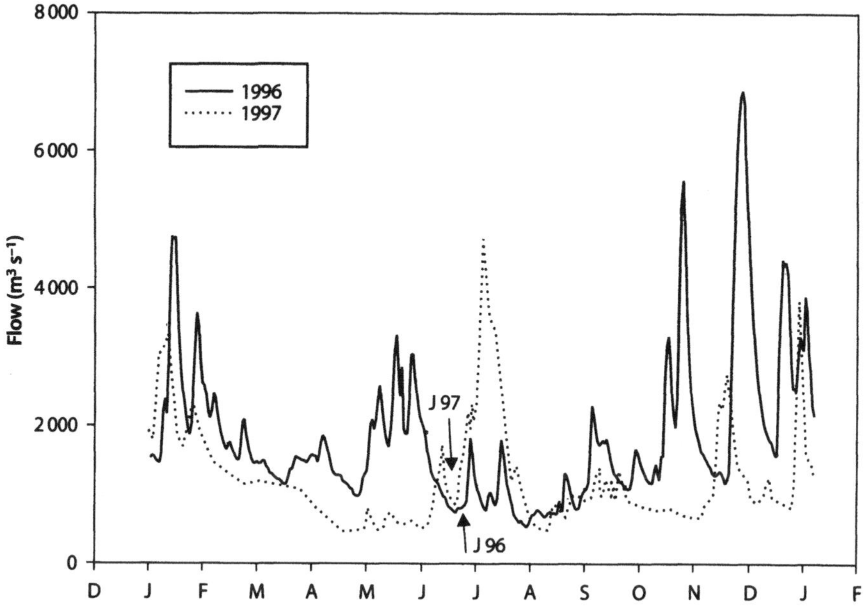

Fig. 22.5. Discharge curves of the Po River in 1996 and 1997. The two *arrows* indicate the two-year cruise period

Fig. 22.6. Calculated increments (μM) in *DOC* concentrations over the background winter value of 76 μM in June 1996 and June 1997

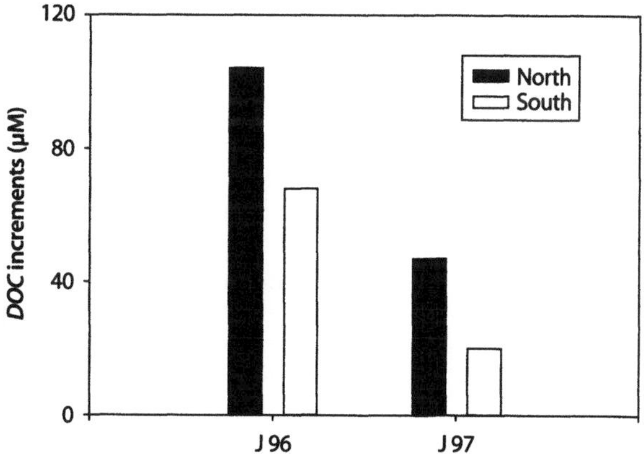

compared to June 1996. This was confirmed by the use of the remote observing vehicle (ROV) during the cruises. The scavenging effect exerted in the water column by mucous macroaggregates may partially explain an inverse relationship between macroaggregates and dissolved *DOC* concentrations. However, these findings point out that high *DOC* levels, as in the case of June 1996, do not necessarily produce macroaggregates, strengthening the importance of the qualitative characteristics of the *DOC* pool in the mass transfer between *DOC*, *COC* and the mucous fraction.

The conclusion by Passow (2000) that TEP-precursors, which consist of strongly sulphated polysaccharides actively involved in aggregation processes, appear to be a distinct group of polysaccharides, whose production and standing stocks are uncoupled from bulk carbohydrates, further remarks the importance of the qualitative characteristics of *DOM* in the formation of aggregates. The higher sulphur level found in macroaggregates in our previous investigations (Pettine et al. 1995) is consistent with a large role played by TEP in the formation of macroaggregate.

22.6 Composition of *DOM*

Dissolved organic matter in the northern Adriatic system is characterized by the presence of a high colloidal fraction with a shift in the molecular weight distribution toward the large size fraction and a high contribution of carbohydrates to dissolved organic matter (Pettine et al. 1999 and 2001).

The average values (µM) calculated from all the data were 67 ±14 for the overall colloidal organic fraction (>1 kDa) and 39 ±13 for the truly dissolved organic fraction (*DOC* > 1 kDa). Differences between the June and February cruises were very small (Fig. 22.7). The colloidal organic carbon pool consisted of high (>10 kDa, *HCOC*) and low (1 to 10 kDa, *LCOC*) molecular weight classes. The *HCOC* fraction accounted for

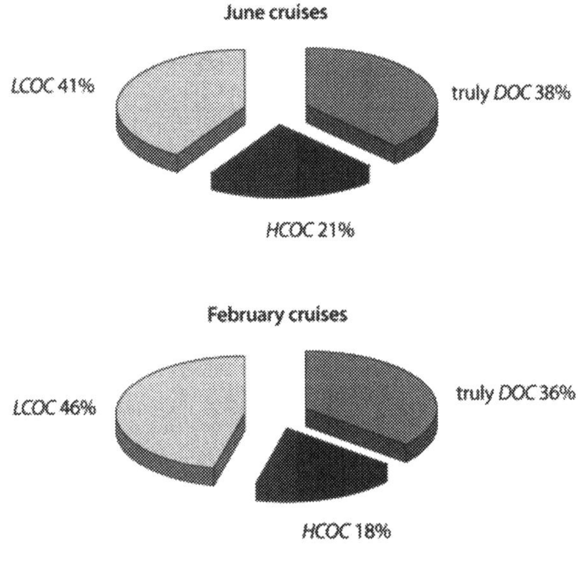

Fig. 22.7. Distribution of dissolved organic matter (<0.7 µm) between the high (*HCOC*) and low (*LCOC*) colloidal classes and truly dissolved components in the June and February cruises

21 ±5 and 18 ±6% of *DOC* in the June and February cruises, respectively. The *LCOC* fraction contributed 41 ±7 and 46 ±6% to the overall *DOC* in the June and February cruises. Both these colloidal classes were tightly correlated to *DOC* (Pettine et al. 2001).

Total dissolved carbohydrate concentrations from the two investigated areas and various cruises ranged from 6.0 to 72.4 µM in terms of carbon (average 18.3 ±12.0) and were tightly correlated to *DOC*. The entire data set (µM) was described by ($p < 0.001$):

$$TDCHO = -10.46\ (\pm 2.22) + 0.28\ (\pm 0.02)\ DOC; \quad n = 155;\ r = 0.74;\ \text{s.d.} = 8.07 \qquad (22.1)$$

The percent contributions of *TDCHO* to *DOC* were 20.3 ±10.7; 15.1 ±4.9; 18.4 ±4.3 and 14.2 ±5.0 in June 1996, February 1997, June 1997 and February 1998, respectively, with average values of 19.1 ±7.3% in the June cruises and 14.5 ±4.9% in the February cruises (Fig. 22.8).

According to the slope of Eq. 22.1, *TDCHO* is responsible for 28% of the *DOC* variations over the study period. This slope is in the high range of values reported for

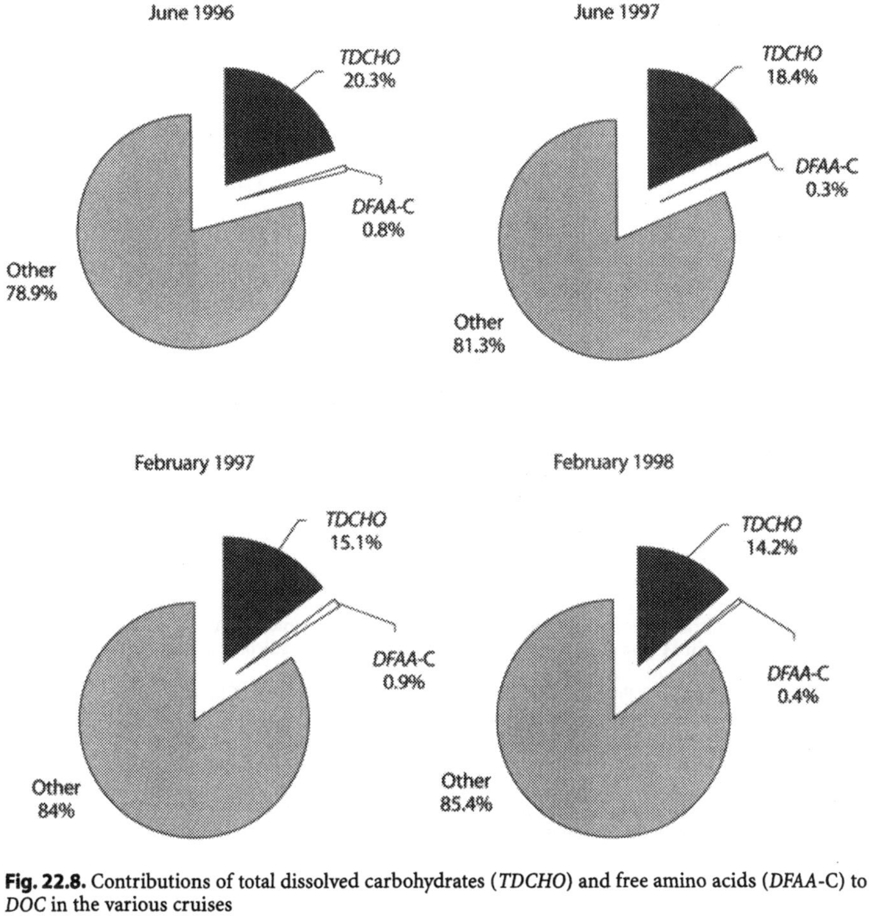

Fig. 22.8. Contributions of total dissolved carbohydrates (*TDCHO*) and free amino acids (*DFAA*-C) to *DOC* in the various cruises

TDCHO vs. *DOC* slopes, which have been found to vary from 0.09 to 0.29 (Burney and Sieburth 1977; Senior and Chevolot 1991; Pakulski and Benner 1994).

Contrary to colloidal and carbohydrate components, *DFAA* varied within the approximate range 100–400 nM reported for marine waters (Thurman 1985) and contributed <1% to *DOC* in the various cruises. However, based on a limited data set available for total dissolved amino acids (*TDAA*), they gave average carbon concentrations (μM) of 3.9 ±0.7, 5.1 ±2.5 and 2.4 ±2.1 in February 1997, June 1997 and February 1998, contributing 4.5, 4.9 and 2.9% to *DOC*, respectively.

Further investigations are needed to improve our knowledge of the qualitative characteristics of *DOM*, with particular emphasis on dominant polymeric compounds that are included in the colloidal fraction and may give rise to coagulation processes leading to the formation of macroaggregates.

Acknowledgements

The studies summarized in this paper were carried out in the framework of the multidisciplinary project on the Adriatic Sea (PRISMA) supported by the Ministry of University and Scientific Research (MURST).

References

Benner RH (1998) Cycling of dissolved organic matter in the ocean. In: Hessen DO, Tranvick L (eds) Ecol Stud 133:317–331

Benner RH, Chin-Leo G, Gardner W, Eadie B, Cotner J (1992) The fates and effects of riverine and shelf-derived DOM on Mississippi River plume/Gulf shelf processes. Proceedings of the workshop "Nutrient enhanced coastal ocean productivity". National Oceanic and Atmospheric Administration, Coastal Ocean Program Office, pp 84–94

Benner RH, Biddanda B, Black B, McCarthy M (1997) Abundance, size distribution and stable carbon and nitrogen isotopic compositions of marine organic matter isolated by tangential-flow ultrafiltration. Mar Chem 57:243–263

Buffle J, Leppard GG (1995) Characterization of aquatic colloids and macromolecules 2. Key role of physical structures on analytical results. Environ Sci Technol 29:2176–2184

Burney CM, Sieburth J (1977) Dissolved carbohydrates in sea water. II. A spectrophotometric procedure for total carbohydrate analysis and polysaccharide estimation. Mar Chem 5:15–28

Chisholm S (1992) What limits phytoplankton growth? Oceanus 35(3):36–46

Cloern JE (1996) Phytoplankton bloom dynamics in coastal ecosystem: A review with some general lessons from sustained investigation of San Francisco Bay, California. Rev Geophys 34:127–168

Copin-Montegut G, Avril B (1993) Vertical distribution and temporal variation of dissolved organic carbon in the north-western Mediterranean Sea. Deep-Sea Res 40(10):1963–1972

Cowie GL, Hedges JI (1992) Improved amino acids quantification in environmental samples: Charge matched recovery standards and reduced analysis time. Mar Chem 37:223–238

Dai M, Buesseler KO, Ripple P, Andrews J, Belastock RA, Gustafsson O, Moran SB (1998) Evaluation at two cross-flow ultrafiltration membranes for isolating marine organic colloids. Mar Chem 62:117–136

Degobbis D, Gilmartin M (1990) Nitrogen, phosphorus and biogenic silicon budgets for the northern Adriatic sea. Oceanol Acta 13:31–45

Degobbis D, Fonda-Umani S, Franco P, Malej A, Precali R, Smodlaka N (1995) Changes in northern Adriatic ecosystem and the hypertrophic appearance of gelatinous aggregates. Sci Total Environ 165:43–58

Degobbis D, Malej A, Fonda Umani S (1999) The mucilage phenomenon in the northern Adriatic sea. A critical review of the present scientific hypotheses. Annali Istituto Superiore di Sanità (ISSN) 35(3):373–381

Fonda-Umani S, Ghirardelli E, Specchi M (1988) Il fenomeno del "mare sporco" nell'Adriatico. In: Brambati A (ed) Consiglio Nazionale delle Ricerche, Progetto Strategico Oceanografia e Tecnologie Marine, pp 37–42

Funari E, Farooq A, Fonda Umani S, Pagnotta R (1999) State of the art and new scientific hypotheses on the phenomenon of mucilages in the Adriatic sea. Annali Istituto Superiore di Sanità (ISSN) 35(3):353–425

Guo L, Coleman CH Jr, Santschi PH (1994) The distribution of colloidal and dissolved organic carbon in the Gulf of Mexico. Mar Chem 45:105–119
Guo L, Santschi PH, Warnken KW (1995) Dynamics of dissolved organic carbon (*DOC*) in oceanic environments. Limnol Oceanogr 40(8):1392–1403
Guo L, Wen LS, Tang D, Santshi PH (2000) Re-examination of cross-flow ultrafiltration for sampling aquatic colloids: Evidence from molecular probes. Mar Chem 69:75–90
Harding LW Jr, Degobbis D, Precali R (1999) Production and fate of phytoplankton: Annual cycles and interannual variability. In: Malone TC, Malej A, Harding LW Jr, Smodlaka N, Turner RE (eds) Ecosystems at the land-sea margin. Drainage basin to coastal sea. American Geophysical Union, Washington DC (Coastal and estuarine studies, 55, pp 131–172)
Keil RG, Kirchman DL (1991) Dissolved combined amino acids in marine waters as determined by a vapor-phase hydrolysis method. Mar Chem 33:242–259
Kemp WP, Faganeli J, Puskaric S, Smith EM, Boynton WR (1999) Pelagic-benthic coupling and nutrient cycling. In: Malone TC, Malej A, Harding LW Jr, Smodlaka N, Turner RE (eds) Ecosystems at the land-sea margin. Drainage basin to coastal sea. American Geophysical Union, Washington DC (Coastal and estuarine studies, 55, pp 295–337)
Kempe S, Pettine M, Cauwet G (1991) Biogeochemistry of European rivers. In: Degens ET, Kempe S, Richey JE (eds) Biogeochemistry of major World rivers. Wiley, New York, pp 169–211
Kepkay PE (1994) Particle aggregation and the biological reactivity of colloids. Mar Ecol Prog Ser 109:293–304
Klinkhammer GP, McManus J, Colbert D, Rudnicki MD (2000) Behaviour of terrestrial dissolved organic matter at the continent-ocean boundary from high-resolution distributions. Geochim Cosmochim Acta 64(16):2765–2774
Lindroth P, Mopper K (1979) High performance liquid chromatographic determination of subpicomole amounts of amino acids by precolumn fluorescence derivatization with o-phthaldehyde. Anal Chem 51(11):1667–1674
Malone TC (1994) Anthropogenic nitrogen loading and assimilation capacity of the Hudson River estuarine system, USA. In: Kennedy VS (ed) The estuary as a filter. Academic Press, New York, pp 291–312
Malone TC, Maley A, Harding LW Jr, Smodlaka N, Turner RE (1999) Ecosystems at the land-sea margin. Drainage basin to coastal sea. American Geophysical Union, Washington DC (Coastal and estuarine studies)
Maske H (1994) Long-term trend in seston and chlorophyll a in a Kiel Bight, Western Baltic. Contin Shelf Res 14:791–801
Marchetti R (1990) Algal bloom and gel production in the Adriatic sea. In: Barth H, Fegan L (eds) Eutrophicated-related phenomena in the Adriatic Sea and in other Mediterranean coastal zones. Commission of the European Communities (Water Pollution Research Report, 16, CEE-EUR 12978, pp 21–42)
Meybeck M (1982) Carbon, nitrogen and phosphorus transport by world rivers. Am J Sci 282:401–450
Mopper K, Schultz CA, Chevolot L, Germain C, Revuelta R, Dawson R (1992) Determination of sugars in unconcentrated seawater and other natural waters by liquid chromatography and pulsed amperometric detection. Environ Sci Technol 26:133–138
Morel FMM, Hudson JM (1985) The geobiological cycle of trace elements in aquatic systems: Redfield revisited. In: Stumm W (ed) Chemical processes in lakes. John Wiley and Sons, New York, pp 251–281
Norrman B (1993) Filtration of water samples for *DOC* studies. Mar Chem 41:239–242
Pakulski DJ, Benner R (1992) An improved method for the hydrolysis and MBTH analysis of dissolved and particulate carbohydrates in seawater. Mar Chem 40:143–160
Pakulski DJ, Benner R (1994) Abundance and distribution of carbohydrates in the ocean. Limnol Oceanogr 39(4):30–940
Parson TR, Maita YR, Lalli CM (1984) A manual of chemical and biological methods for seawater analysis, 2nd edn. Pergamon Press, New York, p 173
Passow U (2000) Formation of transparent exopolymer particles, TEP, from dissolved precursor material. Mar Ecol Prog Ser 192:1–11
Passow U, Alldredge AL (1994) Distribution, size and bacterial colonization on transparent exopolymer particles (TEP) in the ocean. Mar Ecol Progr Ser 113:185–198
Pettine M, Pagnotta R, Liberatori A (1995) Composition of mucilaginous macroaggregates and hypotheses for their formation. Ann Chim 85:431–441
Pettine M, Patrolecco L, Camusso M, Crescenzio S (1998) Transport of carbon and nitrogen to the Northern Adriatic Sea by the River Po. Estuar Coast Shelf S 46:127–142
Pettine M, Patrolecco L, Manganelli M, Capri S, Farrace MG (1999) Characterization of dissolved organic matter in the northern Adriatic Sea. Mar Chem 64:153–169
Pettine M, Capri S, Manganelli M, Patrolecco L, Puddu A, Zoppini A (2001) The dynamics of DOM in the northern Adriatic sea. Estuar Coast Shelf S 52:471–489

Senior W, Chevolot L (1991) Studies of carbohydrates (or carbohydrate-like substance) in an estuarine environment. Mar Chem 32:19–35

Stachowitsch M, Fanuko N, Richter M (1990) Mucus aggregates in the Adriatic Sea: An overview of stages and occurrences. Mar Ecol 11(4):327–350

Thurman EM (1985) Organic Geochemistry of natural waters . Nijhoff/Junk The Hague, The Netherlands, pp 158–163

Tupas LM, Brian NP, Karl DM (1994) Dissolved organic carbon in oligotrophic waters: Experiments on sample preservation, storage and analysis. Mar Chem 45:207–216

Yoro SC, Panahgiotopoulos C, Sempere R (1999) Dissolved organic carbon contamination induced by filters and storage bottles. Wat Res 33(8):1956–1959

Index

Symbols

A

C

D

G

H

J

K

L

M

N

O

P

Q

R

S

T

MIX
Papier aus verantwortungsvollen Quellen
Paper from responsible sources
FSC® C105338

If you have any concerns about our products,
you can contact us on
ProductSafety@springernature.com

In case Publisher is established outside the EU,
the EU authorized representative is:
Springer Nature Customer Service Center GmbH
Europaplatz 3, 69115 Heidelberg, Germany

Printed by Libri Plureos GmbH
in Hamburg, Germany